Management of Phytonematodes: Recent Advances and Future Challenges

Rizwan Ali Ansari • Rose Rizvi • Irshad Mahmood
Editors

Management of Phytonematodes: Recent Advances and Future Challenges

Springer

Editors
Rizwan Ali Ansari
Section of Plant Pathology and Nematology, Department of Botany
Aligarh Muslim University
Aligarh, India

Rose Rizvi
Section of Plant Pathology and Nematology, Department of Botany
Aligarh Muslim University
Aligarh, India

Irshad Mahmood
Section of Plant Pathology and Nematology, Department of Botany
Aligarh Muslim University
Aligarh, India

ISBN 978-981-15-4086-8 ISBN 978-981-15-4087-5 (eBook)
https://doi.org/10.1007/978-981-15-4087-5

This Springer imprint is published by the registered company Springer Nature Singapore Pte Ltd.
The registered company address is: 152 Beach Road, #21-01/04 Gateway East, Singapore 189721, Singapore

Preface

Phytoparasitic nematode causes considerable hurdles in the intensification of agricultural crop produce. Plant parasitic nematodes have caused a greater reduction in the plant growth and yield characters of various crop plants. Severe infestations of the phytonematodes cause greater impairment of plant health which is reflected by poor yield of the crops. The present book, **'Management of Phytonematodes: Recent Advances and Future Challenges'** has been written with the aim to provide a single pot solution related to management of plant parasitic nematodes. Thus, most of the chapters have been taken from the learned researchers, scientists and scholars so that a good and informative book could be prepared. In brief, this book illustrates that biological control, biopesticides, organic additives, manures, phytoextracts, biogenic nanoparticles, etc. are the available options for the sustainable management of phytonematodes. The chapters have been therefore written in such a way that a uniformity and coherence among the chapters could be properly maintained and the maximum knowledge on this aspect could be brought out before the researchers. The updated knowledge for the management phytoparasitic nematodes has been conglomerated.

Moreover, Editorial board is highly grateful to the contributors/authors who took a lot of pain and worked out day and night in the compilation of this book within and beyond the limit, without their support, this book would have just been a dream of the editors.

Furthermore, editors extend hearty thanks to our beloved **Prof. Tariq Mansoor**, Hon'ble Vice Chancellor, Aligarh Muslim University, Aligarh, for being a source of inspiration.

Prof. Akhtar Haseeb, Ex-Vice Chancellor, Narendra Deva University of Agriculture and Technology, Kumarganj, Faizabad, India; Pro-Vice Chancellor, Aligarh Muslim University, Aligarh, India; Prof. Mujeebur Rahman Khan, Dean, Faculty of Agricultural Sciences; Prof. Parvez Qamar Rizvi, Chairperson, Department of Plant Protection; Prof. Saghir A. Ansari, former Dean, Faculty of Agricultural Sciences; Prof. M. Yunus Khalil Ansari, former Chairperson, Department of Botany; Prof. Nafees A. Khan, Chairperson, Department of Botany; Prof. Zaki A. Siddiqui;

Prof. Abrar A. Khan; Prof. Tabreiz A Khan; Prof. Hisamuddin; Prof. Mansoor A. Siddiqui; Prof. Iqbal Ahmad; Prof. A. Malik; Dr. R. U. Khan; Prof. M. S. Ansari (Retd.); Prof. M. Haseeb (Retd.); Prof. S. Asharf (Retd.) and Dr. Shamsh Tabreiz of Aligarh Muslim University, Aligarh, India, receive special thanks for providing us critical suggestion during the write-up of this book.

The young Assistant Professors Dr. Faheem Ahmad, Dr. Mu. Naeem, Dr. Tariq Aftab and Dr. Asim Masood also kept pushing us and helped a lot during the proofreading of the book.

Deepest and heartfelt gratitude is extended to Dr. Aisha Sumbul who always took up each obstacle while at some time we felt to be in a scientific turmoil and translated it into an enjoyable moment. Her hot scientific debate and discussion pertaining to the scientific issues are unforgettable which brought about glorious time memorial from onset of this journey.

Dr. Sartaj A. Tiyagi, Dr. Safiuddin, Dr. Ziaul Haque, Dr. Tahir Mohammad Chauhan, Dr. Ram Pal Singh, Dr. Majhrool Hak Ansari, Dr. Zeba Khan, Mr. Hari Raghu Kumar, Ms. Aiman Zafar and Ms. Geeta Rautela were constantly surrounded us whenever we felt to give up—sincere thanks to all of them. Editors would have not completed this task without endless support, prayers and encouragements of their elders and young people during light and dark situations.

We can never stop thinking about our 'little doctor', Mr. Ayan Mahmood who would practically look up and smile at us with two lovely and twinkling eyeballs, each time muttering words of comfort and encouragement.

It is anticipated that our efforts to forward the readers towards the better state of plant science shall be fruitful.

Aligarh, Uttar Pradesh, India — Rizwan Ali Ansari
Aligarh, Uttar Pradesh, India — Rose Rizvi
Aligarh, Uttar Pradesh, India — Irshad Mahmood

Contents

About the Editors

Dr. Rizwan Ali Ansari is currently working as Assistant Professor in the department of Plant Protection, Faculty of Agricultural Sciences, Aligarh Muslim University, Aligarh, India. He obtained his Ph.D. degree from the same university and has focused his research on understanding the mechanisms involved in the development of disease complexes; and also on utilization of organic additives and biological agents for the management of nematode–fungus disease complexes. He has received a number of prestigious awards from various scientific societies, e.g. the Society of Plant Protection Sciences (SPSS) and Nematological Society of India (NSI), for his outstanding contributions in the field of Plant Pathology/Nematology. He has published several book chapters, research and review articles on the utility of organic additives, biocontrol fungi and bacteria, mycorrhizal fungi and plant growth-promoting bacteria in the sustainable management of plant pathogens. Dr. Ansari has also edited the two-volume book *Plant Health Under Biotic Stress*, which has enjoyed international success.

Dr. Rose Rizvi is currently working as assistant professor in the Department of Botany, Aligarh Muslim University, Aligarh, India. She has thus far published a number of research, review articles in the journals of great repute pertaining to management of plant parasitic nematodes. Dr. Rizvi has also participated in various national and international symposia/conferences/workshops and successfully attended the queries of the delegates related to area of her research. She has been the recipient of various awards such as UGC-BSR Research Fellow by UGC; Norman Ernest Borlaug Research Award by Plant Pathology (Photon Foundation); Junior Scientist Award by NESA; CSIR-Research Associateship by MHRD; Scientist of the Year Award-2018 by NESA and Young Scientist of the Year Award-2018 by IFEE, Kolkata.

Dr. Irshad Mahmood is a professor of Plant Pathology and Nematology who has focused his research on the application of organic additives and potent microorganisms for the sustainable management of phytoparasitic nematodes and plant pathogenic fungi. He has been engaged with teaching programme of undergraduate and postgraduate level students for the last 30 years and has many overseas visits including the United States, France and the UK. He has attended a significant number of national and international conferences pertaining to wide area of agricultural sciences and published more than 150 original research papers, review articles and book chapters in various refereed national and international publication media. He has successfully completed many training courses in various ICAR sponsored research institute in India and also in North Carolina State University, Raleigh, USA. He has guided ten Ph.D., several M. Phil. and large number of M.Sc. dissertations. In addition to academic and research contribution, he has also served the university in various administrative capacity.

Chapter 1
Nanobiotechnology-Driven Management of Phytonematodes

M. I. S. Safeena and M. C. M. Zakeel

Abstract Plant parasitic nematodes are responsible for causing significant damages to various commercial crops. At present, several management strategies are applied such as biological, chemical, organic, cultural, nanobiotechnology to control pathogenic nematodes. Use of nematicides of chemical origin are although effective, on another hand it causes environmental perturbations. The emerging of two novel techniques, nanotechnology and biotechnology has resolved many concerns that prevail with the traditional strategies of nematode managements in plants and environment. Nanotechnology based agricultural systems have developed with a worthy scope to manage phytonematodes using drug-carrier and a controllable drug targeting and releasing system as it can enhance the quality of life and world's economy. Through advancement in nanotechnology, there are a number of state of-the-art techniques available including applications of several types of nanoparticles as protectants and carriers in the form of 'nanonematicide'. Several pathogenic phytonematodes are very effectively managed with the means of nanotechnology. Genetic engineering have evolved as a promising field in the management of plant pathogenic nematodes by the means of gene cloning and gene modification of host plants. Various transgenics plants have been developed so far against plant pathogenic nematodes. The key objective of the genetic manipulation would be to control all possible physiological and biological activities of nematode due to the counter effect of host plants by possessing resistance gene/s on the basis of gene for gene concept. There are several proteinase inhibitors genes which have been identified and transferred into host plants to create resistance against pathogenic nematodes. Nematicidal proteins are also considered as "anti-nematode proteins" can directly inhibit the multiplication of pathogenic nematodes. Protein from *Bacillus thuringiensis*, lectins and some antibodies are regarded as nematicidal proteins.

M. I. S. Safeena (✉)
Department of Biological Sciences, Faculty of Applied Sciences, South Eastern University of Sri Lanka, Sammanthurai, Sri Lanka
e-mail: safeenim@seu.ac.lk

M. C. M. Zakeel
Department of Plant Sciences, Faculty of Agriculture, Rajarata University of Sri Lanka, Anuradhapura, Sri Lanka

R. A. Ansari et al. (eds.), *Management of Phytonematodes: Recent Advances and Future Challenges*, https://doi.org/10.1007/978-981-15-4087-5_1

Similarly, other house-keeping genes have been manipulated through RNA interference technique. There are many benefits from the integration of both disciplines i.e. nanotechnology and biotechnology for the management of pathogenic nematodes. Some important issues are yet to be addressed which needs proper and extensive research

Keywords Nanotechnology · Biotechnology · Phytonematodes · Nematode management · Eco-friendly approaches

1.1 Introduction

Plant parasitic nematodes (PPNs), known as phytonematodes, are invisible to the naked eye due to their small size (300–1000 μm and some nearly 4 mm long and 15–35 μm wide). Phytopathogenic nematodes can be encountered in wide range of agroclimatic conditions (Ansari and Khan 2012a, b; Ali et al. 2015). As plant parasites, they have a solid or hollow spear/stylet, which they use to make perforations to withdraw nutrients from plant cells. Phytonematodes can cause diseases in plants themselves or by associating with other pathogens such as fungi, bacteria, and viruses (Elhady et al. 2017; Adam et al. 2014). The combined pathogenic potential of nematodes then becomes crucial and sometimes appear to be far greater in terms of the quantity of injuries they can produce compared to any of the other pathogens individually (Agrios 2005). Among the nematodes, the most disease-causing in plants belonging to the orders Tylenchida and Dorylaimida. Accordingly, many genera of the above two orders are a source of severe damage to economically important plants (Table 1.1). Nearly 90 different species of root-knot nematodes (RKNs) have been identified, and these belong to the genus *Meloidogyne* (Moens et al. 2009).

Since the current world population is rapidly increasing, the agricultural sector faces a huge challenge to produce adequate food to feed all populations under a safe food production system. The world needs an upsurge of agricultural productivity to feed this significantly increasing population. The annual yield loss in crop production in the agricultural sector is approximately $10 billion in the United States and $230 billion worldwide (Abd-Elhawad and Askary 2015; Batish et al. 2008; Chitwood 2003). Cyst and root-knot plant parasitic nematodes are the most prevailing and highly dangerous plant pathogens, causing a very significant loss of economically importance crop plants, including wheat, tomatoes, potatoes, maize, soybeans, sugar beets, and woody plants like pine (Ali et al. 2017). Thus, agricultural scientists and farming communities are seriously experiencing difficulties in the management of PPNs.

The management of disease causing number of PPS under threshold level in the environment of soil and crops is very vital for the sustainable food production while maintaining the food security. At present, there are many strategies applied to control PPNs (Fig. 1.1).

Table 1.1 Economically important phytonematodes in various crop plants

Genus name	Common name	Crop plant
Anguina	Seed gall nematode, shoot gall nematode, seed and leaf gall nematode	Wheat, sugarcane, etc.
Ditylenchus	Stem or bulb nematode	Alfalfa, onion, narcissus, etc.
Belonolaimus	Sting nematode	Cereals, legumes, cucurbits, etc.
Tylenchorhynchus	Stunt nematode	Tobacco, corn, cotton, etc.
Pratylenchus	Lesion nematode	Almost all crops and trees
Radopholus	Burrowing nematode	Banana, citrus, coffee, sugarcane, etc.
Hoplolaimus	Lance nematode	Corn, sugarcane, cotton, alfalfa, etc.
Rotylenchulus	Reniform nematode	Cotton, papaya, tea, tomato, etc.
Globodera	Round-cyst nematode	Potato
Heterodera	Cyst nematode	Tobacco, soybean, sugar beets, cereals, etc.
Meloidogyne	Root-knot nematode	Almost all crop plants
Criconemella	Ring nematode	Woody plants
Hemicycliophora	Sheath nematode	Various plants
Paratylenchus	Pin nematode	Various plants
Tylenchulus	Citrus nematode	Citrus, grapes, olive, lilac, etc.
Aphelenchoides	Foliar nematode	Chrysanthemum, strawberry, begonia, rice, coconut, etc.
Bursaphelenchus	Red ring nematodes	Pine, coconut palm, etc.
Longidorus	Needle nematode	Some plants
Xiphinema	Dagger nematode	Trees, woody vines, many annuals
Paratrichodorus	Stubby-root nematode	Cereals, vegetables, cranberry, apple
Trichodorus	Stubby nematode	Sugar beet, potato, cereals, apple

(Adopted from Agrios 2005)

The agricultural sector applies various traditional control methods of crop cultivation, such as cultural, organic, chemical, and biological techniques, to reduce the damage caused by PPNs to economically important crops and trees. Most of the time they are integrated approach (Hague and Gowen 1987). While there are several cultural methods practiced (Fig. 1.1), crop rotation is generally used as a cultural method, but it is inadequate to control nematodes (Shojaei et al. 2019).

Since nematodes spend their lives in soil or in the vicinity of roots, delivering chemicals to plants sometimes becomes ineffective. Due to high cost and health hazards, such as toxicity to humans and the environment, contamination of groundwater, and residues in food or food products, some effective nematicides have been banned or are no longer in use in crop farming. Similarly, organophosphate and carbamate compounds (oxamyl, fosthiazate, and ethoprophos) are at risk of removal by EU Instruction 91/414/EEC due to their harmful nature (Clayton et al. 2008). For

Fig. 1.1 Different management strategies applied to plant parasitic nematodes (PPNs)

a lot of reasons, the use of resistant varieties of some crops is limited to the control of nematode infections (Roberts 1992). Although the biological method is comparatively safe and practicable, there is uncertainty in the feasibility of use of biocontrol agents. It is a real challenge to develop a biological control agent that will be positively effective worldwide for any PPNs (Dababat et al. 2015; Martinuz et al. 2012). However, compared to all the above practices, emerging fields such as nanotechnology and biotechnology demonstrate extensive promising pathways for managing PPNs through minimized production inputs and maximized crop yield. In addition, both fields have changed the entire scenario of the agricultural sector with a high potential to conceive products under a healthy and friendly environment.

1.2 Nanotechnology in Nematode Management

Nanotechnology was first introduced by physicist Richard Feynman in 1959 (Feynman 1992). The field of nanotechnolgy has grown extraordinarily from its inception to influence all kinds of organic and inorganic materials to the extreme Nano scales, characteristically less than 100 nm (Abraham et al. 2008). The most striking feature of nanoparticles or nanomaterials is a large surface-to-volume ratio, which offers a crossing layer between the materials themselves and their surrounding environment. In addition, the high surface-to-volume ratio increases the rate of chemical and biochemical activities (Dubchak et al. 2010). Nanoparticles (NPs) are greatly considered and have obtained the attention of scientists due to their uncommon physical and chemical characteristics. A significant number of research has been conducted in recent years to assess the prospective use of nanoparticles in an extensive array of applications, such as in biology, genetic engineering, tissue engineering, agricultural techniques, etc. (Fig. 1.2).

In the twenty-first century, the increasing advancement of nanotechnology in agriculture has gained a substantial consideration worldwide since it can be applied to any system of agriculture involved in crop cultivation through a potential and well-ordered release and targeted supply of agrochemicals toward PPNs. Among the different types of nanoparticles, carbonaceous nanoparticles (CNPs) are the most widely used nanomaterials today due to their striking characteristics and various applications in diverse fields in agriculture (Shojaei et al. 2019). Nanotechnology has emerged as a vibrant technique in agriculture when conventional agricultural

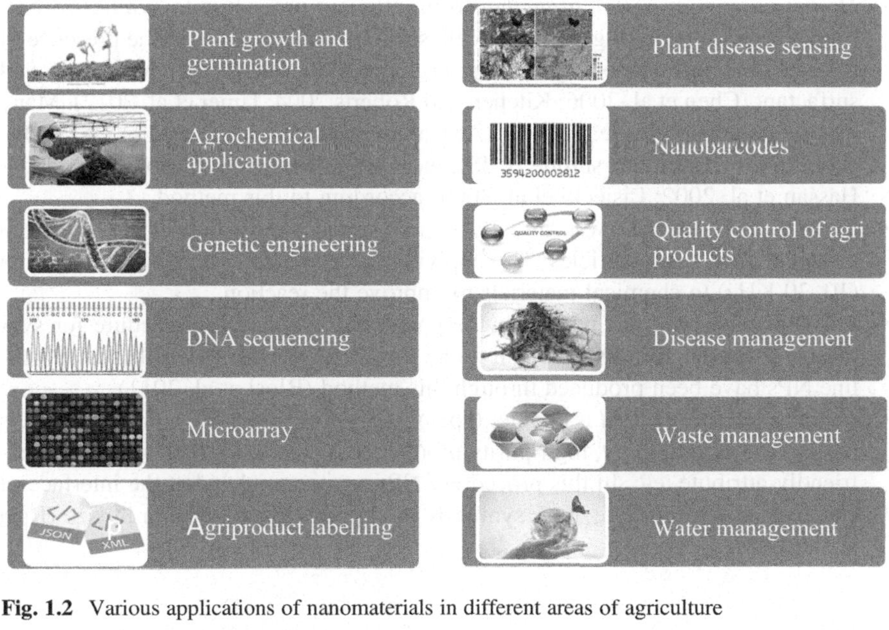

Fig. 1.2 Various applications of nanomaterials in different areas of agriculture

practices have failed to achieve a yield increase for the rapidly growing world population. Furthermore, this provides an ecosystem-friendly technique by reducing the application of pesticides, water usage, and the overall cost of crop production for a sustainable and fastidiousness agricultural system. There are two main features to be considered when nanomaterials are used in plant disease management (Khan and Rizvi 2014):

1. Synthesis of nanomaterials, which deals with the conversion of relevant materials to nanosized particles, between 1 nm and 100 nm.
2. Effective use of nanomaterial for a specific interest or purpose, alone or mixed with some other relevant materials.

The synthesis of specific nanomaterials with the correct and homogeneous size is not easy processes which require unique skill and facilities. Several methods for the synthesis of NPs are employed, such as the following.

1.2.1 Chemical methods

- *Chemical reduction method* is used to synthesize copper nanoparticles by reducing copper salt (Song et al. 2004) in the presence of specific chemicals acting as reducing agents (sodium borohydride (Aslam et al. 2002), isopropyl alcohol and cetyltrimethyl ammonium bromide (CTAB) (Athawale et al. 2005), ascorbate (Wang et al. 2006), polyol (Park et al. 2007), and ascorbic acid (Umer et al. 2012)).
- *Microemulsion or colloidal method* is an efficient procedure to synthesize NPs from microemulsion or group of micelles (oil in water or O/W in the presence of hydrophobic surfactants) by mixing an appropriate quantity of water, oil, and surfactant (Chen et al. 2006; Kitchen and Roberts 2004; Umer et al. 2012). Many metallic nanoparticles (e.g., silver (Ag), aluminum (Al), titanium dioxide (TiO_2), copper (Cu), cadmium sulfide (CdS), etc.) are synthesized (Cason et al. 2001; Hassan et al. 2002; Lisiecki et al. 2000) according to this method.
- *Sonochemical method* was initially demonstrated by Suslick et al. (1996) to produce iron nanoparticles by applying a powerful ultrasound radiation (10–20 KHz) to chemical materials to improve the reaction.
- *Microwave method* has become widespread and is a simple procedure for synthesizing copper NPs (Komarneni 2003; Zhu et al. 2004). Polyol-based crystalline NPs have been produced through this method (Blosi et al. 2011).
- *Electrochemical method* is another type of procedure attracting many researchers because of its simplicity, high-purity product, less cost, user- and environmental-friendly attribute, etc. In this procedure, NPs are accumulated at the interface of electrode and electrolyte, e.g., synthesis of copper NPs 40–60 nm in size (Raja et al. 2008).

- *Solvothermal decomposition* is carried out based on hydrothermal process, in which a heterogenous chemical reaction is allowed to be conducted in a closed vessel in the availability of an aqueous or a nonaqueous solvent at above ambient temperature and > 1 atm pressure (Byrappa and Yoshimura 2001; Byrappa 2005).

1.2.2 Magnetic Nanoparticles

Physical vapor deposition and chemical routes are used to assemble individual atoms into NPs, whereas mechanical abrasion is forced on large particles of materials to break them into NPs (Khan and Rizvi 2014).

1.2.3 Biological Synthesis of NPs

Nanobiotechnology plays a major role in the production of efficient and eco-friendly NPs using "natural bioresources" such as microorganisms and plant natural extracts (Khan et al. 2009):

- *Nanoparticles produced from microorganisms*: Several research studies have shown that large particles of materials are possible to convert into nanoscale particles using "vast and natural factories" of microbes (Khan and Anwer 2011; El-Rafie et al. 2012). There are a significant number of NPs that have been synthesized using different types of microorganisms. For example, (1) silver NPs have been synthesized using *Escherichia coli* (Gurunathan et al. 2009; Manonmani and Juliet 2011), *Fusarium solani* (El-Rafie et al. 2012), and extremophilic yeast strain (Mourato et al. 2011), and (2) gold (Au) NPs have been synthesized using extremophilic yeast (Mourato et al. 2011), extremophilic *Thermomonospora* sp. (Ahmed et al. 2003a, b), mesophilic *Shewanella* sp. (Konishi et al. 2004), *Rhodopseudomonas* sp. (He et al. 2007), *Pseudomonas aeruginosa* (Husseiny et al. 2007), yeast *Pichia jadinii* (Gericke and pinches 2006), and *S. cerevisiae* (Jha et al. 2009).

  Likewise, there are other fungal and bacterial species that have been exploited to synthesize various types of NPs.
- *Nanoparticles from plants*: NPs phytosynthesized using natural extracts of various plants are yet again considered as one of the cost-effective and eco-friendly substrates. One of the widely utilized NPs is silver NPs, which are synthesized with the support of many plant extracts, including *Ocimum tenuiflorum*, *Centella asiatica*, *Syzygium cumini* (Patil et al. 2012), *Acalypha indica* (Krishnaraj et al. 2012), *Camellia sinensis* (Loo et al. 2012), *Urtica dioica* (Nasiri et al. 2014), *Urtica urens* (Nassar 2016), etc., in the presence of a silver nitrate solution.

1.3 Nanomaterials in Plant Disease Management

The application of nanomaterials to control plant diseases is an innovative and smart process. This has made evolvement of number of strategies for application of nanoparticles. Further, this may evidence the exact influence and potential forecasts in forthcoming decade with the advancement of application features of nanotechnology. Moreover, it is assessed that nearly 90% of applied pesticides are wasted or lost throughout or subsequently in the agriculture system (Stephenson 2003; Ghormade et al. 2011). As a consequence, there is an increased enthusiasm to change to the less cost and high performance pesticides, which are minimum or less detrimental to the environment. Hence, the nano size materials in the form of particles, carbon tubes, capsules, cups etc. are used in number of ways to manage many important plant diseases which cause a significant yield loss annually.

Furthermore, nanotechnology can provide benefits to pesticides, like (a) reducing toxicity, (b) enhancing the shelf-life, (c) accelerating the solubility of poorly water-soluble pesticides, (d) improving site-specific uptake into the target pest, etc. (Elizabeth et al. 2018; Hayles et al. 2017). All of these could ensure a green environment with a positive impact. Hence, similar or improved results of the application of chemical pesticides or other cultural or biological methods could be obtained by a direct application of NPs to seeds, roots, and leaves while minimizing the disadvantages of the abovementioned traditional methods. Although a direct application of NPs would be significantly effective, the beneficial microbial population or nontarget organisms that are found surrounding the root zone of plants will be affected much when nanoparticles are introduced straight to the soil. However, a control release of various chemicals for different purposes of plants is possible through NPs as carriers especially when plants are under any stress conditions (Khan et al. 2014) such as flood.

Metal-based nanoprotectants or NPs, such as silver, gold, copper, titanium oxide, zinc oxide (Kah and Hofmann 2014; Gogos et al. 2012; Kim et al. 2018a; Mishra and Singh 2015; Sadeghi et al. 2017; Malerba and Cerana 2016; Rafique et al. 2017), and carbon nanoparticles (Shojaei et al. 2019), and other NPs function as common nanocarriers, like silica (Mody et al. 2014; Barik et al. 2008), chitosan (Malerba and Raffaella 2018; Kashyap et al. 2015; Li et al. 2011), solid lipid nanoparticles (SLN) (Ekambaram et al. 2012; Borel and Sabliov 2014), and layered double hydroxides (LDH) (Xu et al. 2006; Mitter et al. 2017; Bao et al. 2016) nanoparticles have been utilized for plant disease management.

There are two major processes through which the advantages of application of nanomaterial or NPS can be obtained in order to estimate the possibility and usage of NPS in plant disease controlling, espescially regarding the the phytonematodes management:

1. Direct effect on pathogens when NPs are applied alone and used as protectants or applied in the form of nanopesticides by formulating nanomaterials using nanocarriers for insecticides, fungicides, herbicides, and RNA interference (RNAi).

2. Effect of nanomaterials or NPs on the physiological and biochemical activities of pathogens/microorganisms since nanomaterials are highly reactive substances due to their high surface-to-volume ratio.

Many research studies have been carried out under the plant protection strategies to assess the damage caused by the nanomaterial on phytopathogens like bacteria, fungi, viruses, etc. (Table 1.2) and similarly on phytonematodes.

1.4 Nanomaterials in Phytonematode Management

Phytonematodes are traditionally controlled by applying chemical nematicides, employing cultural methods, and cultivating nematode resistance crops. Recently, nanotechnology has been widely developed to improve the agricultural system for controlling phytonematodes and other microorganisms. Nanotechnology-based agricultural systems have developed with a worthy scope to manage phytonematodes using drug carriers and a controllable drug targeting and releasing system (Khot et al. 2012; Mattos et al. 2017; Iavicoli et al. 2016; De Oliveira et al. 2016; Wang et al. 2017; Shen et al. 2017; Liu et al. 2016; Yamamoto and Kuroda 2016; Nakamura et al. 2017). The most crop-damaging nematodes are generally root-knot nematodes, *Meloidogyne* sp., and cyst nematodes (*Heterodera* and *Globodera*), which are sedentary endoparasites affecting many agriculturally important crops from grasses to trees (Oka et al. 2000).

In addition, there are other types of nematodes, such as *Caenorhabditis elegans*, entomopathogenic nematodes, etc., that have been used as experimental organisms to study nanosafety or the effect of NPs on other nontargeted organisms in the surrounding of the treatment area (Kim et al. 2018b; Ma et al. 2018; Taha and Abo-Shady 2016; Kim et al. 2012; Meyer et al. 2010). The nematode *C. elegans* feeds on soil microorganisms and has been used to exemplify the nematode phylum (Boyd and Williams 2003). Hence, *C. elegans* is generally used as a model experimental organism either to prove the activity of various NPs against phytonematodes or to predict the potential application of NPs in other biological aspects.

1.4.1 *Effect of Silver Nanoparticles on Phytonematodes*

Although several methods are used to produce silver nanoparticles (AgNPs), the inclusion of toxic chemical substances cannot be avoided in their common method of synthesis, the chemical approach (Hardman 2006). Therefore, a considerable number of research is being carried out to synthesize them using plant extracts as the biological base so as to maintain "clean," "nontoxic," "harmless," and "eco-friendly green chemistry." There are quite a lot of journal papers that claim to reveal that plant extracts have nematicidal and nematostatic properties (Nour El-Deen and

Table 1.2 Different nanomaterials (protectants and carriers) used in plant disease control

Nanoparticles	Phytopathogens/disease	Reference
Silver nanoparticles	*Alternaria alternate*	Krishnaraj et al. (2012), Bryaskova et al. (2014)
	Sclerotinia sclerotiorum	
	Macrophomina phaseolina	
	Rhizoctonia solani	
	Botrytis cinerea	
	Curvularia lunata	
	Sunn-hemp rosette virus	Jain and Kothari (2014)
	Bean yellow mosaic virus	Elbeshehy et al. (2015)
Titanium dioxide nanoparticles	Bacteria and inactivation of viruses	Sadeghi et al. (2017)
Poly-dispersed gold nanoparticles	Barley yellow mosaic virus	Alkubaisi et al. (2015)
Chitosan	Mosaic virus of alfalfa	Kochkina et al. (1994), Pospieszny et al. (1991), Chirkov (2002)
	Fusarium sp. *Botrytis sp.* Bean mild mosaic virus *Pyricularia grisea*	Kashyap et al. (2015)
	Tobacco mosaic virus Tobacco necrosis virus	Malerba and Raffaella (2018)
	Oleander aphid (*Aphis nerii*) Cotton leafworm (*Spodoptera littoralis*) Root-knot nematode (*Meloidogyne javanica*) Nymphs of the pear psylla (*Cacopsylla pyricola*)	
Zinc nanoparticles	*P. aeruginosa*, *Aspergillus flavus*	Jayaseelan et al. (2012), Rajput et al. (2018)
Ag NPs/PVP (hybrid materials based on polyvinylpyrrolidone with silver nanoparticles)	*Staphylococcus aureus* (gram-positive bacteria), *E. coli* (gram-negative bacteria), *P. aeruginosa* (nonferment gram-negative bacteria), as well as spores of *Bacillus subtilis*,	Azam et al. (2012)
	Candida albicans, *C. krusei*, *C. tropicalis*, *C. glabrata*, and *Aspergillus brasiliensis*	Bryaskova et al. (2011)
CuO NPs	*S. aureus*, *Bacillus subtilis*, *P. aeruginosa*, and *E. coli*	Bryaskova et al. (2011), Azam et al. (2012)
CuSo4 and Na2B4O7	Rust fungi	Singh et al. (2012)
Manganese and zinc	Damping off and charcoal rot diseases in sunflower	Abd El-Hai et al. (2009)

Darwish 2011; Nour El-Deen et al. 2014; Khan et al. 2017; Singh et al. 2017; Cromwell et al. 2014). A potential and effective nematicide can be made from plant extracts when they are formulated into metal-based NPs. Silver NP (AgNP) is one of the most utilized nanomaterial that has emerged as a superior product to control phytonematodes. Silver NPs possess sufficient conductivity, have a good catalytic attribute with pronounced antimicrobial activity, and are chemically stable (Nour El-Deen and Bahig Ahmed El-Deeb 2018; Roh et al. 2009; Chen et al. 2007; Li et al. 2007a, b; Setua et al. 2007).

Lim et al. (2012) have demonstrated that AgNPs cause oxidative stress in the cells of nematodes. Similarly, Nassar (2016) has studied the AgNPs of *Urtica urens* extract associated with rugby (AgNPs-rugby) showed an enhanced nematicidal activity against eggs and second larval stage of *M. incognita* with 11-fold more compare to the plant extract in ethyl acetate (least toxic). The same plant extract (*Urtica urens*) in petroleum ether and in the form of Ag-PE nanoparticles was highly toxic against NPs against both eggs and larva due to their consistent particle size. Another plant-based AgNP has been synthesized through biological and chemical methods after the reaction of silver nitrate with an aqueous solution of ginger extract (*Zingiber officinale*) and sodium borohydride separately. However, the plant based AgNPs has revealed a very significant control of *M. incognita* by reducing number of galls and egg mass and resulted in improved growth and fresh weight of tomato (Nour El-Deen and Bahig Ahmed El-Deeb 2018). Similar research was conducted by Abbasy et al. (2017) to evaluate the nematicidal activities of leaf extracts of *Conyza dioscoridis*, *Melia azedarach*, and *Moringa oleifera* against eggs and second-stage juveniles (J2s) of *Meloidogyne incognita* using crude extracts in different solvents and their Ag nanoformulations. The phytochemical based synthesis of AgNPs showed enriched nematicidal activity affecting J2 and eggs up to the levels of 5 and 2 times respectively while "rugby" was the reference nematicide and that was most toxic against *M. incognita*. The study revealed that the toxicity of all extracts either inhibited nematode activity or caused death, depending on the concentration. Also, both forms of extract (crude and AgNPs) of *C. dioscoridis* showed the highest nematicidal activity among the phytochemical extracts with low LC_{50} value. Nevertheless, the AgNPs showed well-enhanced activity on nematode by increasing certain metabolites 2.5-folds more compared to all other crude extracts.

Green silver NPs (GSNPs), which were formulated from *Ulva lactuca* and *Turbinaria turbinata*, have been applied to eggplants (*Solanum melongena* cv. Login) as a nematicide to control root-knot nematodes (*Meloidogyne javanica*). It has been observed that GSNPs (12.75 mg/100 mL^{-1}) of both algae extracts were active against nematode activity in eggplants. The molecular experiment to evaluate the damage caused by GSNPs to deoxyribonucleic acid (DNA) has been conducted using random amplified polymorphic DNA (RAPD) and expressed sequence tag (EST) markers. Accordingly, the DNA of eggplants has been modified, but differently depending on the concentrations. However, the overall growth rate of the eggplants was remarkably well improved without the phytotoxicity to the plants (Abdellatif et al. 2016).

1.4.2 Other Types of NPs Applied against Phytonematodes

Metal-based types of NPs—gold, platinum, TiO_2, selenium (Se), zinc, copper—and some other synthetic types are also applied to manage nematode attack.

1.4.2.1 Gold NPs

Thakur and Shirkot (2017) and Thakur et al. (2018) have applied gold NPs to *M. incognita*, which was lethal to the nematode without any negative impact on the tomato plants under pot experiment. In fact, the plants had improved in terms of growth and development. The mortality and pathogenic effect of gold nanoparticles (GNPs) have been investigated on entomopathogenic nematodes (*Steinernema feltiae*) by Kucharska et al. (2011). It was shown that the concentration of GNPs and the total duration of the larval stage of nematodes determine the mortality and degree of pathogenicity of nematodes on plants. Generally, engineered nanogold (nAu) particles become insoluble in water once they are released during production or from cosmetic materials or during targeted therapeutic treatments in which nAu is one of the main active materials (Chen et al. 2012; Wang et al. 2011; Zhang et al. 2014; Guix et al. 2008). Bosch et al. (2018) have investigated on the lethal effect of such NPs on *C. elegans* and found that nAu causes internal gonad damage at high concentration, therefore hindering reproduction rather than affecting normal growth. A similar study has been conducted by Panel Laura Gonzalez-Moragas et al. (2017) reveal that treating of *C. elegans* with 11-nm AuNPs caused a higher toxicity compare to that of 150-nm. Chun-Chih Hu et al. (2018) have analyzed the effect of size-tunable gold nanoparticles (Au NPs) with or without 11-mercaptoundecanoic acid (MUA) coating on *C. elegans* as a model experimental nematode. Both the noncoated AuNPs and the MUA-AuNPs were found to be absorbed inside the body, in the intestine and cavities of the nematode. In addition, they affected the growth of axons, and the ratio of MUA to AuNPs was influenced by the body size, mobility, and brood size of the nematode.

1.4.2.2 Copper NPs

A study on the bioaccumulation and toxicological effect of engineered copper nanoparticles (ECuNPs) has proposed a mapping technique to identify the distribution of NPs inside the body of model experimental nematode *C. elegans* using radiation microbeam synchrotron X-ray fluorescence (m-SRXRF) (Gao et al. 2008). The mapping results indicated that ECuNPs has distributed throughout the whole body and the result resembles to the degree of toxicity against *C. elegans*. The effect of CuNPs on the mortality of "entomopathogenic nematodes (EPNs)" *Steinernema feltiae* was verified, and it was confirmed that the mortality of nematodes depends on the concentrations of CuNPs and the length of exposure of

Steinernema larvae to CuNPs (Kucharska et al. 2014). Mohamed et al. (2019) has carried out a similar study using in vitro application of CuNPs to *M. incognita* and it was revealed that the 0.2 g/L of CuNPs was adequate to cause 100% mortality of the nematode. Also, an additional impact of CuNPs was revealed from the experiment, particularly their superior nematicidal efficacy over both silicon carbide NPs (Al Banna et al. 2018) and AgNPs (Taha Entsar 2016).

1.4.2.3 Other NPs

A study has been conducted to evaluate the toxicity of zinc oxide nanoparticles (Pinnacle^{AF} ZnO NP suspension) over an aqueous solution of zinc chloride ($ZnCl_2$) against *C. elegans* (Ma et al. 2009). The types, ZnO NPs and $ZnCl_2$ had very similar effect on experimental parameters like lethality, behavior, reproduction, and transgene expression transgenic strain of *C. elegans*. The findings revealed that there is no significant difference in toxicity caused by ZnO NPs and $ZnCl_2$ against free-living *C. elegans*. Anderson et al. (2018) have focused on an improved way to use NPs to know how plants are modified to increase their resistance against plant pathogens during the application of NPs. Overall, their findings demonstrate that CuO and ZnO NPs alter interkingdom cell signaling processes relevant to crop production. Other nanosized products are cited for enhancing plants' resistance to pathogens. Such products include silica (Suriyaprabha et al. 2014) and elicitors of plant resistance, nanosized glucans, and chitosan (Egusa et al. 2015; Anusuya and Sathiyabama 2015). Likewise, selenium nanoparticles have induced the resistance of tomato to *M. incognita* (Udalova et al. 2018). Nanocapsules of lansiumamide B (NCLB), which have been identified as an innovative nematicide, had efficient and durable effect against phytonematodes (Yin Yan-hua et al. 2012). A similar study by Ardakani (2013) and Kim et al. (2010) showed that silver, silicon oxide, Platinum and titanium oxide had toxicity against the root-knot nematode *M. incognita* under greenhouse condition. Chitosan-based nanoparticles have been applied to control pine wood nematodes and other phytopathogens (Wenlong Liang et al. 2018; Malerba and Cerana 2018; Hassan and Chang 2017).

A comparative study (Ma et al. 2018) of food additive, bulk TiO_2, and nanosized P25 showed high toxicity due to the accumulation of all three components in the body of *C. elegans*. Interestingly, the experiment carried out by Kim et al. (2010) has demonstrated the application of platinum NPs (nano-Pt) functioning as antioxidants and the enhanced scavenging ability of superoxide and hydrogen peroxide in *C. elegans*. As a result, the lifespan of *C. elegans* was extended, and it exhibited strong resistance against excessive oxidative stress.

Carbonaceous nanoparticles (CNPs) are novel NPs and are highly utilized at present as quasi-spherical carbonaceous nanomaterials that are less than 10 nm in size (Mobli et al. 2019; Yang et al. 2009; Baker and Baker 2010; Shojaei et al. 2019). CNPs are considered to be outstanding NPs in this decade because of their unique and versatile characteristics. And they demonstrate a promising role in diverse fields, like agriculture, medicine, biotechnology, material science, etc. (Baptista et al. 2015;

Parisi et al. 2015). A very interesting characteristic of CNPs is the nontoxic carbon as the core molecule, which gives them an extraordinary feature of user-friendliness, allowing them to be utilized in any biological applications. However, although they are extensively used in many divisions of the agricultural sector (Shojaei et al. 2019), according to the latest published data, there is no record yet of any specific and direct application of CNPs for phytonematode management.

Despite the many advantages of nanotechnology relating to plant disease management, there are certain risks associated with the application of this technique in the environment. Importantly, the phytotoxic character of nanomaterials should be assessed carefully before releasing them for commercial purposes. Nanomaterials can not only influence pathogenic nematodes, but even more, they can interfere with the growth, development, and reproduction of host plants. Hence, a detailed study is essential to assess their influence or impact during the seed germination, seedling, and flowering stages of plants.

There are a few previous studies describing unforeseen and controversial results against the use NPs to manage phytonematodes. One of such studies has shown that the dietary exposure of quantum dots (Qdot 625 ITKTM, carboxyl quantum dot NPs) to *C. elegans* has induced multigenerational phenotypic effects due to quantum dot transfer. The result of this study has questioned the "potential safety hazards" of using NPs. In support of this study, Vishnu et al. (2017) have found that various formulas of ZnO nanoparticles have a negative effect on terrestrial plants, aquatic animals and plants, and soil microorganisms. Taha Entsar (2016) has also evaluated the side effect of AgNP on "non-target nematodes," "entomopathogenic nematodes (EPNs)," which are found naturally in the same soil environment and contribute to insect pest control. It was found that the percentage of mortality of EPNs depended on nano-Ag concentrations and exposure time.

1.5 Biotechnology Approaches in Nematode Management

Biotechnology plays a major role in the management of plant diseases caused by pests and pathogens like phytonematodes. Nematodes have a great ability to alter and trigger the plant cell environment by its secretion of three glands located in different parts of the body (Urwin 2007). Barthels et al. (1997) carried out an experiment by selecting initial feeding cells in root with the means of stylet through which secretion from gland cells is injected into the plant cells and thereby adapting the cell environment for their further growth and development. Consequently, there is an intense transformation in the cell development programs and gene expression of root cells (Urwin 2007). Among the major disease causing nematodes, cyst nematodes have taken more influences over the root-knot nematodes regarding these processes through an extensive and swift modification of a single cell by cell dissolution and fusion to end up with a distended syncytium formed due to the union of nearly 200 adjacent cells (Davis et al. 2000; Favery et al. 1998).

Genetic engineering has evolved as a promising field to manage phytonematodes through gene cloning and gene modification of host plants. There are many information on transgenic plants against phytonematodes as to express insecticidal genes such as Bt, trypsin suppressor, lectins, plantibodies, ribosome inactivating proteins, secondary plant metabolites, vegetative insecticidal proteins, etc. (Huang et al. 2018; Hui et al. 2012; Iatsenko et al. 2014a, b; Ali et al. 2017; Wang et al. 2014; Wang et al. 2012; Yu et al. 2015; Urwin et al. 1998; Tamilarasan and Rajam 2013; Yogesh et al. 2017; Banerjee et al. 2017; Davies and Elling 2015; Dutta et al. 2015). Besides, there is a gene transfer technique to transfer especially Bt Cry genes to microbes, such as fungi associated with disease-causing nematodes, which helps the nematodes to complete their life cycle (Cheng et al. 2018; Li et al. 2015).

1.5.1 Developing Host Resistance Transgenic Plants

Plant–nematode interaction provides many avenues to manipulate host plants against nematodes. The main objective of the manipulation of plants by transferring resistance gene/s against PPN in them and controlling all possible physiological and biological activities of nematodes due to the expression of those genes. The basic concept of ‘gene for gene’ interaction plays a significant role corresponding to the ‘resitance gene (R) in the host and an avirulence (Avr) gene’ in the pathogenic nematode. As a result, there will be a series of defense responses (hypersensitive response (HR)) causing the death of cells (necrosis) at the site of infection, or the responses may have an impact on and alter normal physiological and biological reactions, such as reproduction, digestion, metabolic processes, etc., of nematodes. There are many natural resistnace genes found as single gene or poly genic manner which were used to produce transgenic plants against phytonematodes (Fuller et al. 2008). The first R gene (Hs 1^{pro-1}) from wild species of beet was cloned to be applied against *H. schachtii* nematodes (Cai et al. 1997).

However, the R gene that was cloned against *H.schachtii* was failure to function against nematode as expected, due to a lacking character that to fit the pattern of leucine rich repeat (LRR) in the predicted mature protein (Ellis and Jones 1998). The second R gene *Mi* obtained from tomato was cloned in eggplant and tomato to act against three different nematodes (*M. incognita*, *M. javanica*, and *M. arenaria*) (Williamson 1998; Milligan et al. 1998). The expressed amino acid of Mi belonged to a resistance type of plant protein, and its sequence consists of a nucleotide-binding site (NBS) and LRR domains (Williamson 1999). The feasibility of commercially releasing this gene was studied in other crops, such as lettuce and tobacco, using *Agrobacterium*-mediated gene transformation. There was a variation in the segregation of resistance among the crops against the tested nematodes as both tomato and eggplant showed resistance to *M. incognita* (Williamson 1999) but only tomato was resistant to *M. euphorbiae* since R genes are generally effective in only one or a limited species of nematode (Williamson 1998).

Similarly, the transgene Gpa2 from *S. tuberrosum* (Van der Vossen et al. 2000; van der Voort et al. 1999) showed resistance in potato, interestingly activated the female nematode of *Globodera pallida* to become translucent and stagnated. The Hero A gene from tomato demonstrated resistance to *Globodera pallida* and *G. rostochiensis* in potato (Fuller et al. 2008; Sobczak et al. 2005). The gene produced hypersensitive reactions, which caused the degeneration of the surrounding cells in the infected area and in turn made the syncytia abnormal and necrotic. Exactly the same response was shown by two other genes, Rhg1 and Rhg4, from soybean (*Glycine max*) in the genetically engineered soybean plant against *H. glycines* (Kandoth et al. 2011; Liu et al. 2012; Matthews et al. 2013).

1.5.2 Proteinase/Protease Inhibitor Gene/s to Manage Phytonematodes

During plant–nematode interaction, a number of proteinases or proteases are released by the pathogen at the wounding site, and as a countereffect, healthy plants frequently produce protein-based proteinase inhibitors (PIs) to minimize the damage. Phytonematodes have the ability to synthesize four different categories of proteinases (cysteine, serine, metalloproteinases, and aspartic). Several PIs have been studied so far to improve host resistance to nematodes (Table 1.3). Another promising PI, cystatin, has demonstrated enhanced resistance to nematodes in various crops (Urwin et al. 1997, 1998; Chan et al. 2010, 2015; Green et al. 2012; Tripathi et al. 2015; Papolu et al. 2016). Generally, the gene product of Oc-IΔD86 retards the reproductive success of many nematodes (Urwin et al. 1995, 1997, 2000, 2003; Lilley et al. 2004; Vain et al. 1998; Atkinson et al. 2004; Vieira et al. 2015; Papolu et al. 2016).

The fusion of CpTI and Oc-IIΔ86 genes provided an additional resistance property to *Arabidopsis* to manage *G. pallida* and *H. schachtii* effectively (Hepher and Atkinson 1992; Urwin et al. 1998). Further, SpTI-1, CpTI, and PIN2 also had shown significant resistance to nematodes by influencing their sexual fate, fertility, growth, and development (Vishnudasan et al. 2005; Hepher and Atkinson 1992; Cai et al. 2003). Similarly, CeCPI hindered sex determination and gall formation in *M. incognita* (Chan et al. 2010, 2015), while CCII impeded the reproductive success and feeding behavior of *R. similis*, *Helicotylenchus multicinctus*, and *Meloidogyne* sp. (Roderick et al. 2012; Tripathi et al. 2015). In recent times, a dual approach for creating resistance, e.g., CpTI and Oc-IΔD86 (Urwin et al. 1998) and CeCPI and PjCHI-1 (Chan et al. 2015), i.e., "dual proteinase inhibitor," has been used to manage phytonematodes without disturbing soil quality. The combination of two different resistance genes creates a targeted resistance environment to *G. pallida* without causing any harmful effect to nontarget nematodes in the soil atmosphere (Green et al. 2012).

Table 1.3 Various protease inhibitor (PI) genes as a progressive way of developing transgenic plants resistant to phytonematodes

PIs gene and source of origin	Transgenic crop	Targeted phytonematode
CpTI—*Vigna unguiculata*	Potato	*G. pallida* and *M. incognita* Hepher and Atkinson (1992)
SpTI-1—*Ipomoea batatas*	Sugar beet	*H. schachtii* Cai et al. (2003)
PIN2—*Solanum tuberosum*	Wheat	*H. avenae* Vishnudasan et al. (2005)
Oc-IΔD86—*Oryza sativa*	Potato	*G. pallida* and *M. incognita* Urwin et al. (1995, 2003) and Lilley et al. (2004)
	Arabidopsis thaliana	*H. schachtii*, *M. incognita*, and *R. reniformis* Urwin et al. (1997, 2000)
	Rice	*M. incognita* Vain et al. (1998)
	Musa acuminate	*R. similis* Atkinson et al. (2004)
	Lilium longiflorum	*Pratylenchus penetrans* Vieira et al. (2015)
	Solanum melongena	*M. incognita* Papolu et al. (2016)
CeCPI—*Colocasia esculenta*	Tomato	*M. incognita* Chan et al. (2010, 2015)
CCII—*Zea mays*	*Musa spp.*	*R. similis, Helicotylenchus multicinctus*, and *Meloidogyne spp.* Roderick et al. (2012), Tripathi et al. (2013) and Tripathi et al. (2015)

1.5.3 Nematicidal Proteins

They are considered as “anti-nematode proteins,” which can directly inhibit the growth and development of nematodes. Protein from *Bacillus thuringiensis*, lectins, and some antibodies are regarded as nematicidal proteins. Although Bt toxin was first used as an antinematode protein by Marroquin et al. (2000), Cheng et al. (2011) revealed in detail that a prismatic and irregular-shaped parasporal crystals from *Bacillus thuringiensis* had the potential to control phytonematodes because of their high toxicity.

Accordingly, the nematicidal activity of Cry 1 Ea 11 from *B. thuringiensis* BRC-XQ12 was tested against the pine wood nematode *Bursaphelenchus xylophilus*, and it was found that BRC-XQ12 had the most toxic insecticidal crystal proteins (ICPs) against nematodes with LC50 equal to 32.13 μg/ml (Huang et al. 2018). Fascinatingly, a similar concept has been applied to fungi, on which *Bursaphelenchus xylophilus* (pine wood nematode (PWN)) depends to complete its life cycle. Here, Bt Cry gene was transferred to the genome of fungus eaten by PWN using *Agrobacterium*-mediated gene transformation. The result of this study showed that Cry5Ba3Θ retarded the growth and fitness of the PWN (Cheng et al. 2018). However, a serious limitation is encountered in using Cry genes against the very popular parasitic RKN (*M. incognita*) and CN (*H. schachtii*) basically due to the dissimilarity in the ability of their stylet to take up different sizes of toxic proteins expressed by Cry genes. Apparently, a larger size of toxic protein, 50 kDA, can be sieved through the "molecular sieve" (stylet) of RKN (Sobczak et al. 1999; Li et al. 2007a, b, 2008) but not 25 kDA through that of CN (Urwin et al. 1998). This has caused a major challenge of applying the technique to suit all kinds of pathogenic nematodes.

Lectins have the ability to bind with glycans or free sugar or glycoproteins or glycolipids, thereby hindering intestinal digestion pathways (Peumans and Van Damme 1995; Vasconcelos and Oliveira 2004). CaMV35S promoter driving *Galanthus nivalis* lectin or agglutinin (GNA) has extensively been utilized to control root-knot, cyst, and lesion nematodes in several economically important crops like potato, *Brassica napus*, etc. (Burrows et al. 1998; Ripoll et al. 2003).

The formation of syncytium is a vital step in the life cycle of parasitic nematodes since it creates a supportive environment in the host plants so they can feed on them. The secretion of pharyngeal glands by the nematodes induces the plant cells to redifferentiate to form syncytia. "Plantibodies" are antibodies expressed in host plants that function against proteins in pharyngeal secretions and create resistance in the host against RKN and CN by suppressing the formation of syncytia. However, there is a limited study being reported on the application of plantibodies to manage phytonematodes (Fioretti et al. 2002; Sharon et al. 2002).

1.5.4 Housekeeping Genes and RNA Interference (RNAi) in Transgenic Developments

The first genome sequence of *C. elegans* and other plant parasitic nematodes have unveiled many unanswered questions in proteomics, genomics, and transcriptional processes regarding the molecular basis behind the pathogenicity of nematodes. The exposed information provided the means to identify biologically essential genes that would be the basis and targets for RNA interference (Rosso et al. 2009; Thorat et al. 2017).

1.5.4.1 Housekeeping Genes

Any living organisms possess functionally characterized genes that are responsible for many basic tasks. These kinds of genes are called "House-keeping genes" (Tamilarasan and Rajam 2013; Dutta et al. 2014). Plant parasitic nematodes also have such genes, which are arbitrarily expressed and involved in several physiological and biological processes during growth and development. This has opened avenues to manipulate such genes through RNAi techniques in order to hinder the pathogenic ability of parasitic phytonematodes (Banerjee et al. 2017b). The first two of such genes (for splicing and for integrase activities) that were genetically engineered in tobacco plants against *M. incognita* clearly showed that the transcript dsRNA of both genes under promoter control was depleted in the female adult, and therefore a significant reduction in the number and size of the galls of *M. incognita* was observed in transgenic tobacco (Yadav et al. 2006). Three other genes (RPS-3a, RPS-4, and SPK-1) from *Heterodera glycines* that were engineered in soybean reduced the infection of *H. glycines* by 80–88% (Klink and Matthews 2009). A similar result was observed with the PRP 17 gene, which reduced infection by 53% and reproduction by 79%; meanwhile, Cpn 1 showed 95% reduction of the egg mass of *H. glycines* (Li et al. 2010). Although, there is a high potential to use housekeeping genes to control nematode through RNAi techniques, it is subject to the great risk of using them since they are mostly conserved across the plant and animal kingdoms. Hence, they may target or affect any beneficial organisms, including the host plant.

1.5.4.2 RNA Interference (RNAi) Technique to Suppress Nematode

RNAi has been emerged as a very valuable technique and becoming an interesting field of study for gene-silencing intended at useful analysis of number of genes by overpowering their expression in PPNs. In this strategy, the pathogenic nematodes take in "double-stranded RNA" (dsRNA) or "short interfering RNAs" (siRNAs) from the plants expressing these RNAs, which elicit a systemic RNAi response in nematodes (Fig.1.3).

RNAi is considered as an obvious method to slience the effector genes in nemattodes (Gheysen and Vanholme 2007; Lilley et al. 2007; Fuller et al. 2008; Rosso et al. 2009; Maule et al. 2011; Tamilarasan and Rajam 2013). Lilley et al. (2012) have reviewed numerous methods, from in vitro assays with *C. elegans* to delivering RNAi *in planta*, to reduce cyst nematodes. Similarly, Youssef et al. (2013) have confirmed the efficiency of RNAi technique by silencing the *H. glycines* gene HgALD (responsible for encoding fructose-1, 6-diphosphate aldolase) to provide energy for the mobility of nematodes during the infection phase in host plants, and this resulted in 58% reduction of female plants. In recent time, Tripathi et al. (2017) have reviewed the application of RNAi for improving nematode resistance by the suppression of important effector proteins. RNAi-mediated crop security against nematode give the impression to be most promising than other existing methods, in terms of effectiveness, constancy and its capability to overwhelm gene expression in a controlled manner.

However, during an effective plant–nematode interaction, nematodes are somehow able to suppress defense-related genes, the overexpression of which leads to enhanced resistance (Ali et al. 2013). Therefore, to overcome this problem, specific promoters that have the ability to express in a controlled manner at the feeding site only could be used (Siddique et al. 2009, 2011). Nevertheless, silencing the genes of host plants or using constitutive promoters to overcontrol the delivery of the genes or

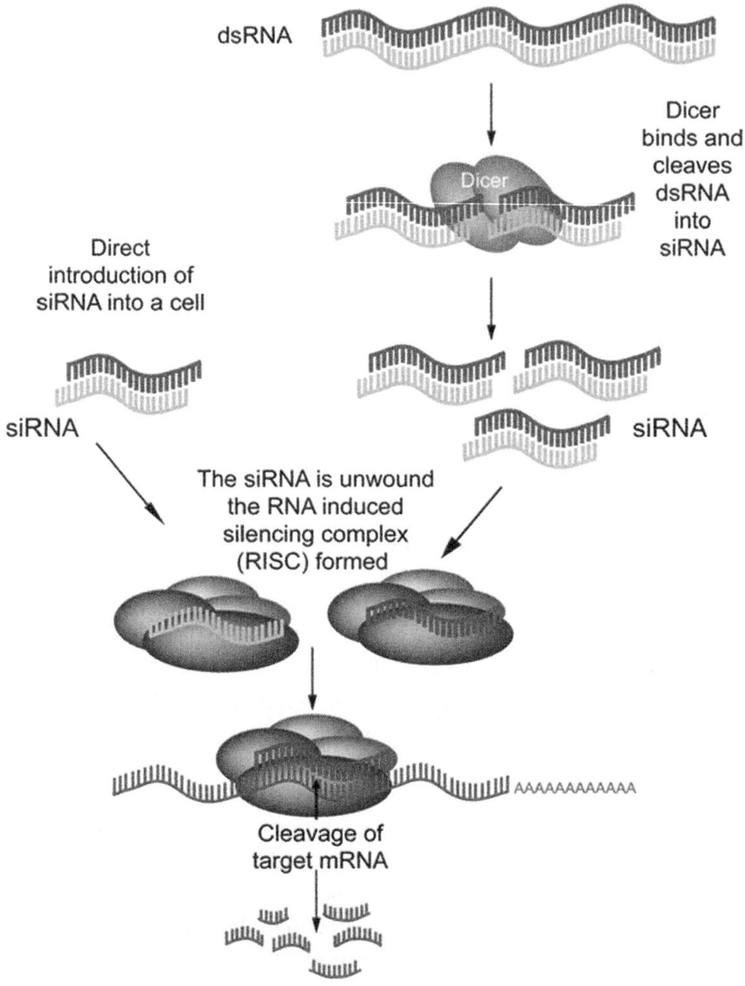

Fig. 1.3 Schematic description of RNA interference of gene silencing in nematode (source: http://www.landesbioscience.com/curie/chapter/4738/)

suppress the genes (Ali and Abbas 2016) will cause a negative impact on the host plants by disturbing their normal physiological processes. Aside from the "CaMV-35S" promoter, quite a few syncytium-related promoters could be applied to enhance the defense-associated genes in feeding sites in order to increase resistance (Ali et al. 2013, 2014; Ali and Abbas 2016). However, a genome of a host plant with all possible genes to enhance resistance against pathogenic nematode may bring the exclusive "immunity" against nematodes. Sometimes pathogenic nematodes are smart and retard the defense system of host plants (Kyndt et al. 2012; Ali et al. 2015). This may perhaps be the stimulating window of information for additional future studies to elucidate on how nematodes are able to conquer systemic plant defense mechanisms. It is concluded that the use of different transgenic strategies has shown good promise for nematode resistance. They have been helpful for the reduction of nematodes.

1.6 Conclusions and Future Prospects

Many literature reviews, inside analysis of facts and data have shown that there is a significant vision to utilize nanotechnology in vast, prospective ways relevant to plant disease identification and management, especially the control of pathogenic phytonematodes. It has been revealed that the application of various types of nanoparticles and nanocarriers and considering other relevant aspects on this technique play a vital role in managing the damage caused by nematodes, with special reference to root-knot, cyst, and other considerable disease-causing nematodes. Various methods are employed to synthesize nanomaterials in order to attain very effective results. The direct application of nanoparticles expressively reduces the damage caused by phytonematodes in several economically important crops. Furthermore, other nanomaterials (nanocapsules, nanotubes, etc.) are also being used for transferring and control releasing of highly active components of biopesticides, organic pesticides, host-resistance-inducing chemicals, and inhibitors to manage pathogenic nematodes in soil and plants. Besides disease control, metal-based (Ag, Cu, Ti, Se, Au, cobalt (Co)) nanosensors or enzyme-based biosensors are extensively used in disease diagnosis and in residue analysis of pesticides.

The interaction mechnaisms between plant and nematode provides sufficient opportunities to involve biotechnology in the management of plant diseases, including the effect of phytonematodes. Apparently, host plants should possess adequate resistance against nematodes in order to minimize the damage caused by pathogens. There are several potential ways to apply biotechnology to manage phytonematodes. Host plants are genetically engineered by cloning resistance gene/s from other organisms or by enhancing the expression of their own resistance gene using some versatile "promoters." As a consequence of this, enormous yield losses in crop plants caused by nematodes could be minimized. Besides gene/s from other or same species of plants responsible for inducing protease inhibitors, housekeeping genes of nematodes are diversely manipulated to enhance resistance in host plants. The recent involvement of targeted "gene silencing" of nematode effector proteins through *in planta* RNAi technology shows a pronounced prospective for plant resistance against numerous species of nematodes.

References

Abbasy MA, Abdel-Rasoul MA, Nassar AMK, Soliman MB (2017) Nematicidal activity of silver nanoparticles of botanical products against root-knot nematode, Meloidogyne incognita. Arch Phytopathol Plant Prot 50(17–18):909–926. https://doi.org/10.1080/03235408.2017.1405608

Abd El-Hai KM, El-Metwally MA, El-Baz SM, Zeid AM (2009) The use of antioxidants and microelements for controlling damping-off caused by *Rhizoctonia solani* and charcoal rot caused by *Macrophomina phasoliana* on sunflower. Plant Pathol J 8(3):79–89

Abd-Elhawad MM, Askary TH (2015) Impact of phytonematodes on agriculture economy. In: Askary TH, Martinelli PRP (eds) Biocontrol agents of phytonematodes. Wallingford, CABI, pp 3–49

Abdellatif KF, Abdelfattah RH, Mostafa El-Ansary MS (2016) Green nanoparticles engineering on root-knot nematode infecting eggplants and their effect on plant DNA modification. Iran. J Biotechnol 14(4):250–259. https://doi.org/10.15171/ijb.1309

Abraham A, Kannangai K, Sridharan G (2008) Nanotechnology: a new frontier in virus detection in clinical practice. Indian J Med Microbiol 26(4):297

Adam M, Westphal A, Hallmann J, Heuer H (2014) Specific microbial attachment to root knot nematodes in suppressive soil. Appl Environ Microbiol 80:2679–2686

Agrios GN (2005) Plant pathology, 5th edn. Elsevier Academic Press, Burlington, MA

Ahmed A, Mukherjee P, Senapati S, Mandal MI, Khan R, Kumar R, Sastry M (2003a) Extracellular biosynthesis of silver nanoparticles using the fungus *Fusarium oxysporum*. Colloids Surf B Biointerfaces 28:313–318

Ahmed A, Senapati S, Khan R, Kumar R, Sastry M (2003b) Extracellular biosynthesis of gold nanoparticles by a novel extremophilic actinomycete, *Thermomonospora* sp. Langmuir 19:3550–3553

Al Banna L, Salem N, Ghrair A, Habash S (2018) Impact of silicon carbide nanoparticles on hatching and survival of soil nematodes *Caenorhabditis elegans* and *Meloidogyne incognita*. Appl Ecol Environ Res 16(3):2651–2662

Ali MA, Abbas A (2016) Analysis of reporter proteins GUS and DsRed driven under the control of CaMV35S promoter in syncytia induced by beet cyst nematode *Heterodera schachtii* in *Arabidopsis* roots. Adv Life Sci 3:89–96

Ali MA, Plattner S, Radakovic Z, Wieczorek K, Elashry A, Grundler FM et al (2013) An Arabidopsis ATPase gene involved in nematode-induced syncytium development and abiotic stress responses. Plant J 74:852–866

Ali MA, Wieczorek K, Kreil DP, Bohlmann H (2014) The beet cyst nematode *Heterodera schachtii* modulates the expression of WRKY transcription factors in syncytia to favour its development in Arabidopsis roots. PLoS one 9:e102360

Ali MA, Abbas A, Azeem F, Javed N, Bohlmann H (2015) Plant-nematode interactions: from genomics to metabolomics. Int J Agric Biol 17:1071–1082. https://doi.org/10.17957/IJAB/15.0037

Ali MA, Azeem F, Abbas A, Joyia FA, Li H, Dababat AA (2017) Transgenic strategies for enhancement of nematode resistance in plants. Front Plant Sci 8:750. https://doi.org/10.3389/fpls.2017.00750

Alkubaisi NAO, Aref NMMA, Hendi AA (2015) Method of inhibiting plant virus using gold nanoparticles. US Patents US9198434B11

Anderson AJ, Joan EM, Astrid RJ, David WB (2018) CuO and ZnO nanoparticles modify Interkingdom cell signaling processes relevant to crop production. J Agric Food Chem 66:6513–6524. https://doi.org/10.1021/acs.jafc.7b01302

Ansari RA, Khan TA (2012a) Parasitic association of root-knot nematode, *Meloidogyne incognita* on guava. e-J Sci Technol 5:65–67

Ansari RA, Khan TA (2012b) Diversity and community structure of phytonematodes associated with guava in and around Aligarh, Uttar Pradesh, India. Trends Biosci 5(3):202–204

Anusuya S, Sathiyabama M (2015) Protection of turmeric plants from rhizome rot disease under field conditions by β-D-glucan nanoparticle. Int J Biol Macromol 77:9–14

Ardakani AS (2013) Toxicity of silver, titanium and silicon nanoparticles on the root-knot nematode, Meloidogyne incognita, and growth parameters of tomato. Nematology 15(6):1–7. https://doi.org/10.1163/15685411-00002710

Aslam M, Gopakumar G, Shoba TL, Mulla IS, Vijayamohanan K (2002) Formation of Cu and Cu_2O nanoparticles by variation of the surface ligand: preparation, structure and insulating-to-metallic transition. J Colloid Interf Sci 255:79–90

Athawale AA, Katre PP, Kumar M, Majumdar MB (2005) Synthesis of CTAB-IPA reduced copper nanoparticles. Mater Chem Phys 91:507–512

Atkinson HJ, Grimwood S, Johnston K, Green J (2004) Prototype demonstration of transgenic resistance to the nematode *Radopholus similis* conferred on banana by a cystatin. Transgenic Res 13:135–142

Azam AS, Ahmed M, Oves MS, Khan SS, Memic HA (2012) Antimicrobial activity of metal oxide nanoparticles against gram-positive and gram-negative bacteria: a comparative study. Int J Nanomedicine 7:6003–6009

Baker SN, Baker GA (2010) Luminescent carbon nanodots: emergent nanolights. Angew Chem Int Ed 49(38):6726–6744

Banerjee S, Banerjee A, Gill SS, Gupta OP, Dahuja A, Jain PK, Sirohi A (2017) RNA interference: a novel source of resistance to combat plant parasitic nematodes. Front Plant Sci 8:834

Banerjee S, Banerjee A, Gill SS, Gupta OP, Dahuja A, Jain PK, Sirohi A (2017b) RNA Interference: a novel source of resistance to combat plant parasiti nematodes. Front Plant Sci 8:834

Bao W, Wang J, Wang Q, O'Hare D, Wan Y (2016) Layered double hydroxide Nanotransporter for molecule delivery to intact plant cells. Sci Rep 6:26,738. https://doi.org/10.1038/srep26738

Baptista FR et al (2015) Recent developments in carbon nanomaterial sensors. Chem Soc Rev 44 (13):4433–4453

Barik T, Sahu B, Swain V (2008) Nanosilica—from medicine to pest control. Parasitol Res 103:253–258

Barthels N, der Leevan FM, Klap J, Goddijn OJM, Karimi M, Puzio P, Grundler FMW, Ohl SA, Lindsey K, Robertson L, Robertson WM, Montagu M, Gheysenyalf G, Sijmons PC (1997) Regulatory sequences of Arabidopsis drive reporter gene expression in nematode feeding structures. Plant Cell 9(21):19–2134

Batish DR, Singh HP, Kohli SK, Kaur S (2008) Eucalyptus essential oil as a natural pesticide. For Ecol Manag 256:2166–2174

Blosi M, Albonetti S, Dondi M, Martelli C, Baldi G (2011) Microwave-assisted polyol synthesis of cu nanoparticles. J Nanopart Res 13:127–138

Borel T, Sabliov C (2014) Nanodelivery of bioactive components for food applications: types of delivery systems, properties, and their effect on ADME profiles and toxicity of nanoparticles. Annu Rev Food Sci Technol 5:197–213

Bosch S, Botha TL, Jordaan A, Maboeta M, Wepener V (2018) Sublethal effects of ionic and nanogold on the nematode *Caenorhabditis elegans*. J Toxicol 2018:6218193. https://doi.org/10.1155/2018/6218193

Boyd WA, Williams PL (2003) Comparison of the sensitivity of three nematode species to copper and their utility in aquatic and soil toxicity tests. Environ Toxicol Chem 22:2768–2774

Bryaskova R, Pencheva D, Nikolov S, Kantardjiev T (2011) Synthesis and comparative study on the antimicrobial activity of hybrid materials based on silver nanoparticles (AgNPs) stabilized by polyvinylpyrrolidone (PVP). J Chem Biol 4:185–191. https://doi.org/10.1007/s12154-011-0063-9

Bryaskova R, Pencheva D, Nikolov S, Kantardjiev T (2014) Synthesis and comparative study on the antimicrobial activity of hybrid materials based on silver nanoparticles (AgNPs) stabilized by polyvinylpyrrolidone (PVP). J Chem Biol 4:185–191

Burrows PR, Barker ADP, Newell CA, Hamilton WDO (1998) Plant-derived enzyme inhibitors and lectins for resistance against plant parasitic nematodes in transgenic crops. Pestic Sci 52:176–183

Byrappa K (2005) Hydrothermal processing. In: Kirk RE, Othmer DF (eds) Encyclopedia of chemical technology. Wiley, London, UK

Byrappa K, Yoshimura M (2001) Hydrothermal technology: principles and applications. In: Byrappa K, Yoshimura M (eds) Handbook of hydrothermal technology. Noyes publications, New Jersey

Cai DG, Kleine M, Kifle S, Harloff HJ, Sandal NN, Marcker KA et al (1997) Positional cloning of a gene for nematode resistance in sugar beet. Science 275:832–834. https://doi.org/10.1126/science.275.5301.832

Cai D, Thurau T, Tian Y, Lange T, Yeh KW, Jung C (2003) Sporamin mediated resistance to beet cyst nematodes (*Heterodera schachtii* Schm) is dependent on trypsin inhibitory activity in sugar beet (*Beta vulgaris* L) hairy roots. Plant Mol Biol 51:839–849

Cason JP, Miller ME, Thompson JB, Roberst CB (2001) Solvent effects on copper nanoparticles growth behavior in AOT reverse micelle systems. J Phys Chem B 105:2297–2302

Chan Y-L, Yang A-H, Chen J-T, Yeh K-W, Chan M-T (2010) Heterologous expression of taro cystatin protects transgenic tomato against *Meloidogyne incognita* infection by means of interfering sex determination and suppressing gall formation. Plant Cell Rep 29:231–238

Chan Y-L, He Y, Hsiao T-T, Wang C-J, Tian Z, Yeh K-W (2015) Pyramiding taro cystatin and fungal chitinase genes driven by a synthetic promoter enhances resistance in tomato to root-knot nematode *Meloidogyne incognita*. Plant Sci 231:74–81

Chen L, Zhang D, Chen J, Zhou H, Wan H (2006) The use of CTAB to control the size of copper nanoparticles and the concentration of alkyl thiols on their surface. Mater Sci Eng A 415:156–161

Chen M, Feng Y-G, Wang X, Li T-C, Zhang J-Y, Qian D-J (2007) Silver nanoparticles capped by oleylamine: formation, growth, and self-organization. Langmuir 23(10):5296–5304. https://doi.org/10.1021/la700553d

Chen H, Li B, Ren X, Li S, Ma Y, Cui S, Gu Y (2012) Multifunctional near-infrared-emitting nano-conjugates based on gold clusters for tumor imaging and therapy. Biomaterials 33:8461–8476

Cheng F, Zhang D, Liu Y, Wang Z, Cheng J, He Xiao MQ (2011) Nematicidal activities and identification of a biocontrol bacterium YC-10 strain. ACTA Phytopathol Sinica 41:203–209

Cheng C, Qin J, Wu C, Lei M, Wang Y, Zhang L (2018) Suppressing a plant-parasitic nematode with fungivorous behavior by fungal transformation of a Bt cry gene. Microb Cell Factories 17:116. https://doi.org/10.1186/s12934-018-0960-5

Chirkov S (2002) The antiviral activity of chitosan (review). Appl Biochem Microbiol 38:1–8

Chitwood DJ (2003) Research on plant—parasitic nematode biology conducted by the United States Department of Agriculture, Agricultural Research Service. Pest Manage Sci 59:748–753

Clayton R, Storey M, Parker B, Ballingall M, Davies K (2008) Impact of reduced pesticide availability on control of potato cyst nematodes and weeds in potato crops. Potato Council Ltd, Oxford

Cromwell WA, Yang J, Starr JL, Jo Y-K (2014) Nematicidal effects of silver nanoparticles on root-knot nematode in Bermuda grass. J Nematol 46(3):261–266

Dababat A, Imren M, Erginbas-Orakci G, Ashrafi S, Yavuzaslanoglu E, Toktay H (2015) The importance and management strategies of cereal cyst nematodes, *Heterodera* sp, in Turkey. Euphytica 202:173–188. https://doi.org/10.1007/s10681-014-1269-z

Davies LJ, Elling AA (2015) Resistance genes against plant-parasitic nematodes: a durable control strategy? Nematology 17:249–263

Davis EL, Hussey RS, Baum TJ, Bakker J, Schots A (2000) Nematode parasitism genes. Annu Rev Phytopathol 38:365–396. https://doi.org/10.1146/annurev.phyto.38.1.365

De Oliveira JL, Campos EV, Bakshi M, Abhilash PC, Fraceto LF (2016) Application of nanotechnology for the delivery to intact plant cells. Sci Rep 6:6. https://doi.org/10.1016/j.enmm.2017.12.006

Dubchak S, Ogar A, Mietelski JW, Turnau K (2010) Influence of silver and titanium nanoparticles on arbuscular mycorrhiza colonization and accumulation of radio caesium in *Helianthus annuus*. Spanish J Agric Res 8:103–108

Dutta TK, Banakar P, Rao U (2014) The status of RNAi based-transgenic research in plant nematology. Front Microbiol 5:760

Dutta TK, Papolu PK, Banakar P, Choudhary D, Sirohi A, Rao U (2015) Tomato transgenic plants expressing hairpin construct of a nematode protease gene conferred enhanced resistance to root-knot nematodes. Front Microbiol 6:260. https://doi.org/10.3389/fmicb.2015.00260

Egusa M, Matsui H, Urakami T, Okuda S, Ifuku S, Nakagami H, Kaminaka H (2015) Chitin nanofiber elucidates the elicitor activity of polymeric chitin in plants. Front Plant Sci 6:1098

Ekambaram P, Sathali AAH, Priyanka K (2012) Solid lipid nanoparticles: a review. Sci Rev Chem Commun 2:80–102

Elbeshehy EKF, Elazzazy AM, Aggelis G (2015) Silver nanoparticles synthesis mediated by new isolates of *Bacillus* sp nanoparticle characterization and their activity against bean yellow mosaic virus and human pathogens. Front Microbiol 6:453

Elhady A, Giné A, Topalovic O, Jacquiod S, Sørensen SJ, Sorribas FJ (2017) Microbiomes associated with infective stages of root-knot and lesion nematodes in soil. PLoS One 12(5): e0177145. https://doi.org/10.1371/journal.pone.0177145

Elizabeth AW, Aflaq H, Karishma TM, Neena M, Hanu RP (2018) Review on nanotechnology for plant disease management. Agronomy 8:285. https://doi.org/10.3390/agronomy8120285

Ellis J, Jones D (1998) Structure and function of proteins controlling strain-specific pathogen resistance in plants. Curr Opin Plant Biol 1:288–293

El-Rafie MH, Shaheen TI, Mohamed AA, Hebeish A (2012) Bio-synthesis and applications of silver nanoparticles onto cotton fabrics. Carbohydr Polym 90:915–920

Favery B, Lecomte P, Gil N, Bechtold N, Bouchez D, Dalmasso A, Abad P (1998) A plant gene involved in early developmental steps of nematode feeding cells. EMBO J 17:6799–6811

Feynman (1992) There's plenty of room at the bottom [data storage]. J Microelectromech Syst 1 (1):60–66

Fioretti L, Porter A, Haydock PJ, Curtis R (2002) Monoclonal antibodies reactive with secreted-excreted products from the amphids and the cuticle surface of *Globodera pallida* affect nematode movement and delay invasion of potato roots. Int J Parasitol 32:1709–1718

Fuller VL, Lilley CJ, Urwin PE (2008) Nematode resistance. New Phytol 180:27–44

Gao Y, Liu N, Chen C, Luo Y, Li Y, Zhang Z, Zhao Y, Zhao B, Iidad A, Zhifang Chai J (2008) Mapping technique for biodistribution of elements in a model organism *Caenorhabditis elegans* after exposure to copper nanoparticles with microbeam synchrotron radiation X-ray fluorescence. Anal At Spectrom 23:1121–1124

Gericke M, Pinches A (2006) Biological synthesis of metal nanoparticles. Hydrometallurgy 83:132–140

Gheysen G, Vanholme B (2007) RNAi from plants to nematodes. Trends Biotechnol 25:89–92

Ghormade V, Deshpande MV, Paknikar KM (2011) Perspectives for nano-biotechnology enabled protection and nutrition of plants. Biotechnol Adv 29:792–803

Gogos A, Knauer K, Bucheli TD (2012) Nanomaterials in plant protection and fertilization: current state, foreseen applications, and research priorities. J Agric Food Chem 60:9781–9792

Gonzalez-Moragas L, Berto P, Vilches C, Quidant R, Kolovou A, Santarella-Mellwig R, Schwab Y, Stürzenbaum S, Roiga A, Laromaine A (2017) In vivo testing of gold nanoparticles using the *Caenorhabditis elegans* model organism. Acta Biomater 53:598–609. https://doi.org/10.1016/j.actbio.2017.01.080

Green J, Wang D, Lilley CJ, Urwin PE, Atkinson HJ (2012) Transgenic potatoes for potato cyst nematode control can replace pesticide use without impact on soil quality. PLoS One 7:e30973

Guix C, Carbonell J, Comenge J (2008) Nanoparticles for cosmetics how safe is safe? Contrib Sci 4:213–217

Gurunathan S, Kalishwaralal K, Vaidyanathan R, Venkataraman D, Pandian SRK (2009) Biosynthesis, purification and characterization of silver nanoparticles using *Escherichia coli*. Colloid Surf B Biointerfaces 74:328–335

Hague NGM, Gowen SR (1987) Chemical control of nematodes. In: Brown RH, Kerry BR (eds) Principles and practice of nematode control in crops. Academic Press, Sydney, pp 131–178

Hardman R (2006) A toxicologic review of quantum dots: toxicitydepends on physicochemical and environmental factors. Environ Health Perspect 114:165–172

Hassan O, Chang T (2017) Chitosan for eco-friendly control of plant disease. Asian J Plant Pathol 11:53–70. https://doi.org/10.3923/ajppaj.2017.53.70

Hassan PA, Raghavan SR, Kaler EW (2002) Microstructural changes in SDS micelles induced by hydrotropic salt. Langmuir 18:2543–2548

Hayles J, Johnson L, Worthley C, Losic D (2017) Nanopesticides: a review of current research and perspectives. New Pestic Soil Sens:193–225. https://doi.org/10.1016/B978-0-12-804299-1.00006-0

He SY, Guo ZR, Zhang Y, Zhang S, Wang J, Gu N (2007) Biosynthesis of gold nanoparticles using the bacteria *Rhodopseudomonas capsulata*. Matter Lett 61:3984–3987

Hepher A, Atkinson HJ (1992) Nematode control with proteinase inhibitors. EP 0502730 B1

Hu C-C, Wu G-H, Lai S-F, Shanmugam MM, Hwu Y, Wagner OI, Yen T-J (2018) Toxic effects of size-tunable gold nanoparticles on *Caenorhabditis elegans* development and gene regulation. Sci Rep 8:15245

Huang T, Lin Q, Qian X, Zheng Y, Yao J, Wu H, Li M, Jin X, Pan X, Zhang L, Guan X (2018) Nematicidal activity of *Cry1Ea11* from *Bacillus thuringiensis* BRC-XQ12 against the pine wood nematode (*Bursaphelenchus xylophilus*). Phytopathology 108:44–51. https://doi.org/10.1094/PHYTO-95-47-0179-R

Hui F, Scheib U, Hu Y, Sommer RJ, Aroian RV, Ghosh P (2012) Structure and glycolipid binding properties of the nematicidal protein *Cry5B*. Biochemistry 51:9911–9921

Husseiny MI, El-Aziz BY, Mahmoud MA (2007) Biosynthesis of gold nanoparticles using *Pseudomonas aeruginosa*. Spectrochim Acta A: Mol Biomol Spectosc 67:1003–1006

Iatsenko I, Boichenko I, Sommer RJ (2014a) *Bacillus thuringiensis* DB27 produces two novel proteins, *Cry21Fa1* and *Cry21Ha1*, which act synergistically against nematodes. Appl Environ Microbiol 80:3266–3275

Iatsenko I, Nikolov A, Sommer RJ (2014b) Identification of distinct *Bacillus thurigiensis*4A4 nematicidal factors using the model nematodes *Pristionchus pacificus* and *Caenorhabditis elegans*. Toxin (Basel) 6:2050–2063

Iavicoli I, Leso V, Beezhold DH, Shvedova AA (2016) Nanotechnology in agriculture: opportunities. Int J Mol Sci 17(7):996

Jain D, Kothari S (2014) Green synthesis of silver nanoparticles and their application in plant virus inhibition. J Mycol Plant Pathol 44:21

Jayaseelan C, Rahuman AA, Kirthi AV, Marimuthu S, Santhoshkumar T, Bagavan A, Gaurav K, Karthik L, Rao KVB (2012) Novel microbial route to synthesize ZnO nanoparticles using *Aeromonas hydrophila* and their activity against pathogenic bacteria and fungi. Spectrochim Acta A Mol Biomol Spectrosc 90:78–84

Jha AK, Prasad K, Prasad K (2009) A green low-cost biosynthesis of Sb_2O_3 nanoparticles. Biochem Eng J 43:303–306

Kah M, Hofmann T (2014) Nanopesticide research: current trends and future priorities. Environ Int 63:224–235

Kandoth PK, Ithal N, Recknor J, Maier T, Nettleton D, Baum TJ et al (2011) The soybean *Rhg1* locus for resistance to the soybean cyst nematode *Heterodera glycines* regulates the expression of a large number of stress and defense-related genes in degenerating feeding cells. Plant Physiol 155:1960–1975

Kashyap PL, Xiang X, Heiden P (2015) Chitosan nanoparticle based delivery systems for sustainable agriculture. Int J Boil Macromol 77:36–51

Khan MR, Anwer A (2011) Fungal bioinoculants for disease management. In: Paul M, Clinton M, Ahmed I (eds) Microbes and microbial technology. Springer, USA, pp 447–488

Khan MR, Rizvi TF (2014) Nanotechnology: scope and application in plant disease management. Plant Pathol J 13:214–231. https://doi.org/10.3923/ppj.2014.214.231

Khan MR, Altaf S, Mohidin FA, Khan U, Anwar A (2009) Biological control of plant nematodes with phosphate solubilizing microorganisms. In: Kahan MS, Zaidi A (eds) Phosphate solubilizing microbes for crop improvement. Novo Science Publisher Inc., New York, pp 395–426

Khan MR, Haque Z, Kausar N (2014) Management of the root-knot nematode *Meloidogyne graminicola* infesting rice in the nursery and crop field by integrating seed priming and soil application treatments of pesticides. Crop Prot 63:15–25

Khan A, Mohd A, Moh T, Bushra R, Kavita P, Mansoor AS (2017) Phytochemical investigation, nematostatic and nematicidal potential of weeds extract against the root-knot nematode *Meloidogyne incognita* in vitro. Asian J Biol Sci 10:38–46

Khot LR, Sankaran S, Majaa JM, Ehsania R, Schuster EW (2012) Applications of nanomaterials in agricultural production and crop protection. Crop Prot 35:64–67. https://doi.org/10.1016/j.cropro.2012.01.007

Kim J, Shirasawa T, Miyamoto Y (2010) The effect of TAT conjugated platinum nanoparticles on lifespan in a nematode *Caenorhabditis elegans* model. Biomaterials 22:5849–5854. https://doi.org/10.1016/j.biomaterials.2010.03.077

Kim SW, Nam SH, An YJ (2012) Interaction of silver nanoparticles with biological surfaces of *Caenorhabditis elegans*. Ecotoxicol Environ Saf 77:64–70. https://doi.org/10.1016/j.ecoenv.2011.10.023

Kim DY, Kadam A, Shinde S, Saratale RG, Patra J, Ghodake G (2018a) Recent developments in nanotechnology transforming the agricultural sector: a transition replete with opportunities. J Sci Food Agric 98(3):849–864. https://doi.org/10.1002/jsfa.8749

Kim SW, Moon J, An YJ (2018b) Matricidal hatching can induce multi-generational effects in nematode *Caenorhabditis elegans* after dietary exposure to nanoparticles. Environ Sci Pollut Res 25:36394–36402. https://doi.org/10.1007/s11356-01803535-4

Konishi Y, Ohno K, Saitoh N, Nomura T, Nagamine S (2004) Microbial synthesis of gold nanoparticles by metal reducing bacterium. Trans Mater Res Soc Jpn 29:2341–2343

Kitchen CL, Roberts CB (2004) Copper nanoparticles synthesis in compressed liquid and supercritical fluid reverse micelle systems. Ind Eng Chem Res 43:6070–6981

Klink VP, Matthews BF (2009) Emerging approaches to broaden resistance of soybean to soybean cyst nematode as supported by gene expression studies. Plant Physiol 151:1017–1022

Kochkina Z, Pospeshny G, Chirkov S (1994) Inhibition by chitosan of productive infection of T-series bacteriophages in the *Escherichia coli* culture. Mikrobiologiia 64:211–215

Komarneni S (2003) Nanophase materials by hydrothermal, microwave hydrothermal and microwave-solvothermal methods. Curr Sci 85:1730–1734

Krishnaraj C, Ramachandran R, Mohan K, Kalaichelvan P (2012) Optimization for rapid synthesis of silver nanoparticles and its effect on phytopathogenic fungi. Spectrochim Acta Part A Mol Biomol Spectrosc 93:95–99

Kucharska K, Tumialis D, Pezowicz E, Skrzecz I (2011) The effect of gold nanoparticles on the mortality and pathogenicity of entomopathogenic nematodes from *Owinema bio-preparation* Insect Pathogens Entomopathogenic Nematodes, IOBC/WPRS Bull 66:347–349

Kucharska K, Pezowicz E, Tumialis D, Kucharski D, Zajde B (2014) Nanoparticles of copper and entomopathogenic nematodes *Steinernema feltiae* (Filipjev, 1934) in reducing the number of the lesser mealworm beetle *Alphitobius diaperinus* (panzer, 1797). Anim Sci 53:29–35

Kyndt T, Nahar K, Haegeman A, De Vleesschauwer D, Hofte M, Gheysen G (2012) Comparing systemic defense-related gene expression changes upon migratory and sedentary nematode attack in rice. Plant Biol 14:73–82. https://doi.org/10.1111/j.1438-8677.2011.00524.x

Li XQ, Wei JZ, Tan A, Aroian RV (2007a) Resistance to root-knot nematode in tomato roots expressing a nematicidal *Bacillus thuringiensis* crystal protein. Plant Biotechnol J 5:455–464

Li ZZ, Chen JF, Liu F, Liu AQ, Wang Q, Sun HY, Wen LX (2007b) Study of UV-shielding properties of novel porous hollow silica nanoparticle carriers for avermectin. Pest Manag Sci 63 (3):241–246

Li X-Q, Tan A, Voegtline M, Bekele S, Chen C-S, Aroian RV (2008) Expression of *Cry5B* protein from *Bacillus thuringiensis* in plant roots confers resistance to root-knot nematode. Biol Control 47:97–102

Li J, Todd TC, Oakley TR, Lee J, Trick HN (2010) Host-derived suppression of nematode reproductive and fitness genes decreases fecundity of *Heterodera glycines*. Ichinohe Planta 232:775–785

Li M, Huang Q, Wu Y (2011) A novel chitosan-poly (lactide) copolymer and its submicron particles as imidacloprid carriers. Pest Manag Sci 67:831–836

Li J, Zou C, Xu J, Ji X, Niu X, Yang J, Huang X, Zhang K-Q (2015) Molecular mechanisms of nematode-nematophagous microbe interactions: basis for biological control of plant-parasitic nematodes. Annu Rev Phytopathol 53:67–95

Liang W, Yu A, Wang G, Zheng F, Jia J (2018) Chitosan-based nanoparticles of avermectin to control pine wood nematodes. Int J Biol Macromol 112:258–226. https://doi.org/10.1016/j.ijbiomac.2018.01.174

Lilley CJ, Urwin PE, Johnston KA, Atkinson HJ (2004) Preferential expression of a plant cystatin at nematode feeding sites confers resistance to *Meloidogyne incognita* and *Globodera pallida*. Plant Biotechnol J 2:3–12

Lilley CJ, Bakhetia M, Charlton WL, Urwin PE (2007) Recent progress in the development of RNA interference for plant parasitic nematodes. Mol Plant Pathol. 8:701–711

Lilley CJ, Davies LJ, Urwin PE (2012) RNA interference in plant parasitic nematodes: a summary of the current status. Parasitology 139:630–640

Lim DH, Jang J, Kim S, Kang T, Lee K, Choi IH (2012) The effects of sub-lethal concentrations of silver nanoparticles on inflammatory and stress genes in human macrophages using cDNA microarray analysis. Biomaterials 33:4690–4699. https://doi.org/10.1016/j.biomaterials.2012.03.006

Lisiecki I, Filankembo A, Sack-Kongehl H, Weiss K, Pileni MP (2000) Structural investigations of copper nanorods by high-resolution TEM. Phys Rev B 61:4968–4974

Liu S, Kandoth PK, Warren SD, Yeckel G, Heinz R, Alden J et al (2012) A soybean cyst nematode resistance gene points to a new mechanism of plant resistance to pathogens. Nature 492:256–260

Liu BX, Wang Y, Yang F, Wang X, Shen H, Cui HX, Wu DC (2016) Construction of a controlled-release delivery system for pesticides using biodegradable PLA-based microcapsules. Colloids Surf B. Biointerfaces 144:38–45

Loo YY, Chieng BW, Nishibuchi M, Radu S (2012) Synthesis of silver nanoparticles by using tea leaf extract from *Camellia sinensis*. Int J Nanomedicine 27:4263–4267

Ma H, Bertsch PM, Glenn TC, Kabengi NJ, Williams PL (2009) Toxicity of manufactured zinc oxide nanoparticles in the nematode *Caenorhabditis elegans*. Environ Toxicol Chem 28:1324–1330. https://doi.org/10.1016/j.enmm.2017.12.006

Ma W, Saccardo A, Roccatano D (2018) Modular assembly of proteins on nanoparticles. Nat Commun 9:1489. https://doi.org/10.1038/s41467-018-03931-4

Malerba M, Cerana R (2016) Chitosan effects on plant systems. Int J Mol Sci 17(7):996. https://doi.org/10.3390/ijms17070996

Malerba M, Cerana R (2018) Recent advances of chitosan applications in plants. Polymers 10:118. https://doi.org/10.3390/polym10020118

Malerba M, Raffaella C (2018) Recent Advances of Chitosan Applications in Plants. Polymers 10:118. https://doi.org/10.3390/polym10020118

Manonmani V, Juliet V (2011) Biosynthesis of Ag nanoparticles for the detection of pathogenic bacteria. In: Proceedings of the 2nd international conference on innovation, management and service, 16–18 Sep, Singapore, pp 307–311

Marroquin LD, Elyassnia D, Griffitts JS, Feitelson JS, Aroian RV (2000) *Bacillus thuringiensis* (Bt) toxin susceptibility and isolation of resistance mutants in the nematode *Caenorhabditis elegans*. Genetics 155:1693–1699

Martinuz A, Schouten A, Sikora RA (2012) Post-infection development of *Meloidogyne incognita* on tomato treated with the endophytes Fusarium oxysporum strain Fo162 and *Rhizobium* strain G12. BioControl 58:95–104. https://doi.org/10.1094/PHYTO-04-11-0120R.P

Matthews BF, Beard H, Macdonald MH, Kabir S, Youssef RM, Hosseini P et al (2013) Engineered resistance and hypersusceptibility through functional metabolic studies of 100 genes in soybean to its major pathogen the soy bean cyst nematode. Planta 237:1337–1357

Mattos BD, Tardy BL, Magalhães WLE, Rojas OJ (2017) Controlled release for crop and wood protection: recent progress toward sustainable and safe nanostructured biocidal systems. J Control Release 28:139–150. https://doi.org/10.1016/j.jconre

Maule AG, Mcveigh P, Dalzell JJ, Atkinson L, Mousley A, Marks NJ (2011) An eye on RNAi in nematode parasites. Trends Parasitol 27:505–513

Meyer JN, Lord CA, Yang XY, Turner EA, Badireddy AR (2010) Intracellular uptake and associated toxicity of silver nanoparticles in *Caenorhabditis elegans*. Aquat Toxicol 100 (2):140–150. https://doi.org/10.1016/j.aquatox.2010.07.016

Milligan SB, Bodeaua J, Yaghoobia J, Kaloshianb I, Zabelc P, Williamson VM (1998) The root knot nematode resistance gene *Mi* from tomato is a member of the leucine zipper nucleotide binding leucine-rich repeat family of plant genes. Plant Cell 10:1307–1319

Mishra S, Singh H (2015) Biosynthesized silver nanoparticles as a nanoweapon against phytopathogens: exploring their scope and potential in agriculture. Appl Microbiol Biotechnol 99:1097–1107

Mitter N, Worrall EA, Robinson KE, Li P, Jain RG, Taochy C, Fletcher SJ, Carroll BJ, Lu GQ, Xu ZP (2017) Clay nanosheets for topical delivery of RNAi for sustained protection against plant viruses. Nat Plants 3:16,207. https://doi.org/10.1038/nplants.2016.207

Mobli H, Aghbashlo M, Rashid SA, Tan T (2019) Applications of nanotechnology and carbon nanoparticles. In: Rashid SA et al (eds) Agriculture synthesis, technology and applications of carbon nanomaterials. Elsevier, Amsterdam. https://doi.org/10.1016/B978-0-12-815757-2.00011-5

Mody VV, Cox A, Shah S, Singh A, Bevins W, Parihar H (2014) Magnetic nanoparticle drug delivery systems for targeting tumor. Appl Nanosci 4:385–392

Moens M, Perry R, Starr J (2009) *Meloidogyne* species- a diverse group of novel and important plant parasites. In: Perry RN, Moens M, Starr JL (eds) Root-knot nematodes. CABI, Wallingford, UK, p 483

Mohamed EA, Elsharabasy SF, Abdulsamad D (2019) Evaluation of in vitro nematicidal efficiency of copper nanoparticles against root-knot nematode *Meloidogyne Incognita*. South Asian J Parasitol 2(1):1–6

Mourato A, Gadanho M, Lino AR, Tenreiro R (2011) Biosyntheis of crystalline silver and gold nanoparticles by extremophilic yeasts. Bioinorg Chem Applic 8:225. https://doi.org/10.1155/2011/546074

Nakamura K, Matsubara H, Akagi S, Sarashina T, Ejiri K, Kawakita N, Yoshida M, Miyoshi T, Watanabe A, Nishii N, Ito H (2017) Nanoparticle-mediated drug delivery system for pulmonary arterial hypertension. J Clin Med 6(5):48. https://doi.org/10.3390/jcm6050048PMCID

Nasiri M, Azizi K, Hamzehzarghani H, Ghaderia R (2014) Studies on the nematicidal activity of stinging nettle (*Urtica dioica*) on plant parasitic nematodes. Arch Phytopathol Plant Protect 47:591–599

Nassar AMK (2016) Effectiveness of silver nanoparticles of extracts of *Urtica urens* (Urticaceae) against root-knot nematode *Meloidogyne incognita*. Asian J Nematol 5:14–19

Nour El-Deen AH, Darwish H (2011) Nematicidal activity of certain Egyptian weeds and bald cypress callus extracts against *Meloidogyne incognita* infecting eggplant under greenhouse conditions. Egypt J Agronematol 10(2):242–254

Nour El-Deen AH, Cseh E, Darwesh HY (2014) Evaluation of certain Hungarian plant extracts for their nematicidal properties against root-knot nematode, *Meloidogyne incognita in*-vitro. Int J Adv Res 2(8):443–448

Nour El-Deen AH, El-Deeb BA (2018) Effectiveness of silver nanoparticles against root-knot nematode, meloidogyne incognita infecting tomato under greenhouse conditions. J Agric Sci 10:2. https://doi.org/10.5539/jas.v10n2p148

Oka Y, Nacar S, Putievsky E, Ravid U, Yaniv Z, Spiegel Y (2000) Nematicidal activity of essential oils and their components against the root-knot nematode. Phytopathology 90:710–715

Papolu PK, Dutta TK, Tyagi N, Urwin PE, Lilley CJ, Rao U (2016) Expression of a cystatin transgene in eggplant provides resistance to root-knot nematode *Meloidogyne incognita*. Front Plant Sci 7:1122

Parisi C, Vigani M, Rodríguez-Cerezo E (2015) Agricultural nanotechnologies: what are the current possibilities? Nano Today 10(2):124–127

Park BK, Jeong S, Kim D, Moon J, Lim S, Kim JS (2007) Synthesis and size control of monodisperse copper nanoparticles by polyol method. J Colloid Interface Sci 311:417–424

Patil RS, Kokate MR, Kolekar SS (2012) Bioinspired synthesis of highly stabilized silver nanoparticles using Ocimum tenuiflorum leaf extract and their antibacterial activity. Spectrochim Acta Part A: Mol Biomol Spectrosc 91:234–238

Peumans WJ, Van Damme EJM (1995) The role of lectins in plant defense. Histochem J 27:253–271

Pospieszny H, Chirkov S, Atabekov J (1991) Induction of antiviral resistance in plants by chitosan. Plant Sci 79:63–68

Rafique M, Sadaf I, Rafique MS, Tahir MB (2017) A review on green synthesis of silver nanoparticles and their applications. Artif Cells Nanomed Biotechnol 45:1272–1291

Raja M, Subha J, Ali FB, Ryu SH (2008) Synthesis of copper nanoparticles by electro reduction process. Mater Manuf Process 23:782–785

Rajput VD, Minkina TM, Behal A, Sushkova SN, Mandzhieva S, Singh R, Gorovtsov A, Tsitsuashvili VS, Purvis WO, Ghazaryan KA (2018) Effects of zinc-oxide nanoparticles on soil plants animals and soil organisms: a review. Environ Nanotechnol Monit Manag 9:76–84

Ripoll C, Favery B, Lecomte P, Van Damme E, Peumans W, Abad P et al (2003) Evaluation of the ability of lectin from snowdrop (*Galanthus nivalis*) to protect plants against root-knot nematodes. Plant Sci 164:517–523

Roberts PA (1992) Current status of the availability, development, and use of host plant-resistance to nematodes. J Nematol 24:213–222

Roderick H, Tripathi L, Babirye A, Wang D, Tripathi J, Urwin PE et al (2012) Generation of transgenic plantain (*Musa* spp) with resistance to plant pathogenic nematodes. Mol Plant Pathol 13:842–851

Roh JY, Sim SJ, Yi J, Park K, Chung KH, Ryu DY (2009) Ecotoxicity of silver nanoparticles on the soil nematode *Caenorhabditis elegans* using functional ecotoxicogenomics. Environ Sci Technol 43(10):3933–3940

Rosso MN, Jones JT, Abad P (2009) RNAi and functional genomics in plant parasitic nematodes. Annu Rev Phytopathol 47:207–232

Sadeghi R, Rodriguez RJ, Yao Y, Kokini JL (2017) Advances in nanotechnology as they pertain to food and agriculture: benefits and risks. Annu Rev Food Sci Technol 8:467–492

Setua P, Chakraborty A, Seth D, Bhatta MU, Satyam PV, Sarkar N (2007) Synthesis, optical properties, and surface enhanced Raman scattering of silver nanoparticles in nonaqueous methanol reverse micelles. N J Phys Chem C 111:3901–3907

Sharon E, Spiegel Y, Salomon R, Curtis RH (2002) Characterization of *Meloidogyne javanica* surface coat with antibodies and their effect on nematode behavior. Parasitology 125:177–185

Shen Y, Wang Y, Zhao X, Sun C, Cui B, Gao F, Zeng Z, Cui H (2017) Preparation and physicochemical characteristics of thermo-responsive Emamectin benzoate microcapsules. Polymers 9(9):418. https://doi.org/10.3390/polym9090418

Shojaei TR, Salleh MAM, Tabatabaei M, Hossein M, Aghbashlo M, Rashid SA, Tan T (2019) Applications of nanotechnology and carbon nanoparticles in agriculture. In: Rashid SA et al (eds) Agriculture synthesis, technology and applications of carbon nanomaterials. Elsevier, Amsterdam, pp 247–277. https://doi.org/10.1016/B978-0-12-815757-200011-5

Siddique S, Endres S, Atkins JM, Szakasits D, Wieczorek K, Hofmann J et al (2009) Myo-inositol oxygenase genes are involved in the development of syncytia induced by *Heterodera schachtii* in Arabidopsis roots. New Phytol 184:457–472

Siddique S, Wieczorek K, Szakasits D, Kreil DP, Bohlmann H (2011) The promoter of a plant defensin gene directs specific expression in nematode induced syncytia in Arabidopsis roots. Plant Physiol Biochem 49:1100–1107

Singh NG, Anil K, Sewak R, Vinod K (2017) Evaluation of nematicidal activity of ethanolic extracts of medicinal plants to *Meloidogyne incognita* (Kofoid and white) Chitwood under lab conditions. Int J Pure Appl Biosci 5(1):827–831

Sobczak M, Golinowski WA, Grundler FMW (1999) Ultrastructure of feeding plugs and feeding tubes formed by *Heterodera schachtii*. Nematology 1:363–374

Sobczak M, Avrova A, Jupowicz J, Phillips MS, Ernst K, Kumar A (2005) Characterization of susceptibility and resistance responses to potato cyst nematode (*Globodera* spp) infection of tomato lines in the absence and presence of the broad-spectrum nematode resistance Hero gene. Mol Plant Microbe Interact 18:158–168

Song X, Sun S, Zhang W, Yin Z (2004) A method for the synthesis of spherical copper nanoparticles in the organic phase. J Colloid Interface Sci 273:463–469

Stephenson GR (2003) Pesticide use and world food production: risks and benefits. ACS Publications, Washington, DC

Suriyaprabha R, Karunakaran G, Kavitha K, Yuvakkumar R, Rajendran V, Kannan N (2014) Application of silica nanoparticles in maize to enhance fungal resistance. IET Nanobiotechnol 8:133–137

Suslick KS, Hyeon T, Fang M, Cichowlas AA (1996) Sonochemical preparation of nanostructured catalyst. In: Moser WR (ed) Advanced catalysts and nanostructured materials: modern synthetic methods. Academic press, New York, pp 197–212

Taha EH, Abo-Shady NM (2016) Effect of silver nanoparticles on the mortality pathogenicity and reproductivity of entomopathogenic nematodes. Int J Zool Res 12:47–50

Taha Entsar H (2016) Nematicidal effects of silver nanoparticles on root-knot nematodes (*Meloidogyne incognita*) in laboratory and screenhouse. J Plant Prot Pathol 7(5):333–337

Tamilarasan S, Rajam MV (2013) Engineering crop plants for nematode resistance through host-derived RNA interference. Cell Dev Biol 2:2. https://doi.org/10.4172/2168-9296.1000114

Thakur RK, Shirkot P (2017) Potential of biogold nanoparticles to control plant pathogenic nematodes. J Bioanal Biomed 9:220–222. https://doi.org/10.4172/1948-593X.1000182

Thakur RK, Dhirta B, Shirkot P (2018) Studies on effect of gold nanoparticles on *Meloidogyne incognita* and tomato plants growth and development. bioRxiv. doi: https://doi.org/10.1101/428144

Thorat YE, Banerjee S, Kumar A, Manimaran B, Jain PK, Pankaj AS (2017) In silico analysis of root-knot nematode, Meloidogyne incognita house-keeping gene. Indian J Nematol 47 (1):121–126

Tripathi L, Tripathi JN, Roderick H, Atkinson HJ (2013) Engineering nematode resistant plantains for sub-Saharan Africa. Acta Hortic 974:99–107

Tripathi L, Babirye A, Roderick H, Tripathi JN, Changa C, Urwin PE et al (2015) Field resistance of transgenic plantain to nematodes has potential for future African food security. Sci Rep 5:8127

Tripathi L, Atkinson H, Roderick H, Kubiriba J, Tripathi JN (2017) Genetically engineered bananas resistant to *Xanthomonas* wilt disease and nematodes. Food Energy Secur 6:36–47

Udalova ZV, Folmanis GE, Khasanov FK, Zinovieva SV (2018) Selenium nanoparticles—an inducer of tomato resistance to the root-knot nematode *Meloidogyne incognita* (Kofoid et white, 1919) Chitwood 1949. Biochem Biophys Mol Biol 482:264–267

Umer NS, Naveed R, Rafique M (2012) Selection of a suitable method for the synthesis of copper nanoparticles. Nano 07. https://doi.org/10.1142/S1793292012300058

Urwin PE (2007) Biotechnological control of plant parasitic nematodes. In: Xu Z et al (eds) Biotechnology and sustainable agriculture 2006 and beyond. Springer, Dordrecht, pp 181–190

Urwin PE, Atkinson HJ, Waller DA, Mcpherson MJ (1995) Engineered *Oryzacystatin-I* expressed in transgenic hairy roots confers resistance to *Globodera pallida*. Plant J 8:121–131

Urwin PE, Lilley CJ, Mcpherson MJ, Atkinson HJ (1997) Resistance to both cyst and root-knot nematodes conferred by transgenic Arabidopsis expressing a modified plant cystatin. Plant J 12:455–461

Urwin PE, Mcpherson MJ, Atkinson HJ (1998) Enhanced transgenic plant resistance to nematodes by dual proteinase inhibitor constructs. Planta 204:472–479

Urwin PE, Levesley A, Mcpherson MJ, Atkinson HJ (2000) Transgenic resistance to the nematode *Rotylenchulus reniformis* conferred by *Arabidopsis thaliana* plants expressing proteinase inhibitors. Mol Breed 6:257–264

Urwin PE, Green J, Atkinson HJ (2003) Expression of a plant cystatin confers partial resistance to *Globodera* full resistance is achieved by pyramiding a cystatin with natural resistance. Mol Breed 12:263–269

Vain P, Worland B, Clarke MC, Richard G, Beavis M, Liu H et al (1998) Expression of an engineered cysteine proteinase inhibitor (*Oryzacystatin-IID86*) for nematode resistance in transgenic rice plants. Theor Appl Genet 96:266–271

Van der Vossen EA, Van Der Voort JN, Kanyuka K, Bendahmane A, Sandbrink H, Baulcombe DC et al (2000) Homologues of a single resistance-gene cluster in potato confer resistance to distinct pathogens: a virus and a nematode. Plant J 23:567–576

Vasconcelos IM, Oliveira JT (2004) Antinutritional properties of plant lectins. Toxicon 44:385–403

Vieira P, Wantoch S, Lilley CJ, Chitwood DJ, Atkinson HJ, Kamo K (2015) Expression of a cystatin transgene can confer resistance to root lesion nematodes in *Lilium longiflorum* cv 'Nellie white'. Transgenic Res. 24:421–432

Vishnudasan D, Tripathi MN, Rao U, Khurana P (2005) Assessment of nematode resistance in wheat transgenic plants expressing potato proteinase inhibitor (*PIN2*) gene. Transgenic Res. 14:665–675

Vishnu R, Tatiana M, Svetlana S, Viktoriia T, Saglara M, Andrey G, Dina N, Natalya G (2017) Effect of nanoparticles on crops and soil microbial communities. J Soils Sediments:2179–2187. https://doi.org/10.1007/s11368-017-1793-2

van der Voort JR, Kanyuka K, Van der Vossen E, Bendahmane A, Mooijman P, Klein-Lankhorst R, Stiekema W, Baulcombe D, Bakker J (1999) Tight physical linkage of the nematode resistance gene *Gpa2* and the virus resistance gene Rx on a single segment introgressed from the wild species *Solanum tuberosum* subsp andigena C 1673 into cultivated potato. MPMI 12:197–206

Wang YH, Chen PL, Liu MH (2006) Synthesis of well-defined copper nanocubes by a one-pot solution process. Nanotechnology 17:6000–6006

Wang Y, Chen J, Irudayaraj J (2011) Nuclear targeting dynamics of gold nanoclusters for enhanced therapy of HER2+ breast cancer. ACS Nano 5:9718–9725

Wang S, Ghosh AK, Bongio N, Stebbings KA, Lampe DJ, Jacobs-Lorena M (2012) Fighting malaria with engineered symbiotic bacteria from vector mosquitoes. Proc Natl Acad Sci USA 109:12,734–12,739

Wang PX, Zhang CY, Gao MM, Guo SX, Zhu YG, Zheng JS, Zhu L, Ruan LF, Peng DH, Sun M (2014) Complete genome sequence of *Bacillus thuringiensis* YBT – 1518, a typical strain with high toxicity to nematodes. J Biotechnol 171:1–2

Wang L, Hu C, Shao L (2017) The antimicrobial activity of nanoparticles: present situation and prospects for the future. Int J Nanomedicine 12:1227–1249. https://doi.org/10.2147/IJN.S121956

Williamson VM (1998) Root-knot nematode resistance genes in tomato and their potential for future use. Annu Rev Phytopathol 36:277–293

Williamson VM (1999) Plant nematode resistance genes. Curr Opin Plant Biol 2:327–331

Xu Z, Zeng Q, Lu M, Yu A (2006) Inorganic nanoparticles as carriers for efficient cellular delivery. Chem Eng Sci 61:1027–1040. https://doi.org/10.1016/j.ces.2005.06.019

Yadav BC, Veuthambi K, Subramaniam K (2006) Host generated double stranded RNA induces RNAi in plant parasitic nematodes and protects the host from infection. Mol Biochem Parasitol 148:219–222

Yamamoto E, Kuroda K (2016) Colloidal Mesoporous Silica Nanoparticles. Bull Chem Soc Jpn 89:501–539

Yang ST, Wang X, Wang H, Lu F, Luo PG, Cao L, Meziani MJ, Liu JH, Liu Y, Chen M, Huang YP, Sun YP (2009) Carbon dots as nontoxic and high-performance fluorescence imaging agents. J Phys Chem C 113:18110. https://doi.org/10.1021/jp9085969

Yan-Hua Y, Guo Q-m, Yun H, Wang L-j, Wan S-q (2012) Preparation, characterization and nematicidal activity of Lansiumamide B nano-capsules. J Integr Agric 11(7):1151–1158

Yogesh ET, Banarjee S, Kumar A, Manimaran B, Jain PK, Pankaj SA (2017) *In silico* analysis of root-knot nematode, *Meloidogyne incognita* house-keeping gene. Indian J Nematol 47:12

Youssef RM, Kim KH, Haroon SA, Matthews BF (2013) Posttranscriptional gene silencing of the gene encoding aldolase from soy bean cyst nematode by transformed soybean roots. Exp Parasitol 134:266–274

Yu Z, Xiong J, Qq Z, Luo H, Hu S, Xia L, Sun M, Li L, Yu Z (2015) The diverse nematicidal properties and biocontrol efficacy of *Bacillus thurigiensisCry6A* against root-knot nematode *Meloidogyne hapla*. J Invertebr Pathol 125:73–80

Zhang XD, Luo Z, Chen J, Shen X, Song S (2014) Ultra small au(10-12)(SG)(10-12) Nanomolecules for high tumor specificity and Cancer radiotherapy. Adv Mater 26:4565–4568

Zhu HT, Zhang CY, Yin YS (2004) Rapid synthesis of copper nanoparticles by sodium hypophosphite reduction in ethylene glycol under microwave irradiation. J Cryst Growth 270:722–728

Chapter 2
Bioprospecting Compost for Long-Term Control of Plant Parasitic Nematodes

Judy Rouse-Miller, Ezra S. Bartholomew, Chaney C. G. St. Martin, and Piterson Vilpigue

Abstract Many of the microbial antagonists of plant-parasitic nematodes (PPNs) have been found in compost. As such, compost represents a great potential source for bioprospecting microbes capable of controlling PPNs. This is particularly so since most of the estimated billions of microbes per gram of compost (>50K species/g) have not been discovered and/or are not yet cultivable. Moreover, the composting process allows for some degree of manipulation to continuously produce desired microbial species that are often capable of surviving under challenging environmental conditions, such as high soil temperature. Compost also contains a rich diversity of nematode antagonistic compounds (microbial and non-microbial sources), such as humic acids, phenolics and fatty acids, and have been reported to enhance soil resident microbial antagonists, increase plant tolerance and resistance and alter soil physiology profiles, making it unsuitable for nematode survival and activity. Due to its multiple suppression mechanisms, in this chapter, compost is argued to be a great potential source for research aimed at extracting the maximum commercial value from its genetic and biochemical resources, thus making it a more holistic and sustainable approach for managing nematodes rather than a single-type approach, such as the use of synthetic pesticides.

J. Rouse-Miller
Department of Life Science, The University of the West Indies, St. Augustine, Republic of Trinidad and Tobago

E. S. Bartholomew
Beijing Key Laboratory of Growth and Developmental Regulation for Protected Vegetable Crops, College of Horticulture, China Agricultural University, Beijing, People's Republic of China

C. C. G. St. Martin (✉)
Inter-American Institute for Cooperation on Agriculture, Couva, Republic of Trinidad and Tobago
e-mail: chaney.stmartin@iica.int

P. Vilpigue
Department of Food Production, The University of the West Indies, St. Augustine, Republic of Trinidad and Tobago

R. A. Ansari et al. (eds.), *Management of Phytonematodes: Recent Advances and Future Challenges*, https://doi.org/10.1007/978-981-15-4087-5_2

Keywords Organic amendment · Nematophagous antagonists · Microbial diversity · Resource management

2.1 Introduction

Bioprospecting compost is being advanced as a sustainable approach for managing PPNs mainly because of its capacity to exhibit all the known mechanisms of nematode suppression by organic amendments. According to Oka (2010), these include (1) the introduction or enhancement of antagonistic microorganisms; (2) the release of pre-existing nematicidal compounds in soil amendments; (3) the generation of nematicidal compounds, such as ammonia and fatty acids, during degradation; (4) the increase in plant tolerance and resistance; and (5) changes in soil physiology that are unsuitable for nematode behaviour. A characteristic of high input agriculture is the application of fertilisers and synthetic pesticides, which have disrupted soil microbial ecology, minimising the role of beneficial microbes in the rhizosphere. Coupled with the negative effect of chemical residues on human and environmental health, there is now the realisation that a switch from synthetic to biological inputs is needed to sustain crop productivity (Bhatia and Mukherjee 2018). For these reasons, compost provides a distinct advantage over traditional single-type approaches, such as the addition of nematicides. Additionally, it fits well with other traditional control methods, such as crop rotation and the use of resistant crop cultivars (Lopes et al. 2019).

Further, as a source for bioprospecting, compost better lends itself to manipulation and management to obtain desired microbes for PPN control than other in situ environments, which arguably have similar or lower microbial diversity than compost but greater protection and conservation concerns. Though important, the effectiveness of compost against PPNs does not guarantee the long-term sustainability of this management approach. In contrast, the long-term control of PPNs is heavily dependent on the capacity of compost or compost-derived products to develop suppressive soils or plant growth substrate (St. Martin 2015). According to St. Martin and Ramsubhag (2015), the sustainability of such an approach is mainly achieved by altering the microbial profile and activity of the rhizosphere and/or the soil as a whole. This is particularly important since nematode populations are highly influenced by many biological control agents found in soils, which include nematophagous fungi, bacteria, nematodes, mites and earthworms (Stirling 2018). To achieve this, more research on the microbial (structural and functional) ecology of compost is needed, particularly since compost has been described to be part of a new paradigm of disease control, which introduces partially or uncharacterised microbial communities usually with no known activity (Litterick et al. 2004). Such an understanding will assist in improving the sustainable effectiveness of compost and inform the development of various compost-based or -derived biocontrol products, which can be used with other control strategies, including the use of inorganic amendments, the application of botanical compounds with the potential to repel nematodes and the selection or breeding of resistant and tolerant cultivars (Oka et al. 2000; Zakaria et al. 2013; Abd-Elgawad and Askary 2018).

This chapter reviews biological antagonists found in compost, nematode antagonists and their modes of action against nematodes. Additionally, the effect of compost in combination with nematicidal plant extracts and microbial fungal and microbial isolates on nematode populations and the concept of engineering compost as a strategy for the management of PPNs are also advanced.

2.2 Biological Antagonists in Compost

The effects of nematode antagonistic organisms in the regulation of plant parasitic nematode populations have been extensively studied and their modes of parasitism well established (Ansari and Khan 2012a, b; Devi 2018; Zhou et al. 2019). Recently, an assessment of compost bacterial and fungal communities was conducted that provided unique insight into microbial dynamics across different compost recipes and preparation techniques and through time as compost cures (Neher et al. 2013). From studies by Neher et al. (2013) and others (Malandraki et al. 2008; Daami-remadi et al. 2012; El Khaldi et al. 2015), it is evident that compost is a rich source of diverse microbes with the potential to control several PPNs using different modes of action. For example, some of the most studied nematophagous bacterial species from the genera *Bacillus*, *Pseudomonas*, *Stenotrophomonas* and *Streptomyces* have been reported as the main taxa responsible for the suppressive effects of compost (Kwok et al. 1987; Hoitink 1990; Kouki et al. 2012; Khaldi et al. 2015). Likewise, *Trichoderma* spp., a major taxa found in compost, is one of the most studied nematophagous fungi (Hoitink and Fahy 1986; Kwok et al. 1987; Malandraki et al. 2008; Daami-remadi et al. 2012). Interestingly, to the best of the authors' knowledge, viruses have never been extracted from compost for use in controlling nematodes. However, Heringa et al. (2010) found that a five-strain bacteriophage mixture isolated from sewage effluent and applied to dairy manure compost significantly reduced in *Salmonella enterica* within 4 h compared with controls. These findings illustrate the potential use of viruses in the control of pathogens, which might include PPNs. Moreover, the activity rather than the relative diversity of specific or groups of microbial taxa may be a more relevant metric in predicting the potential of compost for suppressing nematodes. To this end, researchers have recommended the complimentary use of culture-based and molecular techniques to better understand the functional and metabolic capacity of microbial communities. This presents tremendous opportunities to better understand compost-nematode interaction, particularly under field conditions.

2.3 Nematode Microbial Antagonists

2.3.1 Fungal Antagonists and Mode of Action

Nematode antagonistic fungi infect and consume either one or more stages of living nematodes (eggs, juveniles, vermiform adults and sedentary females) and drastically decrease the population densities of nematodes. These fungi can be classified into nematophagous fungi and endophytic fungi and include more than 200 species of taxonomically different groups (Cumagun and Moosavi 2015). On the basis of the mechanisms that they use to attack nematodes, nematophagous fungi can be categorised into four major groups: nematode-trapping fungi, endoparasitic fungi, egg- and female-parasitic fungi and toxin-producing fungi (Moosavi and Zare 2012). Some of the most studied nematophagous fungi include *Paecilomyces lilacinus*, *Trichoderma harzianum*, *Hirsutella* spp., *Verticillium chlamydosporium*, *Arthrobotrys dactyloides*, *A. oligospora*, *Myrothecium verrucaria* and *Rhizophagus irregularis* (Khan et al. 2004). Nematophagous fungi *Arthrobotrys oligospora*, *Agaricus bisporus*, and *Dactylaria dasguptae* and an unidentified *Dacylalla* species were identified in mushroom compost, and these microbes were effective at reducing the nematode population by 35% and trapped 80% of the nematode species (Koning et al. 1996). In another study, Kumar et al. (2011) observed the occurrence and colonisation of 12 species of nematophagous fungi in compost, both endoparasitic fungi (*Harposporium anguillulae* and *Meristacrum osteospermum*) and predatory parasitic fungi species (*Dactylaria brochopaga*, *Arthrobotrys cladodes*, *A. conoides*, *A. dactyloides*, *A. oligospora*, *Monacrosporium ellipsosporum*, *M. eudermatum*, *M. gephyropagum*, *M. phymatopagum* and *Stylopaga hadra*). In a field experiment, a strain of *A. oligospora* incorporated into compost blocks was efficient at reducing nematode populations and increasing tomato seedling growth (Duponnois et al. 1996). Nematode-trapping fungi such as *A. oligospora*, in response to signals from bacteria that are under attack by bacteriovorus nematodes, switch from saprophytic to predator mode, producing structures that trap and kill nematodes (Bordallo et al. 2002; Kiontke and David 2013; Wang et al. 2014; Liang et al. 2019) and other fungi such as *Verticillium chlamydosporium* and *Trichoderma* spp. parasitise nematode eggs (Sharon et al. 2001; Bordallo et al. 2002; Szabó et al. 2012). Szabó et al. (2012) identified among several tested paired combinations a combination of nematode-trapping and egg-parasitising fungi *Monacrosporium cionopagum and Trichoderma harzianum*, respectively, as effective for reducing nematode populations. The recognition of this synergistic effect of soil microbes has led to numerous investigations on the use of microbes for the control of nematodes in agricultural systems (Zakaria et al. 2013; Bhatia and Mukherjee 2018).

Endophytic fungi colonise the interior plant tissues where they grow and live symbiotically within the plant tissues without causing disease symptoms (Schulz and Boyle 2006; Sikora et al. 2008). Endophytic fungi can be divided into mycorrhizal fungi, *Fusarium* endophytes and *Neotyphodium* endophytes (Porras-Alfaro and Bayman 2011). The most studied endophytes associated with plant roots are arbuscular mycorrhizal (AM) fungi, which have an obligate symbiotic association with their plant hosts (Veresoglou and Rillig 2011). Several studies have reported that compost addition enhanced AM root colonisation, spore production and the development of AM extra-radical hyphae (Labidi et al. 2007; Valarini et al. 2009; Tanwar et al. 2013; Cavagnaro, 2015). Recently, Yang et al. (2018) concluded that 1-year compost addition significantly enhanced AM root colonisation, extra-radical hyphae density and spore density without affecting their community composition in the soil. Plants endophytically colonised by nematophagous fungi also showed enhanced defence responses against PPNs (Maciá-Vicente et al. 2008). Lopez-Llorca et al. (2010) reported the production of proteases in roots of plants colonised endophytically by a nematophagous fungi. These proteases are produced by nematophagous fungi on their nematode host, and the expression of these proteins in the absence of their host would imply that plants colonised by endophytic fungi may have a repellent effect against nematodes and could be protected before nematode contact. Fungi also produce primary and secondary metabolites with nematicidal activity. Degenkolb and Vilcinskas (2016a) have reviewed reports on numerous such compounds found in nematophagous ascomycetes. Additionally, some 102 non-nematicidal and nematicidal secondary metabolites produced by non-nematophagous ascomycetes and basidiomycetes and nematophagous basidio-mycetes are, according to Degenkolb and Vilcinskas (2016b), worth evaluating for use as biocontrol agents against nematodes.

2.3.2 Bacterial Antagonists and Mode of Action

Antagonistic bacteria have been repeatedly used for the control of nematodes (Khan et al. 2008; Siddiqui et al. 2000; Yoon et al. 2012; Abd-Elgawad and Askary 2018). Nematophagous bacteria are ubiquitous with wide host ranges (Maheshwari et al. 2013; Trivedi and Malhotra 2013). They either suppress the growth of PPNs by producing antibiotics and toxins that inhibit nematode reproduction, juvenile hatch and survival or cause direct death of nematodes by lysis of the surface walls (Siddiqui and Mahmood 1999; Ansari et al. 2017a). Endophytic bacteria colonise the interior plant tissue without causing disease symptoms. These bacteria have been detected in numerous plant species and in almost all plant parts, including above and below ground. Most endophytic bacteria may occupy both the rhizosphere and endorhiza simultaneously, where they protect plants against pathogens (Hallmann et al. 2009; Ansari et al. 2017b). Some of the most studied nematophagous bacteria are *Pasteuria* sp., *Pseudomonas* sp., *Bacillus thuringiensis* and *Streptomyces avermitilis* (Wislocki et al. 1989). *Pasteuria* sp. currently appears to have the greatest potential to be an economically and environmentally practical biological control agent for several important genera of PPNs (Chen et al. 1996; Sharma 1992; Oostendorp et al. 1990). In a study by Werner et al. (2000), they concluded that where *P. penetrans* was naturally present in soils, the balance of bacteria to root-knot nematodes (RKNs) does not allow for the suppression of the RKNs. However, they found that when an 'exotic' isolate of *P. penetrans* was added to soil already infected with *P. penetrans*, there was more infection of stage-2 juveniles (J2) and females and suppression of the RKN populations. This suggests that the relative amount of bacteria present is critical, and this is further supported by Stirling et al. (2017), who demonstrated, using a bioassay, endospore attachment and parasitism of *Pratylenchus zeae* in soils with high levels of *P. thornei* from two sugar cane sites. They report though that the soil at one site was non-suppressive against RKNs infecting sugarcane in the field, and they speculated that this may be due to the length of time this site had been tilled compared to the other.

Several strains of *Paenibacillus ehimensis* have also been reported as antagonistic against plant-parasitic nematodes (Aktuganov et al. 2008). *P. ehimensis* strain RS820 found in the rhizosphere soil of tomatoes showed an enhanced antagonistic activity against *M. incognita.* These *Paenibacillus* spp. were reported to produce several hydrolytic enzymes that played important roles in the biocontrol of plant-parasitic nematodes (Yang et al. 2004). Other bacteria producing nematoxic enzymes such as chitinases, proteases and gelatinase can kill nematode or damage the eggshells (Tian et al. 2007). Chitinase produced by *P. illinoisensis* KJA-424 was found to degrade *M. incognita* eggshells, thus inhibiting hatching (Jung et al. 2002), while the proteases from *Bacillus nematocida* and *Brevibacillus laterosporus* and gelatinase produced by *Lysobacter capsici* YS1215 also had nematicidal effects on the J2 stage of root-knot nematodes (RKNs) (Niu et al. 2006; Lee et al. 2013). Similarly, *Stenotrophomonas maltophilia* was found to reduce the nematode population density of potatoes infected with the trichodorid nematodes *Paratrichodorus pachydermus* and *Trichodorus primitivus* as a result of its proteolytic and chitinolytic activities (Hayward et al. 2010). Yang et al. (2006) isolated a new species, *Stenotrophomonas koreensis* sp. nov, from compost, and Wang et al. (2014) demonstrated the ability of three *Stenotrophomonas maltophilia* isolates from cow dung to induce the nematode trapping fungus *Arthrobotrys oligospora.* Bacterial strains also produce volatile organic substances effective against PPNs and induce systemic resistance to PPNs in plants. Zhai et al. (2018) identified seven volatile organic compounds from a strain of *Pseudomonas putida* 1A00316 isolated from Antarctic soil which, among them, exhibited nematicide fumigant and repellent modes of activity against *M. incognita.* This isolate was also reported by Tang et al. (2014) to limit the population increase of *M. incognita* in vitro and in pot experiments. They also reported that tomato plants treated with the strain 1A00316 increased the production of defense enzymes, phenylalanine ammonia lyase, polyphenol oxidase and peroxidase, thereby displaying induced systemic resistance to *M. incognita.*

2.3.3 Other Antagonistic Agents

Predaceous nematodes and mites had been identified as a biocontrol agent decades ago (Cobb 1917; Linford and Oliveira 1938). However, their potential in controlling

PPNs has only begun to be studied in recent years. Among a broad range of soil predators that prey on nematodes, predatory nematodes are the most important (Bilgrami and Brey 2005). They eat all types of nematodes and have established themselves as an important part of the soil food web. Predatory nematodes comprise four main taxonomic groups, namely mononchids, dorylaimids, aphelenchids and diplogasterids (Cobb 1917). Increased populations of predaceous nematodes have been shown to decrease the incidence of RKNs (Akhtar and Mahmood 1993). Further suppression of the development of root-knot symptoms in tomato and chilli was observed when compost was incorporated into the soil (Akhtar 1995). Various nematophagous mites have also been identified in soil ecosystems, and nematodes are a preferred food regime for many mites (Beaulieu and Walter 2007; Pakyari and Maghsoudlo 2011; Walter and Proctor 2013; Szafranek et al. 2013). Many nematophagous mites develop more rapidly, with higher reproductive rates, when nematodes are included in their diet (Walter et al. 1987). According to Van de Bund (1972), the addition of stable compost to field soil caused a two-third decrease in PPN populations, while at the same time omnivorous arthropods increased four- to sevenfold. In addition to their biocontrol potential against PPNs, nematophagous nematodes and mites also stimulate the cycling of plant nutrients, which in turn may help plants better withstand any nematode damage (Yeates and Wardle 1996; Stirling 2011). At present, it is difficult to commercially produce predaceous nematodes and mites. However, adding these natural enemies to compost can create environmental conditions that are unfavourable to nematodes. Therefore, organic amendments hold the key to the success of such a strategy for nematode control.

2.4 Nematode Antagonists and Plant Nematicides for the Control of PPNs

The integration of a bacterial or fungal antagonist with a known plant-derived nematicidal product against PPNs in soils of cultivated plants has recently been reviewed by Abd-Elgawad and Askary (2018). Overall, this integrated approach resulted in reduction in nematode populations. Several cultivated and uncultivated plants are known to produce secondary metabolites, which exhibit nematicidal activity (Renčo et al. 2014; Khan et al. 2017), of which neem (*Azadirachta indica*) is the best known example and most used. Neem products, including leaf, seed kernel, seed powders, seed extracts, oil, sawdust and particularly oilcake, have been reported to be effective for the control of several nematode species (Akhtar and Mahmood 1996: Akhtar 1998; Ansari and Mahmood 2019a, b). Neem leaf compost has also been shown to reduce the population of several significant PPN species (*Meloidogyne*, *Heterodera* and *Tylenchus* spp.) (Olabiyi and Oladeji 2014). The components of neem include nimbin, salanin, thionemone, azadirachtin and various flavonoids, with known nematicidal action (Lim and Lim 2014). Other plant extracts from castor (*Ricinus communis*), groundnut (*Arachis hypogea*) and mahua

(*Madhuca indica*) have also been reported to be effective at reducing nematode populations (Akhtar and Anver 1990; Akhtar and Alam 1991). Marigold extracts (*Tagetes* sp.) are also effective at controlling nematodes (Bridge 2002). Many species of marigold, including the African marigold *T. erecta*, the French marigold *T. patula* and the South American marigold *T. minuta*, can effectively control nematodes on agricultural crops when they are grown in rotation, interplanted with the crop or used as compost (Siddiqui and Alam 1988; Akhtar and Mashkoor Alam 1992; Akhtar 1998; Grubišić et al. 2018). Several members of the Brassicaceae family are also effective at controlling plant-parasitic nematodes (McSorley and Frederick 1995; Mojtahedi 2010; El-Nagdi and Youssef 2019; Dutta et al. 2019). Nematicidal activity of seaweed extracts was demonstrated in in vitro assays by Ara et al. (1997), Ngala et al. (2016) and Khan et al. (2015). Improved growth of roots and reduction in the number of root nematodes in greenhouse-grown tomato plants treated with foliar applications and/or soil drenches of seaweed extract was demonstrated by Featonby-Smith and Staden (1983) and Crouch and Van Staden (1993). More recently, Sultana et al. (2011) reported reduced gall formation and endoparasitic nematode penetration in the roots of tomato plants when seaweed extract was used as a soil amendment.

2.5 Compost, Microbial Antagonists and Biological Nematicides for the Control of PPNs

Composts have been shown to improve soil quality due to increased organic matter content and soil microbial populations (Chang et al. 2007). Microbes in composts improve soil quality and enhance plant resistance by directly eliminating plant-parasitic nematodes or by increasing the populations of beneficial fungi/bacteria and other nematode antagonistic agents (Chen et al. 2000; Bulluck et al. 2002; Usman and Siddiqui 2013; Ansari and Mahmood 2017a, b). Compost contains several microbes that are antagonistic to nematodes, and compost combined with plant-derived nematicidal compounds can have a significant impact on PPN population development (Chikaoka et al. 1982; Dickson and Hewlett 1989; Dutta et al. 2019). Several studies have reported reductions in plant-parasitic nematode numbers following compost application (Marull et al. 1997; Akhtar and Malik 2000). Arancon et al. (2003) reported a reduction in PPNs and an increase in fungivorous and, to some extent, bacterivorous nematode populations in soils treated with vermicompost and inorganic fertilisers of field-grown grape and strawberry; the converse was true for soils treated with inorganic fertiliser only. Additionally, *Meloidogyne incognita* infested soils treated with *Verticillium chlamydosporium* and *Photorhabdus luminescens* fungal and bacterial nematode antagonists respectively combined with compost gave maximum reduction in nematode egg production, female numbers and juveniles in the soil and a reduction in gall formation on cucumber roots compared with control and single treatments of fungus, bacteria or compost (Zakaria et al. 2013). Composting of plants with known nematicidal was reported by Olabiyi and Oladeji (2014) to significantly reduce the populations of PPNs (*Meloidogyne*, *Heterodera* and *Tylenchus* spp.) in soils of okra fields treated with neem-leaf-based compost, cassava-peel-based compost or *Tithonia*-leaf-based compost, with *Trichoderma harzianum* used as the decomposer. *Trichoderma viride* integrated with compost also significantly improved plant performance and reduced root gall formation in gotu kola (*Centella asiatica*) (Shamalie et al. 2012).

2.6 Bioprospecting Compost for Sustainable Nematode Control

Agronomic functions of compost include improved plant growth, nutrient and water retention and disease suppression (Guo et al. 2019; Ansari et al. 2019). The composition of compost and compost end products is affected by many factors that interact with each other; these include the nature of the feedstock, the composting process used, compost additives, the microbial community, particle size, temperature, pH, C/N ratio, moisture content and oxygen content (St. Martin and Ramsubhag 2015). These factors can also be manipulated to obtain a compost with the capacity to develop soils suppressive to PPNs. In the preceding sections, we outlined studies that demonstrated that fungal and bacterial antagonists of nematodes reduce nematode population in in vitro assays, and in greenhouse experiments, we also highlighted studies that used plant extracts with nematicidal properties to reduce nematode populations. In other studies, compost alone or in combination with nematophagous fungi and bacteria and compost derived from plant materials with known nematicidal properties significantly enhanced the suppression of PPNs. We therefore suggest that the composting parameters be manipulated to achieve a product that can be applied to soils infected with PPNs. The use of feedstock such as neem, marigold and castor, which have known nematicidal substances, would add to the effect of controlling nematodes. Additionally, the composting system is known to affect the suppressive potential of compost as it affects the quantity, type and activity of microbes present in the compost (St. Martin and Ramsubhag 2015). Nematophagous fungi and bacteria have been identified in compost, and known fungal and bacteria isolates that exhibit various modes of action against PPNs can also be included as needed. The compost can be engineered to exhibit multiple mechanisms that inhibit PPNs.

2.7 Conclusions and Future Prospects

Chemical nematicides for the control of PPNs in agricultural systems are being replaced by eco-friendly biocontrol products. There are several natural enemies of nematodes, which include nematophagous bacteria and fungi, mites and predaceous nematodes. However, nematophagous fungi and bacteria are the most studied for the biocontrol of nematodes. These organisms utilise several mechanisms to reduce nematode infections in plants, including the induction of systemic resistance, parasitism and the production of secondary metabolites with nematicide activity. Additionally, many plants produce nematicidal secondary metabolites in seeds and above- and below-ground organs, and these have been used as biocontrol products. Compost can be optimised as a biocontrol product for the development of nematode suppressive soils. Several nematophagous microbes have been isolated from compost. The choice of the feedstock and composting processes should be carefully determined to yield a compost product that can develop and sustain nematode suppressive soils. Field trials are needed to demonstrate the effectiveness of the compost product to consistently control nematode populations.

References

Abd-Elgawad MMM, Askary TH (2018) Fungal and bacterial nematicides in integrated nematode management strategies. Egypt J Biol Pest Control 28:74. https://doi.org/10.1186/s41938-018-0080-x

Akhtar M (1995) Biological control of the root-knot nematode *Meloidogyne incognita* in tomato by the predatory nematode *Mononchus aquaticus*. Int Pest Control 37(1):18–19

Akhtar M (1998) Biological control of plant-parasitic nematodes by neem products in agricultural soil. Appl Soil Ecol 7(3):219–223. https://doi.org/10.1016/S0929-1393(97)00031-0

Akhtar M, Anver AY (1990) Effects of organic amendments to soil as nematode suppressants. Int Nematol Net Newslett 7:21–22

Akhtar M, Mahmood I (1993) Effect of *Mononchus aquaticus* and organic amendments on *Meloidogyne incognita* development on chili. Nematol Mediterr 21:251–252

Akhtar M, Mahmood I (1996) Control of plant-parasitic nematodes with organic and inorganic amendments in agricultural soil. Appl Soil Ecol 4(3):243–247. https://doi.org/10.1016/S0929-1393(96)00114-X

Akhtar M, Malik A (2000) Roles of organic soil amendments and soil organisms in the biological control of plant-parasitic nematodes: a review. Bioresour Technol 74(1):35–47. https://doi.org/10.1016/S0960-8524(99)00154-6

Akhtar M, Mashkoor Alam M (1992) Effect of crop residues amendments to soil for the control of plant-parasitic nematodes. Bioresour Technol 41(1):81–83. https://doi.org/10.1016/0960-8524(92)90102-4

Akthar M, Alam M (1991) Integrated control of plant-parasitic nematodes on potato with organic amendments, nematicide and mixed cropping with mustard. Nematol Mediterr 19:169–171

Aktuganov G, Melentjev A, Galimzianova N, Khalikova E, Korpela T, Susi P (2008) Wide-range antifungal antagonism of *Paenibacillus ehimensis* IB-X-b and its dependence on chitinase and β-1,3-glucanase production. Can J Microbiol 54(7):577–587. https://doi.org/10.1139/w08-043

Ansari RA, Khan TA (2012a) Parasitic association of root-knot nematode, *Meloidogyne incognita* on guava. e-J Sci Technol 5:65–67

Ansari RA, Khan TA (2012b) Diversity and community structure of phytonematodes associated with guava in and around Aligarh, Uttar Pradesh, India. Trends Biosci 5(3):202–204

Ansari RA, Mahmood I (2017a) Optimization of organic and bio-organic fertilizers on soil properties and growth of pigeon pea. Sci Hortic 226:1–9

Ansari RA, Mahmood I (2017b) Determination of disease incidence caused by *Meloidogyne* spp. and or *Fusarium udum* on pigeonpea in Aligarh district: a survey. Trends Biosci 10 (24):5239–5243

Ansari RA, Mahmood I (2019a) Plant health under biotic stress: volume 2: Microbial interactions. Springer, Singapore. https://doi.org/10.1007/978-981-13-6040-4

Ansari RA, Mahmood I (2019b) Plant health under biotic stress: volume 1: Organic strategies. Springer, Singapore. https://doi.org/10.1007/978-981-13-6043-5

Ansari RA, Rizvi R, Sumbul A, Mahmood I (2017a) PGPR: current vogue in sustainable crop production. In: Kumar V, Kumar M, Sharma S, Prasad R (eds) Probiotics and plant health. Springer, Singapore, pp 455–472

Ansari RA, Mahmood I, Rizvi R, Sumbul A (2017b) Siderophores: Augmentation of soil health and crop productivity. In: Kumar V, Kumar M, Sharma S, Prasad R (eds) Probiotics in agroecosystem. Springer, Singapore, pp 291–312

Ansari RA, Sumbul A, Rizvi R, Mahmood I (2019) Organic Soil Amendments: Potential tool for soil and plant health management. In: Ansari RA, Mahmood I (eds) Plant health under biotic stress. Springer, Singapore, pp 1–35

Ara J, Ehteshamul-Haque S, Sultana V, Ghaffar A, Qasim R (1997) Use of Sargassum species for the control of Meloidogyne javanica in okra. Nematol Medit 25(1):125–128

Arancon NQ, Galvis P, Edwards C, Yardim E (2003) The trophic diversity of nematode communities in soils treated with vermicompost: the 7th international symposium on earthworm ecology · Cardiff · Wales · 2002. Pedobiologia 47(5):736–740. https://doi.org/10.1078/0031-4056-00752

Beaulieu F, Walter DE (2007) Predation in suspended and forest floor soils: observations on Australian mesostigmatic mites. Acarologia 47(1–2):43–54

Bhatia C, Mukherjee PK (2018) Chapter 12—Microbial technologies for sustainable crop production. In: Prasad R, Gill SS, Tuteja N (eds) Crop improvement through microbial biotechnology. Elsevier, Amsterdam, the Netherlands, pp 255–261; ISBN: 9780444639875. https://doi.org/10.1016/B978-0-444-63987-5.00012-8

Bilgrami AL, Brey C (2005) Potential of predatory nematodes to control plant parasitic nematodes. In: Grewal P, Ehlers R, Shapiro-llan D (eds) Nematodes as biocontrol agents. CABI International, Wallingford, pp 447–464. https://doi.org/10.1079/9780851990170.0447

Bordallo JJ, Lopez-Llorca LV, Jansson H-B, Salinas J, Persmark L, Asensio L (2002) Colonization of plant roots by egg-parasitic and nematode-trapping fungi. New Phytol 154(2):491–499. https://doi.org/10.1046/j.1469-8137.2002.00399.x

Bridge J (2002) Nematode management in sustainable and subsistence agriculture. Annu Rev Phytopathol 34(1):201–225. https://doi.org/10.1146/annurev.phyto.34.1.201

Bulluck LR, Barker KR, Ristaino JB (2002) Influences of organic and synthetic soil fertility amendments on nematode trophic groups and community dynamics under tomatoes. Appl Soil Ecol 21(3):233–250. https://doi.org/10.1016/S0929-1393(02)00089-6

Cavagnaro TR (2015) Biologically regulated nutrient supply systems. Adv Agron 129:293–321. https://doi.org/10.1016/bs.agron.2014.09.005

Chang EH, Chung RS, Tsai YH (2007) Effect of different application rates of organic fertilizer on soil enzyme activity and microbial population: original article. Soil Sci Plant Nutr 53 (2):132–140. https://doi.org/10.1111/j.1747-0765.2007.00122.x

Chen ZX, Dickson DW, McSorley R, Mitchel DJ, Hewlett TE (1996) Suppression of *Meloidogyne arenaria* race 1 by soil application of endospores of *Pasteuria penetrans*. J Nematol 28 (2):159–168

Chen J, Abawi GS, Zuckerman BM (2000) Efficacy of *Bacillus thuringiensis*, *Paecilomyces marquandii*, and *Streptomyces costaricanus* with and without organic amendments against *Meloidogyne hapla* infecting lettuce. J Nematol 32(1):70–77

Chikaoka I, Onbayash NP, Suina S (1982) The effect of green manure, marigold, groundnut, sorghum and watermelon on the population dynamics of *Pratylenchus penetrans* and *Meloidogyne incognita*. Jap J Nematol 11:19–23

Cobb NA (1917) The Mononchus: a genus of free living predatory nematodes. Soil Sci 3:431–486
Crouch IJ, Van Staden J (1993) Effect of seaweed concentrate from Ecklonia maxima (Osbeck) Papenfuss on Meloidogyne incognita infestation on tomato. J Appl Phycol. https://doi.org/10.1007/BF02182420
Cumagun CJR, Moosavi MR (2015) Significance of biocontrol agents of phytonematodes. In: Askary T, Martinelli P (eds) Biocontrol agents of phytonematodes. CABI International, Boston, MA, pp 50–78. https://doi.org/10.1079/9781780643755.0050
Daami-remadi M, Dkhili I, Jabnoun-Khiareddine H, El Mahjoub M (2012) Biological control of potato leak with antagonistic fungi isolated from compost teas and solarized and non-solarized soils. Pest Technol 6:32–40
Degenkolb T, Vilcinskas A (2016a) Metabolites from nematophagous fungi and nematicidal natural products from fungi as an alternative for biological control. Part I: metabolites from nematophagous ascomycetes. Appl Microbiol Biotechnol 100(9):3799–3812. https://doi.org/10.1007/s00253-015-7233-6
Degenkolb T, Vilcinskas A (2016b) Metabolites from nematophagous fungi and nematicidal natural products from fungi as alternatives for biological control. Part II: metabolites from nematophagous basidiomycetes and non-nematophagous fungi. Appl Microbiol Biotechnol 100 (9):3813–3824. https://doi.org/10.1007/s00253-015-7234-5
Devi G (2018) Utilization of nematode destroying fungi for management of plant-parasitic nematodes-a review. Biosci Biotechnol Res Asia 15:377–396. https://doi.org/10.13005/bbra/2642
Dickson DW, Hewlett TE (1989) Effects of bahiagrass and nematicides on *Meloidogyne arenaria* on peanut. J Nematol 21(4S):671–676
Duponnois R, Mateille T, Sene V, Sawadogo A, Fargette M (1996) Effect of different west African species and strains of *Arthrobotrys nematophagous* fungi on *Meloidogyne* species. Entomophaga 41(3/4):474–483
Dutta TK, Khan MR, Phani V (2019) Plant-parasitic nematode management via biofumigation using brassica and non-brassica plants: current status and future prospects. Curr Plant Biol 17:17–32. https://doi.org/10.1016/j.cpb.2019.02.001
El Khaldi R, Daami-remadi M, Hamada W et al (2015) Plant pathology & microbiology the potential of Serratia marcescens: an indigenous strain isolated from date palm compost as biocontrol agent of Rhizoctonia solani on potato. https://doi.org/10.4172/2157-7471.S3-006
El-Nagdi WMA, Youssef MMA (2019) Brassica vegetable leaf residues as promising biofumigants for the control of root knot nematode, Meloidogyne incognita infecting cowpea. Agric Eng Int CIGR J 21(1):134–139
Featonby-Smith BC, Van Staden J (1983) The effect of seaweed concentrate on the growth of tomato plants in nematode-infested soil. Sci Hortic (Amsterdam). https://doi.org/10.1016/0304-4238(83)90134-6
Grubišić D, Uroić G, Ivošević A, Grdiša M (2018) Nematode control by the use of antagonistic plants. Agriculturae Conspectus Scientificus 83(4):269–275
Guo X-x, Liu H-t, Wu S-b (2019) Humic substances developed during organic waste composting: formation mechanisms, structural properties, and agronomic functions. Sci Total Environ 662:501–510. https://doi.org/10.1016/j.scitotenv.2019.01.137
Hallmann J, Davies KG, Sikora R (2009) Biological control using microbial pathogens, endophytes and antagonists. In: Hallmann J, Davies KG, Sikora R (eds) Root-knot nematodes. CABI, Wallingford, pp 380–411. https://doi.org/10.1079/9781845934927.0380
Hayward AC, Fegan N, Fegan M, Stirling GR (2010) Stenotrophomonas and Lysobacter: ubiquitous plant-associated gamma-proteobacteria of developing significance in applied microbiology. J Appl Microbiol 108(3):756–770. https://doi.org/10.1111/j.1365-2672.2009.04471.x
Heringa SD, Kim J, Jiang X, Doyle MP, Erickson MC (2010) Use of a mixture of bacteriophages for biological control of Salmonella enterica strains in compost. Appl Environ Microbiol 76 (15):5327–5332. https://doi.org/10.1128/AEM.00075-10
Hoitink HA (1990) United States Patent [19]. 286–287

Hoitink HAJ, Fahy PC (1986) Basis for the control of Soilborne plant pathogens with composts. Annu Rev Phytopathol 24:93–114

Jung WJ, Jung SJ, An KN, Jin YL, Park RD, Kim KY et al (2002) Effect of chitinase-producing *Paenibacillus illinoisensis* KJA-424 on egg hatching of root-knot nematode (*Meloidogyne incognita*). J Microbiol Biotechnol 12(6):865–871

Khan A, Williams K, Nevalainen H (2004) Effects of protease and chitinase on the eggshell structures and hatching of juveniles. Biol Control 31:346–352. https://doi.org/10.1016/j.biocontrol.2004.07.011

Khan Z, Kim SG, Jeon YH, Khan HU, Son SH, Kim YH (2008) A plant growth promoting rhizobacterium, *Paenibacillus polymyxa* strain GBR-1, suppresses root-knot nematode. Bioresour Technol 99(8):3016–3023. https://doi.org/10.1016/j.biortech.2007.06.031

Khan SA, Abid M, Hussain F (2015) Nematicidal activity of seaweeds against Meloidogyne javanica Pakistan. J Nematol 33(2):195–203

Khan A, Asif M, Tariq M, Rehman B, Parihar K, Siddiqui MA (2017) Phytochemical investigation, nematostatic and nematicidal potential of weeds extract against the root-knot nematode, Meloidogyne incognita in vitro. Asian J Biol Sci 10:38–46

Kiontke K, David HA (2013) Current biology primer fitch nematodes. 23(19):PR862–R864. https://doi.org/10.1016/j.cub.2013.08.009

Koning G, Hamman B, Eicker A (1996) The efficacy of nematophagous fungi on predaceous nematodes in soil compared with saprophagous nematodes in mushroom compost. South Afr J Bot 62(1):49–53. https://doi.org/10.1016/S0254-6299(15)30578-0

Kouki S, Saidi N, RA Ben et al (2012) Control of Fusarium wilt of Tomato Caused by Fusarium oxysporum F. Sp. Radicis-Lycopersici using mixture of vegetable and Posidonia oceanica compost. 2012. https://doi.org/10.1155/2012/239639

Kumar N, Singh RK, Singh KP (2011) Occurrence and colonization of nematophagous fungi in different substrates, agricultural soils and root galls. Arch Phytopathol Plant Prot 44 (12):1182–1195. https://doi.org/10.1080/03235408.2010.484945

Kwok O, Fahy P, Hoitink H, Kuter G (1987) Interactions between bacteria and Trichoderma hamatum in suppression of Rhizoctonia damping-off in bark compost media. Phytopathology 77:1206. https://doi.org/10.1094/Phyto-77-1206

Labidi S, Nasr H, Zouaghi M, Wallander H (2007) Effects of compost addition on extra-radical growth of arbuscular mycorrhizal fungi in *Acacia tortilis* ssp. *raddiana* savanna in a pre-Saharan area. Appl Soil Ecol 35(1):184–192. https://doi.org/10.1016/j.apsoil.2006.04.009

Lee YS, Park YS, Anees M, Kim YC, Kim YH, Kim KY (2013) Nematicidal activity of *Lysobacter capsici* YS1215 and the role of gelatinolytic proteins against root-knot nematodes. Biocontrol Sci Technol 23(12):1427–1441. https://doi.org/10.1080/09583157.2013.840359

Liang LM, Zou CG, Xu J, Zhang KQ (2019) Signal pathways involved in microbe-nematode interactions provide new insights into the biocontrol of plant-parasitic nematodes. Philos Trans R Soc Lond Ser B Biol Sci 374(1767):20180317. https://doi.org/10.1098/rstb.2018.0317

Lim TK, Lim TK (2014) *Azadirachta indica. Edible medicinal and non medicinal plants, A. Juss.* C. Springer, Dordrecht, pp 409–455. https://doi.org/10.1007/978-94-017-8748-2_30

Linford MB, Oliveira JM (1938) Potential agents of biological control of plant-parasitic nematodes (Abstract). Phytopathology 28:14

Litterick AM, Harrier L, Wallace P, Watson CA, Wood M (2004) The role of uncomposted materials, composts, manures, and compost extracts in reducing pest and disease incidence and severity in sustainable temperate agricultural and horticultural crop production-a review. Crit Rev Plant Sci 23:453–479

Lopes EA, Dallemole-Giaretta R, dos Santos Neves W, Parreira DF, Ferreira PA (2019) Eco-friendly approaches to the management of plant-parasitic nematodes. In: Ansari R, Mahmood I (eds) Plant health under biotic stress. Springer, Singapore. https://doi.org/10.1007/978-981-13-6043-5_9

Lopez-Llorca LV, Gómez-Vidal S, Monfort E, Larriba E, Casado-Vela J, Elortza F et al (2010) Expression of serine proteases in egg-parasitic nematophagous fungi during barley root colonization. Fungal Genet Biol 47(4):342–351. https://doi.org/10.1016/j.fgb.2010.01.004
Maciá-Vicente JG, Jansson HB, Abdullah SK, Descals E, Salinas J, Lopez-Llorca LV (2008) Fungal root endophytes from natural vegetation in Mediterranean environments with special reference to Fusarium spp. FEMS Microbiol Ecol 64(1):90–105. https://doi.org/10.1111/j.1574-6941.2007.00443.x
Maheshwari DK, Shukla S, Aeron A, Kumar T, Jha CK, Patel D et al (2013) Rhizobacteria for management of nematode disease in plants. In: Maheshwari DK (ed) Bacteria in agrobiology: disease management. Springer, Heidelberg, pp 379–404. https://doi.org/10.1007/978-3-642-33639-3_14
Malandraki I, Tjamos SE, Pantelides IS, Paplomatas EJ (2008) Thermal inactivation of compost suppressiveness implicates possible biological factors in disease management. Biol Control 44:180–187. https://doi.org/10.1016/j.biocontrol.2007.10.006
Marull J, Pinochet J, Rodríguez-Kábana R (1997) Agricultural and municipal compost residues for control of root-knot nematodes in tomato and pepper. Compost Sci Util 5(1):6–15. https://doi.org/10.1080/1065657X.1997.10701860
McSorley R, Frederick JJ (1995) Responses of some common cruciferae to root-knot nematodes. J Nematol 27(4S):550–554. http://www.ncbi.nlm.nih.gov/pubmed/19277321%0A http://www.pubmedcentral.nih.gov/articlerender.fcgi?artid=PMC2619653
Mojtahedi H (2010) Managing *Meloidogyne chitwoodi* on potato with rapeseed as green manure. Plant Dis 77(1):42. https://doi.org/10.1094/pd-77-0042
Moosavi MR, Zare R (2012) Fungi as biological control agents of plant-parasitic nematodes. In: Mérillon JM, Ramawat KG (eds) Plant defence: biological control. Springer, Heidelberg, pp 67–107. https://doi.org/10.1007/978-94-007-1933-0_4
Neher DA, Weicht TR, Bates ST, Leff JW, Fierer N (2013) Changes in bacterial and fungal communities across compost recipes, preparation methods, and composting times. PLoS One 8(11):e79512. https://doi.org/10.1371/journal.pone.0079512
Ngala BM, Valdes Y, dos Santos G, Perry RN, Wesemael WML (2016) Seaweed-based products from *Ecklonia maxima* and *Ascophyllum nodosum* as control agents for the root-knot nematodes *Meloidogyne chitwoodi* and *Meloidogyne hapla* on tomato plants. J Appl Phycol 28(3):2073–2082. https://doi.org/10.1007/s10811-015-0684-4
Niu Q, Huang X, Zhang L, Li Y, Li J, Yang J, Zhang K (2006) A neutral protease from *Bacillus nematocida*, another potential virulence factor in the infection against nematodes. Arch Microbiol 185(6):439–448. https://doi.org/10.1007/s00203-006-0112-x
Oka Y (2010) Mechanisms of nematode suppression by organic soil amendments—a review. Appl Soil Ecol 44(2):101–115. https://doi.org/10.1016/j.apsoil.2009.11.003
Oka Y, Koltai H, Bar-Eyal M, Mor M, Sharon E, Chet I, Spiegel Y (2000) New strategies for the control of plant-parasitic nematodes. Pest Manag Sci 56:983–988. https://doi.org/10.1002/1526-4
Olabiyi TI, Oladeji OO (2014) Assessment of four compost types on the nematode population dynamics in the soil sown with okra. Int J Org Agric Res Dev 9(9):146–155. http://orgprints.org/28203
Oostendorp M, Dickson DW, Mitchell DJ (1990) Host range and ecology of isolates of *Pasteuria* spp. from the Southeastern United States. J Nematol 22(4):525–531
Pakyari H, Maghsoudlo M (2011) Development and life table of *Tyrophagus putrescentiae* (Astigmata: Acaridae) on mushroom and Phytonematode. Acad J Entomol 4:59–63
Porras-Alfaro A, Bayman P (2011) Hidden fungi, emergent properties: endophytes and microbiomes. Annu Rev Phytopathol 49(1):291–315. https://doi.org/10.1146/annurev-phyto-080508-081831
Renčo M, Sasanelli N, Maistrello L (2014) Plants as natural sources of nematicides. In: Davis LM (ed) Nematodes: comparative genomics, disease management and ecological importance, 1st edn. NOVA Science Publisher, New York, pp 115–141
Schulz B, Boyle C (2006) What are endophytes? In: Schulz B, Boyle C, Sieber T (eds) Microbial root endophytes. Springer, Berlin, Heidelberg, pp 1–13. https://doi.org/10.1007/3–540-33526-9_1

Shamalie BVT, Fonseka RM, Rajapaksha RGAS (2012) Effect of trichoderma viride and carbofuran (Curator®) on management of root knot nematodes and growth parameters of gotukola (Centella asiatica L.). Trop Agric Research 23:61–69. https://doi.org/10.4038/tar.v23i1.4632

Sharma RD (1992) Biocontrol efficiency of *Pasteuria penetrans* against *Meloidogyne javanica*. Ciencia Bio Ecol E Syst 12:43–47

Sharon E, Bar-Eyal M, Chet I, Herrera-Estrella A, Kleifeld O, Spiegel Y (2001) Biological control of the root-knot nematode *MeloidogUSyne javanica* by *Trichoderma harzianum*. Phytopathology 91(7):687–693

Siddiqui MA, Alam MM (1988) Control of plant-parasitic nematodes with *Tagetes tenuifolia*. Rev Nematol 11:369–370

Siddiqui ZA, Mahmood I (1999) Role of bacteria in the management of plant parasitic nematodes: a review. Bioresour Technol 69(2):167–179. https://doi.org/10.1016/S0960-8524(98)00122-9

Siddiqui IA, Qureshi SA, Sultana V, Ehteshamul-Haque S, Ghaffar A (2000) Biological control of root rot-root knot disease complex of tomato. Plant and Soil 227(1–2):163–169. https://doi.org/10.1023/A:1026599532684

Sikora RA, Pocasangre L, zum Felde A, Niere B, Vu TT, Dababat AA (2008) Mutualistic endophytic fungi and in-planta suppressiveness to plant parasitic nematodes. Biol Control 46 (1):15–23. https://doi.org/10.1016/j.biocontrol.2008.02.011

Stirling GR (2011) Biological control of plant-parasitic nematodes: an ecological perspective, a review of progress and opportunities for further research. In: Davies K, Spiegel Y (eds) Biological control of plant-parasitic nematodes. Springer, Cham, Switzerland, pp 1–38. https://doi.org/10.1007/978-1-4020-9648-8_1

Stirling GR (2018) Chapter 5 Biological control of plant-parasitic nematodes. In: Poinar GO Jr, Jansson H-B (eds) Diseases of nematodes, vol 11. CRC Press, Boca Raton, FL, p 160

Stirling GR, Wong E, Bhuiyan S (2017) *Pasteuria*, a bacterial parasite of plant-parasitic nematodes: its occurrence in Australian sugarcane soils and its role as a biological control agent in naturally-infested soil. Austr Plant Pathol 46:563–569. https://doi.org/10.1007/s13313-017-0522-z

St. Martin CCG (2015) Enhancing soil suppressiveness using compost and compost tea. In: Meghvansi M, Varma A (eds) Organic amendments and soil suppressiveness in plant disease management. Soil biology, vol 46. Springer, Cham. https://doi.org/10.1007/978-3-319-23075-7_2

St. Martin CCG, Ramsubhag A (2015) Potential of compost for suppressing plant diseases. In: Ganesan S, Kurucheve Vadivel JJ (eds) Sustainable crop disease management using natural products. CABI, Boston, MA, pp 345–388

Sultana V, Baloch GN, Ara J, et al (2011) Seaweeds as an alternative to chemical pesticides for the management of root diseases of sunflower and tomato. Journal of Applied Botany and Food Quality 84 :162–168

Szabó M, Csepregi K, Gálber M, Virányi F, Fekete C (2012) Control plant-parasitic nematodes with Trichoderma species and nematode-trapping fungi: the role of chi18–5 and chi18–12 genes in nematode egg-parasitism. Biol Control 63(2):121–128. https://doi.org/10.1016/j.biocontrol.2012.06.013

Szafranek P, Lewandowski M, Kozak M (2013) Prey preference and life tables of the predatory mite Parasitus bituberosus (Acari: Parasitidae) when offered various prey combinations. Exp Appl Acarol 61(1):53–67. https://doi.org/10.1007/s10493-013-9701-y

Tang JP, Zhang Z, Jing X, Yu Z, Zhang J, Shao Z et al (2014) Mechanism of antagonistic bacteria Pseudomonas putida 1A00316 from the South Pole soil against Meloidogyne incognita. Chin J App Environ Biol 20:1046–1051. https://doi.org/10.3724/SP.J.1145.2014.05009

Tanwar A, Aggarwal A, Yadav A, Parkash V (2013) Screening and selection of efficient host and sugarcane bagasse as substrate for mass multiplication of *Funneliformis mosseae*. Biol Agric Hortic 29(2):107–117. https://doi.org/10.1080/01448765.2013.771955

Tian B, Yang J, Zhang KQ (2007) Bacteria used in the biological control of plant-parasitic nematodes: populations, mechanisms of action, and future prospects. FEMS Microbiol Ecol 61(2):197–213. https://doi.org/10.1111/j.1574-6941.2007.00349.x

Trivedi PC, Malhotra A (2013) Bacteria in the management of plant-parasitic nematodes. In: Maheshwari DK (ed) Bacteria in agrobiology: disease management. Springer, Heidelberg, pp 349–377. https://doi.org/10.1007/978-3-642-33639-3_13

Usman A, Siddiqui MA (2013) Integrated approaches of phytonematodes management by organic soil amendments and ploughing. Pak J Nematol 31:157–163

Valarini PJ, Curaqueo G, Seguel A, Manzano K, Rubio R, Cornejo P, Borie F (2009) Effect of compost application on some properties of a volcanic soil from central South Chile. Chilean J Agric Res 69(3):416–425. https://doi.org/10.4067/S0718-58392009000300015

Van de Bund CF (1972) Some observations on predatory action of mites on nematodes. Zasz Probl Postepow Nauk Roln 129:103–110

Veresoglou SD, Rillig MC (2011) Suppression of fungal and nematode plant pathogens through arbuscular mycorrhizal fungi. Biol Lett 8:214–217

Walter DE, Proctor HC (2013) Mites: ecology, evolution & behaviour: life at a microscale, 2nd edn. Springer, Cham, Switzerland. https://doi.org/10.1007/978-94-007-7164-2

Walter DE, Hunt HW, Elliott ET (1987) The influence of prey type on the development and reproduction of some predatory soil mites. Pedobiologia 30:419–424

Wang X, Li GH, Zou CG, Ji X, Liu T, Zhao PJ, Liang LM, Xu JP, An ZQ, Zheng X, Qin YK, Tian MQ, Xu YY, Ma YC, Yu ZF, Huang XW, Liu SQ, Niu XM, Yang JK, Huang Y et al (2014) Bacteria can mobilize nematode-trapping fungi to kill nematodes. Nat Commun 5:5776. https://doi.org/10.1038/ncomms6776

Werner M, Blok V, Daudi A, Bala G, Davies K, Netscher C, Sawadogo A et al (2000) The importance of tropical root-knot nematodes (Meloidogyne spp.) and factors affecting the utility of *Pasteuria penetrans* as a biocontrol agent. Nematology 2(8):823–845

Wislocki PG, Grosso LS, Dybas RA. (1989) Environmental Aspects of Abamectin Use in Crop Protection. In: Campbell W.C. (eds) Ivermectin and Abamectin. Springer, New York, pp 182–200

Yang J, Kharbanda PD, Mirza M (2004) Evaluation of *Paenibacillus polymyxa* PKB1 for biocontrol of pythium disease of cucumber in a hydroponic system. Acta Hortic 635:59–66. https://doi.org/10.17660/ActaHortic.2004.635.7

Yang H-C, Im W-T, Kang MS, Shin D-Y, Lee S-T (2006) *Stenotrophomonas koreensis sp.* nov., isolated from compost in South Korea. Int J Syst Evol Microbiol 56:81–84. https://doi.org/10.1099/ijs.0.63826-0

Yang W, Gu S, Xin Y, Bello A, Sun W, Xu X (2018) Compost addition enhanced hyphal growth and sporulation of arbuscular mycorrhizal fungi without affecting their community composition in the soil. Front Microbiol 9:169. https://doi.org/10.3389/fmicb.2018.00169

Yeates GW, Wardle DA (1996) Nematodes as predators and prey: relationships to biological control and soil processes. Pedobiologia 40(1):43–50

Yoon GY, Lee YS, Lee SY, Park RD, Hyun HN, Nam Y, Kim KY (2012) Effects on *Meloidogyne incognita* of chitinase, glucanase and a secondary metabolite from Streptomyces cacaoi GY525. Nematology 14(2):175–184. https://doi.org/10.1163/138855411X584124

Zakaria HM, Kassab AS, Shamseldean MM, Oraby MM, El-Mourshedy MMF (2013) Controlling the root-knot nematode, Meloidogyne incognita in cucumber plants using some soil bioagents and some amendments under simulated field conditions. Ann Agric Sci 58(1):77–82. https://doi.org/10.1016/j.aoas.2013.01.011

Zhai Y, Shao Z, Cai M, Zheng L, Li G, Huang D et al (2018) Multiple modes of nematode control by volatiles of *Pseudomonas putida* 1A00316 from Antarctic soil against *Meloidogyne incognita*. Front Microbiol 9:253. https://doi.org/10.3389/fmicb.2018.00253

Zhou D, Feng H, Schuelke T, De Santiago A, Zhang Q, Zhang J, Luo C, Wei L (2019) Rhizosphere microbiomes from root knot nematode non-infected plants suppress nematode infection. Microb Ecol 78:470–481. https://doi.org/10.1007/s00248-019-01319-5

Chapter 3
Plant-Growth-Promoting Rhizobacteria (PGPR)-Based Sustainable Management of Phytoparasitic Nematodes: Current Understandings and Future Challenges

Rizwan Ali Ansari, Rose Rizvi, Aisha Sumbul, and Irshad Mahmood

Abstract Undoubtedly, phytoparasitic nematodes cause great damage to important agricultural crops, which signifies great monetary loss. Nematicides are used to kill the plant parasitic nematodes. These chemicals have caused greater losses to our biodiversity which are untargeted leading to a great perturbation of ecosystem ecology. The impact of these chemicals on human health cannot be ignored. PGPR uses various mechanisms to manage the plant nematodes. They are also known as plant growth enhancer, phytohormone producer, siderophore producer leading to enhanced plant health. They are also helpful in the enhancement of quantum of resistance of the plants against various pathogens including plant parasitic nematodes. Inoculation of suitable rhizobacteria not only enhances the plant growth and yield characters of plants but also restrict the multiplication of pathogens and pest populations. PGPR is one of the best alternatives which could be used against plant nematodes for bringing down their population below threshold level. There are several mechanisms implicated in the management of phytoparasitic nematodes and such mechanisms have been described in the chapter.

Keywords Phytohormone · Siderophore · Plant health · Phosphate solubilisation · Enzymes

---

R. A. Ansari (✉)
Section of Plant Pathology and Nematology, Department of Botany, Aligarh Muslim University, Aligarh, India

Present Address: Department of Plant Protection, Faculty of Agricultural Sciences, Aligarh Muslim University, Aligarh, India

R. Rizvi · I. Mahmood
Section of Plant Pathology and Nematology, Department of Botany, Aligarh Muslim University, Aligarh, India

A. Sumbul
Section of Plant Pathology and Nematology, Department of Botany, Aligarh Muslim University, Aligarh, India

Present Address: Department of Biosciences, Jamia Millia Islamia, New Delhi, India

R. A. Ansari et al. (eds.), *Management of Phytonematodes: Recent Advances and Future Challenges*, https://doi.org/10.1007/978-981-15-4087-5_3

3.1 Introduction

When we read newspapers, magazines, articles etc., there is frequent caution about the hazardous effects of chemical pesticides, including nematicides, and their consequence on the ecosystem's productivity. In order to intensify the crop produce, growers use the chemical fertilizers, pesticides, nematicides, etc. frequently without caring the catastrophic effects of such chemicals. Non-judicious use of synthetic fertilizers or pesticides causes greater reduction in biota and also offer the development of new strains of various pathogens (Ansari and Mahmood 2017b; Ansari et al. 2019; Ansari and Mahmood 2019a, b). The current situation has indeed reached at alarming stage, thus, there must be some way out to find appropriate alternatives. Biocontrol and biofertilzers are the appropriate option which may help in the protection of plant pathogens and pest making the environment more efficient and hostile. Among several biocontrol microorganisms, PGPR can ameliorate plant growth and yield by depleting pathogen population. Various mechanisms either singly or in multiple are operated for registration of good plant health. Although there are several factors which are responsible for better bacterial colonisation to the root system. These abiotic factors also play a crucial role in the establishment of beneficial phytobiomes. It has been suggested that PGPR can enhance plant growth and yield parameters (Prasad et al. 2019).

In addition, soil is the place where microbial activities, including PGPR activities are considered highly efficient. PGPR have a great potential to manage pest and pathogens effectively and are therefore considered to be an important factor in the intensification of sustainable agriculture (Ansari and Mahmood 2017a; Ansari et al. 2017a; Mahmood et al. 2019). PGPR accelerate the synthesis of various important plant-growth-promoting organic molecules, such as phytohormones, antibiotics, enzymes etc., which offer better plant health (Ansari et al. 2017a). In addition, PGPR play driving role in soil health augmentation, leading to improved plant growth, productivity and yield (Ansari et al. 2017b). Various plant pathogens, including phytoparasitic nematodes, have been found to be effectively controlled by these important microbes (Almaghrabi et al. 2013; Xiang et al. 2018; Viljoen et al. 2019). PGPR are a small portion of rhizobacteria, which can promote plant heath directly as biofertilisers or rhizoremediators or phytostimulators and stress controllers or indirectly as inhibitors of plant pathogens, including fungi, bacteria, viruses and nematodes (Lugtenberg and Kamilova 2009; Antoun 2013; Mhatre et al. 2018). PGPR are found among both Gram-negative and Gram-positive bacteria; however, Gram-negative bacteria, like pseudomonads, *Burkholderia*, *Arthrobacter*, *Serratia*, *Achromobacter*, *Rhizobium* spp., which are capable of nitrogen fixation; *Azospirillum* spp.; *Azotobacter* spp.; and *Diazotrophs* spp., show plant-growth-promoting activity (Antoun 2013). Gram-positive bacterial isolates of *Brevibacterium*, *Corynebacterium*, *Micrococcus*, *Paenibacillus*, *Sarcina*, *Bacillus* and *Pseudomonas* have also been shown to promote plant health (Antoun 2013; Kloepper et al. 2004). Among the PGPR, *Pseudomonas* spp. and *Bacillus* spp. are the two important genera that have been extensively studied (Podile and Kishore 2007). The present chapter brings the latest information on the role of PGPR on plant health amelioration, and plant parasitic nematode management. In order to provide straight information, recent studies on these aspects have been consolidated and presented in a simplified and coherent manner.

3.2 Mechanisms of PGPR

PGPR is used as alternative in the management of phytonematodes. Application of suitable strains help the plants to defend themselves from wide range of pathogens attack (Zandi and Basu 2016; Ijaz et al. 2019). PGPR enhance plant growth and yield through either direct or indirect mechanisms (Glick 1995). Direct mechanisms include different processes like phosphate solubilisation, nitrogen fixation, siderophore (iron-chelating compound) production, hydrogen cyanide (HCN), ammonia, vitamin and plant hormone (auxin, cytokinin and gibberellin) production; on the other hand, indirect mechanisms involve secondary metabolite production, which hampers soil pathogen proliferation, cell-wall-degrading enzyme synthesis, competition, induction of resistance etc., minimising the deleterious effects of pathogens (Glick et al. 1999; Prasad et al. 2019).

3.2.1 *Phosphate Solubilisation*

Phosphorus is a good nutrient for plants to help with its growth and developments. The phosphorus which are present in the environment is not easily accessible to the plants. Deprived inorganic phosphate (orthophosphate) in soil significantly hampers crop production (Miller et al. 2010; Wang et al. 2017). To obviate the situation, phosphate-solubilising bacteria (rhizosphere-colonising bacteria and endophytes) can help in the liberation of organic phosphates (Otieno et al. 2015). Phosphate-solubilising bacteria have been extensively studied in many cases, and it has been found that they improve plant growth and yield. Joe et al. (2016) conducted an experiment with two bacteria, *Acinetobacter* sp. and *Bacillus* sp., which were isolated from the rhizosphere of *Phyllanthus amarus*. The bacteria had shown salt tolerance and phosphate-solubilising character. Finally, they reported that the application of these two bacteria promoted vigour index, phosphorus content, the percentage of germination, plant biomass, phenolic content, and also the antioxidative activity of uninoculated control. Inagaki et al. (2015) reported the findings that the application of different phosphate-solubilising bacteria in acidic sandy soil enhanced the phosphorus content in the leaf tissue of maize. Delfim et al. (2018) indicated that the use of phosphate-solubilising bacteria *Bacillus thuringiensis* increased the availability of soil phosphorus. In brief, wheat plant was inoculated with *B. thuringiensis* at 20th day and was reinoculated 46 days after sowing. The inoculation of these phosphate-solubilising bacteria registered 11% improvement in phosphorus in the rhizosphere at Z46 (Ultisol) and 34% and 67% in aerial tissues at Z46 (Andisol and Ultisol), respectively. On the other hand,

75% enhancement of phosphorus was observed in root tissues at Z87 (Ultisol). Acid phosphatase activity, microbial biomass and root biomass were significantly increased.

3.2.2 Siderophore Production

Siderophore is a iron-chelating agents which are low in molecular weight (Chu et al. 2010; Hider and Kong 2010; Goswami et al. 2016; Ansari et al., 2017a). Iron is considered to be one of the most important elements used for the development and normal functioning of plants and a wide range of soil microorganisms. Large amount of iron is available in the soil but is not accessible form to plants due to its complex nature. These siderophores produced from PGPR assist well in the fulfilment of iron to plants by solubilising and making it available to the plants (Wandersman and Delepelaire 2004; Arora et al. 2013; Singh et al. 2017a, b). These iron-chelating agents (siderophores) contain a variety of chemical structures that can bind with metal cations (Chu et al. 2010; Hider and Kong 2010; Verma et al. 2011; Ghavami et al. 2017). PGPR have also been used in the acquisition of nitrogenase co-factors, molybdenum (Mo) and vanadium (V). Besides, siderophore production in *A. vinelandii* was tested under a variety of trace metal environments, and increased siderophore production was recorded under Fe limitation; on the other hand, under Mo limitation, only catechol-type siderophore production was found to be significantly enhanced (McRose et al. 2017). Siderophore-producing strains of bacteria possess good quantum of plant growth promotion and biocontrol features (Kumar et al. 2016, Kumar et al., 2017a,b; Bindu and Nagendra 2016). Sheirdil et al. (2019) identified some strains of PGPR through 16Sr RNA gene sequencing fatty acid profile and biolog and thereafter selected ten potential strains of PGPR on the basis of their ACC deaminase activity, siderophore production, P solubilisation and the production of indole acetic acid for the plant growth promotion of wheat. They further reported that inoculation of these PGPR significantly enhanced the plant growth and yield characters over control.

3.2.3 Plant Hormone Production

Plant hormone regulates various metabolic and biochemical reaction which are inevitable for plant growth (Waadt et al. 2015; Wani et al. 2016; Ibrahim et al. 2019). Various plant hormones affect the biochemical, physiological and various functioning of plants, including stress management. PGPR also help in the alleviation of various abiotic stresses, such as drought, salinity, heat, cold, flooding, ultraviolet radiation etc., which are common issues in the current era, and the whole world is struggling with these factors. The common phytohormones can be identified as auxin, cytokinin, ethylene, gibberellins and abscisic acid; however, some newly discovered phytohormones have also been included, such as brassinosteroids, jasmonates and strigolactones, which play a significant role in the development of stress-tolerant crop plants (Egamberdieva et al. 2017; Abd-Allah et al. 2018). Brilli et al. (2019) inoculated *Pseudomonas chlororaphis* subsp. *aureofaciens* strain M71 to tomato with the objective of assessing water tolerance level. The production of phytohormones like abscisic acid (ABA) and indoleacetic acid (IAA) was significantly enhanced, contributing a lot in shaping the leaf without alteration in photosynthesis. IAA is directly involved in cell differentiation, cell division and cell elongation of crop plants and thus is a key hormone of plant bodies (Bhardwaj et al. 2014). Maximum PGPR considerably secrete some organic compounds that are directly involved in plant growth and yield promotion leading to ameliorated plant health (Kumar et al. 2015). The amount of IAA concentration varies from species to species, strains to strains. It has been observed most of the time that *Pseudomonas* sp. is the most potent IAA producer among all genera; however, *Pseudomonas putida* is considered to be more superior in terms of production of IAA than *P. fluorescens* (Bharucha et al. 2013; Reetha et al. 2014; Kumar et al. 2015). Ethylene has an important place in the promotion of plant growth and development. The hormone can work efficiently even at very low concentrations (Abeles 1992). Ethylene concentration may effectively control plant growth and senescence (Nazar et al. 2014). Important hormones, i.e. abscisic acids (ABAs), are the molecules that are considerably involved in the alleviation of several environmental stresses and also have significant impact on plants' defence system against a wide range of plant pathogens (Alazem and Lin 2017; Davies and Zhang 1991). Two rhizobacteria, *Bacillus licheniformis* Rt4M10 and *Pseudomonas fluorescens* Rt6M10, isolated from the rhizosphere of grapevines produced ABA, IAA and gibberellins. The concentration of ABA was recorded to be higher in 45-day-old *Vitis vinifera* plants inoculated with *B. licheniformis* and *P. fluorescens* than in control plants (Salomon et al. 2014). Application of *Bacillus licheniformis* SA03 with *Chrysanthemum* plants grown under saline-alkaline conditions significantly alleviated the saline-alkaline stress leading to improved photosynthesis and biomass (Zhou et al. 2017).

3.2.4 Ammonia and Hydrogen Cyanide Production

Likewise, HCN and ammonia production is considered to be an important growth-promoting trait of strains. HCN is also considered to be involved in phytopathogen management in agroecosystems (Rijavec and Lapanje 2016). PGPR-mediated HCN production and synthesis vary considerably and depend upon the genus prevailing in the area; their efficacious nature suggest that such PGPR can be used as biological fertilisers or biocontrol in the intensification of crop production under a climate change scenario (Agbodjato et al. 2015; Rijavec and Lapanje 2016). A large number of research suggested that PGPR-producing HCN can be used for the growth promotion and yield enhancement of various horticultural crops (Rijavec and Lapanje 2016; Kumar et al. 2016). Heydari et al. (2008) isolated cyanogenic strain of *Pseudomonas fluorescence*, which exhibited biocontrol activity, leading to enhanced length of the stems and roots and enhanced germination rate in rye, wild barley and wheat. In another study, Kumar et al. (2012a,b) isolated 40 fluorescent *Pseudomonas* strains from a diverse range of soil. Among the seven strains, P1, P10, P13, P18, P21, P28 and P38, that were further selected for trial depending on their possessing of single or multiple PGPR traits, P38 was found to be a good producer of HCN. In addition, ammonia production by PGPR helps in the promotion of root and shoot elongation and the improvement of plant growth and yield performance (Marques et al. 2010). Many of the PGPR strains have been discovered so far to have both characters, i.e. production of HCN and production of ammonia. The synergistic effects of these two aspects considerably enhance the physiological and biochemical properties of plants (Agbodjato et al. 2015; Kumar et al. 2016).

3.2.5 Enzyme Production

Plant contains a wide range of enzymes, which regulate the various life cycles of plants and also help in the promotion of plant growth (Brilli et al. 2019). Samaddar et al. (2019) reported that the normal plant growth was arrested with the increased concentrations of salinity stress resulting increased stress levels, disrupted the photosynthetic variables, and also affected the antioxidant enzymatic traits in bacterial non-inoculated control plants. Further, the inoculation of *Pseudomonas* spp. considerably alleviated stress ethylene emission and exhibited enhanced plant growth and yield. In the same experiment, catalase activity was significantly higher in the *Pseudomonas* spp. inoculated plants and also neutralised the hydrogen peroxide ions formed due to oxidative stress in plants grown under salinity stress. Besides, the plant resistance against *M. javanica* infesting tomato cv. CALJN3 was improved using salicylic acid and *Pseudomonas fluorescens* CHAO. The results indicated that salicylic acid and *Pseudomonas fluorescens* CHAO elicitors induced the removal of high concentrations of toxic reactive oxygen species by scavenging antioxidant enzymes, especially that of superoxide dismutase, peroxidase and catalase. Application of these elicitors at different time schedules registered significant diminution in the number of galls, egg masses or eggs of *M. javanica* infected tomato plants over control (Nikoo et al. 2014).

3.2.6 Nitrogen Fixation

Nitrogen fixation is also an important phenomenon in various strains of PGPR, which contributes a lot to plant growth promotion (Prasad et al. 2019). Nitrogen although present in the atmosphere in gaseous form about 78% but not directly available to the plants and therefore needs some route through which it might be fixed. The nitrogen fixation is not uncommon among prokaryotes with strains in both bacteria and archaea (Dekas et al. 2009; Das et al. 2015). In addition, the synergistic action of *Rhizobium tropici* strain CIAT 899 and *Paenibacillus polymyxa* strain DSM 36 resulted in higher nodulation, leghaemoglobin, concentration, nitrogenase and nitrogen fixation efficiency and thereby improved the plant health status of common bean. PGPR benefits on specific nodulation were evident on accumulated plant nitrogen (Figueiredo et al. 2008). Yadegari et al. (2010) evaluated the individual and synergistic effects of *Pseudomonas fluorescens* P-93 and *Azospirillum lipoferum* S-21, as well as two highly effective *Rhizobium* strains, in relation to the improvement of plant growth and yield-contributing character. The application of PGPR and *Rhizobium* enhanced the nodulation and plant growth and yield of kidney bean.

3.3 Mechanisms of Biocontrol of Plant Pathogenic Nematodes

PGPR adopt various mechanisms to alleviate the pernicious effects of phytonematodes and avoid greater damage to the growth and yield-contributing characters of plants (Table 3.1). Amaki et al. (2019) reported that the application of *Bacillus* sp. strains AT-332 (NITE BP-1095) and AT-79 (NITE BP-1094) isolated from nature showed biocontrol activity against plant pathogenic activity, also improved the plant health. The *Bacillus* sp. strains AT-332 and AT-79 were found to be effective in the management of a wide range of plant pathogens that cause damage to plants. Ramamoorthy et al. (2001) reported that consortia of different PGPR strains resulted in improved efficacy by inducing systemic resistance against a wide range of plant pathogens. They further illustrated that seed treatment with PGPR resulted in cell wall structural modifications and physiological and biochemical changes, leading to enhanced synthesis of proteins and chemicals involved in plant defence mechanisms. In another study, microbial strains (*Pseudomonas aeruginosa* (B1), MTCC7195, and *Burkholderia gladioli* (B2), MTCC10242) were used against plant parasitic nematodes. The application of these microbial agents showed a favourable response in terms of antioxidant enhancement, which helps in the defence expression of *Lycopersicon esculentum* to alleviate oxidative stress generated under nematode infection (Khanna et al. 2019). Application of the liquid-based *B. subtilis* formulations in soil registered greater reduction in the reproduction of *M. javanica*, leading to promoted plant growth and yield of tomato plant (Lopes et al. 2019b). Turatto et al. (2018) evaluated anti-nematodal activity of five PGPR isolated from garlic rhizosphere. They reported that the potentiality of these PGPR arrested hatching (*Meloidogyne javanica*) and motility (*Ditylenchus* spp.). It was observed that isolates CBSAL02 (*Bacillus*) and CBSAL05 (*Pseudomonas*) significantly impaired the hatching of *M. javanica* eggs by 74% and 54.77%, respectively. Likewise, the motility of another important nematode, i.e. *Ditylenchus* spp., was reduced by 55.19% and 53.53%, respectively. In addition, various biocontrol mechanisms are well known, such as antibiosis and lytic enzyme production, and induced systemic resistance (ISR) helps in the restriction of plant pathogens (Kumar et al. 2011). PGPR possess a characteristic that can help in the acceleration of nutrient uptake and in the enhancement of the plant growth and yield attributes of

Table 3.1 PGPR-mediated plant parasitic nematode management

Country	Crops/in vitro	PPN	PGPR	Mode of action	References
China	In vitro test	*M. incognita*	*Pseudomonas putida* MCCC 1A00316	Antibiosis, some active compounds	Zhai et al. (2019)
India	*Lycopersicon esculentum*	*M. incognita*	*Pseudomonas aeruginosa* and *Burkholderia gladioli*	Antibiosis	Khanna et al. (2019)
Brazil	Tomato	*M. javanica*	*Bacillus subtilis* isolate 34	Secretion of some enzymes like protease and chitinase, lytic enzymes and phyto-hormones; siderophore production	Lopes et al. (2019a)
South Africa	Carrot	*Meloidogyne incognita*	*Bacillus firmus* T11, *Bacillus aryabhattai* A08, *Paenibacillus barcinonensis* A10, *Paenibacillus alvei* T30 and *Bacillus cereus* N10w	Production of extracellular toxic com-pounds or enzymes against nematodes	Viljoen et al. (2019)
Italy	Tomato	*Meloidogyne incognita*	*Bacillus firmus* strain 1-1582	Alteration in soil-beneficial microbial populations	d'Errico et al. (2019)
China	Tomato	*M. incognita*	*Bacillus cereus* BCM2	Increased stimulation of root exudates, which helps the plant to keep the nema-todes away	Li et al. (2019)
China	Cucumber	*Meloidogyne incognita*	*Bacillus subtilis* strain Bs-1	Production of some volatile compounds	Cao et al. (2019)
USA	Corn	*Meloidogyne incognita*	*Burkholderia renojensis*, *Streptomyces avermitilis* and *Bacillus firmus*		Aljaafri et al. (2019)
Egypt	*Solanum lycopersicum*	*Meloidogyne incognita*	*Bacillus amyloliquefaciens* subsp. plantarum SA5 and *Lysinibacillus sphaericus* Amira strain	Toxin and enzyme production	Abdel-Salam et al. (2018)
China	Tomato	*Meloidogyne incognita*	*Bacillus cereus* BCM2	Nutrient and space competition	Wang et al. (2018)

India	Gerbera	*Meloidogyne incognita*	*Bacillus subtilis strain* Bbv 57 (KF718836)	Antibiosis, enzyme production	Ramyabharathi et al. (2018)
China	Wheat cv. Aikang 58	*Heterodera avence*	*B. cereus* XZ 24-2-1, *B. cereus* XZ-33-3, *B. weihenstephanensis* MH-58-60-01 and *B. thuringiensis* MH 032-003	Excretion of hydrolytic enzymes	Ahmed et al. (2018)
Egypt	Tomato	*Meloidogyne incognita*	*Bacillus thuringiensis* and *Bacillus subtilis*	Production of metabolites	Khalil and El-Naby (2018)
Korea	Tomato	*Meloidogyne* spp.	*Pseudomonas chlororaphis* O6	HCN production	Kang et al. (2018)
China	In vitro	*Meloidogyne incognita*	*Pseudomonas putida* 1A00316	Volatile organic compounds	Zhai et al. (2018)
Brazil	In vitro	*Meloidogyne javanica* and *Ditylenchus* spp.	Rhizobacteria (CBSAL02, CBSAL05, CBSAL14, CBSAL18 and CBSAL21)	Production of toxic metabolites, secondary metabolites and extracellular enzymes	Turatto et al. (2018)
Iran	Tomato	*Meloidogyne javanica*	*Paenibacillus polymyxa* (Prazmowski) Ash et al. (NCIM 2188), *Bacillus subtilis* (Ehrenberg) Cohn (MCC 0067), *Pseudomonas striata* Chester (NCIM 2847) and *P. fluorescens* (Flugge) Migula (333-S)	Production of toxic metabolites	Sohrabi et al. (2018)
Iran	Pistachio	*Meloidogyne incognita*	*Pseudomonas fluorescens strains* VUPF5, VUPF52; *Bacillus cereus* strain PRC95; and *Bacillus subtilis* strain PRC96	Development of ISR	Zeynadini-Riseh et al. (2018)

plants, leading to reduced multiplication of plant pathogens (Kloepper and Schroth 1981; Liu et al. 2016). Moreover, PGPR also assist in plant growth promotion by acting as biofertilisers, rhizoremediators, phytostimulators and stress alleviators and also restrict the reproduction of a wide range of plant pathogens like fungi, bacteria, viruses and plant parasitic nematodes (Antoun 2013; Lugtenberg and Kamilova 2009). Important PGPR like fluorescent and non-fluorescent pseudomonads, *Arthrobacter*, *Burkholderia*, *Achromobacter*, *Rhizobium* spp., *Serratia*, which are capable of nitrogen fixation; *Azospirillum* spp.; *Azotobacter* spp.; and *Diazotrophs* spp. (Antoun 2013). Various isolates of *Brevibacterium*, *Corynebacterium*, *Micrococcus*, *Paenibacillus*, *Sarcina*, *Bacillus* and *Pseudomonas* have also shown plant-growth-promoting activities (Antoun 2013; Kloepper et al. 2004). However, a wide range of bacterial genera are found to be plant growth enhancers; *Bacillus* and *Pseudomonas* spp. are the predominant genera that are currently being exploited at a large scale (Podile and Kishore 2007).

Liu et al. (2012) reported that dual inoculations of *Glomus versiforme* and *G. mosseae* and *Bacillus polymyxa* and *Bacillus* sp. exhibited greater management of *Meloidogyne incognita* and promoted the plant growth and yield productivity of *Lycopersicon esculentum*, which might be due to the stimulation of plant-growth-promoting molecules by the rhizobacteria, helping in the reduction of nematode infections. Anwar-ul-Haq et al. (2011) assessed the effectiveness of *Bacillus* spp., *Azotobacter* spp., *Pseudomonas putida* and *P. fluorescens* against *Meloidogyne incognita* infection on tomato root cultivar 'Money Maker' in the green house at 30 ± 4 °C. Tomato plants treated with *P. fluorescens* considerably reduced nematode-related parameters, leading to ameliorated plant growth and yield. PGPR such as *P. putida* and *Bacillus* spp. also enhanced the plant-growth and yield-related performances of the tomato plants. Siddiqui and Singh (2005) conducted experiments with the objective of assessing fly ash amendments at different dose levels, i.e. 0, 20 and 40% + soil, and *Pseudomonas striata* and a nodule-forming bacteria, *Rhizobium* sp., against the reproduction of *Meloidogyne incognita* infecting pea. The inoculation of second-stage juveniles of *M. incognita* reduced the rate of transpiration as well as plant growth and yield attributes of pea. On the other hand, the inoculation of *Rhizobium* sp. and *P. striata* improved transpiration from first week onwards with or without nematode-inoculated plants. Moreover, the addition of 40% fly ash registered considerable reduction in nematode multiplication. However, the highest reduction in nematode population was recorded in the *Rhizobium* sp. inoculated plants as it caused greater pernicious effect on galling and nematode multiplication than *P. striata*. The inoculation of both rhizobacteria registered highest improvement in the reduction of nematode population as compared to their individual effects, which suggested that both organisms have a synergistic role in the management of plant nematodes. This study also elaborated that both PGPR had the ability to release some nematotoxic compounds, which became lethal to nematodes life cycle. Khan et al. (2016) described that the inoculation of *Pseudomonas aeruginosa*, *P. fluorescens*, *P. stutzeri* and *P. striata* enhanced root nodulation and reduced the *Meloidogyne incognita* infesting mungbean. The application of rhizobacteria such as *Pseudomonas aeruginosa*,

P. fluorescens, *P. stutzeri* and *P. striata* improved the plant growth and yield characters of mungbean and also reduced nematode-related parameters such as galling, egg masses and fecundity, which might be due to the production of siderophores, HCN, IAA and NH_3. Pankaj et al. (2010) reported that *Gluconacetobacter diazotrophicus* Co99–70, *Bacillus* sp. RKB-91 and *Pseudomonas* sp. RKP-33 potentially controlled the severity of *Meloidogyne graminicola* of rice cv. Basmati-370. Seed inoculation with *G. diazotrophicus* Co99-70 was found to be highly effective in the diminution of galling and in lowering nematode multiplication. Restriction in the multiplication of root-knot nematodes was found to be due to antibiosis against nematodes and volatile fatty acids produced by *G. diazotrophicus*. These bacteria also promoted the root and shoot growth of seedlings. These organic molecules may also reduce egg hatching ability by intervening in embryogenesis. Padgham and Sikora (2007) evaluated the performance of *Bacillus megaterium*, which was isolated from a rice-growing region of Taiwan. Root dip and soil-drenching methods successfully impaired nematode-related variables (*Meloidogyne graminicola*) and enhanced crop health. *Bacillus megaterium* possesses endospore-forming ability, which creates a conducive environment for the biological control of *Meloidogyne graminicola*. It was further justified that the dry spores of *Bacillus* could be applied directly through seed treatments as these bacteria quickly colonise the roots and help in the enhancement of plant growth. Huang et al. (2010) conducted an in vitro test followed by pot experiments and reported that *Bacillus megaterium* YMF3.25 inhibited the egg hatching and reduced the infection of *Meloidogyne incognita* through the production of various organic compounds that are nematicidal in nature. After gas chromatography-mass spectrometry (GCMS), benzeneacetaldehyde, 2-nonanone, decanal, 2-undecanone and dimethyl disulphide were found to be nematostatic to the juveniles and eggs at the concentration of 0.5 mmol. In addition, some organic compounds such as nonane; phenylethanone; 3,5-dimethoxy-toluene; phenol; 2,3-dimethyl-butanedinitrile; and 1-ethenyl-4-methoxy-benzene also exhibited nematicidal activities. Chen et al. (2000) reported that some organics were incorporated into soil with or without fumigants (methyl bromide). Three weeks later, *Meloidogyne hapla*, *Bacillus thuringiensis*, *Paecilomyces marquandii* and *Streptomyces costaricanus* were inoculated to soil individually. *B. thuringiensis* plus *S. costaricanus* inoculated plants enhanced lettuce head weight in unfumigated organic soil. All other amendments were also found to be effective against *M. hapla*, which reduced root galling in soil treated with or without fumigants. In this trial, the organic residue might be the carbon source for the proliferation of biocontrol microorganisms, which reduced the nematode population. The organic compounds released after the decomposition of the organic matter was also considered to be the reason behind nematode population reduction.

Mendoza et al. (2008) elaborated the mode of action of antagonist bacteria *Bacillus firmus* against *Meloidogyne incognita*. Significant rates of paralysis and mortality were detected after the incubation of three nematode species in low concentrations of pure culture filtrates following the removal of the bacterial cells. The same culture filtrates also significantly reduced the hatching of *Meloidogyne incognita*. Pure bacterial cell suspensions added to sand also reduced the survival of

R. similis in bioassays by 41% over the control. The mode of action responsible for nematode paralysis and mortality was therefore demonstrated to be closely associated with the production of bioactive compounds (secondary metabolites) by the bacteria. Mashela and Nthangeni (2002) evaluated the effectiveness of *Ricinus communis* with and without *Bacillus* species for the suppression of *Meloidogyne incognita*. Application of organic materials and *Bacillus* suppressed nematode population, which might be due to enhanced concentrations of chemicals during microbial decomposition or enhanced nematophagous microbial populations. Ricin, a compound isolated from *Ricinus communis*, could also be the possible reason behind the considerable reduction of nematode population. Serfoji et al. (2010) conducted a glasshouse experiment for the effectiveness of *Glomus aggregatum* and *Bacillus coagulans* against the *Meloidogyne incognita* infesting tomato cv. Pusa Ruby. *G. aggregatum* alone and in combination with *B. coagulans* exhibited maximum plant biomass of tomato cv. Pusa Ruby and decreased root-knot nematode population, which might be due to increased beneficial microbial population. Xiong et al. (2015) reported the biocontrol effectiveness of *Bacillus firmus* YBf-10 against *Meloidogyne incognita*. Results further revealed that the inoculation of *B. firmus* YBf-10 caused lethal activity, inhibition of egg hatch and mortality of *M. incognita*. In addition, pot trials revealed that soil drenching with YBf-10 considerably controlled nematode reproduction by efficiently reducing the damage by *M. incognita* to tomato plants. Nematode-related parameters such as root galls, egg masses and total nematode population were found to be significantly checked, which might be due to the secondary metabolites produced by YBf-10 leading to improved plant growth and yield attributes. Kavitha et al. (2007) conducted an experiment to assess the effectiveness of some biocontrol microorganisms such as *Pseudomonas fluorescens*, *Bacillus subtilis* and *Trichoderma viride* against *Meloidogyne incognita* infesting tropical sugar beet cv. Indus. 2.5 kg/ha of biocontrol was applied and compared with carbofuran (1 kg a.i./ha). The application of these antagonists improved the plant growth and yield performance of sugar beet and also caused greater reduction in nematode population over control. *P. fluorescens* registered maximum improvement in the plant growth and yield variables of sugar beet. Besides, some well-studied plant defence enzymes, such as peroxidase, polyphenol oxidase and phenylalanine ammonia-lyase, were found to be increased, being higher in plants treated with *P. fluorescens*, followed by *T. viride* and *B. subtilis*, which might be due to the induction of resistance to plant nematodes by the crop plants. Moreover, a study published in the *Journal of Phytopathology* in 2015 reported that *B. pumilus* L1 produced protease and chitinase enzymes, which were found to be nematostatic as they inhibited second-stage juveniles. Hatching and mortality rate were correspondingly increased with increasing concentrations of crude enzymes and time, which might be due to the partial destruction of the eggshell and juvenile body. The pot experiment also demonstrated that the application of a biocontrol agent to potted soil caused significant reduction in the number of galls and egg masses of *M. arenaria*, leading to enhanced plant growth of tomato, which could be due to the presence of certain plant-growth-promoting molecules (Lee and Kim 2016). Moghaddam et al. (2014) reported that two strains of *Bacillus pumilus*, ToIrFT-KC806241 and

TolrMA-KC806242, were isolated and characterised from tomato fields through morphological and molecular-based techniques. The inoculation of Tolr-MA against *M. javanica* considerably reduced nematode-related parameters such as the number of galls and eggs. Tolr-MA had the ability to produce proteolytic enzymes, which might be the source of the suppression of egg hatching. Siddiqui et al. (2007) used some biocontrol PGPR, such as *Pseudomonas putida*, *P. alcaligenes*, *Paenibacillus polymyxa*, *Bacillus pumilus* and *Rhizobium* sp. for the management of *Meloidogyne javanica* on lentil. It was seen that *Pseudomonas putida* registered greater suppression on the egg hatching and penetration of *M. javanica*, followed by other rhizobacteria like *P. alcaligenes*, *P. polymyxa* and *B. pumilus*, leading to improved plant growth and yield-related characters. Interestingly, the plant growth character was found to be considerably improved. In nematode-inoculated plant, *Rhizobium* sp. of lentil strain was the agent that registered the highest improvement in the plant's health as compared to other PGPR. Combinatorial effect of *Rhizobium* sp. with any other PGPR caused a considerable reduction in nematode galling and egg masses. Highest reduction was recorded in *Rhizobium* plus *P. putida* inoculated-plant which might be due to production of siderophores and other nematostatic organic molecules by the rhizobacteria as per analysis performed with the help of SDS-PAGE. Tong-Jian et al. (2013) tested the effectiveness of *Bacillus cereus* X5 against *Meloidogyne* sp. in vitro by examining the mortality and egg hatching of second-stage juveniles. The biofumigation of the *Meloidogyne* sp. infested soil with some organic additives such as chicken manure, pig manure and rice straw alone or in combination with *B. cereus* X5 was also found to show great nematicidal activity. The application of bio-organics like *B. thuringiensis* BTG or *T. harzianum* SQR-T037 enhanced the plant's biomass and reduced nematode galling and population. *B. cereus* X5 also exhibited great quantum of effectiveness in the management of root-knot nematodes, which might be due to the production of nematicidal compounds, and the production of plant-growth-promoting molecules could be reason behind enhanced plant growth and yield. Mahdy et al. (2000) tested the biocontrol potentiality of *Bacillus cereus* S18 against three species of root-knot nematodes, *Meloidogyne incognita*, *M. javanica* and *M. arenaria*, infesting tomato. A drastic reduction trend line of galls and the number of egg masses was recorded after the treatment of the plants with *Bacillus cereus* S18, which might be due to the congestion in embryogenesis, egg hatching, mortality etc., leading to improved plant growth and yield attributes of tomato. Siddiqui and Mahmood (2001) assessed the effectiveness of *Pseudomonas fluorescens*, *Azotobacter chroococcum* and *Azospirillum brasilense*, singly as well as in combination with *Rhizobium* sp. and *Glomus mosseae*, for the plant growth promotion of *Cicer arietinum*. All rhizobacteria and plant symbionts significantly enhanced the plant growth and yield characters of chickpea on one hand, while on the other hand, the application of these microorganisms drastically reduced the root galls, egg masses and reproducing ability of *Meloidogyne javanica*, which might be due to the release of some toxic chemicals inhibitory to nematodes. Oyekanmi et al. (2007) reported that the inoculation of soybean plants with *Glomus mosseae* (200 spores/plant), *Bradyrhizobium japonicum* (10^6 cells/plant) and *Trichoderma pseudokoningii*

(6.8×10^7 spores/plant) caused greater reduction in *Meloidogyne incognita* population and enhanced plant growth and yield characters. The application of these microorganisms solely, dually or in consortium registered greater improvement in plant health, and the microorganisms were assumed to be working synergistically. The inoculation of these microbes also proved to be antagonistic to root-knot nematodes as there were reduced root galls, egg masses and other nematode-related parameters, which might be due to the combinatorial application of these microorganisms. Different cultivar response to root-knot nematodes were the robust reason behind the poor reproduction of nematodes.

Bhat et al. (2012) reported that *Meloidogyne incognita* race-1 inoculation hampered the growth, nodulation, the nitrogen contents of root and shoot as well as leghaemoglobin, bacteroid and nitrogenase activity, exhibiting poor plant health of blackgram. However, the application of *Bradyrhizobium* inhibited *Meloidogyne incognita* race-1 activity because in the treated plant, there were lesser number of galls and egg masses, which might be due to the inhibition of egg hatching and secondary infection. Saikia et al. (2013) reported that the rhizobacteria *Bacillus megaterium* (ATCC No. 14581) and *Pseudomonas fluorescens* (ATCC No. 13525), fungi *Trichoderma viride* (MTCC No. 167) and *Paecilomyces lilacinus* (PDBC PL55) and plant symbiont *Glomus intraradices* registered greater management of *M. incognita* infesting *Withania somnifera* cv. Poshita. Except plant symbiont (*G. fasciculatum*), all rhizospheric microbes exhibited greater reduction in nematode population, which might be due to various reasons like competition, hyperparasitism, induced resistance etc. Rhizobacteria generally produces some organic molecules, such as phenolic compounds, organic acids and secondary metabolites, which have already been proven to be nematostatic, resulting in the enhancement of plant growth and yield characters. Tian et al. (2014) reported the effectiveness of endophytic bacterium *Sinorhizobium fredii* Sneb183 against soybean cyst nematodes (*Heterodera glycines*). The inoculation of *Sinorhizobium fredii* Sneb183 inhibited the penetration of juveniles and their further development inside the roots of soybean, which might be due to change in root exudates or the production of toxic chemicals. Reimann et al. (2008) reported that the inoculation of *Rhizobium etli* G12 to tomato plants resulted in induced resistance against *Meloidogyne incognita*. The application of *R. etli* accelerated the colonisation of *Glomus intraradices* on tomato. In addition, the application of these two microorganisms significantly reduced nematode penetration and stopped the reproduction of *Meloidogyne incognita*. However, the sole application of both beneficial microorganisms registered greater enhancement in plant growth and reduced nematode development and population, and their combinatorial application significantly enhanced the rate of reduction in nematode multiplication, which might be due to the reduced rate of secondary infection. Ugwuoke and Eze (2010) reported that the inoculation of *Glomus geosporum* and *Rhizobium* (IRJ 21774) registered a significant increase in the root nodulation of *Vigna unguiculata* L. Walp). The inoculation of these microorganisms also reduced root gallings and other nematode-related parameters, leading to enhanced plant growth and yield parameters.

3.4 Conclusions and Future Prospects

A concluding remark could be drawn here that PGPR could be an alternative approach in the management of phytoparasitic nematodes. PGPR are the currently available option, which can be potentially isolated from a diverse range of soil and exploited for the purpose. Suitable and putative strains of PGPR, like *Rhizobium*, *Pseudomonas*, *Bacillus* etc., can be commercialised at a large scale. The biopesticides extracted from these biocontrol bacteria could also be used in the management of various soil-borne plant pathogens and pests, including phytonematodes. The application of PGPR not only improves soil health but also provides a good source of nutrition to plants by obviating phytonematode stress. A single mechanism is not the reason behind the biocontrol of these plant parasitic nematodes. Multiple mechanisms are operated simultaneously, which help in the reduction of nematode population and thereby enhance plant health. However, significant research has been conducted that advocated that PGPR could be exploited, commercialised and used at a large scale for the betterment of agroecosystems. However, a major hurdle in the commercialisation of these PGPR might be the isolation and identification of suitable strains. Identification of potential strains and their proper exploitation will be an option in the 'next generation agriculture'. Potent strain selection could be done through genomics and metagenomics studies, which will indeed unravel many hidden facts about the biomanagement of these phytonematodes.

References

Abd-Allah EF, Alqarawi AA, Hashem A, Radhakrishnan R, Al-Huqail AA, Al-Otibi FON et al (2018) Endophytic bacterium Bacillus subtilis (BERA 71) improves salt tolerance in chickpea plants by regulating the plant defense mechanisms. J Plant Interact 13(1):37–44

Abdel-Salam MS, Ameen HH, Soliman GM, Elkelany US, Asar AM (2018) Improving the nematicidal potential of *Bacillus amyloliquefaciens* and *Lysinibacillus sphaericus* against the root-knot nematode *Meloidogyne incognita* using protoplast fusion technique. Egypt J Biol Pest Control 28(1):31

Abeles, F. B. M. (1992). Ethylene in plant biology, Frederick B. Abeles, Page W. Morgan, Mikal E. Saltveit

Agbodjato NA, Noumavo PA, Baba-Moussa F, Salami HA, Sina H, Sèzan A et al (2015) Characterization of potential plant growth promoting rhizobacteria isolated from Maize (*Zea mays* L.) in central and Northern Benin (West Africa). *Appl Environ Soil Sci* 2015:1–15

Ahmed S, Liu Q, Jian H (2018) Biocontrol potential of Bacillus isolates against cereal cyst nematode (*Heterodera avenae*). Pak J Nematol 36(2):163–176

Alazem M, Lin NS (2017) Antiviral roles of abscisic acid in plants. Front Plant Sci 8:1760

Aljaafri WA, Husain ER, Al-Fadhal FA (2019) Examination of *Burkholderia renojensis, Streptomcyes avermentilis*, and *Bacillus firmus* to management of *Meloidogyne incognita* on corn. J Pharm Sci Res 11(3):942–947

Almaghrabi OA, Massoud SI, Abdelmoneim TS (2013) Influence of inoculation with plant growth promoting rhizobacteria (PGPR) on tomato plant growth and nematode reproduction under greenhouse conditions. Saudi J Biol Sci 20(1):57–61

Amaki Y, Tanaka K, Tanaka M, Takahashi A (2019) U.S. Patent Application No. 10/219,517

Ansari RA, Mahmood I (2017a) Optimization of organic and bio-organic fertilizers on soil properties and growth of pigeon pea. Sci Hortic 226:1–9

Ansari RA, Mahmood I (2017b) Determination of disease incidence caused by *Meloidogyne* spp. and or *Fusarium udum* on pigeonpea in Aligarh district: a survey. Trends Biosci 10 (24):5239–5243

Ansari RA, Mahmood I (2019a) Plant health under biotic stress: volume 2: Microbial interactions. Springer, Singapore. https://doi.org/10.1007/978-981-13-6040-4

Ansari RA, Mahmood I (2019b) Plant health under biotic stress: volume 1: Organic strategies. Springer, Singapore. https://doi.org/10.1007/978-981-13-6043-5

Ansari RA, Rizvi R, Sumbul A, Mahmood I (2017a) PGPR: current vogue in sustainable crop production. In: Kumar V, Kumar M, Sharma S, Prasad R (eds) Probiotics and plant health. Springer, Singapore, pp 455–472

Ansari RA, Mahmood I, Rizvi R, Sumbul A (2017b) Siderophores: Augmentation of soil health and crop productivity. In: Kumar V, Kumar M, Sharma S, Prasad R (eds) Probiotics in agroecosystem. Springer, Singapore, pp 291–312

Ansari RA, Sumbul A, Rizvi R, Mahmood I (2019) Organic Soil Amendments: Potential tool for soil and plant health management. In: Ansari RA, Mahmood I (eds) Plant health under biotic stress. Springer, Singapore, pp 1–35

Antoun H (2013) Plant-growth-promoting rhizobacteria. In: Maloy S, Hughes K (eds) Brenner's encyclopedia of genetics. Academic Press, San Diego, CA, pp 353–355. https://doi.org/10.1016/B978-0-12-374984-0.01169-4

Anwar-ul-Haq M, Anwar SA, Shahid M, Javed N, Khan SA, Mehamood K (2011) Management of root knot nematode Meloidogyne incognita by plant-growth promoting Rhizobacteria on tomato. Pak J Zool 43(6):1027–1031

Arora NK, Tewari S, Singh R (2013) Multifaceted plant-associated microbes and their mechanisms diminish the concept of direct and indirect PGPRs. In: Arora NK (ed) Plant microbe symbiosis: fundamentals and advances. Springer, New Delhi, pp 411–449

Bhardwaj D, Ansari MW, Sahoo RK, Tuteja N (2014) Biofertilizers function as key player in sustainable agriculture by improving soil fertility, plant tolerance and crop productivity. Microb Cell Factories 13(1):66

Bharucha U, Patel K, Trivedi UB (2013) Optimization of indole acetic acid production by *Pseudomonas putida* UB1 and its effect as plant growth-promoting rhizobacteria on mustard (*Brassica nigra*). Agric Res 2(3):215–221

Bhat MY, Wani AH, Fazal M (2012) Effect of *Paecilomyces lilacinus* and plant growth promoting rhizobacteria on *Meloidogyne incognita* inoculated black gram, *Vigna mungo* plants. J Biopest 5 (1):36

Bindu P, Nagendra PG (2016) Siderophore production by Pseudomonas aeruginosa isolated from the paddy fields of Kuttanad, Kerala. Int J Sci Res 5:1577–1581

Brilli F, Pollastri S, Raio A, Baraldi R, Neri L, Bartolini P et al (2019) Root colonization by *Pseudomonas chlororaphis* primes tomato (*Lycopersicum esculentum*) plants for enhanced tolerance to water stress. J Plant Physiol 232:82–93

Cao H, Jiao Y, Yin N, Li Y, Ling J, Mao Z et al (2019) Analysis of the activity and biological control efficacy of the *Bacillus subtilis* strain Bs-1 against *Meloidogyne incognita*. Crop Prot 122:125

Chen J, Abawi GS, Zuckerman BM (2000) Efficacy of Bacillus thuringiensis, Paecilomyces marquandii, and Streptomyces costaricanus with and without organic amendments against Meloidogyne hapla infecting lettuce. J Nematol 32(1):70

Chu BC, Garcia-Herrero A, Johanson TH, Krewulak KD, Lau CK, Peacock RS et al (2010) Siderophore uptake in bacteria and the battle for iron with the host; a bird's eye view. Biometals 23(4):601–611

Das NP, Kumar A, Singh PK (2015) Cyanobacteria, pesticides and rice interaction. Biodivers Conserv 24(4):995–1005

Davies WJ, Zhang J (1991) Root signals and the regulation of growth and development of plants in drying soil. Annu Rev Plant Biol 42(1):55–76

Dekas AE, Poretsky RS, Orphan VJ (2009) Deep-sea archaea fix and share nitrogen in methane-consuming microbial consortia. Science 326(5951):422–426

Delfim J, Schoebitz M, Paulino L, Hirzel J, Zagal E (2018) Phosphorus availability in wheat, in volcanic soils inoculated with phosphate-solubilizing Bacillus thuringiensis. Sustainability 10 (1):144

d'Errico G, Marra R, Crescenzi A, Davino SW, Fanigliulo A, Woo SL, Lorito M (2019) Integrated management strategies of Meloidogyne incognita and Pseudopyrenochaeta lycopersici on tomato using a Bacillus firmus-based product and two synthetic nematicides in two consecutive crop cycles in greenhouse. Crop Prot 122:159

Egamberdieva D, Wirth SJ, Alqarawi AA, Abd-Allah EF, Hashem A (2017) Phytohormones and beneficial microbes: essential components for plants to balance stress and fitness. Front Microbiol 8:2104

Figueiredo MVB, Martinez CR, Burity HA, Chanway CP (2008) Plant growth-promoting rhizobacteria for improving nodulation and nitrogen fixation in the common bean (*Phaseolus vulgaris* L.). World J Microbiol Biotechnol 24(7):1187–1193

Ghavami N, Alikhani HA, Pourbabaei AA, Besharati H (2017) Effects of two new siderophore-producing rhizobacteria on growth and iron content of maize and canola plants. J Plant Nutr 40 (5):736–746

Glick BR (1995) The enhancement of plant growth by free-living bacteria. Can J Microbiol 41 (2):109–117

Glick BR, Patten CL, Holguin G, Penrose DM (1999) Penrose biochemical and genetic mechanism used by plant growth promoting bacteria. Imperial College Press, London, UK

Goswami D, Thakker JN, Dhandhukia PC (2016) Portraying mechanics of plant growth promoting rhizobacteria (PGPR): a review. Cogent Food Agric 2(1):1127500

Heydari S, Rezvani-Moghadam P, Arab M (2008) Hydrogen cyanide production ability by *Pseudomonas fluorescence* bacteria and their inhibition potential on weed germination. Paper presented at Competition for Resources in a changing world: New drive for rural development, Tropentag, Hohenheim. http://www.tropentag.de/2008/abstracts/full/676.pdf. Accessed 26 Aug 2015

Hider RC, Kong X (2010) Chemistry and biology of siderophores. Nat Prod Rep 27(5):637–657

Huang Y, Xu C, Ma L, Zhang K, Duan C, Mo M (2010) Characterisation of volatiles produced from Bacillus megaterium YFM3. 25 and their nematicidal activity against *Meloidogyne incognita*. Eur J Plant Pathol 126(3):417–422

Ibrahim M, Agarwal M, Yang JO, Abdulhussein M, Du X, Hardy G, Ren Y (2019) Plant growth regulators improve the production of volatile organic compounds in two rose varieties. Plan Theory 8(2):35

Ijaz M, Tahir M, Shahid M, Ul-Allah S, Sattar A, Sher A et al (2019) Combined application of biochar and PGPR consortia for sustainable production of wheat under semiarid conditions with a reduced dose of synthetic fertilizer. *Braz J Microbiol* 50:1–10

Inagaki AM, Guimarães VF, do Carmo Lana M, Klein J, Peres Rodrigues da Costa AC, Ormond Sobreira Rodrigues LF, Rampim L (2015) Maize initial growth with the inoculation of plant growth-promoting bacteria (PGPB) under different soil acidity levels. Aust J Crop Sci 9 (4):271–280

Joe MM, Devaraj S, Benson A, Sa T (2016) Isolation of phosphate solubilizing endophytic bacteria from Phyllanthus amarus Schum & Thonn: evaluation of plant growth promotion and antioxidant activity under salt stress. J Appl Res Med Aromatic Plants 3(2):71–77

Kang BR, Anderson AJ, Kim YC (2018) Hydrogen cyanide produced by Pseudomonas chlororaphis O6 exhibits nematicidal activity against Meloidogyne hapla. Plant Pathol J 34 (1):35

Kavitha J, Jonathan EI, Umamaheswari R (2007) Field application of *Pseudomonas fluorescens, Bacillus subtilis* and *Trichoderma viride* for the control of *Meloidogyne incognita* (Kofoid and white) Chitwood on sugarbeet. J Biol Control 21(2):211–215

Khalil MS, El-Naby SSA (2018) The integration efficacy of formulated abamectin, *Bacillus thuringiensis* and *Bacillus subtilis* for managing *Meloidogyne incognita* (Kofoid and white) Chitwood on tomatoes. J Biopest 11(2):146–153

Khan MR, Mohidin FA, Khan U, Ahamad F (2016) Native Pseudomonas spp. suppressed the root-knot nematode in in vitro and in vivo, and promoted the nodulation and grain yield in the field grown mungbean. Biol Control 101:159–168

Khanna K, Jamwal VL, Kohli SK, Gandhi SG, Ohri P, Bhardwaj R et al (2019) Role of plant growth promoting Bacteria (PGPRs) as biocontrol agents of Meloidogyne incognita through improved plant defense of Lycopersicon esculentum. Plant Soil 436(1–2):325–345

Kloepper JW, Schroth MN (1981) Relationship of in vitro antibiosis of plant growth-promoting rhizobacteria to plant growth and the displacement of root microflora. Phytopathology 71 (10):1020–1024

Kloepper JW, Ryu CM, Zhang SA (2004) Induced systemic resistance and promotion of plant growth by *Bacillus* spp. Phytopathology 94:1259–1266. https://doi.org/10.1094/PHYTO.2004.94.11.1259

Kumar A, Prakash A, Johri BN (2011) Bacillus as PGPR in crop ecosystem. In: Maheshwari DK (ed) Bacteria in agrobiology: crop ecosystems. Springer, Berlin, Heidelberg, pp 37–59

Kumar GP, Kishore N, Amalraj ELD, Ahmed SMH, Rasul A, Desai S (2012a) Evaluation of fluorescent *Pseudomonas* spp. with single and multiple PGPR traits for plant growth promotion of sorghum in combination with AM fungi. Plant Growth Regul 67(2):133–140

Kumar P, Dubey RC, Maheshwari DK (2012b) Bacillus strains isolated from rhizosphere showed plant growth promoting and antagonistic activity against phytopathogens. Microbiol Res 167 (8):493–499

Kumar V, Kumar A, Pandey KD, Roy BK (2015) Isolation and characterization of bacterial endophytes from the roots of Cassia tora L. Ann Microbiol 65(3):1391–1399

Kumar A, Singh R, Yadav A, Giri DD, Singh PK, Pandey KD (2016) Isolation and characterization of bacterial endophytes of Curcuma longa L. 3 Biotech 6(1):60

Kumar A, Singh AK, Kaushik MS, Mishra SK, Raj P, Singh PK, Pandey KD (2017a) Interaction of turmeric (Curcuma longa L.) with beneficial microbes: a review. 3 Biotech 7(6):357

Kumar V, Menon S, Agarwal H, Gopalakrishnan D (2017b) Characterization and optimization of bacterium isolated from soil samples for the production of siderophores. Resource-Efficient Technol 3(4):434–439

Lee YS, Kim KY (2016) Antagonistic potential of Bacillus pumilus L1 against root-knot nematode, *Meloidogyne arenaria*. J Phytopathol 164(1):29–39

Li X, Hu H, Li J, Wang C, Chen S, Yan SZ (2019) Effects of the endophytic bacteria Bacillus cereus BCM2 on tomato root exudates and *Meloidogyne incognita* infection. Plant Dis 103:1551–1558

Liu R, Dai M, Wu X, Li M, Liu X (2012) Suppression of the root-knot nematode [*Meloidogyne incognita* (Kofoid & White) Chitwood] on tomato by dual inoculation with arbuscular mycorrhizal fungi and plant growth-promoting rhizobacteria. Mycorrhiza 22(4):289–296

Liu K, Garrett C, Fadamiro H, Kloepper JW (2016) Induction of systemic resistance in Chinese cabbage against black rot by plant growth-promoting rhizobacteria. Biol Control 99:8–13

Lopes E, Ribeiro R, Xavier A, Alves R, Castro M, Martins M, Almeida L, Mizobutsi E, Santos Neto J (2019a) Effect of Bacillus subtilis on Meloidogyne javanica and on tomato growth promotion. J Exp Agric Int 35(1):1–8. https://doi.org/10.9734/jeai/2019/v35i130197

Lopes P, Koschorreck K, Nedergaard Pedersen J, Ferapontov A, Lörcher S, Skov Pedersen J et al (2019b) Bacillus Licheniformis CotA Laccase mutant: ElectrocatalyticReduction of O2 from 0.6 V (SHE) at pH 8 and in seawater. ChemElectroChem 6(7):2043–2049

Lugtenberg B, Kamilova F (2009) Plant-growth-promoting rhizobacteria. Annu Rev Microbiol 63:541–556. https://doi.org/10.1146/annurev.micro.62.081307.162918

Mahdy M, Hallmann J, Sikora RA (2000) Biological control of different species of the root-knot nematode *Meloidogyne* with the plant health-promoting rhizobacterium *Bacillus cereus* S18. Mededelingen-Faculteit Landbouwkundige en Toegepaste Biologische Wetenschappen, Universiteit Gent 65(2b):545–549

Mahmood I, Rizvi R, Sumbul A, Ansari RA (2019) Potential role of plant growth promoting Rhizobacteria in alleviation of biotic stress. In: Ansari RA, Mahmood I (eds) Plant health under biotic stress. Springer, Singapore, pp 177–188

Marques AP, Pires C, Moreira H, Rangel AO, Castro PM (2010) Assessment of the plant growth promotion abilities of six bacterial isolates using Zea mays as indicator plant. Soil Biol Biochem 42(8):1229–1235

Mashela PW, Nthangeni ME (2002) Efficacy of *Ricinus communis* fruit meal with and without *Bacillus* species on suppression of *Meloidogyne incognita* and growth of tomato. J Phytopathol 150(7):399–402

McRose DL, Baars O, Morel FM, Kraepiel AM (2017) Siderophore production in Azotobacter vinelandii in response to Fe-, Mo-and V-limitation. Environ Microbiol 19(9):3595–3605

Mendoza AR, Kiewnick S, Sikora RA (2008) In vitro activity of Bacillus firmus against the burrowing nematode Radopholus similis, the root-knot nematode Meloidogyne incognita and the stem nematode Ditylenchus dipsaci. Biocontrol Sci Tech 18(4):377–389

Mhatre PH, Karthik C, Kadirvelu K, Divya KL, Venkatasalam EP, Srinivasan S et al (2018) Plant growth promoting rhizobacteria (PGPR): a potential alternative tool for nematodes bio-control. Biocatal Agric Biotechnol 17:119–128

Miller SH, Browne P, Prigent-Combaret C, Combes-Meynet E, Morrissey JP, O'Gara F (2010) Biochemical and genomic comparison of inorganic phosphate solubilization in Pseudomonas species. Environ Microbiol Rep 2(3):403–411

Moghaddam MR, Mahdikhani Moghaddam E, Baghaee Ravari S, Rouhani H (2014) The first report of Bacillus pumilus influence against Meloidogyne javanica in Iran. J Crop Prot 3(1):105–112

Nazar R, Khan MIR, Iqbal N, Masood A, Khan NA (2014) Involvement of ethylene in reversal of salt-inhibited photosynthesis by sulfur in mustard. Physiol Plant 152(2):331–344

Nikoo FS, Sahebani N, Aminian H, Mokhtarnejad L, Ghaderi R (2014) Induction of systemic resistance and defense-related enzymes in tomato plants using Pseudomonas fluorescens CHAO and salicylic acid against root-knot nematode Meloidogyne javanica. J Plant Prot Res 54 (4):383–389

Otieno N, Lally RD, Kiwanuka S, Lloyd A, Ryan D, Germaine KJ, Dowling DN (2015) Plant growth promotion induced by phosphate solubilizing endophytic Pseudomonas isolates. Front Microbiol 6:745

Oyekanmi EO, Coyne DL, Fagade OE, Osonubi O (2007) Improving root-knot nematode management on two soybean genotypes through the application of *Bradyrhizobium japonicum*, *Trichoderma pseudokoningii* and *Glomus mosseae* in full factorial combinations. Crop Prot 26(7):1006–1012

Padgham JL, Sikora RA (2007) Biological control potential and modes of action of *Bacillus megaterium* against *Meloidogyne graminicola* on rice. Crop Prot 26(7):971–977

Pankaj K, Bansal RK, Nandal SN (2010) Biocontrol of *Meloidogyne graminicola* using rhizobacteria on rice seedlings. Nematol Mediterr 38(2):115–119

Podile AR, Kishore GK (2007) Plant growth-promoting rhizobacteria. In: Gnanamanickam SS (ed) Plant-associated bacteria. Springer, Dordrecht, pp 195–230

Prasad M, Srinivasan R, Chaudhary M, Choudhary M, Jat LK (2019) Plant growth promoting rhizobacteria (PGPR) for sustainable agriculture: perspectives and challenges. In: Singh AK, Kumar A, Singh PK (eds) PGPR Amelioration in Sustainable Agriculture. Woodhead Publishing, Duxford, UK, pp 129–157

Ramamoorthy V, Viswanathan R, Raguchander T, Prakasam V, Samiyappan R (2001) Induction of systemic resistance by plant growth promoting rhizobacteria in crop plants against pests and diseases. Crop Prot 20(1):1–11

Ramyabharathi SA, Meena KS, Rajendran L, Karthikeyan G, Jonathan EI, Raguchander T (2018) Biocontrol of wilt-nematode complex infecting gerbera by *Bacillus subtilis* under protected cultivation. Egypt J Biol Pest Control 28(1):21

Reetha S, Bhuvaneswari G, Thamizhiniyan P, Mycin TR (2014) Isolation of indole acetic acid (IAA) producing rhizobacteria of *Pseudomonas fluorescens* and *Bacillus subtilis* and enhance growth of onion (*Allim cepa* L.). Int J Curr Microbiol App Sci 3(2):568–574

Reimann S, Hauschild R, Hildebrandt U, Sikora RA (2008) Interrelationships between rhizobium etli G12 and Glomus intraradices and multitrophic effects in the biological control of the root-knot nematode *Meloidogyne incognita* on tomato. J Plant Dis Prot 115(3):108–113

Rijavec T, Lapanje A (2016) Hydrogen cyanide in the rhizosphere: not suppressing plant pathogens, but rather regulating availability of phosphate. Front Microbiol 7:1785

Saikia SK, Tiwari S, Pandey R (2013) Rhizospheric biological weapons for growth enhancement and *Meloidogyne incognita* management in Withania somnifera cv. Poshita. *Biol Control* 65 (2):225–234

Salomon MV, Bottini R, de Souza Filho GA, Cohen AC, Moreno D, Gil M, Piccoli P (2014) Bacteria isolated from roots and rhizosphere of *Vitis vinifera* retard water losses, induce abscisic acid accumulation and synthesis of defense-related terpenes in *in vitro* cultured grapevine. Physiol Plant 151(4):359–374

Samaddar S, Chatterjee P, Choudhury AR, Ahmed S, Sa T (2019) Interactions between *Pseudomonas* spp. and their role in improving the red pepper plant growth under salinity stress. Microbiol Res 219:66–73

Serfoji P, Rajeshkumar S, Selvaraj T (2010) Management of root-knot nematode, *Meloidogyne incognita* on tomato cv Pusa ruby. By using vermicompost, AM fungus, *Glomus aggregatum* and mycorrhiza helper bacterium, *Bacillus coagulans*. J Agric Technol 6(1):37–45

Sheirdil R, Hayat R, Zhang XX, Abbasi N, Ali S, Ahmed M et al (2019) Exploring the potential soil bacteria for sustainable wheat (*Triticum aestivum* L.) production. Sustainability 11:3361

Siddiqui ZA, Mahmood I (2001) Effects of rhizobacteria and root symbionts on the reproduction of *Meloidogyne javanica* and growth of chickpea. Bioresour Technol 79(1):41–45

Siddiqui ZA, Singh LP (2005) Effects of fly ash, *Pseudomonas striata* and *Rhizobium* on the reproduction of nematode *Meloidogyne incognita* and on the growth and transpiration of pea. J Environ Biol 26(1):117–122

Siddiqui ZA, Baghel G, Akhtar MS (2007) Biocontrol of Meloidogyne javanica by rhizobium and plant growth-promoting rhizobacteria on lentil. World J Microbiol Biotechnol 23(3):435–441

Singh R, Pandey DK, Kumar A, Singh M (2017a) PGPR isolates from the rhizosphere of vegetable crop *Momordica charantia*: characterization and application as biofertilizer. Int J Curr Microbiol App Sci 6(3):1789–1802

Singh VK, Singh AK, Kumar A (2017b) Disease management of tomato through PGPB: current trends and future perspective. 3 Biotech 7(4):255

Sohrabi F, Sheikholeslami M, Heydari R, Rezaee S, Sharifi R (2018) Evaluation of four rhizobacteria on tomato growth and suppression of root-knot nematode, Meloidogyne javanica under greenhouse conditions, a pilot study. Egypt J Biol Pest Control 28(1):56. https://doi.org/10.1186/s41938-018-0059-7

Tian F, Wang Y, Zhu X, Chen L, Duan Y (2014) Effect of *Sinorhizobium fredii* strain Sneb183 on the biological control of soybean cyst nematode in soybean. J Basic Microbiol 54 (11):1258–1263

Tong-Jian XIAO, Fang CHEN, Chao GAO, Qing-Yun ZHAO, Qi-Rong SHEN, Wei RAN (2013) *Bacillus cereus* X5 enhanced bio-organic fertilizers effectively control root-knot nematodes (*Meloidogyne* sp.). Pedosphere 23(2):160–168

Turatto MF, Dourado FDS, Zilli JE, Botelho GR (2018) Control potential of *Meloidogyne javanica* and *Ditylenchus* spp. using fluorescent *Pseudomonas* and *Bacillus* spp. Braz J Microbiol 49 (1):54–58

Ugwuoke KI, Eze SC (2010) The effect of mycorrhiza (Glomus geosporum), rhizobium and Meloidogyne incognita on growth and development of cowpea (Vigna unguiculata L. Walp). Res J Agric Biol Sci 6(6):937–941

Verma VC, Singh SK, Prakash S (2011) Bio-control and plant growth promotion potential of siderophore producing endophytic *Streptomyces* from *Azadirachta indica* A. Juss. *J Basic Microbiol* 51(5):550–556

Viljoen JJ, Labuschagne N, Fourie H, Sikora RA (2019) Biological control of the root-knot nematode *Meloidogyne incognita* on tomatoes and carrots by plant growth-promoting rhizobacteria. Trop Plant Pathol:1–8

Waadt R, Hsu PK, Schroeder JI (2015) Abscisic acid and other plant hormones: methods to visualize distribution and signaling. BioEssays 37(12):1338–1349

Wandersman C, Delepelaire P (2004) Bacterial iron sources: from siderophores to hemophores. Annu Rev Microbiol 58:611–647

Wang D, Lv S, Jiang P, Li Y (2017) Roles, regulation, and agricultural application of plant phosphate transporters. Front Plant Sci 8:817

Wang C, Hu HJ, Li X, Wang YF, Tang YY, Chen SL, Yan SZ (2018) Effects of varying environmental factors on the biological control of *Meloidogyne incognita* in tomato by *Bacillus cereus* strain BCM2. Biocontrol Sci Tech 28(4):359–376

Wani SH, Kumar V, Shriram V, Sah SK (2016) Phytohormones and their metabolic engineering for abiotic stress tolerance in crop plants. Crop J 4(3):162–176

Xiang N, Lawrence KS, Donald PA (2018) Biological control potential of plant growth-promoting rhizobacteria suppression of Meloidogyne incognita on cotton and *Heterodera glycines* on soybean: a review. J Phytopathol 166(7–8):449–458

Xiong J, Zhou Q, Luo H, Xia L, Li L, Sun M, Yu Z (2015) Systemic nematicidal activity and biocontrol efficacy of Bacillus firmus against the root-knot nematode *Meloidogyne incognita*. World J Microbiol Biotechnol 31(4):661–667

Yadegari M, Rahmani HA, Noormohammadi G, Ayneband A (2010) Plant growth promoting rhizobacteria increase growth, yield and nitrogen fixation in *Phaseolus vulgaris*. J Plant Nutr 33 (12):1733–1743

Zandi P, Basu SK (2016) Role of plant growth-promoting rhizobacteria (PGPR) as biofertilizers in stabilizing agricultural ecosystems. In: Nandwani D (ed) Organic farming for sustainable agriculture. Springer, Cham, pp 71–87

Zeynadini-Riseh A, Mahdikhani-Moghadam E, Rouhani H, Moradi M, Saberi-Riseh R, Mohammadi A (2018) Effect of some probiotic bacteria as biocontrol agents of Meloidogyne incognita and evaluation of biochemical changes of plant defense enzymes on two cultivars of pistachio. J Agric Sci Technol 20(1):179–191

Zhai Y, Shao Z, Cai M, Zheng L, Li G, Huang D et al (2018) Multiple modes of nematode control by volatiles of *Pseudomonas putida* 1A00316 from Antarctic soil against *Meloidogyne incognita*. Front Microbiol 9:253

Zhai Y, Shao Z, Cai M, Zheng L, Li G, Yu Z, Zhang J (2019) Cyclo (l-pro–l-Leu) of *Pseudomonas putida* MCCC 1A00316 isolated from Antarctic soil: identification and characterization of activity against *Meloidogyne incognita*. Molecules 24(4):768

Zhou C, Li F, Xie Y, Zhu L, Xiao X, Ma Z, Wang J (2017) Involvement of abscisic acid in microbe-induced saline-alkaline resistance in plants. Plant Signal Behav 12(10):1143

Chapter 4
Organic Additives and Their Role in the Phytoparasitic Nematodes Management

Marwa M. El-Deriny, Dina S. S. Ibrahim, and Fatma A. M. Mostafa

Abstract Application of organic additives is an agricultural strategy for improving the physicochemical properties of the soil, the temperature and humidity conditions, soil structure which are valuable for plant development. Soil amendments like animal and green compost, undecomposed or deteriorated materials, plant extracts, essential oils, oil cakes and chitin, etc. are examples of organic additives. Plant parasitic nematodes cause greater damage to plants health. The application of organic additives is markedly important and can be used in the control of this tiny pest. PPN can survive in wide range of soil and multiply and incite significant losses to various crops. Also, organic additives may improve the soil conditions, such as temperature, pH, salinity, oxygen, etc. which are unsuitable for nematode multiplication. The outcome relies on many factors, such as the material used, the rate of use, the processing/composting of materials, agronomic practices, crop rotation and soil type, the climate, and other environmental factors.

Keywords Nematode · Organic additives · Management · Chitin · Agroecosystem

4.1 Introduction

Highly expensive nematicides (soil fumigants or non-fumigant) are considered good options for controlling plant parasitic nematodes (PPN) in developed countries (Ansari and Khan 2012a, b; Ansari and Mahmood 2017b). The toxicity of nematicides to humans and animals and the disturbance of soil ecosystems have become serious concerns (Sparks 2003). In addition, nematicides often do not show the

M. M. El-Deriny (✉) · D. S. S. Ibrahim
Department of Nematodes Diseases and Central Lab of Biotechnology, Plant Pathology Institute, Agriculture Research Center (ARC), Cairo, Egypt

F. A. M. Mostafa
Agricultural Zoology Department, Faculty of Agriculture, Mansoura University, Mansoura, Egypt

R. A. Ansari et al. (eds.), *Management of Phytonematodes: Recent Advances and Future Challenges*, https://doi.org/10.1007/978-981-15-4087-5_4

consistent results to many years in suppression of nematode population. Therefore, in recent years, several nematicides have been either dragged out from the market or banned due to their highly toxic nature to the plants and other biota. Also, the economic cost of chemicals is a major obstacle for many farmers. Therefore, there is a need for alternate management options against PPNs (Bale et al. 2008). In order to curtail the use of chemical nematicides, development of ecofriendly module is the need of hour which can only protect the crops from pathogens attack. Compared to other pathogens, PPN is sometimes difficult to control because it lives in the soil and has a great ability to attack almost every kind of the plant.

4.2 Organic Additives

Soil organic amendments can be successfully employed for the control of PPNs. The first study on the nematicidal effects of organic additives was done by Linford et al. (1938), who examined the effect of mixing chopped pineapple (*Ananas comosus*) leaves into soil for the management of *Meloidogyne* spp. infesting cowpea (*Vigna unguiculata*). They observed that the population of free-living nematodes grew while that of *Meloidogyne* spp. was suppressed and suggested that microbial species antagonistic to nematodes may be supported by increasing organic additives. Thereafter, various organic additives have been used on various crops to decrease the number of phytonematodes (Akhtar and Malik 2000; Faruk et al. 2001; Olabiyi and Oladeji 2014; Ravindra et al. 2014) and to enhance the structure and fertility of soil.

4.2.1 Definition of Organic Additives

Soil organic matter is any substance that was initially created by living organisms in the soil (Stolt and Lindbo 2010). The quality of organics depends on a range of substances from flawless plant material to the ultimately decomposed organic materials. Tissues of plants have a wide range of organic components, which typically decompose at different rates (Ribeiro et al. 2012; Ansari and Mahmood 2017a, b).

4.2.2 Classification of Organic Additives

We can split organic additives into two wide categories: (i) additives that are cultivated and developed in the soil, for example, cover crops, green manure, crop residues, industrial wastes (oilseed cakes), or town wastes and (ii) additives transported from somewhere else into the field, such as composted yard materials or animal waste and animal compost, etc. (Mokrini et al. 2018; Ansari et al. 2019).

4.2.3 Types of Organic Amendments

Many types of organic amendments, such as nematicidal plants, solid compost, compost teas, protein-rich wastes, and animal and green manure have been evaluated for their effect on suppressing PPNs (Oka 2010; Ansari et al. 2019).

4.2.4 Organic Amendments

The use of organic amendments can reduce diseases caused by nematodes directly by affecting soil properties and indirectly by improving plant growth, changing root physiology, and enhancing the population of antagonistic microorganisms and their productivity (Adegbite and Adesiyan 2005; Rizvi et al. 2012a, b; Tiyagi et al. 2011). The wise use of organic materials may be helpful not only in the production of crops and the health of soil but also in the enhancement of crop chemical fertilizers. The use of organic additives would help in plant metabolism through the supply of many important micronutrients in the early development stage of plants (Mahmood et al. 2007). Utilizing fresh plant materials in soil that has practically zero decomposable materials causes fast increase of microbes, organisms, and different microorganisms, which effectively decay the tissues of plants. As the most readily available energy sources (carbohydrates, fats, proteins) in fresh plants are used up, these microorganisms become comparatively inefficient. Application of organic additives also improved the content of chlorophyll (Mahmood et al. 2007). Vegetative growth and photosynthesis enhancing would lead to the accumulation of more carbohydrates and other metabolites and thus an increase in biomass. The incorporation of plants in soil enhancement microbial activity is known to lead to increased conversion of nitrogen (N) to nitrate form (Gunner 1963), which in turn appears to be responsible for stimulating the activity of nitrate reductase. The application of additives induces the substrate (nitrate) for the enzyme (nitrate reductase) to improve its activity, which ultimately leads to increased plant metabolic activity and hence plant biomass. Root nodulation also increased in soil amended with plant parts, which may be due to the better growth of plants and subsequently the suppression of nematode populations. More studies revealed that the incidence of organic amendment decomposition leading to the control of phytonematodes may be a result of more than one factor: (1) changes in the physical and biological properties of the soil (Ramesh et al. 2009), (2) toxicants released or produced during microbial decomposition (Matthiessen and Kirkegaard 2006; Montasser et al. 2012), and (3) nematotoxic substances present in amendments emitted as a result of decomposition (Khan et al. 1974). Moreover, Southey (1978) observed that organic manure may suppress nematode population and subsequently enhance crop tolerance level. Proper decomposition of organic residues in soil plays a major role in changing the properties of soil and developing an unfavorable condition for nematode reproduction. Organic soil additives also improve soil structure and consequently promote root growth of host plants (Craswell and Lefroy 2001; Young and Crawford 2004). The use of compost prevents the increase of root-knot nematodes (RKNs) by producing toxic root secretions and by incorporating them into the soil, where organic matter increases, thus providing a nutritional source for antagonistic organisms (Valenzuela and Smith 2002; Li et al. 2018). Therefore, decomposed materials ultimately serve as nutrients for plants and improve crop yields (Akhtar and Alam 1993a, b). Many scientists have confirmed that organic matter contains substances toxic to nematodes (Alam et al. 1979, 1980; Badra and Elgindi 1979; El-Deriny 2016; Mohamed and Dalia 2017; Lear 1959; Singh and Sitaramaiah 1973; Sitaramaiah and Singh 1978; Srivastava et al. 1971). Addition of sufficient quantity of organic matter into the soil suppresses the activity of nematodes directly. After further investigation on phytonematodes, several studies have been conducted on plant species that have been discovered to contain nematotoxic compounds, which can be used to control root-knot nematodes (Taba et al. 2008; Ahmad 2009; Ahmad et al. 2010; Douda et al. 2010; El-Deriny 2009). Plant tissues used as organic additives to control nematodes, especially those with high nitrogen/carbon ratios, have been observed to exhibit nematicidal activity, mainly through the production of ammonia from plant parts during their decomposition in the soil or through an increase in the population of antagonistic microorganisms (Oka et al. 2006). Several secondary metabolites from plant parts, such as alkaloids, phenolic compounds, and terpenoids, also appear to have nematicidal activity (Thoden et al. 2009; Mostafa et al. 2016).

4.2.4.1 Plant Extracts

The use of plants and their products is one of the safest and most economical methods for the management of plant parasitic nematodes. These methods are low cost and easy to apply and also have the ability to improve soil texture and fertility (Feizi et al. 2014). Several plant-based products have been involved in plant and nematode interactions. These components involve nematotoxicants, hatching stimulants or inhibitors, attractants, and repellents (Chitwood 2002). Substances that are toxic to parasitic nematodes have been identified in several plants (Akhtar and Malik 2000; Oka 2010). Different species of plant extracts have been evaluated in different studies for their nematicidal properties (Akhtar 1999; Alam et al. 2002). Among plant extracts, cruciferous plants (*Brassica carinata*, *Brassica rapa*), velvet bean (*Mucuna* spp.), sunn hemp (*Crotalaria juncea*), *Eucalyptus*, marigold (*Tagetes* spp.), Rakshak gold (a neem-based product), castor bean (*Ricinus communis*), and *Chrommelina* have been used successfully for controlling root-knot nematodes (*Meloidogyne* spp.) (Wang et al. 2002b; Alam et al. 2002; Hooks et al. 2010; Mokrini et al. 2010; Umar et al. 2010; Ansari et al. 2016). Several studies showed decreasing densities of root-knot nematodes after the application of various types of organic plants. Leaf extract from *Crotalaria virgulata* reduced significantly root galls caused by root-knot nematodes (*Meloidogyne incognita)* in tomato (Jourand et al. 2004). Plant extracts of thyme (*Thymus vulgaris*), *Eucalyptus spp.*, sweet wormwood (*Artemisia absinthium*), fennel (*Foeniculum vulgare*), and peppermint (*Mentha piperita or Mentha balsamea Wild*) reduced the hatching activity of *Meloidogyne incognita* (Ibrahim et al. 2006). The performance of other organic amendments has been improved when used in integration with neem extracts (Oka et al. 2007). Aqueous extract of *Justicia adhatoda* L. and *Lantana camara* L. were effective and decreased the penetration of nematodes in the roots of tomato (Ahmad et al. 2013). Therefore, aqueous concentration treatment might have repellent action or may play a direct role in plant defense mechanisms.

4.2.4.2 Dried Leaf Powder

The population of root-knot nematodes (*M. incognita*) in the roots of cucumber and in the soil was significantly suppressed when seed powder of *Trigonella foenum-graecum* (70.1%), and dried leaf powder of *Catharanthus roseus* (72.6%) were incorporated into the soil (Mostafa et al. 2016). On the other hand, dried peels of *Punica granatum* as well as dried seed powder of *T. foenum-graecum* application achieved the lowest reproduction factor of *Rotylenchulus reniformis* infecting cantaloupe under greenhouse conditions. Under field conditions, a significant reduction in root-knot nematode population (*Meloidogyne* spp.) infecting cucumber was listed when dried leaf powder of *Gomphrena globosa* (79.3%), *Nerium oleander* (78.0%), and *Calendula officinalis* (74.9%) was used. However, the greatest reduction in reniform nematodes infecting cantaloupe was achieved with dried leaf powder of *N. oleander* (67.9%) and dried seed of *Brassica juncea* (69.1%) (El-Deriny 2016). Because an appropriate level of humidity contributed well to the decomposition of plant residues, lethal nematotoxic components are released into the soil to suppress nematode activity (Ahmad 2009). Therefore, the application of organic soil additives at higher level of moisture was slightly better than the use of the same organic additives at low moisture level. Integration of wild spinach powder with fresh chopped leaves of different plants, viz., Mexican poppy, Indian mallow, trailing *Eclipta*, ivy gourd, black pigweed, and wild eggplant, suppressed the pathogenic effect of *M. incognita* infected tomato cv. K-21 (Asif et al. 2016).

4.2.4.3 Neem Preparations

Neem (*Azadirachta indica*) preparations are one of the most promising examples for controlling phytonematodes, including root-knot (Chitwood 2002; Adegbite and Adesiyan 2005; Bharadwaj and Sharma 2007; Oka et al. 2007; Ntalli et al. 2009; Mokrini et al. 2010; Yadav et al. 2018). There are several studies reporting the nematicidal activity, and it has been used as oil cakes, plant extracts, or whole plant materials (Ferraz and de Freitas 2004; Oka 2010; Stirling 1991). Several researchers reported high effectivity of the use of different aqueous extracts of neem preparations against RKNs (Aziz et al. 1995; Khurma and Singh 1997; Akhtar 2000a, b). Neem (*Azadirachta indica*) extracts were also found to be highly effective in reducing the number of rice root-knot nematodes (*M. graminicola*) and in improving plant growth (Dongre and Sobita 2013).

4.2.4.4 Nematostatic Compounds

In the last decade, there was an increasing interest in the discovery of active ingredients from plants or plant products that have nematotoxic effect (Chitwood 2002; Du et al. 2011; Khan et al. 2017). Plant extracts that have alkaloids and flavonoids as active ingredients were observed to have ovicidal property against *Meloidogyne* spp. (Adegbite 2003). A number of nematotoxic ingredients, including salicylic and phenolic acids were found in the aqueous extract of *Argemone mexicana* in order to inhibit *M. javanica*. Plants rich in terpenoids, such as *Salvia officinalis*, *Azadirachta indica*, *Origanum majorana*, *Ocimum basilicum*, and *Lantana camara*, caused significant reduction in *Rotylenchulus reniformis* and *M. incognita* infesting cantaloupe and cucumber, respectively (El-Deriny 2016).

4.2.4.5 Antagonistic Plants

Antagonistic plants release some nematotoxic substances which suppress the population of phytonematodes. Antagonistic crops such as partridge pea, neem, marigold, asparagus, sunn hemp, sesame castor bean, and rape seed have been studied and applied in the management of nematodes. Sunn hemp (*Crotalaria* spp.) as a cover crop is often cultivated for intercrops and soil amendment or direct seeding and is considered as an antagonistic crop for phytonematodes, especially *Meloidogyne* spp. (Wang et al. 2002b). The number of *M. incognita* was suppressed by previous cover crops of *C. juncea* in north Florida (Wang et al. 2004). The use of *Crotalaria* spp. was recommended as precrops by Germani and Plenchette (2004) as long as green manure while at the same time suppressing the number of *Meloidogyne* spp. and improving arbuscular mycorrhizal fungi level. Root-lesion and root-knot nematodes have been shown to be suppressed by marigolds (*Tagetes* spp.) after Kimpinski et al. (2000) observed a reduction in the population of *Pratylenchus penetrans* when marigolds were applied as a cover crop followed by potato, resulting in a significantly higher average yield. In Japan, growing vegetables continuously led to the increase of phytonematodes, particularly *P. coffeae* and *M. incognita*; therefore, integrating marigold with major crops for one season was very important to achieve changes in the cropping system (Yamada 2001). In Morocco, the use of fresh marigold as an amendment was helpful for the management of root-knot nematodes in the protected cultivation of vegetables (Sikora et al. 2005). The inhibition of nematodes by *Crotalaria* spp., including sunn hemp (*Crotalaria juncea*) and marigold (*Tagetes* spp.), has been studied properly (Hooks et al. 2010; Wang et al. 2001, 2002b). Phenolic and tannin compounds produced from plant residues may have nematotoxic effect (Kokalis-Burelle et al. 1994; Mian and Rodriguez-Kabana 1982b).

Brassica Crops

The potential use of *Brassica* crops for the management of sugar beet cyst nematodes (*Heterodera schachtii*) is a common strategy in some parts of the United States and in Northwestern Europe (Hafez and Sundararaj 2004; Muller 1999). Rizvi et al. (2012a, b) observed significant reduction in several phytonematodes such as *Rotylenchulus reniformis*, *Meloidogyne incognita*, *Helicotylenchus indicus*, and *Tylenchorhynchus brassicae* by using some nematotoxic plants such as *Calotropis procera*, *Argemone mexicana*, *Eichhornia echinulata*, and *Solanum xanthocarpum*. Dutta et al. (2019) reported that mixing chopped brassicaceous plant materials into the soil can be useful as an attractive alternative for controlling PPNs.

4.2.4.6 Essential Oils

Several essential oils from medicinal and botanical plants have been found to be effective for controlling several phytonematodes, including root-knot nematodes in vegetables (Oka et al. 2000; Park et al. 2005; Sivakumar and Gunasekaran 2011). The essential oils of *Foeniculum vulgare*, *Mentha rotundifolia*, and *Azadirachta indica* were nematotoxic in vitro, and the essential oils of *Origanum vulgare* and *Coridothymus capitatus* significantly reduced root galling caused by RKNs of cucumber seedlings when mixed with sandy soil (Akhtar 2000a, b; Oka et al. 2000). Meanwhile, the essential oils of four medicinal plants, *Mentha spicata*, *Thymus vulgaris*, *Majorana hortensis*, and *Mentha longifolia* increased nematode mortality (Abd-Elgawad and Omer 1995). Volatiles of clove oil reduced the number of *Meloidogyne incognita* in vitro (Meyer et al. 2008). Similarly, neem-oil-based formulation controlled *Meloidogyne incognita* associated with tomato and chickpea (Javed et al. 2008; Akhtar and Mahmood 1997). Even with lower concentrations, the oils of *Eucalyptus citriodora* and *Ocimum basilicum* had significant impact on the control of *Meloidogyne incognita*.

Oilseed Cakes

Oilseed cakes such as mustard oil cake was found affecting the mortality of *M. incognita* (Goswami 1993). Radwan et al. (2009) noticed the potential use of oilseed cakes in amended soil and found a reduction in root galls in tomato caused by *M. incognita*. The populations of phytonematodes *Rotylenchulus reniformis, Meloidogyne incognita, Helicotylenchus indicus,* and *Tylenchorhynchus brassicae* were significantly suppressed by oilseed cakes of castor bean (*Ricinus communis*), neem (*Azadirachta indica*), duan (*Eruca sativa*), and mustard (*Brassica campestris*) (Tiyagi and Alam 1995). Significant reduction in plant parasitic nematodes infesting eggplant (*Solanum melongena*) was observed due to addition of oil cakes of fennel, sesame, and anise leading to enhanced plant growth and yield contributing characters (El-Sherif et al. 2010). Jojoba oil cakes can significantly inhibit *M. incognita* reproduction, gall formation, and egg mass production (Ashoub et al. 2010). In 33 studies, 91% gave positive results with neem oil cakes (Muller and Gooch 1982). The integration of *Pasteuria penetrans* with organic amendments of oil cakes, namely castor, neem, *Citrullus*, and mustard, against root-knot-nematode-infected chili gave greater reduction in nematode population (85.74%) and, in addition, greater enhancement of plant growth parameters (Chaudhary and Kaul 2013).

4.2.4.7 Animal Manure

Numerous studies have reported nematode suppression in soil and the increment of microorganisms after application of animal manure (Oka 2010; Kaplan and Noe 1993). Adding beef manure into soil around potato roots increased the number of bacterial-feeding nematodes (Kimpinski et al. 2003). Chicken and steer manure suppressed the population of citrus nematodes and potato cyst nematodes and increased their yields (Gonzalez and Canto-Sanenz 1993). The incorporation of cattle manure (70 T/ha) improved potato growth and decreased the multiplication rate of *M. javanica* (El Hajji and Horrigne-Raouani 2012).

Composted Animal Manure

Composted animal manure has been evaluated in several studies for its capacity to control phytonematodes (Akhtar and Mahmood 1996; Renčo et al. 2009, 2011; D'Addabbo et al. 2011). A positive relation between the application of compost and the control of PPN was reported in several studies (McSorley and Gallaher 1996; Everts et al. 2006). Different composts prepared from poultry, sheep, cattle, and horse manure showed nematicidal activity against *Meloidogyne incognita* and reduced galling index in tomato roots due to the presence of predatory nematodes (Kerkeni et al. 2007). The level of nematode suppression depends on several factors, i.e., the nematode species, the maturity of the final product, the application rate, and the type of composting process (Rivera and Aballay 2008). Greatest suppression of *Rotylenchulus reniformis* was obtained with lower carbon/nitrogen (C/N) ratio in soil amended with composts (Ismail et al. 2006). However, a reduction in *M. incognita* population infesting chickpea (*Cicer arietinum*) was observed when compost of cattle manure was applied in a one-season field trial by about 90% (Akhtar 2000a, b).

The effect of organic materials can be modified based on many factors, such as the type of organic matter and its by-products, the time of application, and the species of nematode; nevertheless, there is a common conception that organic materials supplying nutrients to plants consequently increase yields (McSorley and Gallaher 1996; McSorley and Gallaher 1997). Besides, there is a difference in the impact of these applications, depending on whether the experiments were carried out in the field, in microplots, or in a greenhouse (McSorley and Gallaher 1997). Furthermore, the type of cultivar was more important than the choice of organic additives for the management of root-knot nematodes. Large amounts of organic matter are useful for the management of phytonematodes (Whitehead 1997).

4.2.4.8 Chitin and Nitrogenous Amendments

Chitin is widely spread in nature, being the second most common polysaccharide after cellulose (Muzzarelli 1977). Most interesting among nitrogenous organic additives for the management of phytonematodes are those that have chitin or similar mucopolysaccharides. Microbial origin (e.g., cell walls of some fungi as a source for chitinous materials) and crustacean chitin are used in pharmaceutical industries and as waste products of the food. Tylenchoid egg shells consist of chitin in their second layer (Bird and McClure 1976). Chitin is depolymerized through chitinase activity when it is used in the soil. The management of *H. glycines* (Rodriguez-Kfibana et al. 1984) and *M. arenaria* (Godoy et al. 1983a, b; Mian et al. 1982) by chitin amendments was observed in studies at Auburn University. There were several species of fungi isolated from the eggs of *H. glycines* and *Meloidogyne* spp. that were found in chitin-treated soil and are known as egg parasites (Godoy et al. 1983a, b; Rodriguez-Kfibana et al. 1984).

4.3 Mechanism of Organic Additives

Many scientists have conducted several studies to explain how organic additives affect plants in the presence of nematodes, and they revealed that decomposing materials can release nematotoxic compounds, enhance plant tolerance to nematode infection, and induce nematodes' natural enemies (Akhtar and Malik 2000; Oka 2010; Stirling 1991; Thoden et al. 2011; Ansari and Mahmood 2019a, b). There are multiple mechanisms implicated simultaneously, therefore, it is difficult to know which are important as multiple mechanisms may work together (Akhtar and Malik 2000).

4.3.1 Mechanism of Chitin

Chitin and other additives may have multiple modes of action. Because chitin has a low C:N ratio of 6:4 (Rodriguez-Kabana 1986), it decomposes quickly in soil and releases significant amounts of ammonia (Mian et al. 1982). It is interesting to observe that chitinous amendments resulted in impressive reduction in the number of *Heterodera glycines* (Rodriguez-Kabana et al. 1984) or *M. arenaria* (Godoy et al. 1983a, b; Mian et al. 1982). Similar results were achieved in the population control of nematodes under the same conditions using other materials (Kokalis-Burelle et al. 1994). Reduction in *Meloidogyne arenaria* population by chicken manure to ammoniacal nitrogen initially followed by possible suppression by microorganisms were referred by Kaplan and Noe (1993). Integrating urea with chitin decreased phytotoxicity compared to the use of urea alone and the application of additional ammoniacal nitrogen (Rodriguez-Kabana et al. 1989). In field tests in California, Westerdahl et al. (1992) conducted that amendment of soil with 1–2 mt/ha of a chitin product generally suppressed number of phytonematodes relative to an unamended control. Although these observations refer toward releasing of ammonia as a mode of action for chitin, the additional possibility of biological control cannot be eliminated, because reduction of nematode were noticed in a second crop following chitin amendment, long after any short-term effects from ammonia in the first crop would have dissipated (Rodriguez-Kabana et al. 1987).

4.3.2 Ammonia (NH_3)

Oka and Yermiyahu (2002) reported that ammonia is more toxic to nematodes than ammonium ion (NH4+). Under acidic soil conditions, ammonia is ionized to NH4+ (Oka et al. 2007; Rodriguez-Kabana et al. 1989). Therefore, the increase of pH may be better for soil as ammonium can be converted to ammonia (Oka 2010). From this point, Zasada (2005) studied why organic materials can cause a reduction in nematode population. Rodriguez-Kabana et al. (1987) noticed that in soils with pH 5.5–6.5 in Alabama, higher concentrations of chitin were needed than alkaline soils in Israel. Under alkaline conditions, nitrification can be caused missing of ammonia (Oka 2010). On the other hand, Oka et al. (2007) reported that mixing amendments with neem extracts in soil with pH 8.5 can lengthen the activity of ammonia because neem can stop nitrification.

4.3.3 Urea

Urea was effective at lower amounts and is more effective than several plant materials in suppressing the number of root-knot nematodes (Chavarria-Carvajal and Rodriguez-Kabana 1998) because it is considered a more credible source of ammonia than various types of additives. Rodriguez-Kabana and other workers reported that urea and ammonia in soil at amounts as low as 300–400 mg/kg were influential against nematodes (Eno et al. 1955; Rodriguez-Kabana 1986; Rodriguez-Kabana et al. 1981, 1989; Rodriguez-Kabana and King 1980).

4.3.4 Plant Tolerance

One of the surprising qualities of plants grown in mulched plots or pots is that, although healthy, they have the same knots compared to unmulched control plants (Watson 1954). Amendments can provide nutrients and water, which improve plant health and yield, consequently bringing negative effects on nematodes (Akhtar and Malik 2000; McSorley and Gallaher 1995; Noling 1999). Kimpinski et al. (2003) observed in a field study that manure and compost amendments did not cause a reduction in the number of plant parasitic nematodes, but potato yield increased by 27%. Similar trend is observed with McSorley and Gallaher (1995) who concluded that crop yields were greatly improved by compost amendments in split of high population levels of *M. incognita.* Thoden et al. (2011) added that nematode fecundity could be supported by nutrient-enriched root systems. While the examples discuss improved plant tolerance to nematodes (as defined by Cook and Evans 1987; i.e., nematode numbers not affected). Stirling (1991) recorded that the use of organic additives with high levels of phenols may increase plant resistance to nematodes, resulting in lower numbers of plant parasitic nematodes. Some plant extracts, fungi, and bacteria may catalyze plant resistance to phytonematode (Oka 2010; Thoden et al. 2011), but it is not defined how much this mechanism can suppress plant parasitic nematodes by the addition of organic additives.

4.3.5 Toxins from Specific Plants

Ferraz and de Freitas (2004) isolated nematotoxic compounds from plant species. Additives from some plants, such as *Mucuna* spp. and *Ricinus communis,* may have some effect against phytonematodes (Oka 2010; Ritzinger and McSorley 1998; Stirling 1991). Cruciferous plants have had a good effect against plant parasitic nematodes, as well as other pathogens, when used as decomposed products (Akhtar and Malik 2000; Oka 2010; Zasada and Ferris 2004). These plants when used under plastic in biosolarization (Ros et al. 2008) or biofumigation (Bello 1998) may be more effective. This technique could also be effective with other plant species (Piedra Buena et al. 2007).

4.3.6 Stimulation Nematodes Natural Enemies

Wide range of organic materials addition may increase the population of beneficial organisms (Akhtar and Malik 2000; Chavarria-Carvajal et al. 2001; Oka 2010; Riegel et al. 1996; Stirling 1991). Odour-Owino (2003) and Wang et al. (2002a) achieved that some fungi which can parasitize on plant parasitic nematodes. The use of organic materials to control nematode and development of biological control is not new, Linford and co-workers reported it since 1930s (Linford 1937; Linford et al. 1938).

4.3.7 Habitat Modification

"Suppressed of soil pathogens may be as a result to structure and ecology" (Muller and Gooch 1982). In one study, a decrease in *Paratrichodorus minor* population levels was noticed with the doubling of soil organic matter content (McSorley and Gallaher 1996). Oka (2010) concluded that organic amendments to soil may affect nematodes directly by modifying factors such as soil structure, particle accumulation, pH, salinity, carbon dioxide level, oxygen level, and chemical content. It is conceivable to greatly change the biological properties of the soil and its structure (Ansari et al. 2017a, b). For example, the number of juveniles and root galls of *Meloidogyne* spp. in roots and the soil were reduced with higher application rates (50–100%) of composts in pots (Nico et al. 2004), but these application rates are found in pots that have more compost than soil. Numerous studies are needed to show how soil modification affects the presence of nematodes and improves plant performance.

4.3.8 Basic Results of Disintegration

Organic acids, nitrogen compounds, and other compounds could be produced from plant residues. These compounds may have a negative effect on phytonematodes (Oka 2010; Thoden et al. 2011). A common compound is ammonia, which is a by-product of organic matter decomposition (Rodriguez-Kabana 1986; Rodriguez-Kabana et al. 1987). Mian and Rodriguez-Kabana (1982c) observed that decreasing of *M. arenaria* galling by increasing %N of 15 different amendments. Plant materials with a C:N ratio in the range of 15–20 were the most effective. In pot experiments, Oka and Yermiyahu (2002) noticed that concentrations of ammonia produced from composts were more than the lethal level required for *M. javanica* deactivation. Although the C:N ratios of tested oil cakes were low (C:N = 7.0–7.1), it had phytotoxic effect but suppressed galling form by nematode (Mian and Rodriguez-Kabana 1982a, c). Rodriguez-Kabana and coworkers pioneered work with different sources of amendments to produce additional C sources and improve the phytotoxic effects of rapid ammonia generated from organic matters with very low C:N ratios (Mian and Rodriguez-Kabana 1982a, c; Rodriguez-Kabana and King 1980; Rodriguez-Kabana et al. 1987). Based on pot experiments, Castagnone-Sereno and Kermarrec (1991) pioneered the opinion that the use of a sewage sludge (very low C:N = 5.8) caused faster decomposition and released maximum levels of ammoniacal N within 7 days after application. Nematicidal effect usually does not occur from organic materials with C:N >20, possibly because of unsuitable concentrations of released ammonia and different toxins, which have caused slow decomposition, while materials with low C:N (ca. <10) can cause phytotoxicity (Rodriguez-Kabana et al. 1987).

4.3.9 The Hypothesis of Chitin

In early 1980s, many scientists documented the potential use of chitin as organic amendment to control phytonematodes (Godoy et al. 1983a, b; Mian et al. 1982; Rodriguez-Kabana et al. 1983, 1984). However, the direct action is unclear (Duncan 1991). Rodriguez-Kabana and coworkers observed that increased levels of chitinolytic fungi in soil resulted from the use of chitin, which parasitized plant parasitic nematode eggs (Rodriguez-Kabana et al. 1983, 1984, 1987). Rodriguez-Kabana (1986) pioneered the thought that chitin has a low C:N ratio of 6:4, so it produces significant amounts of ammonia as it decomposes quickly in soil (Mian et al. 1982). While chitin amendments resulted in an impressive suppression of *H. glycines (*Rodriguez-Kabana et al. 1984) or *M. arenaria* (Godoy et al. 1983a, b; Mian et al. 1982), under comparatively similar conditions using other materials (Kokalis-Burelle et al. 1994). In field tests in California, Westerdahl et al. (1992) noticed reduced numbers of nematode after adding 1–2 mt/ha of chitin product to soil compared to an unamended control. Kaplan and Noe (1993) attributed the suppression in the number of *Meloidogyne arenaria* by application of chicken litter to ammoniacal nitrogen initially, followed by a possible reduction by microorganisms. Integrating urea with chitin inhibited phytotoxicity and caused additional ammoniacal nitrogen compared urea alone (Rodriguez-Kabana et al. 1989).

4.3.10 The Guess Work of Linford

It is *said that* the *number of phytonematodes can be reduced by using organic additive in soil to stimulate nematode antagonists. The concept was a result of a* Linford 1937 *study. Linford* noted that when plant materials of pineapple was added to soil, the galling from root-knot nematodes (*Meloidogyne* spp.) was reduced, and also revealed that natural enemies may have suppressed root knot population. There are a many evidence suggesting that the use of organic additives can stimulate organisms to attack phytonematodes (Stirling 1991). Even today, it is hard to confirm whether the reason for nematode death after adding organic additives is the increase in either the parasites or predators of nematodes. Some experiments disagree with Linford's hypothesis. For example, the nematode-parasitic fungus *Hirsutella rhossiliensis* decreased in number after adding organic amendments, while bacterivorous nematodes increased (Jaffee et al. 1994). Jaffee (2006) conducted that potential use of nematode-trapping fungi for preying on nematode were opposite and did not follow classic predator-prey dynamics. Nevertheless, the addition of organic amendments to soil is documented to stimulate a wide range of nematode antagonists (Akhtar and Malik 2000; Oka 2010; Riegel et al. 1996; Stirling 1991; Wang et al. 2001, 2002a)

4.4 Conclusions and Future Prospects

This chapter clarifies the potential use of organic additive to improve plant health by reducing nematode population. Also, organic additives have the direct effect on promotion of the physiological and chemical properties of soil. Use of organic additives would be more effective in the suppression of nematode population without harming the agroecosystem. Marginal farmers cannot buy expensive chemical fertilizers for various purposes. Such organic additives could help the purpose of alternative sources of nutrition supply in crop production. These additives are ecofriendly, locally available and help in the improvement of soil health. Nevertheless, experiments are still needed to investigate the active ingredients in these amendments that can be used as potential bio-nematicides.

References

Abd-Elgawad MM, Omer EA (1995) Effect of essential oils of some medicinal plants on phytonematodes. Anzeiger für Schädlingskunde, Pflanzenschutz, Umweltschutz 68(4):82–84

Adegbite AA (2003) Comparative effects of Carbofuran and water extract of *Chromolaena odorata* on growth, yield and food components of root-knot nematode- infested soybean (*Glycine max* (L.) Merrill). Ph.D. thesis, University of Ibadan, Nigeria

Adegbite AA, Adesiyan SO (2005) Root extracts of plants to control root-knot nematode on edible soybean. World J Agric Res 1:18–21

Ahmad F (2009) Studies on the management of root-knot nematode (*Meloidogyne incognita*) with organic soil amendments. Ph.D. thesis, Aligarh Muslim University, Aligarh, p. 262

Ahmad F, Rather MA, Siddiqui MA (2010) Nematicidal activity of leaf extracts from *Lantana camara* L. against *Meloidogyne incognita* (Kofoid and white) Chitwood and its use to manage roots infection of *Solanum melongena* L. Braz Arch Biol Technol 53:543–548

Ahmad F, Siddiqui MA, Babalola OO (2013) Characterization of nematicidal activity of plant residues and their application with moisture approach against *Meloidogyne incognita* in tomato. African J Agric Res 8(1):93–101

Akhtar M (1999) Plant growth and nematode dynamics in response to soil amendments with neem products, urea and compost. Bioresour Technol 69:181–183

Akhtar M (2000a) Effect of organic and urea amendments in soil on nematode communities and plant growth. Soil Biol Biochem 32:573–575

Akhtar M (2000b) Nematicidal potential of the neem tree *Azadirachta indica* (a. Juss). Int Pest Manag Rev 5:57–66

Akhtar M, Alam MM (1993a) Utilization of waste material in nematode control: a review. Bioresour Technol 45:1–7

Akhtar M, Alam MM (1993b) Utilization of waste materials in nematode control: a review. Bioresour Technol 45:1–7

Akhtar M, Mahmood I (1997) Control of root-knot nematode *Meloidogyne incognita* in tomato plants by seed coating with Suneem and neem oil. J Pest Sci 22:37–38
Akhtar M, Mahmood I (1996) Control of plant-parasitic nematodes with organic and inorganic amendments in agriculture soil. Appl Soil Ecol 4:243–247
Akhtar M, Malik A (2000) Roles of organic soil amendments and soil organisms in the biological control of plant-parasitic nematodes: a review. Bioresour Technol 74:35–47
Alam MM, Ahmad M, Khan AM (1980) Effect of organic amendments on the growth and chemical composition of tomato, eggplant and chilli and their susceptibility to attack by *Meloidogyne incognita*. Plant and Soil 57:231–236
Alam MM, Khan AM, Saxena SK (1979) Mechanism of control of plant parasitic nematodes as a result of the application of organic amendments to the soil. V. Role of phenolic compounds. Ind J Nematol 9:136–142
Alam S, Akhter N, Begun F, Banu MS, Islam MR, Chowdhary AN, Alam MS (2002) Anti-fungal activities (*in vitro*) of some plant extracts and smoke on four fungal pathogens of different hosts. Pak J Biol Sci 5:307–309
Ansari RA, Mahmood I (2017b) Determination of disease incidence caused by *Meloidogyne* spp. and or *Fusarium udum* on pigeon pea in Aligarh district: a survey. Trends Biosci 10 (24):5239–5243
Ansari RA, Mahmood I (2019a) Plant health under biotic stress: volume 2: microbial interactions. Springer, Singapore. https://doi.org/10.1007/978-981-13-6040-4
Ansari RA, Mahmood I (2019b) Plant health under biotic stress: volume 1: organic strategies. Springer, Singapore. https://doi.org/10.1007/978-981-13-6043-5
Ansari RA, Khan TA (2012a) Parasitic association of root-knot nematode, *Meloidogyne incognita* on guava. e-J Sci Technol 5:65–67
Ansari RA, Khan TA (2012b) Diversity and community structure of phytonematodes associated with guava in and around Aligarh, Uttar Pradesh, India. Trends Biosci 5(3):202–204
Ansari RA, Mahmood I (2017a) Optimization of organic and bio-organic fertilizers on soil properties and growth of pigeon pea. Sci Hortic 226:1–9
Ansari RA, Mahmood I, Rizvi R, Sumbul A (2017b) Siderophores: augmentation of soil health and crop productivity, In *Probiotics in agroecosystem* (pp. 291–312). Springer, Singapore
Ansari RA, Rizvi R, Sumbul A, Mahmood I (2017a) PGPR: current vogue in sustainable crop production, In *Probiotics and plant health* (pp. 455–472). Springer, Singapore
Ansari RA, Sumbul A, Rizvi R, Mahmood I (2019) Organic soil amendments: potential tool for soil and plant health management. In: Ansari RA, Mahmood I (eds) Plant health under biotic stress. Springer, Singapore, pp 1–35
Ansari T, Asif M, Siddiqui MA (2016). Potential of botanicals for root knot nematode management on tomato. Lambert Academic Publishing, pp 1–121. ISBN No. 9783659910920.
Ashoub AH, Abdel-Bary NA, Yassin MY, AbdEl-Razek GM (2010) Nematicidal potentiality of certain medicinal plants used as soil amendments against *Meloidogyne incognita*. Egypt J Agronematol 9:57–67
Asif M, Khan A, Tariq M, Siddiqui MA (2016) Sustainable management of root knot nematode *Meloidogyne incognita* through organic amendment on *Solanum lycopersicum* L. Asian J Biol 1 (1):1–8
Aziz I, Ahmad R, Inam-ul-Haq M (1995) Effect of insecticides and leaf extracts on egg hatching and larval mortality of root-knot nematode (*Meloidogyne javanica*). Pak J Phytopathol 7:65–67
Badra T, Elgindi DM (1979) The relationship between phenolic content and *Tylenchulus semipenetrans* populations in nitrogen-amended citrus plants. Rev Nematol 2:161–164
Bale JS, van Lenteren JC, Bigler F (2008) Biological control and sustainable food production. Philos Trans R Soc Lond Ser B Biol Sci 363(1492):761–776
Bello A (1998) Biofumigation and integrated crop management. In: Bello A, Gonzalez JA, Arias M, Rodriguez-Kabana R (eds) Alternatives to methyl bromide for the southern European countries. Graficas Papallona S.C.V, Valencia, Spain, pp 99–126
Bharadwaj A, Sharma S (2007) Effect of some plant extracts on the hatch of *Meloidogyne incognita* eggs. Int J Botany 3:312–316

Bird AF, McClure MA (1976) The tylenchoid (Nematoda) egg shell: structure, composition and permeability. Parasitology 72:19–28

Castagnone-Sereno P, Kermarrec A (1991) Invasion of tomato roots and reproduction of *Meloidogyne incognita* as affected by raw sewage sludge. Supl J Nematol 23:724–728

Chaudhary KK, Kaul RK (2013) Efficacy of *Pasteuria penetrans* and various oil seed cakes in management of *Meloidogyne incognita* in Chilli pepper (*Capsicum annuum* L.). J Agric Sci Technol 15:617–626

Chavarria-Carvajal JA, Rodriguez-Kabana R (1998) Changes in soil enzymatic activity and control of *Meloidogyne incognita* using four organic amendments. Nematropica 28:7–18

Chavarria-Carvajal JA, Rodriguez-Kabana R, Kloepper JW, Morgan-Jones G (2001) Changes in populations of microorganisms associated with organic amendments and benzaldehyde to control plant parasitic nematodes. Nematropica 31:165–180

Chitwood DJ (2002) Phytochemicals based strategies for nematode control. Annu Rev Phytopathol 40:221–249

Cook R, Evans K (1987) Resistance and tolerance. In: Brown RH, Kerry BR (eds) Principles and practice of nematode control in crops. Academic Press, Sydney, pp 179–231

Craswell ET, Lefroy RDB (2001) The role and function of organic matter in tropical soils. Nutr Cycl Agroecosys 61:7–18

D'Addabbo T, Papajová I, Sasanelli N, Radicci V, Renčo M (2011) Suppression of root-knot nematodes in potting mixes amended with different composted biowastes. Helminthologia 48:278–287

Dongre M, Sobita S (2013) Efficacy of certain botanical extracts in the management of *Meloidogyne graminicola* of rice. IntJAgrSciRes 3:91–98

Douda O, Zouhar M, Maza-kova J, Nova-kova E, Pavela R (2010) Using plant essences as alternative mean for northern root-knot nematode (Meloidogyne hapla) management. J Pest Sci 83:217–221

Du SS, Zhang HM, Bai CQ, Wang CF, Liu QZ, Liu ZL, Deng ZW (2011) Nematicidal flavone-C-glycosides against the root-knot nematode (*Meloidogyne incognita*) from *Arisaema erubescens* Tubers. Molecules 16:5079–5086. https://doi.org/10.3390/molecules16065079

Duncan LW (1991) Current options for nematode management. Annu Rev Phytopathol 29:469–490

Dutta TK, Khan MR, Phani V (2019) Plant-parasitic nematode management via biofumigation using brassica and non-brassica plants: current status and future prospects. Curr Plant Biol 17:17–32

El Hajji L, Horrigne-Raouani N (2012) Efficiency of Solarization and incorporation of cattle manure in management of *Meloidogyne javanica* (Treub.) Chitwood and *Globodera pallida* (Stone) Behrens associated with potato. Pest Technol 6:75–78

El-Deriny MM (2009). Studies on certain nematode pests parasitizing some ornamental plants. M. Sc. thesis, Fac. Agric., Mansoura Univ., 135 pp

El-Deriny MM (2016). Integrated control of certain plant parasitic nematodes infecting cucurbitaceae plants. PhD. Thesis, Fac. Agric., Mansoura Univ., 155 pp

El-Sherif AG, Refaei AR, El-Nagar ME, Salem MMH (2010) Impact of certain oil-seed cakes or powder in comparison with oxamyl or urea on *Meloidogyne incognita* infecting eggplant. Arch Phytopathol Plant Protect 43:88–94

Eno CF, Blue WG, Good JM (1955) The effect of anhydrous ammonia on nematodes, fungi, bacteria, and nitrification in some Florida soils. Proc Soil Sci Soc Am 19:55–58

Everts KL, Sardanelli S, Kratochvil RJ, Armentrout DK, Gallagher LE (2006) Root knot and root lesion nematode suppression by cover crops, poultry litter, and poultry litter compost. Plant Dis 90:487–492

Faruk MI, Bari MA, Nahar MS, Rahman MA, Hossain MM (2001) Management of root knot nematode (*Meloidogyne* spp.) of tomato with two organic amendments and a nematicide. Bangladesh J Plant Pathol 17:27–30

Feizi A, Mahdikhani-Moghadam E, Azizi M, Roohani H (2014) Inhibitory effect of *Allium cepa var. aggregatum*, *Salvia officinalis* and *Kelussia odoratissima* essence on the root-knot nematode (*Meloidogyne javanica*) and extraction of active ingredients. J Plant Prot 28:220–225

Ferraz S, de Freitas LG (2004) Use of antagonistic plants and natural products. In: Chen ZX, Chen SY, Dickson DW (eds) Nematology-advances and perspectives, volume II: nematode management and utilization. CABI Publishing, Cambridge, MA, pp 931–977

Germani G, Plenchette C (2004) Potential of Crotalaria species as green manure crops for the management of pathogenic nematodes and beneficial mycorrhizal fungi. Plant and Soil 266:333–342

Godoy G, Rodriguez-Kabana R, Shelby RA, Morgan-Jones G (1983a) Chitin amendments for control of *Meloidogyne arenaria* in infested soil. II. Effects on microbial population. Nematropica 13:63–74

Godoy G, Rodrlguez-Kztbana R, Morgan-Jones G (1983b) Fungal parasites of *Meloidogyne arenaria* in an Alabama soil. A mycological survey and greenhouse studies. Nematropica 13:201–213

Gonzalez A, Canto-Sanenz M (1993) Comparison of five organic amendments for the control of *Globodera pallid* in microplots in Peru. Nematropica 23:133–139

Goswami BK (1993) Effect of different soil amendments with neem cake on root knot nematode and soil mycoflora in cowpea rhizosphere. Indian J Plant Prot 21:87–89

Gunner HB (1963) Nitrification by Arthrobacter globiformis. Nature 197:1127–1128

Hafez SL, Sundararaj P (2004) Biological and chemical management strategies in the sugar beet cyst nematode management. Proceedings of the winter commodity Schools-2004, University of Idaho, USA, pp 243–248

Hooks CRR, Wang KH, Ploeg A, McSorley R (2010) Using marigold (*Tagetes* spp.) as a cover crop to protect crops from plant-parasitic nematodes. Appl Soil Ecol 46:307–320

Ibrahim SK, Traboulsi AF, El-Hajj S (2006) Effect of essential oils and plant extracts on hatching, migration and mortality of *Meloidogyne incognita*. Phytopathol Mediterr 45:238–246

Ismail AE, Rawia AE, El-Nagdi WMA (2006) Effect of different composts, biofertilizers and olive pomace as soil amendments on Rotylenchulus reniformis, growth and chemical analysis of jasmine. Egypt J Appl Sci 2:909–916; J Plant Pathol 17:27–30

Jaffee BA (2006) Interactions among a soil organic amendment, nematodes, and the nematode-trapping fungus *Dactylellina candidum*. Phytopathology 96:1388–1396

Jaffee BA, Ferris H, Stapleton JJ, Norton MVK, Muldoon AE (1994) Parasitism of nematodes by the fungus *Hirsutella rhossiliensis* as affected by certain organic amendments. J Nematol 26:152–161

Javed N, Gowen SR, El-Hassan SA, Inam-ul-Haq M, Shahina F, Pembroke B (2008) Efficacy of neem (*Azadirachta indica*) formulations on biology of root- knot nematodes (*Meloidogyne javanica*) on tomato. Crop Prot 27:36–43

Jourand P, Rapior S, Fargette M, Mateille T (2004) Nematostatic effects of a leaf extract from *Crotalaria virgulata* subsp. *grantiana* on *Meloidogyne incognita* and its use to protect tomato roots. J Nematol 6:79–84

Kaplan M, Noe JP (1993) Effects of chicken excrement amendments on *Meloidogyne arenaria*. J Nematol 25:71–77

Kerkeni A, Horrigue-Raouani N, Ben Khedher M (2007) Effet suppressif de cinq extraits de compost vis-à-vis du nematode àgalles *Meloidogyne incognita*. Nematol Mediterr 35:15–21

Khan A, Asif M, Tariq M, Rehman B, Parihar K, Siddiqui MA (2017) Phytochemical investigation, nematostatic and nematicidal potential of weeds extract against the root-knot nematode, *Meloidogyne incognita* in vitro. Asian J Biol Sci 10:38–46. https://doi.org/10.3923/ajbs.2017.38.46

Khan MW, Alam MM, Khan AM, Saxena SK (1974) Effect of water soluble fractions of oil-cakes and bitter principles of neem on some fungi and nematodes. Acta Bot Indica 2:120–128

Khurma UR, Singh A (1997) Nematicidal potential of seed extracts: *in vitro* effects on juvenile's mortality and egg hatching of M incognita and M. javanica. Nematol Mediterr 25:49–54

Kimpinski J, Arsenault WJ, Gallant CE, Sanderson JB (2000) The effect of marigolds (*Tagetes* spp.) and other cover crops on *Pratylenchus penetrans* and on following potato crops. Suppl J Nematol 32(4S):531–536

Kimpinski J, Gallant CE, Henry R, Macleod JA, Sanderson JB, Sturz AV (2003) Effect of compost and manure soil amendments on nematodes and on yields of potato and barley: a 7-year study. J Nematol 35(3):289–293

Kokalis-Burelle N, Rodriguez-Kabana R, Weaver CF, King PS (1994) Evaluation of powdered pine bark for control of *Meloidogyne arenaria* and *Heterodera glycines* on soybean. Plant and Soil 162:163–168

Lear B (1959) Application of castor pomace and cropping of castor beans to soil to reduce nematode populations. Plant Dis Rep 43:459–460

Li T, Wang H, Xia X, Cao S, Yao J, Zhang L (2018) Inhibitory effects of components from root exudates of Welsh onion against root knot nematodes. PLoS One 13(7):e0201471. https://doi.org/10.1371/journal.pone.0201471

Linford MB (1937) Stimulated activity of natural enemies of nematodes. Science 85:123–124

Linford MB, Yap F, Oliveira JM (1938) Reduction of soil populations of the root-knot nematode during decomposition of organic matter. Soil Sci 45:127–141

Mahmood I, Tiyagi SA, Azam MF (2007) Efficacy of latex bearing plants for the management of plant-parasitic nematodes and soil-inhabiting fungi on chickpea and mungbean. Environ Biol Conserv 12:23–27

Matthiessen JN, Kirkegaard JA (2006) Biofumigation and enhanced biodegradation: opportunity and challenge in soilborne pest and disease management. Crit Rev Plant Sci 25(3):235–265

McSorley R, Gallaher RN (1995) Cultural practices improve crop tolerance to nematodes. Nematropica 25:53–60

McSorley R, Gallaher RN (1996) Effect of yard waste compost on nematode densities and maize yield. J Nematol 28:655–660

McSorley R, Gallaher RN (1997) Effect of compost and maize cultivars on plant-parasitic nematodes. Suppl J Nematol 29:731–736

Melakeberhan H (2006) Fertiliser use efficiency of soybean cultivars infected with *Meloidogyne incognita* and *Pratylenchus penetrans*. Nematology 8:129–137

Meyer SL, Lakshman DK, Zasada IA, Vinyard BT, Chitwood DJ (2008) Dose–response effects of clove oil from *Syzygium aromaticum* on the root-knot nematode *Meloidogyne incognita*. Pest Manag Sci: formerly Pestic Sci 64(3):223–229

Mian IH, Rodriguez-Kabana R (1982a) Soil amendments with oil cakes and chicken litter for control of *Meloidogyne arenaria*. Nematropica 12:205–220

Mian IH, Rodriguez-Kabana R (1982b) Organic amendments with high tannin and phenolic contents for control of *Meloidogyne arenaria* in infested soil. Nematropica 12:221–234

Mian IH, Rodriguez-Kabana R (1982c) Survey of the nematicidal properties of some organic materials available in Alabama as amendments to soil for control of *Meloidogyne arenaria*. Nematropica 12:235–246

Mian IH, Godoy G, Shelby RA, Rodriguez-Kabana R, Morgan-Jones G (1982) Chitin amendments for control of *Meloidogyne arenaria* in infested soil. Nematropica 12:71–84

Mohamed SK, Dalia MD (2017) Soil amendments the alternative approach in modern agriculture. Agri Res Tech Open Access J 4(5):555646. https://doi.org/10.19080/ARTOAJ.2017.04.555646

Mokrini F, Abbad Andaloussi F, El Aissami A (2010) Effet de quelquesextraits des plantescontreles nematodes àgalles (*Meloidogyne incognita*) associés à la culture de la tomate dans la région du Gharb. In: Proceeding 7^{ème}Congrès de l'Association Marocaine de Protection des Plantes, Rabat, vol 2, pp 427–436

Mokrini F, Janati S, Houari A, Essarioui A, Bouharroud R, Mimouni A (2018) Management of plant parasitic nematodes by means of organic amendment. Rev Mar Sci Agron Vét 6 (3):337–344

Montasser SA, Mostafa MA, Anany AE, Bashandy AS (2012) Influence of Decomposing of some Green Leaf Weeds on Sunflower Plants Infected with *Meloidogyne incognita*. Egypt J Agronematol 11(2):193–371

Mostafa FAM, Refaei AR, Khalil AE, El-Deriny MM (2016) Potential use of botanicals rich in alkaloids for controlling *Meloidogyne incognita* and *Rotylenchulus reniformis* infecting cucurbits. Egypt J Agronematol 15(1):29–43

Muller J (1999) The economic importance of Heterodera schachtii in Europe. Helminthologia 36:205–213

Muller R, Gooch PS (1982) Organic amendments in nematode control: an examination of the literature. Nematropica 12:319–326

Muzzarelli RA (1977) Chitin. Pergamon Press, New York

Nico AI, Rafael RM, Jimenez-Diaz M, Castillo P (2004) Control of root-knot nematodes by composted agro-industrial wastes in potting mixtures. Crop Prot 23:581–587

Noling JW (1999) Plant resistance and soil amendments in Florida tomato and pepper. In: Obenauf GL (ed) 2002 annual research conference on methyl bromide alternatives and emissions reductions. Methyl Bromide Alternatives Outreach, Fresno, CA, pp 34-1 to 34-4

Ntalli NG, Menkissoglu-Spiroudi U, Giannakou IO, Prpphetou-Athanasiadou DA (2009) Efficacy evaluation of neem (*Azadirachta indica* A. Juss) formulation against root-knot nematode, *Meloidogyne incognita*. J Crop Prot 28:489–494

Odour-Owino P (2003) Integrated management of root-knot nematodes using agro-chemicals, organic matter and the antagonistic fungus, Paecilomyces lilacinus in natural field soil. Nematol Mediterr 31:121–123

Oka Y (2010) Mechanisms of nematode suppression by organic soil amendments—a review. Appl Soil Ecol 44:101–115

Oka Y, Yermiyahu U (2002) Suppressive effects of composts against the root-knot nematode *Meloidogyne javanica* on tomato. Nematology 4:891–898

Oka Y, Nacar S, Putieusky E, Ravid U, Zohara Y, Spiegal Y (2000) Nematicidal activity of essential oils and their components against the root knot nematode. Phytopathology 90:710–715

Oka Y, Tkachi N, Shuker S, Yermiyahu U (2007) Enhanced nematicidal activity of organic and inorganic ammonia-releasing amendments by *Azadirachta indica* extracts. J Nematol 39:9–16

Oka Y, Tkachi N, Shuker S, Rosenberg R, Suriano S, Fine P (2006) Laboratory studies on the enhancement of nematicidal activity of ammonia releasing fertilizers by alkaline amendments. Nematology 8:335–346

Olabiyi TI, Oladeji OO (2014) Assessment of four compost types on the nematode population dynamics in the soil sown with okra. Int J Org Agric Res Dev 9:146–155

Park IK, Park JY, Kim KH, Choi KS, Choi IH, Kim CS, Shin SC (2005) Nematicidal activity of plant essential oils and their components from garlic (*Allium sativum*) and cinnamon (*Cinnamomum verum*) oils against the pine wood nematode (*Bursaphelenchus xylophilus*). Nematology 7:767–774

Piedra Buena A, Garcia-Alvarez A, Diez-Rojo MA, Ros C, Fernandez P, Lacasa A, Bello A (2007) Use of pepper crop residues for the control of root-knot nematodes. Bioresour Technol 98:2846–2851

Radwan MA, El-Maadawy EK, Kassem SI, Abu-Elamayem MM (2009) Oil cakes soil amendment effects on *Meloidogyne incognita*, root-knot nematode infecting tomato. Arch Phytopathol Plant Prot 42:58–64

Ramesh P, Panwar NR, Singh AB, Ramana S (2009) Effect of organic nutrient management practices on the production potential, nutrient uptake, soil quality, input use efficiency and economics of mustard (*Brassica juncea*). Indian J Agric Sci 79:40–44

Ravindra H, Sehgal M, Pawan AS, Archana BS, Shruti SA, Narasimhamurty HB (2014) Eco-friendly management of root-knot nematodes using acacia compost and bioagents in brinjal. Pak J Nematol 32(1):33–38

Renčo M, Sasanelli N, Kováčik P (2011) The effect of soil compost treatments on potato cyst nematodes *Globodera rostochiensis* and *Globodera pallida*. Helminthologia 3:184–194

Renčo M, Sasanelli N, Šalamún P (2009) The effect of two compost soil amendments based on municipal green and penicillin production wastes, on plant parasitic nematodes. Helminthologia 46:190–197

Ribeiro HB, Ribeiro RCF, Xavier AA, Campos VP, Dias-Arieira CR, Mizobutsi EH (2012) Resíduos de frutos de pequi no controle do nematoide das galhas em tomateiro. Hortic Brasil 30:453–458. https://doi.org/10.1590/S0102-05362012000300016

Riegel C, Fernandez FA, Noe JP (1996) *Meloidogyne incognita* infested soil amended with chicken litter. J Nematol 28:369–378

Ritzinger CHSP, McSorley R (1998) Effect of castor and velvet bean organic amendments on *Meloidogyne arenaria* in greenhouse experiments. J Nematol 30:624–631

Rivera L, Aballay E (2008) Nematicide effect of various organic soil amendments on *Meloidogyne ethiopica* Whitehead, 1968, on potted vine plants. Chilean J Agric Res 68:290–296

Rizvi R, Mahmood I, Tiyagi SA, Khan Z (2012a) Conjoint effect of oil-seed cakes and *Pseudomonas fluorescens* on the growth of chickpea in relation to the management of plant-parasitic nematodes. Braz Arch Biol Technol 55:801–808

Rizvi R, Mahmood I, Tiyagi SA, Khan Z (2012b) Effect of some botanicals for the management of plant-parasitic nematodes and soil-inhabiting fungi infesting chickpea. Turk J Agric For 36:710–719. https://doi.org/10.3906/tar-1201-45

Rodriguez-Kabana R (1986) Organic and inorganic nitrogen amendments to soil as nematode suppressants. J Nematol 18:129–135

Rodriguez-Kabana R, King PS (1980) Use of mixture of urea and blackstrap molasses for control of root-knot nematodes in soil. Nematropica 10:38–44

Rodriguez-Kabana R, Boube D, Young RW (1989) Chitinous materials from blue crab for control of root-knot nematode. I. Effect of urea and enzymatic studies. Nematropica 19:53–74

Rodriguez-Kabana R, Godoy G, Morgan-Jones G, Shelby RA (1983) The determination of soil chitinase activity: conditions for assay and ecological studies. Plant and Soil 75:95–106

Rodriguez-Kabana R, King PS, Pope MH (1981) Combinations of anhydrous ammonia and ethylene dibromide for control of nematodes parasitic of soybeans. Nematropica 11:27–41

Rodriguez-Kabana R, Morgan-Jones G, Chet I (1987) Biological control of nematodes: soil amendments and microbial antagonists. Plant and Soil 100:237–247

Rodriguez-Kabana R, Morgan-Jones G, Ownley-Gintis B (1984) Effects of chitin amendments to soil on *Heterodera glycines*, microbial populations, and colonization of cysts by fungi. Nematropica 14:10–25

Rodriguez-Kfibana R, Morgan-Jones G, Ownley Gintis B (1984) Effects of chitin amendments to soil on *Heterodera glycines*, microbial populations, and colonization of cysts by fungi. Nematropica 14:10–25

Ros M, Garcia C, Hernandez MT, Lacasa A, Fernandez P, Pascual JA (2008) Effects of biosolarization as methyl bromide alternative for *Meloidogyne incognita* control on quality of soil under pepper. Biol Fertil Soils 45:37–44

Shaukat SS, Siddiqui IA, Khan GH, Zaki MJ (2002) Nematicidal and allelopathic potential of *Argemone mexicana*, atropical annual weed. Plant and Soil 245:239–447

Sikora RA, Bridge J, Starr JL (2005) Management practice: an overview of integrated nematode management technologies. In: Luc M, Sikora RA, Bridge J (eds) Plant parasitic nematodes in subtropical and tropical agriculture. CABI Publishing, Wallingford, UK, pp 793–827

Singh RS, Sitaramaiah K (1973) Control of plant parasitic nematodes with organic amendments of soil. Final technical report. Effect of organic amendments, green manures, and inorganic fertilizers on root-knot of vegetable crops. Research Bulletin, Experiment Station and College of Agriculture, Govind Ballabh Pant University of Agriculture and Technology, Panmagar, India

Sitaramaiah K, Singh RS (1978) Effect of organic amendments on phenolic content of soil and plant and response of *Meloidogyne javanica* and its host to related compounds. Plant and Soil 50:671–679

Sivakumar M, Gunasekaran K (2011) Management of root-knot nematode in tomato, chili and brinjal by neem oil formulations. J Biopest 4:198–200

Southey JF (1978) Plant nematology. Ministry of Agriculture, Fisheries and Food, HMSO, London

Sparks DL (2003) Environmental soil chemistry: an overview. In: Sparks DL (ed) Environmental soil chemistry, 2nd edn. Academic Press, New York, pp 1–42

Srivastava AS, Pandey RC, Ram S (1971) Application of organic amendment for the control of root-knot nematode, *Meloidogyne javanica.* (Trueb.). Labdev J Sci Technol 9B(319):203–205

Stirling GR (1991) Biological control of plant-parasitic nematodes. CAB International, Wallingford, UK

Stolt MH, Lindbo DL (2010) Soil organic matter. In: Stoops G et al (eds) Interpretation of micromorphological features of soils and regoliths. Elsevier, Amsterdam, pp 369–396. https://doi.org/10.1016/B978-0-444-53156-8.00017-9

Taba S, Sawada J, Moromizato Z (2008) Nematicidal activity of Okinawa Island plants on the root-knot nematode *Meloidogyne incognita* (Kofoid and White) Chitwood. Plant and Soil 303:207–216

Thoden TC, Hallmann J, Boppré M (2009) Effects of plants containing pyrrolizidine alkaloids on the northern root-knot nematode *Meloidogyne hapla*. Eur J Plant Pathol 123:27–36

Thoden TC, Korthals GW, Termorshuizen AJ (2011) Organic amendments and their influences on plant parasitic and free living nematodes: a promising method for nematode management? Nematology 13:133–153

Tiyagi SA, Alam MM (1995) Efficacy of oil-seed cakes against plant-parasitic nematodes and soil-inhabiting fungi on mungbean and chickpea. Bioresour Technol 51(2–3, 233):–239

Tiyagi SA, Mahmood I, Khan Z, Ahmad H (2011) Biological control of soil-pathogenic nematodes infecting mungbean using *Pseudomonas fluorescens*. Arch Phytopathol Plant Protect 44:1770–1778

Umar I, Muhammad Z, Okusanya BAO (2010) Effect of organic amendments on the control of *Meloidogyne javanica* (Kofoid and White, 1919) (Chitwood, 1949) on Tomato (*Lycopersicon lycopersicum*, Mill). Agri Bus Technol J 8:63–77

Valenzuela H, Smith J (2002) "Tropic sun", sunnhemp. University of Hawaii at Manoa, College of Tropical Agriculture and Human Resources, Hawaii, USA. www.ctahr.hawaii.edu/oc/freepubs/pdf/GreenManureCrops/sunnhemp.pdf

Wang KH, McSorley R, Gallaher RN (2004) Effect of *Crotalaria juncea* amendment on squash infected with *Meloidogyne incognita*. J Nematol 36:290–296

Wang KH, Sipes BS, Schmitt DP (2001) Suppression *of Rotylenchulus reniformis* by *Crotalaria juncea, Brassica napus*, and *Tagetes erecta*. Nematropica 31:237–251

Wang KH, Sipes BS, Schmitt DP (2002a) Management of *Rotylenchulus reniformis* in pineapple, *Ananas comosus*, by intercycle cover crops. J Nematol 34:106–114

Wang KH, Sipes BS, Schmitt DP (2002b) *Crotalaria* as a cover crop for nematode management: a review. Nematropica 32:35–57

Watson JR (1954) Mulches to control root-knot. Proc Florida Acad Sci 7:151–153

Westerdahl BB, Carlson HL, Grant J, Radewald JD, Welch N, Anderson CA, Darso J, Kirby D, Shibuya F (1992) Management of plant-parasitic nematodes with a chitin-urea soil amendment and other materials. Suppl J Nematol 24:669–680

Whitehead AG (1997) Plant nematode control. CABI, Wallingford, p 384

Yadav S, Patil J, Kumar A (2018) *Bio-nematicidal effect of Azadirachta indica*, against *Meloidogyne incognita* in tomato. IJCS 6(3):2757–2761

Yamada M (2001) Methods of control of injury associated with continuous vegetables cropping in Japan crop rotation and several cultural practices. Japan Agric Res Q 35:39–45

Young IM, Crawford JW (2004) Interactions and self-organisation in the soil-microbe complex. Science 304:1634–1637

Zasada I (2005) Factors affecting the suppression of *Heterodera glycines* by N-Viro Soil. J Nematol 37:220–225

Zasada IA, Ferris H (2004) Nematode suppression with brassicaceous amendments: application based upon glucosinolate profiles. Soil Biol Biochem 36:1017–1024

Chapter 5
Metagenomic Insights Into Interactions Between Plant Nematodes and Endophytic Microbiome

M. C. M. Zakeel and M. I. S. Safeena

Abstract Plant-parasitic nematodes are a serious threat to global agriculture. Although the control of plant-parasitic nematodes mainly depends on chemical methods, the need for environmentally friendly alternative approaches for the control of nematodes has been realized in the recent past. In order to find alternative methods to control plant-parasitic nematodes, endophytic microorganisms have been extensively studied in various plant species. These studies have used an array of molecular techniques from simple PCR to next-generation sequencing approaches. Metagenomic studies of endophytes are believed to be the most advanced and effective techniques to disentangle and understand interactions between endophytes and plant-parasitic nematodes.

Keywords Plant-parasitic nematodes · Metagenomics · Endophytic microorganisms · Plant-microbe interaction · Plant defense · Next-generation sequencing

5.1 Introduction

Plant-parasitic nematodes present significant threats to agriculture worldwide. Control strategies of plant-parasitic nematodes mainly depend on synthetic chemical control methods. However, due to the health and environmental consequences of synthetic chemical use in agriculture, the need for viable alternatives was understood by scientific communities, health professionals, environmental activists, and policy makers. Subsequently, studies focusing on the use of plant-derived phytochemicals

M. C. M. Zakeel (✉)
Department of Plant Sciences, Faculty of Agriculture, Rajarata University of Sri Lanka, Anuradhapura, Sri Lanka
e-mail: zakeel@agri.rjt.ac.lk

M. I. S. Safeena
Department of Biological Sciences, Faculty of Applied Sciences, South Eastern University of Sri Lanka, Sammanthurai, Sri Lanka

R. A. Ansari et al. (eds.), *Management of Phytonematodes: Recent Advances and Future Challenges*, https://doi.org/10.1007/978-981-15-4087-5_5

and microbial pesticides were initiated (Whipps and Davies 2000; Pérez et al. 2003; Agbenin et al. 2005; Ntalli and Caboni 2012; Rizvi et al. 2012; Stirling 2017). In this context, endophytic microorganisms, which colonize interior plant tissues, play a pivotal role in plants' defense mechanism (Dong and Zhang 2006; Bae et al. 2011). Endophytes interact with nematodes directly or via the host plant's metabolic pathways. Endophytes can enhance plants to produce certain secondary metabolites and hormones to facilitate defense against plant nematodes (Aly et al. 2011; Khan et al. 2012, 2014; Brader et al. 2014). Composition and diversity of endophytes that interact with nematodes have been studied using different molecular techniques. Of these, culture-independent, metagenomics studies using next-generation sequencing (NGS) platforms would provide novel insights into these interactions between endophytes and plant-parasitic nematodes. Metagenomics is the study of the entire genomes of microorganisms present in open and host-associated microniches using a suite of molecular techniques and bioinformatics tools (Kumar et al. 2015; Thomas et al. 2012). Metagenomic studies provide insights into the microbial diversity, environmental ecology, and functional gene composition of microbial communities (Dong et al. 2018; Handelsman 2007; Kennedy et al. 2008; Kumar et al. 2015; Thomas et al. 2012). Amplicon-based metagenomic approaches (also known as diversity profiling) using 16S ribosomal ribonucleic acid (rRNA) of bacterial and internal transcribed spacer (ITS) regions of fungal communities have been increasingly used for the composition and diversity analysis of soil, rhizosphere, epiphytic, and endophytic microbiomes. This chapter discusses in detail endophytes, their association with and their benefits to plants, and finally how metagenomic studies have been used for deciphering the interactions between endophytes and plant-associated parasitic nematodes.

5.2 Plant Nematodes and Their Impact on Agriculture

Plant nematodes are obligate parasites that cause a huge loss to global agricultural production (Sasser and Freckman 1987). They are considered as major soilborne agricultural pests, responsible for an annual crop loss of over $150 billion worldwide (Abad et al. 2008). This figure may be an underestimate as the loss of agricultural production in developing countries mostly remains unnoticed due to farmers' poor understanding about plant-parasitic nematodes, their impact on crop production, and control strategies. This necessitates an effective control of plant-parasitic nematodes in agriculture (Sasser and Freckman 1987). However, the impact of soil nematodes in natural ecosystems may vary from substantial to no impact. This is because some nematodes are beneficial to the processes of ecosystems, and some have no impact on plant growth or ecosystems, whereas some plant-parasitic nematodes are associated with crop yield reduction (van der Putten et al. 2006). Therefore, control should not introduce a risk to the beneficial nematodes and the natural ecosystem.

Plant nematodes are tiny organisms that develop specific structures for feeding and produce secretory products that help them infect the hosts and absorb nutrients from plant tissues (Williamson 1999; Ansari and Khan 2012a, b; Ansari and Mahmood 2017b; Bernard et al. 2017). Plant-parasitic nematodes differ significantly from other nematodes as they possess dorsal and subventral esophageal glands and a stylet, which are important adaptation characteristics for plant parasitism (Maier et al. 2013; Quentin et al. 2013). A stylet is a needle-like, hollow structure that is protruded into the host to probe the host tissue and release a collection of proteinaceous secretions from the esophageal glands (Bernard et al. 2017). These secretions enable plant nematodes to disintegrate host cells and enter the host tissues. Carbohydrate-catabolizing enzymes are one of the important components of the secretions. Beta-1,4-endoglucanases (cellulases) that are present in the secretion digest plant cell walls, allowing nematodes to enter the tissues (Bernard et al. 2017). Genome sequencing of *Meloidogyne incognita* (root-knot nematodes) has revealed the presence of a set of 61 carbohydrate-active enzymes (CAZymes), which can degrade plant cell walls (Abad et al. 2008). Beta-1,4-endoglucanase genes with catalytic domains of glycosyl hydrolase families G5 and G45 have been identified in many plant nematodes (Smant et al. 1998; Yan et al. 1998; Bernard et al. 2017). Bacterial G5 sequences show homology with G5 sequences of plant-parasitic nematodes in the order Rhabditida, suggesting horizontal gene transfer from bacteria. A deoxyribonucleic acid (DNA) sequence for the G45 domain found in the nematode *Bursaphelenchus xylophilus* has shown a high level of homology to G45 cellulases of ascomycetous fungi, supporting events of horizontal gene transfer from fungi to nematodes (Kikuchi et al. 2004; Palomares-Rius et al. 2014).

Plant-associated nematodes show a variety of lifestyle patterns and interactions with hosts (Jones et al. 2013; Bernard et al. 2017). Some nematodes, which are known as endoparasitic nematodes, invade plant tissues to allow feeding, whereas others (migratory ectoparasitic nematodes) remain external to obtain required nutrients (Bernard et al. 2017). Migratory ectoparasitic nematodes move through the soil and use roots as ephemeral food sources when they encounter them (Jones et al. 2013). The California dagger nematode (*Xiphinema* spp.) is an ectoparasitic nematode that transmits grapevine fanleaf virus, which is attributed to huge losses in global grape production (Villate et al. 2008; Bernard et al. 2017). Endoparasitic nematodes can be of two groups: (1) migratory and (2) sedentary. Migratory endoparasitic nematodes cause serious damage to plants as they enter plants, move around within the roots, and feed on root tissues. Lesion nematodes (*Pratylenchus* spp.), pine wilt nematodes (*Bursaphelenchus xylophilus*), burrowing nematodes (*Radopholus* spp.), and rice root nematodes (*Hirschmanniella* spp.) are some of the economically important migratory endoparasitic nematodes in plants (Moens and Perry 2009; Jones et al. 2013). Sedentary nematodes stop movement once they establish a biotrophic complex feeding structure within the plant tissue (Williamson and Hussey 1996; Jones et al. 2013). The most economically important plant-parasitic nematodes, such as root-knot nematodes and cyst nematodes, display biotrophic interactions (Jones et al. 2013). Semi-endoparasitic nematodes may partially penetrate host tissues at some points of their life cycle and have migratory stages (Jones et al. 2013). The number of plant-parasitic nematodes identified to date has exceeded 4100, and only a fraction of them causes significant yield losses in agriculture (Decraemer and Hunt 2006). Cereal cyst nematodes (*Heterodera* spp.) cause significant crop losses in wheat (*Triticum aestivum*), which is a staple food for more than 50% of the world's population; oat (*Avena sativa*); and barley (*Hordeum vulgare*) (Anon 2012; Bernard et al. 2017). The damage caused by *Heterodera avenae* in wheat may vary between 30 and 100% (Bonfil et al. 2004; Nicol et al. 2007). Root-lesion nematodes *Pratylenchus neglectus* and *P. thornei* and seed gall nematode *Anguina tritici* also cause yield losses in wheat (Bernard et al. 2017). About 20% of yield reduction is attributed to parasitic nematode problems in rice (*Oryza sativa*). There are more than 100 species of nematodes associated with rice diseases, and *M. graminicola* causes the most devastating problem in rice with a yield loss of about 80%. Rice root nematode *Hirschmanniella oryzae* is widely distributed in Asian countries such as Japan, China, Korea, the Philippines, Indonesia, Thailand, Vietnam, India, Sri Lanka, Pakistan, and Nepal, particularly in irrigated rice fields (Bridge et al. 2005; Kyndt et al. 2014). Out of 50 species of nematodes that show parasitic interaction with maize (*Zea mays*), root-knot nematodes (*Meloidogyne* spp.), cyst nematodes (*Heterodera* spp.), and root-lesion nematodes (*Pratylenchus* spp.) are the major nematode pests in maize (Nicol et al. 2011). Potato (*Solanum tuberosum*) and sweet potato (*Ipomoea batatas*) are also important food crops subject to different nematode problems (Turner and Evans 1998). Potato cyst nematodes *Globodera rostochiensis* and *G. pallida*, originated from South America, cause heavy crop loss in potato and various other solanaceous crops (Turner and Evans 1998). These nematodes are quarantine pests in many countries, including the USA (Jones et al. 2017). *Meloidogyne* spp. and stem nematode *Ditylenchus destructor* also affect potatoes and sweet potatoes (Santo et al. 1980).

5.3 Control of Plant Nematodes

The management of plant-parasitic nematodes has become a vital practice in crop production because of the severity of damage and yield lost caused by them. There are numerous practices applied to control nematodes, including cultural, chemical, organic, biological, etc. (Ansari et al. 2017a, b, 2019; Ansari and Mahmood 2017a, 2019a, b; Safeena and Zakeel 2019). All methods have their own merits and demerits, like two faces of a coin. Nevertheless, none of them individually can be concluded as a perfect controlling strategy; rather, it will be more efficient if they are applied in an integrated way. The development of specific and new approaches in cultural practices has promised to fine-tune the control of plant nematodes in cropping systems. Those approaches include the use of species-specific cover crops against certain plant nematodes, the application of bio-fumigants, and the use of trap crops (Westerdahl 2011). Several studies have suggested the efficacy of bio-fumigation to manage nematodes (Mojtahedi et al. 1991; Ploeg and Stapleton 2001; Ploeg 2008). Similarly, trap crops can be employed for controlling sedentary endoparasitic nematodes such as root-knot and cyst nematodes (Koch 1998; Scholte 2000; Westerdahl et al. 2008). The use of resistant cultivars, which hinder the reproduction of nematodes, is another promising strategy for the management of plant nematodes (Noling 2009). This offers a significant benefit by lowering the nematode population in the next round of cropping system. However, resilient cultivars are unfortunately not available for many cropping systems. Physical, chemical, and biological properties of the soil may reduce the devastation instigated by plant-parasitic nematodes (Westphal 2011). Soil fumigations with methyl iodide; propargyl bromide and 1,3-dichloropropene (Rosskopf et al. 2005); calcium cyanamide (D'Addabbo et al. 1996); and methyl bromide have been applied as a pretreatment option, and they have evidenced the strongest activity against soil nematodes (Wang et al. 2006). However, due to the destruction of soil ecosystems by many of these chemicals, the use of fumigants is limited (Dong et al. 1994; van Wesemael et al. 2011). For instance, calcium cyanamide has been banned because of the risks it presents to environmental health (Humpherson-Jones et al. 1992). As an alternative measure to reduce the impact of chemical use to the environment, various physical methods, including steam disinfection, soil solarization, and hot water injection, are applied to treat the soil. However, some specific factors, including the soil type, climatic conditions, and soil moisture content, can affect the success of physical treatments (Dungan et al. 2003).

There are other soil microbes, such as fungi, bacteria, protozoa and predatory nematodes, that can be used as biological control agents by introducing them into the soil at increased numbers (Oka et al. 2000). *Telluria chitinolytica* and *Bacillus cereus* have been isolated from soil added with collagen, which is the main constituent of the cuticle of nematodes (Urwin et al. 1997). In another study, *Bacillus subtilis* and a fungal species, *Paecilomyces lilacinus*, were tested for the control of root-knot nematodes on tomato in pots containing steamed soil (Gautam et al. 1995). Individual application of the fungus *Verticillium chlamydosporium* and the bacterium *Pasteuria penetrans* has reduced root galling in tomato (Frans et al. 1992). Microorganisms living inside plants tissues, aka endophytes, have been well documented as natural control agents for plant nematodes. However, studies focusing on the interactions between different species of endophytes and their host plants, and associated nematodes are scarce.

5.4 Fungal and Bacterial Endophytes

Plant endophytes are microorganisms that reside within plants to complete their entire, or at least part, of their life cycle. Endophytes were first described by the German botanist, Heinrich Friedrich Link, in 1809 (Link 1809). The term "endophyte" was initially used for certain parasitic fungi that were living inside plant tissues (Hardoim et al. 2015). Later, researchers realized the colonization of bacteria within plant tissues too, and therefore the term "endophyte" then encompassed both bacteria and fungi (Chanway 1996; Hallmann et al. 1997). It was believed in the nineteenth century that healthy plants were sterile and devoid of microorganisms (cited in Compant et al. 2012). However, the presence of bacteria and fungi inside vegetable plants was reported in the later part of the nineteenth century, and it was assumed that these organisms are soilborne, migrate into plants, and provide an array of benefits to the host plants for their growth and development (postulated by Galippe, cited in Hardoim et al. 2015). The isolation and discovery of nitrogen-fixing *Rhizobium leguminosarum* from the root nodules of leguminous plants was an important breakthrough in the history of endophyte research (Beijerinck 1888). The development of molecular techniques and the use of "omics" approaches enabled detailed studies of endophytes, confirming the presence of different microorganisms within plant tissues (Knief 2014; Hardoim et al. 2015). Therefore, it is now well known that plant endophytes include fungi, bacteria, archaea, and single cellular eukaryotes such as amoebae and algae (Trémouillaux-Guiller et al. 2002; Müller and Döring 2009).

Endophytes are defined as organisms that live within plants, do not cause any noticeable harm and can be isolated from within plants or from surface-sterilized plant tissues (Hallmann et al. 1997). However, this pragmatic definition leaves many questions and is valid only for culturable bacterial and fungal species. It is hard to distinguish between plant endophytes and phytopathogens because: (i) the assessment of plant pathogenicity can be performed only with culturable endophytes; (ii) pathogenicity assays are generally performed on limited plant species; and (iii) pathogenicity or mutualism depends on many factors, such as microbial and host genotype, environmental circumstances, microbial numbers, and the quorum sensing potential of microbes (Hardoim et al. 2015). Environmental microbiologists explain that less than 2% of bacteria can be cultured in laboratories. Recent molecular techniques have shown that there are a vast majority of uncultivable bacterial and fungal species found within plant species (Rodriguez et al. 2009; Knief 2014; Tian et al. 2015; Wemheuer et al. 2016, 2017; Liu et al. 2017; Bergna et al. 2019). There are some microbes that remain latent within plants and turn pathogenic under certain environmental circumstances (Kloepper et al. 2013). Some microbes that are non-pathogenic in one plant species may be pathogenic in different plant species or to other animals, including humans. Some well-established pathogenic bacterial species in certain plants have proven to be beneficial and growth promoting in other plant hosts (Reiter et al. 2002; Coombs and Franco 2003; Van Overbeek et al. 2014). Therefore, the term "endophyte" should be associated with the habitat and not with the functional role of microbes. Generally, most endophytes are commensals with functions in plants that are known or yet to be known although a few endophytes show mutualistic (positive) or antagonistic (negative) effects on plants (Hallmann et al. 1997). The collection of genomes of endophytic microbes in plants is known as the endophytic microbiome or endophytic microbiota (Hardoim et al. 2015; Mareque et al. 2018).

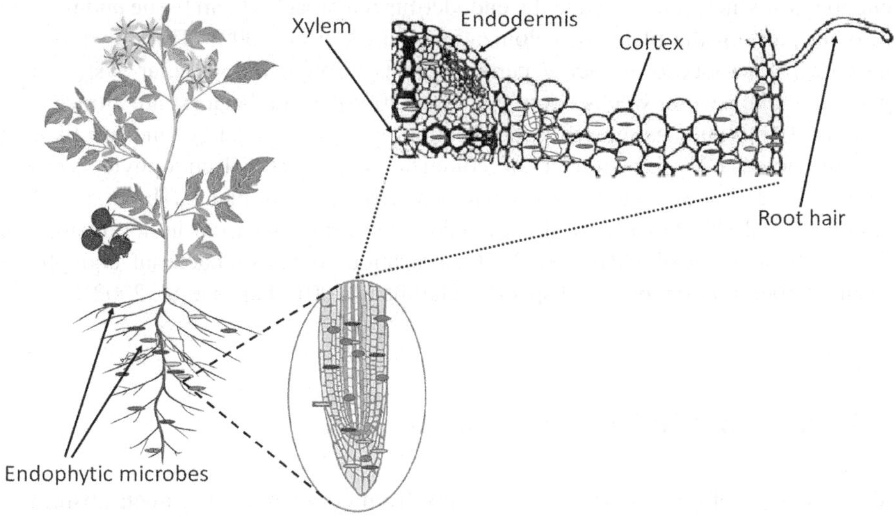

Fig. 5.1 Graphical representation of entry and colonization of endophytes in plant tissues. Endophytes colonize the cortical region and accelerate the metabolic activities involved in plant health ameliorations

5.4.1 Colonization of Endophytes

Many factors such as plant species and tissue type, species and strains of endophytic microbes, and environmental conditions determine the efficacy of colonization of endophytes in plants (Hardoim et al. 2015). Compared to bacterial endophytes, clavicipitaceous and nonclavicipitaceous fungal endophytes exhibit many different behaviors of colonization (Carroll 1988; Stone et al. 2004). Clavicipitaceous fungal endophytes, which are mainly found in grass species, colonize entire plants systemically and proliferate within the plants showing horizontal transmission to leaves and vertical transmission to seeds (Redman et al. 2002; Saikkonen et al. 2002; Tadych et al. 2007). Some clavicipitaceous endophytic fungi, including members of *Ascomycota* and *Basidiomycota*, colonize the aerial tissues of most tree species (Petrini 1986; Gamboa and Bayman 2001; Davis et al. 2003; Murali et al. 2007). Colonization of certain dark, septate endophytic fungi is restricted to root tissues, particularly in cortical cells inter- or intracellularly (O'Dell et al. 1993). Most bacterial endophytes originate from the rhizosphere, and the microbes are attracted to roots by rhizodeposits and root exudates (Compant et al. 2010; Philippot et al. 2013). Colonization of bacterial endophytes starts at root hairs, and then the bacteria enter the roots and move to other parts of the plant (Fig. 5.1). The movement of bacteria from the root surface through the cortical cell layer toward the endodermis may stop at the root endodermis due to its obstructive nature to certain bacteria (Compant et al. 2010; Mercado-Blanco and Lugtenberg 2014). However, bacterial endophytes which can penetrate the endodermis may reach xylem tissue and migrate to other parts of the plant for colonization (James et al. 2002). The movement of bacteria from roots to the aerial parts of plants through the vascular system may generally take several weeks (Compant et al. 2008). Some bacteria are attracted by leaf and stem exudates; however, desiccation, nutrient deficiency, and inactivation by ultraviolet (UV) light hinder the colonization of bacterial endophytes via leaf stomata, hydathodes in stems, and wounds in aerial parts of plants (Hallmann 2001; Compant et al. 2010). Colonization of endophytes may also occur in floral parts and fruits (Compant et al. 2010, 2011). Colonization routes of bacterial endophytes mainly depend on the bacterial species (Hallmann 2001; James et al. 2002).

5.5 Roles of Endophytes in Plants

The functions of endophytes in plants vary from commensalism, amensalism, and parasitism to competition (Reinhold-Hurek and Hurek 1998, 2011; Rodriguez et al. 2009; Hardoim et al. 2015). Most of the endophytes that live inside plants and utilize plant-derived metabolites but do not show any visible effects on plants are known as commensals (Hardoim et al. 2015). Some endophytes change their mode of relationship with the host plant and other endophyte species of endophytes, depending on the host's genotype, the growth stage of the host, and the biotic and abiotic stress conditions of the host plant. For instance, *Fusarium verticillioides*, a beneficial fungal endophyte in maize (*Zea mays*) becomes detrimental and accumulates mycotoxins in hosts when the host plants suffer from abiotic stress and reduce their fitness (Bacon et al. 2008). The composition of endophytes that exhibit different relationship modes with the host such as commensalism, mutualism, parasitism, and competition, is also determined by many factors, including the host's genotype, stress conditions of the host, and the growth stage of the host plant (Hardoim et al. 2015; Albrectsen et al. 2018). In Swedish aspen (*Populus tremula*), which produces defense substances such as salicinoid phenolic glycosides, the composition of fungal endophytes and their relationship modes with the host depend on the host's genotype and interaction with a leaf beetle species (*Chrysomela tremula*) (Albrectsen et al. 2018). Plant endophytes are well known to offer a variety of benefits to their host plants. They protect the host plants from pathogens and arthropod herbivores and promote plant growth through a variety of mechanisms (Rodriguez et al. 2009; Hardoim et al. 2015). Some endophytes are latent pathogens. They live within the host plant and show no visible effects at the beginning although when the host becomes mature or immune compromised due to biotic or abiotic stresses or physiological status, they can turn pathogenic (Scortichini and Loreti 2007). Endophytes also show effects on the competitive response of plants. *Alternaria* endophytes in *Centaurea stoebe* increased the competitive effects of *C. stoebe* over other grass species (Aschehoug et al. 2012).

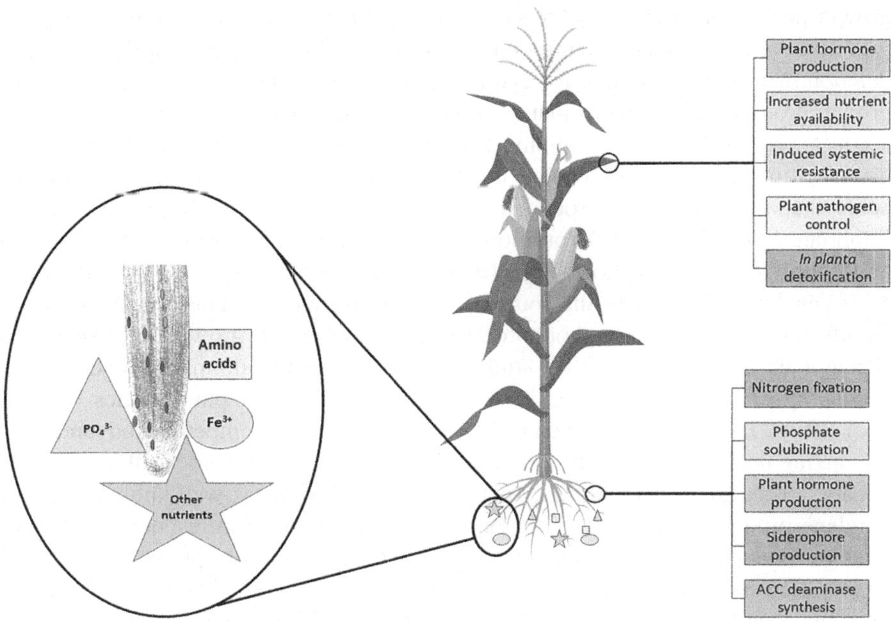

Fig. 5.2 Overview of plant-growth-promoting activities of endophytes. Endophytes improve plant health status through various mechanisms, such as the production of phytohormones, the production of siderophores, increased nitrogen fixation, phosphate solubilization, enzyme synthesis, increased uptake of plant nutrients, induced systemic resistance, etc.

5.5.1 *Plant Growth Promotion*

Although endophytes obtain nutrients from plants, they are also involved in the stimulation of plant growth (Clay 1988; Long et al. 2008; Singh et al. 2011). Benefit-cost balance of endophytic infections in plants is mostly positive or at least neutral, suggesting a beneficial relationship of endophytes with their host plant species (Hardoim et al. 2015). Endophytes associate with growth promotion in plants via a variety of ways, such as the production of plant-growth-promoting hormones, phosphate solubilization, siderophore production, 1-aminocyclopropane-1-carboxylase (ACC) deaminase activity, and nitrogen fixation (Fig. 5.2). Plant growth promotion by endophytes via phytohormone production has been extensively studied (Long et al. 2008; Shi et al. 2009; Khan et al. 2012; Waqas et al. 2012). Two fungal endophytes, which can produce gibberellins (GAs) and indoleacetic acid (IAA), have shown growth promotion in two dwarf mutant rice cultivars that are GA deficient (Waqas et al. 2012). A bacterial endophyte, *Sphingomonas* sp. LK11, isolated from *Tephrosia apollinea* leaves, exhibited a profuse amount of GAs and IAA production and growth promotion in tomato (Khan et al. 2014). An endophytic yeast isolate of *Williopsis saturnus* found in maize roots showed auxin-dependent growth promotion in the host (Nassar et al. 2005). Similarly, cytokinin production was observed in root-colonizing endophytic fungi *Piriformospora indica*, and the deletion of genes responsible for cytokinin biosynthesis showed a complete loss of growth promotion in the host plant (Vadassery et al. 2008).

In *Pisum sativum* L., endophytic bacteria *Pseudomonas* have demonstrated growth stimulation by solubilizing insoluble phosphates with the help of gluconic acids that the endophyte produced (Oteino et al. 2015). Siderophores play an important role in plant growth promotion by sequestering iron from the environment and supplying it to the host plant (Scavino and Pedraza 2013). An endophytic *Streptomyces* sp. found in the roots of a Thai jasmine rice cultivar has shown a significant amount of siderophore production, promoting the growth of the rice plant (Rungin et al. 2012). Another study in rice isolated a dominant siderophore-producing endophytic bacteria, *Pantoea ananatis*, from all plant tissues (Loaces et al. 2011). Bacteria, by producing siderophores, obtain a competitive advantage in colonizing the host and thereby exclude other microbes in the same ecological niche (Loaces et al. 2011). In addition, ACC deaminase activity is well known to lower ethylene production in plants. Ethylene is mainly produced in response to various stresses, both biotic and abiotic, causing growth retardation in plants. Most ACC-deaminase-producing bacteria show synergistic interactions between the ACC deaminase and IAA produced by both bacteria and plants (Glick 2014; Zakeel and Safeena 2019). ACC deaminase production is a widespread feature of the endophytic species of *Burkholderia*, and a study showed growth promotion in tomato by reducing ethylene levels with ACC deaminase activity (Onofre-Lemus et al. 2009). Supplying available forms of nitrogen for plants is another mechanism of plant growth promotion exhibited by endophytic microorganisms. Nitrogen fixation by free-living and symbiotic microorganisms is probably one of the extensively studied fields of science. Various endophytes isolated from different parts of plants have shown nitrogen-fixing potential (Baldani et al. 1997; Reinhold-Hurek and Hurek 1998; Elbeltagy et al. 2001; Cocking 2003). The efficacy of nitrogen fixation in the root nodules of legume–*Rhizobium* association is far higher than that of free-living endophytes. However, *Gluconacetobacter diazotrophicus* in sugarcane and various other host plants showed a comparatively higher nitrogen fixation efficiency (Dong et al. 1994; Bertalan et al. 2009; Eskin et al. 2014).

5.5.2 Improved Plant Fitness and Protection Against Biotic and Abiotic Stresses

Endophytes improve the fitness of host plants to tolerate abiotic stresses, to resist biotic stress due to pathogens and to deter herbivores. Turfgrass with endophytic fungi *Epichloe festucae* has shown improved resistance to leaf spot pathogens *Sclerotina homeocarpa* and *Laetisaria fusiformis* compared to the endophyte-uninfected counterpart turfgrass (Bonos et al. 2005; Clarke et al. 2006). Endophytic actinomycetes in cucumber can control the pathogen *Pythium aphanidermatum* (El-Tarabily et al. 2009). A fungal endophyte, *Fusarium solani* strain Fs–K isolated from tomato roots, has been able to suppress *Fusarium oxysporum* f. sp. *radicis-lycopersici*, which is a root pathogen, and to also induce resistance against *Septoria lycopersici*, a foliar pathogen of tomato, by mediating ethylene production in the host plant (Kavroulakis et al. 2007). Endophytic colonization of *Beauveria bassiana* has protected cotton and tomato seedlings from *Pythium myriotylum* and *Rhizoctonia solani*, both causing seedling damping off (Ownley et al. 2008), and from *Fusarium* sp., which causes wilt in many plant species (Ownley et al. 2010). Similarly, fungal endophytes in *Theobroma cacao* have shown defense against the plant pathogenic *Phytophthora* species (Arnold et al. 2003). *Lecanicillium* spp., endophytic fungi belonging to *Ascomycota*, have exhibited protection of host plants from rust fungi and powdery mildew (Ownley et al. 2010). Endophytes are involved in the induced systemic resistance (ISR) of host plants to tolerate pathogenic stress (Carroll 1991; Robert-Seilaniantz et al. 2011; Zamioudis and Pieterse 2012). Compared to fungal endophytes, bacterial endophytes have been widely reported to be associated with ISR in host plants (Vu et al. 2006; Blodgett et al. 2007; Bae et al. 2011). *Bacillus* and *Pseudomonas* are the most common bacterial genera associated with ISR (Chanway 1998; Kloepper and Ryu 2006), but many other bacterial strains, such as *Burkholderia phytofirmans*, are also capable of triggering ISR (Bordiec et al. 2010; Pavlo et al. 2011). Bacterial outer membrane components such as lipopolysaccharides, bacterial flagella, siderophores, and various secondary metabolites are responsible for the induction of ISR in host plants (van Loon et al. 2008; Bordiec et al. 2010). Fungal endophytes make plants resistant to pathogens, including nematodes infesting crop plants, by means of producing compounds such as toxic peptides, steroids, phenolic compounds, polyketones, alkaloids, flavonoids, quinols, chlorinated compounds, and terpenoids, which inhibit the growth of pathogens and reduce herbivore activities in host plants (Bush et al. 1982; Siegel et al. 1990; Gunatilaka 2006; Higginbotham et al. 2013; Tejesvi et al. 2013). Some fungal endophytes can produce insecticidal, antifungal, antiviral, and bactericidal compounds in plants, which may protect the plants from insects and pathogens (Gunatilaka 2006; Tejesvi et al. 2011).

5.5.3 Secondary Metabolite Production

Secondary metabolites are small organic molecules synthesized by certain organisms that are not necessary for their growth, development, and reproduction (Verpoorte 2000; Monfil and Casas-Flores 2014). These molecules are associated with an array of functions, such as insecticidal, antimicrobial, nematicidal, antiviral, antidiabetic, immunosuppressive, anticancer, and antioxidant activities, in biological systems (Hallmann and Sikora 1996; Tan and Zou 2001; Strobel and Daisy 2003; Gunatilaka 2006; Zhang et al. 2006; Verma et al. 2009; Aly et al. 2010, 2011; Brader et al. 2014). Endophytic actinomycetes are well known for their production of various antimicrobial compounds, such as coronamycin, kakadumycin, and munumbicin (Castillo et al. 2002, 2003; Ezra et al. 2004). Endophytic *Enterobacter* sp. strain 638 produces 4-hydroxybenzoate and 2-phenylethanol, which are antibiotic substances (Taghavi et al. 2010). Endophytic *Streptosporangium oxazolinicum* strain K07-0450^{T} isolated from orchids has been able to produce antitrypanosomal spoxazomicins A to C (Inahashi et al. 2011). An antibacterial compound, multicyclic indolosesquiterpenes, has been found in *Streptomyces* sp. HKI0595, which lives endophytically in the mangrove tree *Kandelia candel* (Ding et al. 2011). Secondary metabolites produced by endophytes are also involved in gene regulation associated with symbiosis, defense against pathogens and herbivores, and signaling mechanisms (Schulz and Boyle 2005; Mousa and Raizada 2013; Nasopoulou et al. 2014). In addition to the production of secondary metabolites that contribute to host plants in many ways, the presence of endophytes in plants influence the hosts to produce various secondary metabolites (Zhang et al. 2006). For example, the interaction of fungal endophyte *Paraphaeosphaeria* sp. with bilberry plant (*Vaccinium myrtillus*) has shown increased biosynthesis of phenolic acids in the host plant (Koskimäki et al. 2009).

5.6 Interaction Between Plant Nematodes and Endophytes

Only a handful of studies have been reported regarding the interactions between plant nematodes and endophytic microbes. Fungal and bacterial endophytes have shown antagonistic interactions with plant nematodes through many different mechanisms, including nematicidal, nematophagous, and parasitoid activities (Bush et al. 1982; Siegel et al. 1990; Dong and Zhang 2006; Gunatilaka 2006). The nematophagous mechanism is mainly exhibited by endophytic and saprophytic fungi (Bordallo et al. 2002; Dong and Zhang 2006). Nematophagous fungi are capable of capturing, trapping and parasitizing or paralyzing nematodes at any stage of their life cycle (Dong and Zhang 2006). Nematophagous fungi use one of the four main types of traps to capture motile nematodes being adhesive knobs, adhesive branches, or constricting rings (Ahrén and Tunlid 2003). Parasitic fungi of nematodes can be endoparasitic or egg and female parasitic (Persmark et al. 1995; Jansson and Lopez-Llorca 2001). Endoparasitic fungi use spores to infect nematodes (Persmark et al. 1995; Jansson and Lopez-Llorca 2001). Toxins produced by fungi to interact with nematodes can be either nematicidal or nematistatic (Kulkarni and Sangit 2000; Luo et al. 2004). Different fungal species of the same genus may exhibit different mechanisms of antagonism against nematodes (Dong and Zhang 2006). Sometimes the same species can interact with different nematodes via different mechanisms (Dong and Zhang 2006). Some species of *Arthrobotrys* showed antagonism against nematodes via egg parasitism (dos Santos et al. 1992; Chen et al. 1996) or toxin production (Anke et al. 1995), whereas others showed a nematode trapping mechanism (Duponnois et al. 1996; Viaene and Abawi 1998). *Meloidogyne arenaria* and *Heterodera glycines* are antagonized by *Humicola fuscoatra* via toxin production and egg parasitism, respectively (Mian et al. 1982; Dong and Zhang 2006). Nematophagous fungi are currently commercially utilized as potential biocontrol agents for nematodes (Jansson et al. 1985; Del Sorbo et al. 2003).

The species of *Trichoderma* are widely studied endophytic fungi (Spiegel and Chet 1998; Sikora et al. 2008; Al-Hazmi and TariqJaveed 2016). *Trichoderma harzianum* has been reported as a potential endophyte for the control of root-knot nematode *Meloidogyne javanica* (Sharon et al. 2001). In addition, there are four species of nematophagous endophytes that are studied extensively, viz., *Phomopsis phaseoli*, *Melanconium botulinum* (Schwarz et al. 2004), *Lecanicillium lecanii* (Monfort et al. 2005), and *Fusarium oxysporum* (Hallmann and Sikora 1996). Some studies have revealed that endophytic *Neotyphodium* sp. infects aerial parts and roots of grasses and produces fungal alkaloids which are translocated to roots (West et al. 1988; Kimmons et al. 1990; Zabalgogeazcoa 2008). *Neotyphodium*-infected plants showed resistance to plant-parasitic nematodes, but the mechanism underlying the protection against nematodes is still unclear (Malinowski and Belesky 2000; Zabalgogeazcoa 2008). Mostly, interactions of endophytes with plant-parasitic nematodes are indirect, where endophytes interact with host plants and make them resistant to plant nematodes through different mechanisms. They are: (1) suppression of invasion by nematodes; (2) enhancement of plant growth; (3) improvement of nutrient uptake; (4) boosting of plant tolerance; and (5) enabling plants to produce nematicidal compounds (Strobel et al. 1982; West et al. 1988; Waceke et al. 2001; Diedhiou et al. 2003). Moreover, *Streptomyces* of actinomycetes has many species with the potential for nematicidal compound production (Samac and Kinkel 2001; El-Nagdi and Youssef 2004). *Streptomyces avermitilis* produced natural avermectins, a group of macrocyclic lactones (Stutzman-Engwall et al. 2005). Semisynthetic and natural avermectins are widely used for nematode control in medicine (ivermectin), veterinary medicine, and agriculture (Campbell 2012; El-Nagdi and Youssef 2004; Huang et al. 2014; Stutzman-Engwall et al. 2005). A fermentation product of avermectins, known as "Abamectin," has been successfully used for the control of root-knot nematode *M. incognita* (El-Nagdi and Youssef 2004; Qiao et al. 2012).

5.6.1 Exploring the Interaction Between Endophytes and Plant Nematodes Using Metagenomics Approach

Endophytes show interactions with host plants and nematodes infecting the host plants, in addition to their interactions with each other (Proença et al. 2010; Tian et al. 2015; Wemheuer et al. 2016, 2017; Su et al. 2017). Some endophytes are predominantly found in nematode-infected plants (Proença et al. 2010; Tian et al. 2015; Su et al. 2017). Su et al. (2017) studied endophytes in banana roots infected with the root-knot nematode *M. javanica* and found that antagonistic endophytes were dominant in highly and moderately affected roots. Root-knot nematode *M. incognita* infected tomato roots showed *Streptomycetales*, *Micromonosporales*, *Rhizobiales*, *Sphingomonadales*, *Burkholderiales*, and *Pseudomonadales* as the dominant orders of endophytic bacteria (Tian et al. 2015). Endophytic diversity and operational taxonomic unit (OTU) richness are higher in nematode-infected plants compared with those in healthy plants (Tian et al. 2015). This suggests that microorganisms infect and subsequently colonize plants during nematode infections (Tian et al. 2015). *Pseudomonas* and *Burkholderia* species isolated from the nematode *Bursaphelenchus xylophilus*, which infects pinewood (*Pinus pinaster*), are the commonly found endophytes in nematode-infected plants (Proença et al. 2010; Tian et al. 2015). The endophytic community composition of root-knot-nematode-affected tomato roots has shown a slight increase in the OTU richness of orders *Micromonosporales* and *Burkholderiales* with a significant decline in the OTU richness of *Pseudomonadales* and *Streptomycetales*, which are known to produce antimicrobial compounds (Tian et al. 2015). However, Su et al. (2017) showed that *Streptomyces* spp., *Bacillus* spp., and *Pseudomonas* spp. were abundant among the endophytic communities of *M. javanica* infected banana roots. A strain of *Streptomyces* sp. has been isolated and identified as a potential biocontrol agent for *M. javanica* (Su et al. 2017).

Different molecular techniques have been in use for the study of the diversity and species richness of endophytes in healthy and nematode-infected plants. Denaturing gradient gel electrophoresis of PCR products of 16S rRNA gene has been used to explore the diversity of endophytes in grasses such as *Lolium perenne*, *Festuca rubra*, and *Dactylis glomerata* (Wemheuer et al. 2016, 2017). Sequencing of the 16S rRNA gene has been used to investigate the diversity of bacterial endophytes in rice (*Oryza sativa*) (Sun et al. 2008), grapevine (*Vitis vinifera*) (Bulgari et al. 2009), and *M. javanica* infected banana roots (Su et al. 2017). A diversity study of culturable fungal endophytes in loblolly pine (*Pinus laeda*) used sequencing of the ITS region (Arnold et al. 2007). Another study has shown the use of polymerase chain reaction–restriction fragment length polymorphism (PCR–RFLP) and sequencing analysis of the rRNA gene of bacterial endophytes isolated from rice roots (Singh et al. 2006). With awareness of the rich diversity of unculturable endophytes and the availability of high-throughput sequencing technologies such as next-generation sequencing (NGS), at present diversity studies of endophytes heavily depend on metagenomics approaches. In particular, PCR amplicons of endophytes are collectively sequenced using NGS platforms. An NGS-based metagenomics approach has been used for the study of endophytes in *Aloe vera* (Akinsanya et al. 2015). Metagenomic studies have been performed on different plant species to explore the diversity of bacterial and fungal endophytes using high-throughput NGS methods (Bullington and Larkin 2015; Tian et al. 2015; Wemheuer et al. 2017).

5.6.2 Next-Generation Sequencing

Next-generation sequencing (NGS) is an ultra-high throughput sequencing method that is quick and cheap and enables parallel sequencing (Türktaş et al. 2015). Until 2004, sequencing of whole genomes involved the cloning and subcloning of DNA fragments, sequencing by the traditional Sanger sequencing approach, assembling the sequences of the subclones and large clones, and reconstructing the entire chromosome (Marra et al. 1997). This "top-down" approach of genome sequencing was revolutionized by the introduction of pyrosequencing with automation for massively parallel sequencing by 454 Life Sciences (Margulies et al. 2005). In contrast to the aforementioned traditional approach, NGS methods do not require cloning; however, construction of a library of DNA fragments or amplicons, which are to be sequenced, is a prerequisite (Margulies et al. 2005). To construct a library, DNA fragments or amplicons are ligated to universal adapter sequences by DNA ligase, or adapter sequences are added to the ends of amplicons by a PCR step (Margulies et al. 2005; Tian et al. 2015; Abdelfattah et al. 2016). NGS is currently used in many fields of research, including medicine, human genetics and genomics, plant genomics, environmental genomics, forensics, etc. Examples of research areas highly demanding the use of NGS methods include RNA sequence analysis to quantify gene expression and discover RNA splice sites and variants, microbial diversity studies, analysis of epigenetic factors, and the study of rare variants (Hurd and Nelson 2009; Trapnell et al. 2009, 2010, 2012; Marroni et al. 2011; Meaburn and Schulz 2012; Finotello and Di Camillo 2015; Abdelfattah et al. 2016). In 2005, 454 pyrosequencing was launched by 454 Life Sciences, and was the first successful NGS platform, albeit with some drawbacks (Liu et al. 2012). After Roche purchased the company in 2007, it produced the 454 GS FLX Titanium system coupled with GS Junior, capable of producing 14-G data output per run in 2008 (Huse et al. 2007; Liu et al. 2012). Sequencing by Oligo Ligation Detection (SOLiD) systems were originally provided by Agencourt before selling the company to Applied Biosystems (ABI) in 2006 (Liu et al. 2012). SOLiD systems were based on a two-base sequencing techniques which provided 99.85% accuracy (Liu et al. 2012). Applied Biosystems released a SOLiD 5500xl platform with 99.99% accuracy, 30-G data output per run, and 85-bp read length in 2010 (Liu et al. 2012). A company known as Solexa produced Genome Analyzer in 2006. The company was sold to Illumina in 2007, and Illumina has been producing different platforms from TruSeq V3 since 2009, HiSeq 2000 since 2010, MiSeq since 2011, and NovaSeq 6000 at present (Liu et al. 2012). Illumina platforms use the chemistry of sequencing by synthesis (Mardis 2008). Ion torrent Proton/PGM (Personal Genome Machine) is another platform mainly used in small labs for clinical purposes (Liu et al. 2012). Ion PGM sequencing uses a different chemistry where pH change, in response to the release of H^+ ions during the addition of nucleotides to DNA molecules is detected to recognize if a base is added (Liu et al. 2012).

Metagenomics of endophytes and their interactions with plant-parasitic nematodes have been mainly studied using amplicon-based NGS methods (Tian et al. 2015). In these techniques, PCR targeting of the ITS regions of fungi or the 16S rRNA gene of bacteria will be performed with total nucleic acids extracted from surface-sterilized plant tissues (Akinsanya et al. 2015; Bullington and Larkin 2015; Tian et al. 2015; Abdelfattah et al. 2016). Roche 454 pyrosequencing and Illumina HiSeq and MiSeq have been widely used for metagenomics studies of endophytes (Lundberg et al. 2013; Akinsanya et al. 2015; Tian et al. 2015; Abdelfattah et al. 2016; Guo et al. 2016; Wemheuer et al. 2017). Different studies have targeted different hypervariable regions of 16S rRNA (Fig. 5.3). For instance, a study of endophytes in tomato plants affected by root-knot nematodes has used primers 338F and 806R for the amplification of the V3–V4 region of the 16S rRNA gene for NGS analysis using the Illumina MiSeq platform (Tian et al. 2015). Another set of primers, 341F and 805R, has also been used to amplify the V3–V4 region (Akinsanya et al. 2015). Guo et al. (2016) used 515F/806R and 926F/1392R (Table 5.1) to target V4 and V6–V8 regions, respectively (Fig. 5.3). Amplification of chloroplast rRNA targets along with the endophytes is a drawback with certain primers. In order to prevent the co-amplification of chloroplast rRNA targets of the host, a nested PCR approach has been used with 799F and 1492R primers for the first round and with 968F and 1401R (Table 5.1) for the nested round (Wemheuer et al. 2017).

Nonetheless, metagenomic studies of fungi have relied on the amplification of the ITS2 regions of the rRNA gene (Ihrmark et al. 2012; Toju et al. 2012; Abdelfattah et al. 2016). In order to amplify the ITS2 region, primers are designed for the ITS3 and ITS4 regions. Moreover, gITS7, fITS7, and fITS9 (Table 5.1) targeting 5.8S regions and ITS4 (Table 5.1) targeting the large subunit (LSU) region of the fungal rRNA gene have been used to amplify the ITS2 region (Ihrmark et al. 2012). Primers designed for the ITS3 and ITS4 regions may co-amplify plant ITS regions (Ihrmark et al. 2012). Therefore, a nested approach is generally used for fungal PCR as well. A study has shown the effectiveness of ITS1F-KYO1 and ITS4R primers (Table 5.1) to amplify a region encompassing ITS1, ITS2, and 5.8S sequences of fungi while excluding the amplification of the host ITS region (Toju et al. 2012).

NGS data are processed and analyzed through a series of steps using bioinformatics pipelines. Firstly, the raw data are processed using software packages according to the standard operating procedure of the software package (Tian et al. 2015). Then the processed sequences are aligned to reference databases (Abdelfattah et al. 2016; Wemheuer et al. 2017). The SILVA database has been generally used for bacterial 16S rRNA sequences and the UNITE dynamic database for fungal ITS sequences (Tian et al. 2015; Abdelfattah et al. 2016; Wemheuer et al. 2017).

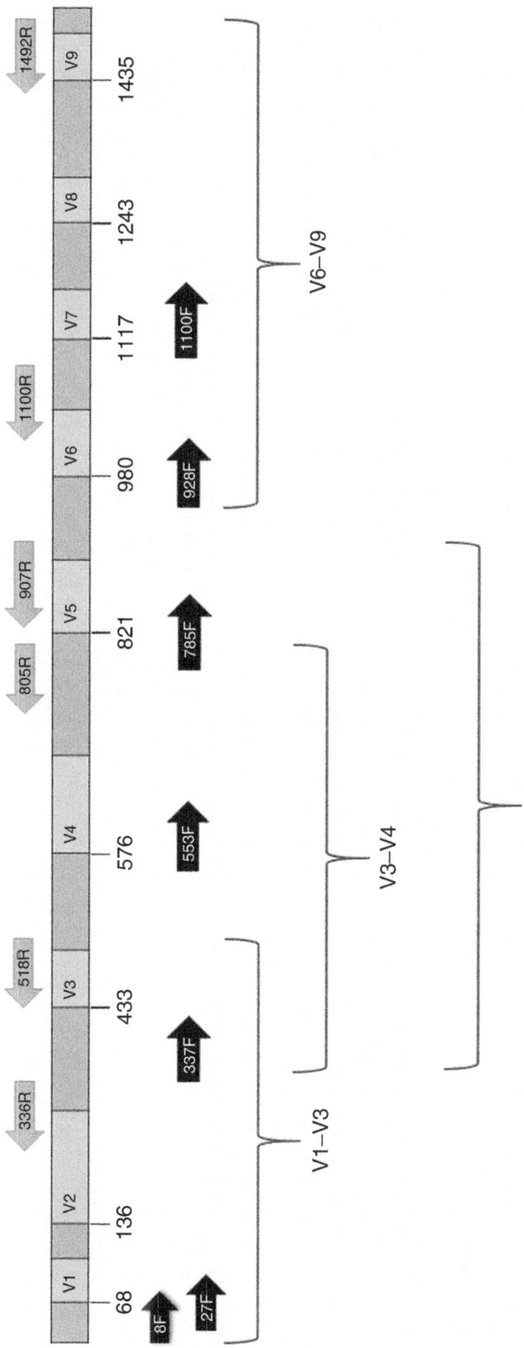

Fig. 5.3 Graphical representation of hypervariable regions of bacterial 16S rRNA gene and primer annealing sites (adapted from EZBioCloud at https://help.ezbiocloud.net/16s-rrna-and-16s-rrna-gene/)

Table 5.1 Primer sequences used in previous metagenomics studies of bacterial and fungal endophytes

Bacterial primers			Fungal primers		
Name	Sequence	Source	Name	Sequence	Source
gITS7	GTGARTCATCGARTCTTTG	Ihrmark et al. (2012)	341F	CCTACGGGNGGCWGCAG	Akinsanya et al. (2015)
fITS7	GTGARTCATCGAATCTTTG	Ihrmark et al. (2012)	805R	GACTACHVGGGTATCTAATCC	Akinsanya et al. (2015)
fITS9	GAACGCAGCRAAIIGYGA	Ihrmark et al. (2012)	338F	ACTCCTACGGGAGGCAGCA	Tian et al. (2015)
ITS4	TCCTCCGCTTATTGATATGC	White et al. (1990)	806R	GGACTACHVGGGTWTCTAAT	Lundberg et al. (2013) Tian et al. (2015)
ITS3	GCATCGATGAAGAACGCAGC	White et al. (1990)	515F	GTGYCAGCMGCCGCGGTAA	Lundberg et al. (2013)
ITS1F-KYO1	CTHGGTCATTTAGAGGAASTAA	Toju et al. (2012)	799F	AACMGGATTAGATACCCKG	Chelius and Triplett (2001)
			1492R	GCYTACCTTGTTACGACTT	Lane (1991)
			968F	AACGCGAAGAACCTTAC	Nübel et al. (1996)
			1401R	CGGTGTGTACAAGACCC	Nübel et al. (1996)

5.7 Conclusions and Future Prospects

Plant endophytes play a vital role in plant defense mechanisms against diseases, disorders and pests including plant-parasitic nematodes. Plant-associated endophytes establish interactions with both the host plants and plant-parasitic nematodes when the nematodes infect the host plants. These interactions either enable plants to enhance immunity against nematodes by means of producing various compounds, which can control nematodes, or directly interfere with nematodes and control their invasion. Studies into these interactions require sophisticated approaches due to the most complex nature of the interactions, often formed between a large number of organisms. A metagenomics approach using NGS platforms is the widely used and most effective technique to study these interactions.

References

Abad P, Gouzy J, Aury J-M, Castagnone-Sereno P, Danchin EGJ, Deleury E, Perfus-Barbeoch L, Anthouard V, Artiguenave F, Blok VC, Caillaud M-C, Coutinho PM, Dasilva C, De Luca F, Deau F, Esquibet M, Flutre T, Goldstone JV, Hamamouch N, Hewezi T, Jaillon O, Jubin C, Leonetti P, Magliano M, Maier TR, Markov GV, McVeigh P, Pesole G, Poulain J, Robinson-Rechavi M, Sallet E, Ségurens B, Steinbach D, Tytgat T, Ugarte E, van Ghelder C, Veronico P, Baum TJ, Blaxter M, Bleve-Zacheo T, Davis EL, Ewbank JJ, Favery B, Grenier E, Henrissat B, Jones JT, Laudet V, Maule AG, Quesneville H, Rosso M-N, Schiex T, Smant G, Weissenbach J, Wincker P (2008) Genome sequence of the metazoan plant-parasitic nematode *Meloidogyne incognita*. Nat Biotechnol 26:909–915

Abdelfattah A, Wisniewski M, Nicosia MGLD, Cacciola SO, Schena L (2016) Metagenomic analysis of fungal diversity on strawberry plants and the effect of management practices on the fungal community structure of aerial organs. PLoS One 11(8):e0160470

Agbenin NO, Emechebe AM, Marley PS, Akpa AD (2005) Evaluation of nematicidal action of some botanicals on *Meloidogyne incognita in vivo* and *in vitro*. J Agric Rural Dev Trop Subtrop 106(1):29–39

Ahrén D, Tunlid A (2003) Evolution of parasitism in nematode-trapping fungi. J Nematol 35 (2):194–197

Akinsanya MA, Goh JK, Lim SP, Ting ASY (2015) Metagenomics study of endophytic bacteria in *Aloe vera* using next-generation technology. Genom Data 6:159–163

Albrectsen BR, Siddique AB, Decker VHG, Unterseher M, Robinson KM (2018) Both plant genotype and herbivory shape aspen endophyte communities. Oecologia 187(2):535–545

Al-Hazmi AS, TariqJaveed M (2016) Effects of different inoculum densities of *Trichoderma harzianum* and *Trichoderma viride* against *Meloidogyne javanica* on tomato. Saudi J Biol Sci 23(2):288–292

Aly AH, Debbab A, Kjer J, Proksch P (2010) Fungal endophytes from higher plants: a prolific source of phytochemicals and other bioactive natural products. Fungal Divers 41(1):1–16

Aly AH, Debbab A, Proksch P (2011) Fungal endophytes: unique plant inhabitants with great promises. Appl Microbiol Biotechnol 90(6):1829–1845

Anke H, Stadler M, Mayer A, Sterner O (1995) Secondary metabolites with nematicidal and antimicrobial activity from nematophagous fungi and Ascomycetes. Can J Bot 73(S1):932–939

Anon (2012) National food security—the wheat initiative—an international research initiative for wheat improvement. In: Second global conference on agricultural research for development, Punta del Este, Uruguay

Ansari RA, Khan TA (2012a) Parasitic association of root-knot nematode, *Meloidogyne incognita* on guava. e-J Sci Technol 5:65–67
Ansari RA, Khan TA (2012b) Diversity and community structure of phytonematodes associated with guava in and around Aligarh, Uttar Pradesh, India. Trends Biosci 5(3):202–204
Ansari RA, Mahmood I (2017a) Optimization of organic and bio-organic fertilizers on soil properties and growth of pigeon pea. Sci Hortic 226:1–9
Ansari RA, Mahmood I (2017b) Determination of disease incidence caused by *Meloidogyne* spp. and or *Fusarium udum* on pigeonpea in Aligarh district: a survey. Trends Biosci 10 (24):5239–5243
Ansari RA, Mahmood I (2019a) Plant health under biotic stress: volume 2: microbial interactions. Springer, Singapore. https://doi.org/10.1007/978-981-13-6040-4. Accessed 20 May 2019
Ansari RA, Mahmood I (2019b) Plant health under biotic stress: volume 1: organic strategies. Springer, Singapore. https://doi.org/10.1007/978-981-13-6043-5. Accessed 20 May 2019
Ansari RA, Mahmood I, Rizvi R, Sumbul A (2017a) Siderophores: augmentation of soil health and crop productivity. In: Kumar V, Kumar M, Sharma S, Prasad R (eds) Probiotics in agroecosystem. Springer, Singapore, pp 291–312
Ansari RA, Rizvi R, Sumbul A, Mahmood I (2017b) PGPR: current vogue in sustainable crop production. In: Kumar V, Kumar M, Sharma S, Prasad R (eds) Probiotics and plant health. Springer, Singapore, pp 455–472
Ansari RA, Sumbul A, Rizvi R, Mahmood I (2019) Organic soil amendments: potential tool for soil and plant health management. In: Ansari RA, Mahmood I (eds) Plant health under biotic stress. Springer, Singapore, pp 1–35
Arnold AE, Mejía LC, Kyllo D, Rojas EI, Maynard Z, Robbins N, Herre EA (2003) Fungal endophytes limit pathogen damage in a tropical tree. Proc Natl Acad Sci 100(26):15649
Arnold AE, Henk DA, Eells RL, Lutzoni F, Vilgalys R (2007) Diversity and phylogenetic affinities of foliar fungal endophytes in loblolly pine inferred by culturing and environmental PCR. Mycologia 99(2):185–206
Aschehoug ET, Metlen KL, Callaway RM, Newcombe G (2012) Fungal endophytes directly increase the competitive effects of an invasive forb. Ecology 93(1):3–8
Bacon CW, Glenn AE, Yates IE (2008) *Fusarium verticillioides*: managing the endophytic association with maize for reduced fumonisins accumulation. Toxin Rev 27(3–4):411–446
Bae H, Roberts DP, Lim H-S, Strem MD, Park S-C, Ryu C-M, Melnick RL, Bailey BA (2011) Endophytic *Trichoderma* isolates from tropical environments delay disease onset and induce resistance against *Phytophthora capsici* in hot pepper using multiple mechanisms. Mol Plant-Microbe Interact 24(3):336–351
Baldani J, Caruso L, Baldani VLD, Goi SR, Döbereiner J (1997) Recent advances in BNF with non-legume plants. Soil Biol Biochem 29(5–6):911–922
Beijerinck MW (1888) Cultur des *Bacillus radicicola* aus den Knöllchen. Bot Ztg 46:740–750
Bergna A, Cernava T, Rändler M, Grosch R, Zachow C, Berg G (2019) Tomato seeds preferably transmit plant beneficial endophytes. Phytobiomes 2(4):183–193
Bernard GC, Egnin M, Bonsi C (2017) The impact of plant-parasitic nematodes on agriculture and methods of control. In: Shah MM, Mahamood M (eds) Nematology-concepts, diagnosis and control. Intechopen, Rijeka, Croatia
Bertalan M, Albano R, de Pádua V, Rouws L, Rojas C, Hemerly A, Teixeira K, Schwab S, Araujo J, Oliveira A (2009) Complete genome sequence of the sugarcane nitrogen-fixing endophyte *Gluconacetobacter diazotrophicus* Pal5. BMC Genomics 10(1):450
Blodgett JT, Eyles A, Bonello P (2007) Organ-dependent induction of systemic resistance and systemic susceptibility in *Pinus nigra* inoculated with *Sphaeropsis sapinea* and *Diplodia scrobiculata*. Tree Physiol 27(4):511–517
Bonfil DJ, Dolgin B, Mufradi I, Asido S (2004) Bioassay to forecast cereal cyst nematode damage to wheat in fields. Precis Agric 5(4):329–344
Bonos SA, Wilson MM, Meyer WA, Reed Funk C (2005) Suppression of red thread in fine fescues through endophyte-mediated resistance. Appl Turfgrass Sci 2(1):1

Bordallo JJ, Lopez-Llorca LV, Jansson HB, Salinas J, Persmark L, Asensio L (2002) Colonization of plant roots by egg-parasitic and nematode-trapping fungi. New Phytol 154(2):491–499

Bordiec S, Paquis S, Lacroix H, Dhondt S, Ait Barka E, Kauffmann S, Jeandet P, Mazeyrat-Gourbeyre F, Clément C, Baillieul F (2010) Comparative analysis of defence responses induced by the endophytic plant growth-promoting rhizobacterium *Burkholderia phytofirmans* strain PsJN and the non-host bacterium *Pseudomonas syringae* pv. *pisi* in grapevine cell suspensions. J Exp Bot 62(2):595–603

Brader G, Compant S, Mitter B, Trognitz F, Sessitsch A (2014) Metabolic potential of endophytic bacteria. Curr Opin Biotech 27:30–37

Bridge J, Plowright RA, Peng DL (2005) Nematode parasites of rice. In: Luo M, Sikora RA, Bridge J (eds) Plant parasitic nematodes in subtropical and tropical agriculture. CABI Publishing, Wallingford, pp 87–130

Bulgari D, Casati P, Brusetti L, Quaglino F, Brasca M, Daffonchio D, Bianco PA (2009) Endophytic bacterial diversity in grapevine (*Vitis vinifera* L.) leaves described by 16S rRNA gene sequence analysis and length heterogeneity-PCR. J Microbiol 47(4):393–401

Bullington LS, Larkin BG (2015) Using direct amplification and next-generation sequencing technology to explore foliar endophyte communities in experimentally inoculated western white pines. Fungal Ecol 17:170–178

Bush LP, Cornelius PL, Buckner RC, Varney DR, Chapman RA, Burrus PB, Kennedy CW, Jones TA, Saunders MJ (1982) Association of N-acetyl loline and N-formyl loline with *Epichloe typhina* in tall fescue. Crop Sci 22(5):941–943

Campbell WC (2012) History of avermectin and ivermectin, with notes on the history of other macrocyclic lactone antiparasitic agents. Curr Pharm Biotechnol 13(6):853–865

Carroll G (1988) Fungal endophytes in stems and leaves: from latent pathogen to mutualistic symbiont. Ecology 69(1):2–9

Carroll GC (1991) Fungal associates of woody plants as insect antagonists in leaves and stems. In: Barbosa P, Krischik VA, Jones CG (eds) Microbial mediation of plant-herbivore interactions. Wiley, New York, pp 253–271

Castillo UF, Strobel GA, Ford EJ, Hess WM, Porter H, Jensen JB, Albert H, Robison R, Condron MAM, Teplow DB (2002) Munumbicins, wide-spectrum antibiotics produced by *Streptomyces* NRRL 30562, endophytic on *Kennedia nigriscansa*. Microbiology 148(9):2675–2685

Castillo U, Harper JK, Strobel GA, Sears J, Alesi K, Ford E, Lin J, Hunter M, Maranta M, Ge H (2003) Kakadumycins, novel antibiotics from *Streptomyces* sp. NRRL 30566, an endophyte of *Grevillea pteridifolia*. FEMS Microbiol Lett 224(2):183–190

Chanway CP (1996) Endophytes: they're not just fungi! Can J Bot 74:321–322

Chanway CP (1998) Bacterial endophytes: ecological and practical implications. In: Endophytism in plant pathology, the 7th international congress of plant pathology, Edinburgh, Scotland, pp 149–170

Chelius MK, Triplett EW (2001) The diversity of archaea and bacteria in association with the roots of *Zea mays* L. Microb Ecol 41(3):252–263

Chen SY, Dickson DW, Mitchell DJ (1996) Pathogenicity of fungi to eggs of *Heterodera glycines*. J Nematol 28(2):148–158

Clarke BB, White JF Jr, Hurley RH, Torres MS, Sun S, Huff DR (2006) Endophyte-mediated suppression of dollar spot disease in fine fescues. Plant Dis 90(8):994–998

Clay K (1988) Fungal endophytes of grasses: a defensive mutualism between plants and fungi. Ecology 69(1):10–16

Cocking EC (2003) Endophytic colonization of plant roots by nitrogen-fixing bacteria. Plant Soil 252(1):169–175

Compant S, Kaplan H, Sessitsch A, Nowak J, Ait Barka E, Clément C (2008) Endophytic colonization of *Vitis vinifera* L. by *Burkholderia phytofirmans* strain PsJN: from the rhizosphere to inflorescence tissues. FEMS Microbiol Ecol 63(1):84–93

Compant S, Clément C, Sessitsch A (2010) Plant growth-promoting bacteria in the rhizo-and endosphere of plants: their role, colonization, mechanisms involved and prospects for utilization. Soil Biol Biochem 42(5):669–678

Compant S, Mitter B, Colli-Mull JG, Gangl H, Sessitsch A (2011) Endophytes of grapevine flowers, berries, and seeds: identification of cultivable bacteria, comparison with other plant parts, and visualization of niches of colonization. Microb Ecol 62(1):188–197

Compant S, Sessitsch A, Mathieu F (2012) The 125th anniversary of the first postulation of the soil origin of endophytic bacteria—a tribute to M.L.V. Galippe. Plant Soil 356(1):299–301

Coombs JT, Franco CMM (2003) Isolation and identification of actinobacteria from surface-sterilized wheat roots. Appl Environ Microbiol 69(9):5603–5608

D'Addabbo T, Filotico A, Sasanelli N (1996) Effect of calcium cyanamide and other ammonia fertilizers on *Meloydogine incognita*. Nematol Mediterr 24(2):209–214

Davis EC, Franklin JB, Shaw AJ, Vilgalys R (2003) Endophytic *Xylaria* (Xylariaceae) among liverworts and angiosperms: phylogenetics, distribution, and symbiosis. Am J Bot 90 (11):1661–1667

Decraemer W, Hunt D (2006) Structure and classification. In: Perry R, Moens M (eds) Plant nematology. CAB International, Oxfordshire, pp 3–32. https://doi.org/10.1079/9781845930561.0000. Accessed 14 Mar 2019

Del Sorbo G, Marziano F, D'Errico FP (2003) Diffusion and effectiveness of the nematophagous fungus *Hirsutella rhossiliensis* in control of the cyst nematode *Heterodera daverti* under field conditions. J Plant Pathol 85(3):219–221

Diedhiou P, Hallmann J, Oerke EC, Dehne HW (2003) Effects of arbuscular mycorrhizal fungi and a non-pathogenic *Fusarium oxysporum* on *Meloidogyne incognita* infestation of tomato. Mycorrhiza 13(4):199–204

Ding L, Maier A, Fiebig H-H, Lin W-H, Hertweck C (2011) A family of multicyclic indolosesquiterpenes from a bacterial endophyte. Org Biomol Chem 9(11):4029–4031

Dong LQ, Zhang KQ (2006) Microbial control of plant-parasitic nematodes: a five-party interaction. Plant Soil 288(1–2):31–45

Dong Z, Canny MJ, McCully ME, Roboredo MR, Cabadilla CF, Ortega E, Rodes R (1994) A nitrogen-fixing endophyte of sugarcane stems (a new role for the apoplast). Plant Physiol 105 (4):1139–1147

Dong M, Yang Z, Cheng G, Peng L, Xu Q, Xu J (2018) Diversity of the bacterial microbiome in the roots of four *Saccharum* species: *S. spontaneum*, *S. robustum*, *S. barberi*, and *S. officinarum*. Front Microbiol 9:267

dos Santos MA, Ferraz S, Muchovej JJ (1992) Evaluation of 20 species of fungi from Brazil for biocontrol of *Meloidogyne incognita* race 3. Nematropica 22(2):183–192

Dungan RS, Ibekwe AM, Yates SR (2003) Effect of propargyl bromide and 1, 3-dichloropropene on microbial communities in an organically amended soil. FEMS Microbiol Ecol 43(1):75–87

Duponnois R, Mateille T, Sene V, Sawadogo A, Fargette M (1996) Effect of different west african species and strains of *Arthrobotrys* nematophagous fungi on *Meloidogyne* species. Entomophaga 41(3–4):475–483

Elbeltagy A, Nishioka K, Sato T, Suzuki H, Ye B, Hamada T, Isawa T, Mitsui H, Minamisawa K (2001) Endophytic colonization and in planta nitrogen fixation by a *Herbaspirillum* sp. isolated from wild rice species. Appl Environ Microbiol 67(11):5285–5293

El-Nagdi W, Youssef M (2004) Soaking faba bean seed in some bio-agents as prophylactic treatment for controlling *Meloidogyne incognita* root-knot nematode infection. J Pest Sci 77 (2):75–78

El-Tarabily KA, Nassar AH, Hardy GESJ, Sivasithamparam K (2009) Plant growth promotion and biological control of *Pythium aphanidermatum*, a pathogen of cucumber, by endophytic actinomycetes. J Appl Microbiol 106(1):13–26

Eskin N, Vessey K, Tian L (2014) Research progress and perspectives of nitrogen fixing bacterium, *Gluconacetobacter diazotrophicus*, in monocot plants. Int J Agron 2014:208383

Ezra D, Castillo UF, Strobel GA, Hess WM, Porter H, Jensen JB, Condron MAM, Teplow DB, Sears J, Maranta M (2004) Coronamycins, peptide antibiotics produced by a verticillate *Streptomyces* sp.(MSU-2110) endophytic on *Monstera* sp. Microbiology 150(4):785–793

Finotello F, Di Camillo B (2015) Measuring differential gene expression with RNA-seq: challenges and strategies for data analysis. Brief Funct Genomics 14(2):130–142

Frans AAMDE, Davies G, Kerry R (1992) The use of *Verticillium chlamydosporium* Goddard and *Pasteuria penetrans* (Thorne) Sayre & Starr alone and in combination to control *Meloidogyne incognita* on tomato plants. Fundam Appl Nematol 15(3):235–242

Gamboa MA, Bayman P (2001) Communities of endophytic fungi in leaves of a tropical timber tree (*Guarea guidonia*: Meliaceae) 1. Biotropica 33(2):352–360

Gautam A, Siddiqui ZA, Mahmood I (1995) Integrated management of *Meloidogyne incognita* on tomato. Nematol Mediterr 23(2):245–247

Glick BR (2014) Bacteria with ACC deaminase can promote plant growth and help to feed the world. Microbiol Res 169(1):30–39

Gunatilaka AAL (2006) Natural products from plant-associated microorganisms: distribution, structural diversity, bioactivity, and implications of their occurrence. J Nat Prod 69(3):509–526

Guo J, Cole JR, Zhang Q, Brown CT, Tiedje JM (2016) Microbial community analysis with ribosomal gene fragments from shotgun metagenomes. Appl Environ Microbiol 82(1):157–166

Hallmann J (2001) Plant interactions with endophytic bacteria. CABI Publishing, New York

Hallmann J, Sikora RA (1996) Toxicity of fungal endophyte secondary metabolites to plant parasitic nematodes and soil-borne plant pathogenic fungi. Eur J Plant Pathol 102(2):155–162

Hallmann J, Quadt-Hallmann A, Mahaffee WF, Kloepper JW (1997) Bacterial endophytes in agricultural crops. Can J Microbiol 43(10):895–914

Handelsman J (2007) Metagenomics and microbial communities. In: Maccarrone Evans M (ed) Encyclopedia of life sciences. Wiley, Hoboken, NJ

Hardoim PR, Van Overbeek LS, Berg G, Pirttilä AM, Compant S, Campisano A, Döring M, Sessitsch A (2015) The hidden world within plants: ecological and evolutionary considerations for defining functioning of microbial endophytes. Microbiol Mol Biol Rev 79(3):293–320

Higginbotham SJ, Arnold AE, Ibañez A, Spadafora C, Coley PD, Kursar TA (2013) Bioactivity of fungal endophytes as a function of endophyte taxonomy and the taxonomy and distribution of their host plants. PLoS One 8(9):e73192

Huang W-K, Sun J-H, Cui J-K, Wang G-F, Kong L-A, Peng H, Chen S-L, Peng D-L (2014) Efficacy evaluation of fungus *Syncephalastrum racemosum* and nematicide avermectin against the root-knot nematode *Meloidogyne incognita* on cucumber. PLoS One 9(2):e89717

Humpherson-Jones FM, Dixon GR, Craig MA, and Ann DM (1992) Control of clubroot using calcium cyanamide: a review. In: Brighton crop protection conference, pests and diseases-1992. British Crop Protection Council, pp 1147–1154

Hurd PJ, Nelson CJ (2009) Advantages of next-generation sequencing versus the microarray in epigenetic research. Brief Funct Genomic Proteomic 8(3):174–183

Huse SM, Huber JA, Morrison HG, Sogin ML, Welch DM (2007) Accuracy and quality of massively parallel DNA pyrosequencing. Genome Biol 8(7):R143

Ihrmark K, Bödeker I, Cruz-Martinez K, Friberg H, Kubartova A, Schenck J, Strid Y, Stenlid J, Brandström-Durling M, Clemmensen KE (2012) New primers to amplify the fungal ITS2 region–evaluation by 454-sequencing of artificial and natural communities. FEMS Microbiol Ecol 82(3):666–677

Inahashi Y, Iwatsuki M, Ishiyama A, Namatame M, Nishihara-Tsukashima A, Matsumoto A, Hirose T, Sunazuka T, Yamada H, Otoguro K, Takahashi Y, Ōmura S, Shiomi K (2011) Spoxazomicins A–C, novel antitrypanosomal alkaloids produced by an endophytic actinomycete, *Streptosporangium oxazolinicum* K07-0460 T. J Antibiot 64(4):303–307

James EK, Gyaneshwar P, Mathan N, Barraquio WL, Reddy PM, Iannetta PPM, Olivares FL, Ladha JK (2002) Infection and colonization of rice seedlings by the plant growth-promoting bacterium *Herbaspirillum seropedicae* Z67. Mol Plant-Microbe Interact 15(9):894–906

Jansson H-B, Lopez-Llorca LV (2001) Biology of nematophagous fungi. In: Misra JK, Horn BW (eds) Trichomycetes and other fungal groups. Science Publishers, Enfield, pp 145–173
Jansson H-B, Jeyaprakash A, Zuckerman BM (1985) Control of root-knot nematodes on tomato by the endoparasitic fungus *Meria coniospora*. J Nematol 17(3):327–329
Jones JT, Haegeman A, Danchin EGJ, Gaur HS, Helder J, Jones MGK, Kikuchi T, Manzanilla-López R, Palomares-Rius JE, Wesemael WML, Perry RN (2013) Top 10 plant-parasitic nematodes in molecular plant pathology. Mol Plant Pathol 14(9):946–961
Jones LM, Koehler AK, Trnka M, Balek J, Challinor AJ, Atkinson HJ, Urwin PE (2017) Climate change is predicted to alter the current pest status of *Globodera pallida* and *G. rostochiensis* in the United Kingdom. Glob Chang Biol 23(11):4497–4507
Kavroulakis N, Ntougias S, Zervakis GI, Ehaliotis C, Haralampidis K, Papadopoulou KK (2007) Role of ethylene in the protection of tomato plants against soil-borne fungal pathogens conferred by an endophytic *Fusarium solani* strain*. J Exp Bot 58(14):3853–3864
Kennedy J, Marchesi JR, Dobson AD (2008) Marine metagenomics: strategies for the discovery of novel enzymes with biotechnological applications from marine environments. Microb Cell Fact 7(1):27
Khan AL, Hamayun M, Kang S-M, Kim Y-H, Jung H-Y, Lee J-H, Lee I-J (2012) Endophytic fungal association via gibberellins and indole acetic acid can improve plant growth under abiotic stress: an example of *Paecilomyces formosus* LHL10. BMC Microbiol 12(1):3
Khan AL, Waqas M, Kang S-M, Al-Harrasi A, Hussain J, Al-Rawahi A, Al-Khiziri S, Ullah I, Ali L, Jung H-Y (2014) Bacterial endophyte *Sphingomonas* sp. LK11 produces gibberellins and IAA and promotes tomato plant growth. J Microbiol 52(8):689–695
Kikuchi T, Jones JT, Aikawa T, Kosaka H, Ogura N (2004) A family of glycosyl hydrolase family 45 cellulases from the pine wood nematode *Bursaphelenchus xylophilus*. FEBS Lett 572 (1–3):201–205
Kimmons CA, Gwinn KD, Bernard EC (1990) Nematode reproduction on endophyte-infected and endophyte-free tall fescue. Plant Dis 74(10):757–761
Kloepper JW, Ryu C-M (2006) Bacterial endophytes as elicitors of induced systemic resistance. In: JES B, JCB C, Thomas NS (eds) Microbial root endophytes. Springer Nature, Cham, Switzerland, pp 33–52
Kloepper JW, McInroy JA, Liu K, Hu C-H (2013) Symptoms of Fern distortion syndrome resulting from inoculation with opportunistic endophytic fluorescent *Pseudomonas* spp. PLoS One 8(3): e58531
Knief C (2014) Analysis of plant microbe interactions in the era of next generation sequencing technologies. Front Plant Sci 5:216
Koch DW (1998) Nematode-resistant oil radish for *Heterodera schachtii* control. II. Sugarbeet-dry bean-corn rotations. J Sugar Beet Res 35(1–2):63–75
Koskimäki JJ, Hokkanen J, Jaakola L, Suorsa M, Tolonen A, Mattila S, Pirttilä AM, Hohtola A (2009) Flavonoid biosynthesis and degradation play a role in early defence responses of bilberry (*Vaccinium myrtillus*) against biotic stress. Eur J Plant Pathol 125(4):629–640
Kulkarni SM, Sangit AD (2000) Cultivation of *Hohenbuehelia atrocaerulea* (Fr.) sing. (Agaricomycetideae): a mushroom with nematicidal potential. Int J Med Mushrooms 2 (2):161–163
Kumar V, Maitra S, Shukla RN (2015) Environmental metagenomics: the data assembly and data analysis perspectives. J Inst Eng (India) Ser A 96(1):71–83
Kyndt T, Fernandez D, Gheysen G (2014) Plant-parasitic nematode infections in rice: molecular and cellular insights. Annu Rev Phytopathol 52:135–153
Lane DJ (1991) 16S/23S rRNA sequencing. In: Stackebrandt E, Goodfellow M (eds) Nucleic acid techniques in bacterial systematics. Wiley, New York, pp 115–175
Link HF (1809) Observationes in ordines plantarum naturales. Dissertatio I. Mag Ges Naturf Freunde Berlin 3:3–42
Liu L, Li Y, Li S, Hu N, He Y, Pong R, Lin D, Lu L, Law M (2012) Comparison of next-generation sequencing systems. BioMed Res Int 2012:251364

Liu H, Carvalhais LC, Crawford M, Singh E, Dennis PG, Pieterse CMJ, Schenk PM (2017) Inner plant values: diversity, colonization and benefits from endophytic bacteria. Front Microbiol 8:2552

Loaces I, Ferrando L, Scavino AF (2011) Dynamics, diversity and function of endophytic siderophore-producing bacteria in rice. Microb Ecol 61(3):606–618

Long HH, Schmidt DD, Baldwin IT (2008) Native bacterial endophytes promote host growth in a species-specific manner; phytohormone manipulations do not result in common growth responses. PLoS One 3(7):e2702

van Loon LC, Bakker PAHM, van der Heijdt WHW, Wendehenne D, Pugin A (2008) Early responses of tobacco suspension cells to rhizobacterial elicitors of induced systemic resistance. Mol Plant-Microbe Interact 21(12):1609–1621

Lundberg DS, Yourstone S, Mieczkowski P, Jones CD, Dangl JL (2013) Practical innovations for high-throughput amplicon sequencing. Nat Methods 10(10):999–1002

Luo H, Mo M, Huang X, Li X, Zhang K (2004) Coprinus comatus: a basidiomycete fungus forms novel spiny structures and infects nematode. Mycologia 96(6):1218–1224

Maier TR, Hewezi T, Peng J, Baum TJ (2013) Isolation of whole esophageal gland cells from plant-parasitic nematodes for transcriptome analyses and effector identification. Mol Plant-Microbe Interact 26(1):31–35

Malinowski DP, Belesky DP (2000) Adaptations of endophyte-infected cool-season grasses to environmental stresses: mechanisms of drought and mineral stress tolerance. Crop Sci 40:923–940

Mardis ER (2008) The impact of next-generation sequencing technology on genetics. Trends Genet 24(3):133–141

Mareque C, da Silva TF, Vollú RE, Beracochea M, Seldin L, Battistoni F (2018) The endophytic bacterial microbiota associated with sweet sorghum (*Sorghum bicolor*) is modulated by the application of chemical N fertilizer to the field. Int J Genomics 2018:7403670

Margulies M, Egholm M, Altman WE, Attiya S, Bader JS, Bemben LA, Berka J, Braverman MS, Chen Y-J, Chen Z, Dewell SB, Du L, Fierro JM, Gomes XV, Godwin BC, He W, Helgesen S, Ho CH, Irzyk GP, Jando SC, Alenquer MLI, Jarvie TP, Jirage KB, Kim J-B, Knight JR, Lanza JR, Leamon JH, Lefkowitz SM, Lei M, Li J, Lohman KL, Lu H, Makhijani VB, McDade KE, McKenna MP, Myers EW, Nickerson E, Nobile JR, Plant R, Puc BP, Ronan MT, Roth GT, Sarkis GJ, Simons JF, Simpson JW, Srinivasan M, Tartaro KR, Tomasz A, Vogt KA, Volkmer GA, Wang SH, Wang Y, Weiner MP, Yu P, Begley RF, Rothberg JM (2005) Genome sequencing in microfabricated high-density picolitre reactors. Nature 437(7057):376–380

Marra MA, Kucaba TA, Dietrich NL, Green ED, Brownstein B, Wilson RK, McDonald KM, Hillier LW, McPherson JD, Waterston RH (1997) High throughput fingerprint analysis of large-insert clones. Genome Res 7(11):1072–1084

Marroni F, Pinosio S, Di Centa E, Jurman I, Boerjan W, Felice N, Cattonaro F, Morgante M (2011) Large-scale detection of rare variants via pooled multiplexed next-generation sequencing: towards next-generation Ecotilling. Plant J 67(4):736–745

Meaburn E, Schulz R (2012) Next generation sequencing in epigenetics: insights and challenges. Semin Cell Dev Biol 23(2):192–199

Mercado-Blanco J, Lugtenberg BJJ (2014) Biotechnological applications of bacterial endophytes. Curr Biotechnol 3(1):60–75

Mian IH, Godoy G, Shelby RA, Rodriguez-Kabana R, Morgan-Jones G (1982) Chitin amendments for control of *Meloidogyne arenaria* in infested soil. Nematropica 12(1):71–84

Moens M, Perry RN (2009) Migratory plant endoparasitic nematodes: a group rich in contrasts and divergence. Ann Rev Phytopathol 47:313–332

Mojtahedi H, Santo GS, Hang AN, Wilson JH (1991) Suppression of root-knot nematode populations with selected rapeseed cultivars as green manure. J Nematol 23(2):170–174

Monfil VO, Casas-Flores S (2014) Molecular mechanisms of biocontrol in *Trichoderma* spp. and their applications in agriculture. In: Gupta VK, Schmoll M, Herrera-Estrella A, Upadhyay RS,

Druzhinina I, Tuohy MG (eds) Biotechnology and Biology of Trichoderma. Elsevier, Amsterdam, pp 429–453

Monfort E, Lopez-Llorca LV, Jansson HB, Salinas J, Park JO, Sivasithamparam K (2005) Colonisation of seminal roots of wheat and barley by egg-parasitic nematophagous fungi and their effects on *Gaeumannomyces graminis* var. *tritici* and development of root-rot. Soil Biol Biochem 37(7):1229–1235

Mousa WK, Raizada MN (2013) The diversity of anti-microbial secondary metabolites produced by fungal endophytes: an interdisciplinary perspective. Front Microbiol 4:65

Müller P, Döring M (2009) Isothermal DNA amplification facilitates the identification of a broad spectrum of bacteria, fungi and protozoa in *Eleutherococcus* sp. plant tissue cultures. Plant Cell Tiss Org Culture 98(1):35–45

Murali TS, Suryanarayanan TS, Venkatesan G (2007) Fungal endophyte communities in two tropical forests of southern India: diversity and host affiliation. Mycol Prog 6(3):191–199

Nasopoulou C, Pohjanen J, Koskimäki JJ, Zabetakis I, Pirttilä AM (2014) Localization of strawberry (*Fragaria* x *ananassa*) and *Methylobacterium extorquens* genes of strawberry flavor biosynthesis in strawberry tissue by *in situ* hybridization. J Plant Physiol 171(13):1099–1105

Nassar AH, El-Tarabily KA, Sivasithamparam K (2005) Promotion of plant growth by an auxin-producing isolate of the yeast *Williopsis saturnus* endophytic in maize (*Zea mays* L.) roots. Biol Fertil Soils 42(2):97–108

Nicol JM, Elekçioğlu IH, Bolat N, Rivoal R (2007) The global importance of the cereal cyst nematode (*Heterodera* spp.) on wheat and international approaches to its control. Commun Agric Appl Biol Sci 72(3):677–686

Nicol JM, Turner SJ, Coyne DL, Den Nijs L, Hockland S, Maafi ZT (2011) Current nematode threats to world agriculture. In: Genomics and molecular genetics of plant-nematode interactions. Springer, Dordrecht, Heidelberg, pp 21–43

Noling JW (2009) Nematode management in tomatoes, peppers and eggplant. University of Florida, IFAS Extension, ENY-032 (December). https://edis.ifas.ufl.edu/ng032. Accessed 14 Apr 2019

Ntalli NG, Caboni P (2012) Botanical nematicides in the mediterranean basin. Phytochem Rev 11 (4):351–359

Nübel U, Engelen B, Felske A, Snaidr J, Wieshuber A, Amann RI, Ludwig W, Backhaus H (1996) Sequence heterogeneities of genes encoding 16S rRNAs in *Paenibacillus polymyxa* detected by temperature gradient gel electrophoresis. J Bacteriol 178(19):5636–5643

O'Dell TE, Massicotte HB, Trappe JM (1993) Root colonization of *Lupinus latifolius* Agardh. And *Pinus contorta* Dougl. By *Phialocephala fortinii* Wang & Wilcox. New Phytol 124(1):93–100

Oka Y, Koltai H, Bar-Eyal M, Mor M, Sharon E, Chet I, Spiegel Y (2000) New strategies for the control of plant-parasitic nematodes. Pest Manage Sci: Formerly Pestic Sci 56(11):983–988

Onofre-Lemus J, Hernández-Lucas I, Girard L, Caballero-Mellado J (2009) ACC (1-aminocyclopropane-1-carboxylate) deaminase activity, a widespread trait in *Burkholderia* species, and its growth-promoting effect on tomato plants. Appl Environ Microbiol 75 (20):6581–6590

Oteino N, Lally RD, Kiwanuka S, Lloyd A, Ryan D, Germaine KJ, Dowling DN (2015) Plant growth promotion induced by phosphate solubilizing endophytic *Pseudomonas* isolates. Front Microb 6:745

Ownley BH, Griffin MR, Klingeman WE, Gwinn KD, Moulton JK, Pereira RM (2008) *Beauveria bassiana*: endophytic colonization and plant disease control. J Invertebr Pathol 98(3):267–270

Ownley BH, Gwinn KD, Vega FE (2010) Endophytic fungal entomopathogens with activity against plant pathogens: ecology and evolution. BioControl 55(1):113–128

Palomares-Rius JE, Hirooka Y, Tsai IJ, Masuya H, Hino A, Kanzaki N, Jones JT, Kikuchi T (2014) Distribution and evolution of glycoside hydrolase family 45 cellulases in nematodes and fungi. BMC Evol Biol 14(1):69

Pavlo A, Leonid O, Iryna Z, Natalia K, Maria PA (2011) Endophytic bacteria enhancing growth and disease resistance of potato (*Solanum tuberosum* L.). Biol Control 56(1):43–49

Pérez MP, Navas-Cortés JA, Pascual-Villalobos MJ, Castillo P (2003) Nematicidal activity of essential oils and organic amendments from *Asteraceae* against root-knot nematodes. Plant Pathol 52(3):395–401

Persmark L, Marban-Mendoza N, Jansson H-B (1995) Nematophagous fungi from agricultural soils of Central America. Nematropica 25(2):117–124

Petrini O (1986) Taxonomy of endophytic fungi of aerial plant tissues. In: Fokkema N, van den Heuvel J (eds) Microbiology of the phyllosphere. Cambridge University Press, Cambridge, pp 175–187

Philippot L, Raaijmakers JM, Lemanceau P, van der Putten WH (2013) Going back to the roots: the microbial ecology of the rhizosphere. Nat Rev Microbiol 11(11):789–799

Ploeg A (2008) Biofumigation to manage plant-parasitic nematodes. In: Ciancio A, Mukerji KG (eds) Integrated management and biocontrol of vegetable and grain crops nematodes. Springer, Dordrecht, the Netherlands, pp 239–248

Ploeg AT, Stapleton JJ (2001) The effects of temperature, time, and amendment of soil with broccoli residues on the infestation of melon by two root-knot nematode species. Plant Prot Q 11:1–3

Proença DN, Francisco R, Santos CV, Lopes A, Fonseca L, Abrantes IMO, Morais PV (2010) Diversity of bacteria associated with *Bursaphelenchus xylophilus* and other nematodes isolated from *Pinus pinaster* trees with pine wilt disease. PLoS One 5(12):e15191

van der Putten WH, Cook R, Costa S, Davies KG, Fargette M, Freitas H, Hol WHG, Kerry BR, Maher N, Mateille T (2006) Nematode interactions in nature: models for sustainable control of nematode pests of crop plants. Adv Agron 89:227–260

Qiao K, Liu X, Wang H, Xia X, Ji X, Wang K (2012) Effect of abamectin on root-knot nematodes and tomato yield. Pest Manag Sci 68(6):853–857

Quentin M, Abad P, Favery B (2013) Plant parasitic nematode effectors target host defense and nuclear functions to establish feeding cells. Front Plant Sci 4:53

Redman RS, Sheehan KB, Stout RG, Rodriguez RJ, Henson JM (2002) Thermotolerance generated by plant/fungal symbiosis. Science 298(5598):1581–1581

Reinhold-Hurek B, Hurek T (1998) Life in grasses: diazotrophic endophytes. Trends Microbiol 6 (4):139–144

Reinhold-Hurek B, Hurek T (2011) Living inside plants: bacterial endophytes. Curr Opin Plant Biol 14(4):435–443

Reiter B, Pfeifer U, Schwab H, Sessitsch A (2002) Response of endophytic bacterial communities in potato plants to infection with *Erwinia carotovora* subsp. *atroseptica*. Appl Environ Microbiol 68(5):2261–2268

Rizvi R, Mahmood I, Tiyagi SA, Khan Z (2012) Effect of some botanicals for the management of plant-parasitic nematodes and soil-inhabiting fungi infesting chickpea. Turk J Agric For 36 (6):710–719

Robert-Seilaniantz A, Grant M, Jones JDG (2011) Hormone crosstalk in plant disease and defense: more than just jasmonate-salicylate antagonism. Annu Rev Phytopathol 49:317–343

Rodriguez RJ, White JF Jr, Arnold AE, Redman RS (2009) Fungal endophytes: diversity and functional roles. New Phytol 182(2):314–330

Rosskopf EN, Chellemi DO, Kokalis-Burelle N, Church GT (2005) Alternatives to methyl bromide: a Florida perspective. Plant Health Prog 6(1):19–48

Rungin S, Indananda C, Suttiviriya P, Kruasuwan W, Jaemsaeng R, Thamchaipenet A (2012) Plant growth enhancing effects by a siderophore-producing endophytic streptomycete isolated from a Thai jasmine rice plant (*Oryza sativa* L. cv. KDML105). Antonie Van Leeuwenhoek 102 (3):463–472

Safeena MIS, Zakeel MCM (2019) Strategies for the management of soil-borne pathogens and crop production under saline environment. In: Ansari RA, Mahmood I (eds) Plant health under biotic stress. Springer, Singapore, pp 213–229

Saikkonen K, Ion D, Gyllenberg M (2002) The persistence of vertically transmitted fungi in grass metapopulations. Proc R Soc Lond B Biol Sci 269(1498):1397–1403

Samac D, Kinkel L (2001) Suppression of the root-lesion nematode (*Pratylenchus penetrans*) in alfalfa (*Medicago sativa*) by *Streptomyces* spp. Plant Soil 235(1):35–44

Santo GS, O'bannon JH, Finley AM, Golden AM (1980) Occurrence and host range of a new root-knot nematode (*Meloidogyne chitwoodi*) in the Pacific northwest. Plant Dis 64(10):951–952

Sasser JN, Freckman DW (1987) A world perspective on nematology: the role of the society. In: Veeck J, Dickson D, Hyattsville M (eds) Vistas on nematology. Society of Nematologists, Hyattsville, MD, pp 7–14

Scavino AF, Pedraza RO (2013) The role of siderophores in plant growth-promoting bacteria. In: Maheshwari D, Saraf M, Aeron A (eds) Bacteria in agrobiology: crop productivity. Springer, Berlin, Heidelberg, pp 265–285

Scholte K (2000) Effect of potato used as a trap crop on potato cyst nematodes and other soil pathogens and on the growth of a subsequent main potato crop. Ann Appl Biol 136(3):229–238

Schulz B, Boyle C (2005) The endophytic continuum. Mycol Res 109(6):661–686

Schwarz M, Köpcke B, Weber RWS, Sterner O, Anke H (2004) 3-Hydroxypropionic acid as a nematicidal principle in endophytic fungi. Phytochemistry 65(15):2239–2245

Scortichini M, Loreti S (2007) Occurrence of an endophytic, potentially pathogenic strain of *Pseudomonas syringae* in symptomless wild trees of *Corylus avellana* L. J Plant Pathol 89 (3):431–434

Sharon E, Bar-Eyal M, Chet I, Herrera-Estrella A, Kleifeld O, Spiegel Y (2001) Biological control of the root-knot nematode *Meloidogyne javanica* by *Trichoderma harzianum*. Phytopathology 91(7):687–693

Shi Y, Lou K, Li C (2009) Promotion of plant growth by phytohormone-producing endophytic microbes of sugar beet. Biol Fertil Soils 45(6):645–653

Siegel MR, Latch GCM, Bush LP, Fannin FF, Rowan DD, Tapper BA, Bacon CW, Johnson MC (1990) Fungal endophyte-infected grasses: alkaloid accumulation and aphid response. J Chem Ecol 16(12):3301–3315

Sikora RA, Pocasangre L, zum Felde A, Niere B, Vu TT, Dababat AA (2008) Mutualistic endophytic fungi and in-planta suppressiveness to plant parasitic nematodes. Biol Control 46 (1):15–23

Singh RK, Mishra RP, Jaiswal HK, Kumar V, Pandey SP, Rao SB, Annapurna K (2006) Isolation and identification of natural endophytic rhizobia from rice (*Oryza sativa* L.) through rDNA PCR-RFLP and sequence analysis. Curr Microbiol 52(5):345–349

Singh LP, Gill SS, Tuteja N (2011) Unraveling the role of fungal symbionts in plant abiotic stress tolerance. Plant Signal Behav 6(2):175–191

Smant G, Stokkermans JPWG, Yan Y, De Boer JM, Baum TJ, Wang X, Hussey RS, Gommers FJ, Henrissat B, Davis EL (1998) Endogenous cellulases in animals: isolation of β-1, 4-endoglucanase genes from two species of plant-parasitic cyst nematodes. Proc Natl Acad Sci 95(9):4906–4911

Spiegel Y, Chet I (1998) Evaluation of *Trichoderma* spp. as a biocontrol agent against soilborne fungi and plant-parasitic nematodes in Israel. Integr Pest Manag Rev 3(3):169–175

Stirling GR (2017) Biological control of plant-parasitic nematodes. In: Tjamos EC, Papavizas GC, Cook RJ (eds) Diseases of nematodes. Springer, Boston, MA, pp 103–150

Stone JK, Polishook JD, White JF (2004) Endophytic fungi. In: Foster MS, Bills GF, Mueller GM (eds) Biodiversity of fungi: inventory and monitoring methods. Elsevier Science, Burlington, MA, pp 241–270

Strobel G, Daisy B (2003) Bioprospecting for microbial endophytes and their natural products. Microbiol Mol Biol Rev 67(4):491–502

Strobel NE, Hussey RS, Roncadori RW (1982) Interactions of vesicular-arbuscular mycorrhizal fungi, *Meloidogyne incognita*, and soil fertility on peach. Phytopathology 72(6):690–694

Stutzman-Engwall K, Conlon S, Fedechko R, McArthur H, Pekrun K, Chen Y, Jenne S, La C, Trinh N, Kim S, Zhang Y-X, Fox R, Gustafsson C, Krebber A (2005) Semi-synthetic DNA shuffling of *aveC* leads to improved industrial scale production of doramectin by *Streptomyces avermitilis*. Metab Eng 7(1):27–37

Su L, Shen Z, Ruan Y, Tao C, Chao Y, Li R, Shen Q (2017) Isolation of antagonistic endophytes from banana roots against *Meloidogyne javanica* and their effects on soil nematode community. Front Microbiol 8:2070

Sun L, Qiu F, Zhang X, Dai X, Dong X, Song W (2008) Endophytic bacterial diversity in rice (*Oryza sativa* L.) roots estimated by 16S rDNA sequence analysis. Microb Ecol 55(3):415–424

Tadych M, Bergen M, Dugan FM, White JF Jr (2007) Evaluation of the potential role of water in spread of conidia of the *Neotyphodium* endophyte of *Poa ampla*. Mycol Res 111(4):466–472

Taghavi S, van der Lelie D, Hoffman A, Zhang Y-B, Walla MD, Vangronsveld J, Newman L, Monchy S (2010) Genome sequence of the plant growth promoting endophytic bacterium *Enterobacter* sp. 638. PLoS Genet 6(5):e1000943

Tan RX, Zou WX (2001) Endophytes: a rich source of functional metabolites. Nat Prod Rep 18 (4):448–459

Tejesvi MV, Kajula M, Mattila S, Pirttilä AM (2011) Bioactivity and genetic diversity of endophytic fungi in *Rhododendron tomentosum* Harmaja. Fungal Divers 47(1):97–107

Tejesvi MV, Segura DR, Schnorr KM, Sandvang D, Mattila S, Olsen PB, Neve S, Kruse T, Kristensen HH, Pirttilä AM (2013) An antimicrobial peptide from endophytic *Fusarium tricinctum* of *Rhododendron tomentosum* Harmaja. Fungal Divers 60(1):153–159

Thomas T, Gilbert J, Meyer F (2012) Metagenomics-a guide from sampling to data analysis. Microb Inform Exp 2(1):3

Tian B-Y, Cao Y, Zhang K-Q (2015) Metagenomic insights into communities, functions of endophytes, and their associates with infection by root-knot nematode, *Meloidogyne incognita*, in tomato roots. Sci Rep 5:17,087

Toju H, Tanabe AS, Yamamoto S, Sato H (2012) High-coverage ITS primers for the DNA-based identification of ascomycetes and basidiomycetes in environmental samples. PLoS One 7(7): e40863

Trapnell C, Pachter L, Salzberg SL (2009) TopHat: discovering splice junctions with RNA-Seq. Bioinformatics 25(9):1105–1111

Trapnell C, Williams BA, Pertea G, Mortazavi A, Kwan G, Van Baren MJ, Salzberg SL, Wold BJ, Pachter L (2010) Transcript assembly and quantification by RNA-Seq reveals unannotated transcripts and isoform switching during cell differentiation. Nat Biotechnol 28(5):511–515

Trapnell C, Roberts A, Goff L, Pertea G, Kim D, Kelley DR, Pimentel H, Salzberg SL, Rinn JL, Pachter L (2012) Differential gene and transcript expression analysis of RNA-seq experiments with TopHat and cufflinks. Nat Protoc 7(3):562–578

Trémouillaux-Guiller J, Rohr T, Rohr R, Huss VAR (2002) Discovery of an endophytic alga in *Ginkgo biloba*. Am J Bot 89(5):727–733

Türktaş M, Kurtoglu KY, Dorado G, Zhang B, Hernandez P, Ünver T (2015) Sequencing of plant genomes: a review. Turk J Agric For 39(3):361–376

Turner SJ, Evans K (1998) The origins, global distribution and biology of potato cyst nematode (*Globodera rostochiensis* (Woll.) and *G. pallida* Stone). In: Marks RJ, Brodie BB (eds) Potato cyst nematodes-biology, distribution and control. CAB International, Wallingford, pp 20–21

Urwin PE, Lilley CJ, McPherson MJ, Atkinson HJ (1997) Resistance to both cyst and root-knot nematodes conferred by transgenic *Arabidopsis* expressing a modified plant cystatin. Plant J 12 (2):455–461

Vadassery J, Ritter C, Venus Y, Camehl I, Varma A, Shahollari B, Novák O, Strnad M, Ludwig-Müller J, Oelmüller R (2008) The role of auxins and cytokinins in the mutualistic interaction between *Arabidopsis* and *Piriformospora indica*. Mol Plant-Microbe Interact 21 (10):1371–1383

Van Overbeek L, van Doorn J, Wichers J, van Amerongen A, van Roermund H, Willemsen P (2014) The arable ecosystem as battleground for emergence of new human pathogens. Front Microbiol 5:104

van Wesemael B, Paustian K, Andrén O, Cerri CE, Dodd M, Etchevers J, Goidts E, Grace P, Kätterer T, McConkey BG (2011) How can soil monitoring networks be used to improve predictions of organic carbon pool dynamics and CO_2 fluxes in agricultural soils? Plant Soil 338(1–2):247–259

Verma VC, Kharwar RN, Strobel GA (2009) Chemical and functional diversity of natural products from plant associated endophytic fungi. Nat Prod Commun 4(11):1511–1532

Verpoorte R (2000) Secondary metabolism. In: Verpoorte R, Alfermann AW (eds) Metabolic engineering of plant secondary metabolism. Kluwer Academic Publishers, Dordrecht, Boston, pp 1–29

Viaene NM, Abawi GS (1998) Fungi parasitic on juveniles and egg masses of *Meloidogyne hapla* in organic soils from New York. J Nematol 30(4S):632–638

Villate L, Fievet V, Hanse B, Delemarre F, Plantard O, Esmenjaud D, Van Helden M (2008) Spatial distribution of the dagger nematode *Xiphinema index* and its associated grapevine fanleaf virus in French vineyard. Phytopathology 98(8):942–948

Vu T, Sikora R, Hauschild R (2006) *Fusarium oxysporum* endophytes induced systemic resistance against *Radopholus similis* on banana. Nematology 8(6):847–852

Waceke JW, Waudo SW, Sikora R (2001) Suppression of *Meloidogyne hapla* by arbuscular mycorrhiza fungi (AMF) on pyrethrum in Kenya. Int J Pest Manage 47(2):135–140

Wang K-H, Mcsorley R, Kokalis-Burelle N (2006) Effects of cover cropping, solarization, and soil fumigation on nematode communities. Plant Soil 286(1–2):229–243

Waqas M, Khan AL, Kamran M, Hamayun M, Kang S-M, Kim Y-H, Lee I-J (2012) Endophytic fungi produce gibberellins and indoleacetic acid and promotes host-plant growth during stress. Molecules 17(9):10,754–10,773

Wemheuer F, Wemheuer B, Kretzschmar D, Pfeiffer B, Herzog S, Daniel R, Vidal S (2016) Impact of grassland management regimes on bacterial endophyte diversity differs with grass species. Lett Appl Microbiol 62(4):323–329

Wemheuer F, Kaiser K, Karlovsky P, Daniel R, Vidal S, Wemheuer B (2017) Bacterial endophyte communities of three agricultural important grass species differ in their response towards management regimes. Sci Rep 7:40,914

West CP, Izekor E, Oosterhuis DM, Robbins RT (1988) The effect of *Acremonium coenophialum* on the growth and nematode infestation of tall fescue. Plant Soil 112(1):3–6

Westerdahl BB (2011) Cultural methods for managing nematodes on vegetables and ornamentals. In: First all African horticultural congress, pp 185–198

Westerdahl BB, Nunez J, and Radewald JD (2008) Trap cropping for managing nematodes on carrots in California. In: Annual international research conference on methyl bromide alternatives and emissions reductions, Orlando, FL

Westphal A (2011) Sustainable approaches to the management of plant-parasitic nematodes and disease complexes. J Nematol 43(2):122–125

Whipps JM, Davies KG (2000) Success in biological control of plant pathogens and nematodes by microorganisms. In: Gurr G, Wratten S (eds) Biological control: measures of success. Springer, Dordrecht, pp 231–269

White TJ, Bruns T, Lee SJWT, Taylor JL (1990) Amplification and direct sequencing of fungal ribosomal RNA genes for phylogenetics. Academic Press Inc., New York

Williamson VM (1999) Plant nematode resistance genes. Curr Opin Plant Biol 2(4):327–331

Williamson VM, Hussey RS (1996) Nematode pathogenesis and resistance in plants. Plant Cell 8 (10):1735–1745

Yan Y, Smant G, Stokkermans J, Qin L, Helder J, Baum T, Schots A, Davis E (1998) Genomic organization of four β-1, 4-endoglucanase genes in plant-parasitic cyst nematodes and its evolutionary implications. Gene 220(1–2):61–70

Zabalgogeazcoa I (2008) Review. Fungal endophytes and their interaction with plant pathogens (Revisión. Los hongos endofíticos y sus interacciones con patógenos de plantas). Span J Agric Res 6:138–146

Zakeel MC, Safeena MI (2019) Biofilmed biofertilizer for sustainable agriculture. In: Ansari RA, Mahmood I (eds) Plant health under biotic stress. Springer, Singapore, pp 65–82

Zamioudis C, Pieterse CMJ (2012) Modulation of host immunity by beneficial microbes. Mol Plant-Microbe Interact 25(2):139–150

Zhang HW, Song YC, Tan RX (2006) Biology and chemistry of endophytes. Nat Prod Rep 23 (5):753–771

Chapter 6
Nanoparticles' Synthesis and Their Application in the Management of Phytonematodes: An Overview

Oluwatoyin Adenike Fabiyi, Ridwan Olamilekan Alabi, and Rizwan Ali Ansari

Abstract One of the most effective novel areas of research is nanotechnology. It is a field that is improving daily, especially in the subject of agriculture. Crop production is hindered by several pests of economic importance, which are responsible for huge crop losses and invariably contributing to global food insecurity with attendant consequences such as malnutrition, starvation, social isolation, and other overlapping issues that cannot be overemphasized. Crops are predisposed to losses from agricultural pests and diseases because the peasant farmer lacks efficient soil, virile seeds, and environmental management techniques. The citizens are subjected to poverty and hunger owing to the devastating effect of huge crop losses in the field and storage. The synthetic-based approach, though highly indispensable in agricultural crop pest management, is laden with severe environmental pollution and has brought various diseases to the human race. This necessitated the development of safer alternative pesticidal substances with the aim of improving crop production. Prodigious attention has been paid by the researcher toward application of nanoparticles (NPs) and their application in sustainable agriculture. NPs are new methods that could be used to redeem the environment from its polluted state. Materials act differently in their nano form, which improves their activity as pesticidal agents. In agriculture, nanomaterials have been used in livestock and crop protection; they exhibit properties like biodegradability, solubility, permeability, and thermal stability. They also possess surface areas that increase their affinity to the target organism. Nanomaterials are available in different forms such as nano-containers, nano-encapsulates, nano-cages, and nano-emulsions for pest management use, and their potency has been established in crop disease control. NPs are

O. A. Fabiyi (✉) · R. O. Alabi
Department of Crop Protection, Faculty of Agriculture, University of Ilorin, Ilorin, Nigeria

R. A. Ansari
Section of Plant Pathology and Nematology, Department of Botany, Aligarh Muslim University, Aligarh, India

Present Address: Department of Plant Protection, Faculty of Agricultural Sciences, Aligarh Muslim University, Aligarh, India

R. A. Ansari et al. (eds.), *Management of Phytonematodes: Recent Advances and Future Challenges*, https://doi.org/10.1007/978-981-15-4087-5_6

good sources of controlled release mechanisms, which is a way of reducing the amount of pesticides or fertilizers injected into the environment while crops are on the field. NP biomarkers can be used in detecting bacteria, viruses, fungi, and nematodes of economic importance in agriculture; as diagnostic tools, nano sensors can indicate certain compounds in plants that are elicited by the plants only in disease conditions. The use of nano sensors as pesticide residue detectors comes up with accurate and reliable information. Nanotechnology consists of two major aspects, that is, the synthesis of nano-sized materials and the application or use of the synthesized nanomaterials for the intended purpose. Phytonematodes cause a wide range of losses to agricultural crops; they are a major threat to world food production as they cause yield losses in all areas of the world. Considerable yield losses are caused annually by several nematode species on crops. In order to increase yield, NPs have been employed in the control of nematodes. This overview highlights synthesis, characterization of NPs, and successes in the area of phytonematode management.

Keywords Nanoparticles · Plants · Phytonematodes · Control · Nanoshapes

6.1 Introduction

Nanoparticles (NPs) are particles that could be obtained by conversion of macromolecules into nano-sized particles or by breaking up of bulk materials into atoms or ions after which they are allowed to condense into NPs whose size ranges from 1 to 100 nm in size (Khan and Rizvi 2014). The NPs' behavior is quite different from the macro material and they exhibit properties that are absent in the bulk material. Nanoparticles have surface-to-volume ratio that is generally high and that helps in the enhancement of their biochemical properties and reactivity (Dubchak et al. 2010). For example, 1 g of gold (Au) converted to nano-sized particles will cover an area of about 100 km^2. A gold nanoparticle of 2.5 nm particle size melts at a lower temperature (~300 °C) compared to a gold slab that melts at 1064 °C (Buffat and Borel 1976). The macro gold has no toxic effect, whereas the gold NPs exhibit toxic effect on bacteria, *Salmonella typhimurium* (Wang et al. 2011). Similarly, the macro silver (Ag) lacks antibacterial and antifungal properties, while the NPs possess all these (Sofi et al. 2012). Synthesis of nanoparticles requires skill and facilities based on the method chosen. The characterization of the resulting NPs is another crucial step that determines the size and uniformity of the particles produced. NPs are characterized by their surface area, dispersity, size, and shape. The homogenization of all these properties is necessary in the application of NPs (Jiang et al. 2009). The common techniques for the characterization of NPs include but are not limited to: energy-dispersive spectroscopy (EDS), transmission electron microscopy (TEM), scanning electron microscopy (SEM), Fourier-transform infrared spectroscopy (FTIR), dynamic light scattering (DLS), ultraviolet-visible (UV-vis) spectrophotometry, and powder X-ray diffraction (XRD; Shahverdi et al. 2007). For measurement of size distribution, surface charge, crystallinity, phase identification, and phase distribution of biosynthesized NPs in liquid medium, dynamic light scattering (DLS) and X-ray diffraction methods are used (Strasser et al. 2010; Gunalan et al. 2011). In UV-visible spectroscopy, light wavelength ranging 300–800 nm is generally used for characterization of metal nanomaterials in the range of 2–100 nm; for gold and silver NPs, absorption measurements range between 500 and 550 and 400 and 450, respectively. Absorption peaks of zinc oxide (ZnO) NPs produced from *Aloe vera* had peaks ranging from 358 to 375 nm. Another important technique, that is, energy dispersive spectroscopy, is applied in nanotechnology to ascertain the elemental composition of NPs (Sun et al. 2000; Feldheim and Foss 2002; Pal et al. 2007). The nature of functional groups clinging on the surface of NPs could be the reason behind stabilization and production of metal NPs. The FTIR is employed in the identification and characterization of functional groups. Generally, functional group bands observed at 3450, 3266, and 2932 cm are assigned to stretching of alcohols and C–H stretching of alkanes, respectively. *Solanum torvum* mediated silver NPs depicted peaks at 1648, 1535, 1450, and 1019 cm and further study revealed that the peak at 1450 cm of carboxylate ions was accountable for the stabilization of the silver NPs produced. The morphological characterization of NPs at the nanometer and nanoscale ranges are measured with SEM and TEM (Eppler et al. 2000; Govindaraju et al. 2010; Dhandapani et al. 2012; Sankar et al. 2014). NPs cannot be viewed under optical microscope; they are smaller in size compared to bacteria cells and virus particles. The shape may be spherical, rod shaped, or polyhedral (Khan and Rizvi 2014).

6.2 Synthesis of Nanoparticles

The method of NPs' synthesis varies widely; this influences the efficiency and properties. The most common ways of NPs' preparation include physical, biological, and chemical.

6.2.1 Physical Method

This involves the use of evaporation condensation, radiolysis, UV irradiation, ultrasonication, and laser ablation. During the process of synthesis, the metal atoms evaporate, followed by condensation; the metallic atoms are then rearranged and aggregated into tiny clusters of metal NPs (Hurst et al. 2006). The cost of production is usually very high with physical synthesis of NPs, because of the need for sophisticated equipment, radiative heating, high power consumption, and chemicals (Khandel et al. 2018).

6.2.2 Chemical Methods

The Chemical methods of NPs' synthesis could be mechanical milling, etching, laser ablation, sputtering, or thermal decomposition. In this process, chemicals are used in the reduction of metal ion in solution. Metal ions may help in either nucleation or aggregation for the formation of small clusters of metals, which depends on the reaction mixture condition. Sodium borohydride, hydrazine, and hydrogen are some of the chemicals employed as reducing agent (Egorova and Revina 2000). Stabilizing agents like cellulose, natural rubber, chitosan, and co-polymer micelles are used with the reducing agents. Furthermore, organic solvents such as ethane, dimethyl, formaldehyde, toluene, and chloroform need to be added to these chemicals because of their hydrophobic nature. Formation of toxic or benign byproducts usually occurs during the process of reaction. The toxicity of these chemicals to the surface of the NPs has made them unpopular in medical applications. In view of the problems associated with the physical and chemical methods of NPs' synthesis, research is redirecting to biological methods of synthesis that are not as expensive and dangerous as former, that is, chemical methods (Patel et al. 2015).

6.2.3 Biological Method

The process includes various plants and microorganisms such as actinomycetes, fungi, and bacteria. The antioxidant and reducing properties present in such agents are responsible for NPs' reduction (Abd El-Rahman and Mohammad 2014). The biogenic synthesis of NPs can be divided into two stages. The first stage is bioreduction where the metal ions are reduced chemically to stable forms via a biological process (Mukunthan and Balaji 2012). Several organisms have the ability to reduce metals, through a process where the metal ion reduction is synchronized with enzyme oxidation, during which a stable and inert metal NPs are produced (Deplanche et al. 2010; Mukunthan and Balaji 2012). The second stage is biosorption involving binding of metal ions in aqueous or soil medium on to the cell wall of the organism, which does not require energy input (Yong et al. 2002). The modified cell walls of plants, fungi, and bacteria contain peptides that automatically bind to metal ions from which stable metal complexes as NPs are formed (Yong et al. 2002). Biologically synthesized NPs are better defined in size and morphology than what has been obtained from physico-chemical methods (Azizi et al. 2013). Invention of efficient and eco-friendly process for the synthesis of NPs is an important and emerging aspect of bio-nanotechnology (Khan et al. 2009; Azizi et al. 2013). Microbially synthesized NPs are eco-friendly, scalable, and compatible with pharmacological application. However, synthesis with microorganisms is more expensive than plant-based synthesis. The requirement of special maintenance and high aseptic conditions makes it unacceptable for large-scale production (Dhuper et al. 2012). Apart from this, plant-based synthesis does not require the use of toxic chemicals, high temperature, and pressures as in the case of the chemical and physical methods. It has a unique advantage over them because it is more eco-friendly, cheaper, and can easily be scaled-up for mass level NPs' synthesis (Shankar et al. 2004). Within the last 30 years, the use of plant or whole plant extract has gained more attention due to its simplicity (Park et al. 2011). Research on plant-mediated synthesis is increasing rapidly because it is a one-step process synthesis.

6.2.3.1 Biosynthesis with Bacteria

As far as formation of metallic NPs are concerned, there are various mechanisms that sometimes significantly vary with each other with different bacteria. Pure gold nanoparticles were synthesized with the bacterium *Delftia acidovorans* (Johnston et al. 2013). Synthesis of Ag NPs is done by reduction of Ag ions with the supernatant culture of *Escherichia coli*, and purified by using sucrose density gradient centrifugation. A uniform distribution of NPs with average size of 50 nm was achieved using TEM characterization; further characterization using UV-vis spectra produced maximum absorbance peak at 420 nm (Gurunathan et al. 2009). An extremophilic yeast strain isolated from acid mine drainage yielded 20 nm and 30–100 nm for Ag and Au NPs, respectively. The bacterium *Bacillus licheniformis* acted as reducing agent in the biosynthesis of silver NPs of 50 nm size; similarly *Aspergillus niger* was employed in the extracellular biosynthesis of silver NPs (Gade et al. 2008; Kalimuthu et al. 2008; Mourato et al. 2011). *Geobacter sulfurreducens* reduced Ag(I) as insoluble Ag(+) ions, through c-type cytochrome mechanism. *Morganella morganii*, a Gram-positive bacteria, produced silver NPs of 30 nm, while 20 nm sized gold NP was synthesized from *Rhodopseudomonas capsulata* (He et al. 2007; Law et al. 2008; Abd et al. 2013). Likewise, culture filtrate of *Lactobacillus* sp. (extracellular) is considered to produce NPs of titanium at room temperature, which are generally spherical and ranges from 40 to 60 nm (Prasad et al. 2007). Sintubin et al. (2009) centered the research on the gold NPs' synthesis by lactic acid producing bacteria. Several members of bacteria were tested to know the potentiality related to Au NPs' synthesis. Bacterial genera such as *Lactobacillus* spp., *Pediococcus pentosaceus*, *Enterococcus faecium*, and *Lactococcus garvieae* were only found to produce the Au NPs. There were two steps that were considered to be helpful in the gold NPs' formation. Firstly, gold ions were observed to get accumulated at the cell wall through biosorption followed by reduction of those ions that formed the metallic NPs. Sintubin et al. (2009) also suggested that the cell wall was the capping agent for the NPs, providing them stability and preventing from aggregation. Enhancement of pH of the media exhibited increased reduction rate of NPs. *Bacillus licheniformis* produces intracellular Ag NPs, which was confirmed by the appearance of dark brown color of culture after addition of silver ions that was found to be dispersed quietly in the solution (Kalimuthu et al. 2008). Pugazhenthiran et al. (2009) reported the synthesis of Ag NPs intracellularly when *Bacillus* sp. were allowed to multiply in the medium containing silver trioxonitrate. This reaction was considered slower process as the incubation time was 7 days and may not be recommended for industrial purpose (Pugazhenthiran et al. 2009). Meanwhile, despite the new extraction step, the method proposed by Kalimuthu et al. (2008) reported the process that was industrially significant, and it took only 24 h for *Bacillus licheniformis* to produce silver NPs. *Stenotrophomonas maltophilia* SELTE02, an isolate of *Astragalus bisulcatus* rhizosphere, exhibited the transformation of selenite to elemental selenium (Di Gregorio et al. 2005). In addition, *Pseudomonas aeruginosa* was isolated from rhizosphere of seleniferous soil. *P. aeruginosa* synthesized nanostructured selenium by the process of biotransformation of selenium oxyanions (intracellularly and extracellularly) to spherical amorphous allotropic elemental red selenium (Yadav et al. 2008). Likewise, synthesis of copper nanoparticles (Cu NPs) was a challenging job for the scientists with the available methods because of non-stability of Cu at the nanometer scale, as it oxidizes rapidly to form copper oxide (Baco-carles et al. 2011). Therefore, the stability of Cu is much more important for the synthesis of Cu NPs. *Morganella morganii* was used to synthesize Cu NPs, which happens to be due to intracellular uptake of the Cu ions together with binding of the ions to either a metal ion reductase or similar protein (Ramanathan et al. 2013). Manonmani and Juliet (2011) reported about biosynthesis of Ag NPs by using *Escherichia coli*, and further characterization of the synthesized NPs was established by UV-visible spectroscopy, FTIR (Fourier-transform infrared spectroscopy), and SEM (scanning electron microscope). *Desulfovibrio desulfuricans* (sulfate reducing bacterium) anaerobically bioreduced and biocrystallized palladium (2+) ions to palladium NPs in the presence of formate as an exogenous electron donor on the cell surface within a minute (Yong et al. 2002) (Table 6.1).

6.2.3.2 Biosynthesis Using Fungi

The production of NPs using fungi as reducing agent has attracted significant attention because of the advantages over the use of bacteria due to various important reasons like ease in scaling up, downstream processing, economic feasibility, and the presence of mycelia that provides enhanced surface areas (Mukherjee et al. 2001). Generally, fungi release more proteinaceous compounds than the bacteria that helps in the amplification of NPs' production. *Fusarium oxysporum* has been widely used in the studies pertaining to synthesis of Ag NPs at a size of 5–15 nm, and it was found that NPs were capped with the help of proteinaceous matter released from fungi. In addition, *Fusarium oxysporum* has also been encountered producing various NPs such as cadmium sulfide (CdS), lead sulfide (PbS), zinc sulfide (ZnS), and molybdenum sulfide (MoS), subject to the addition of suitable appropriate salts in the growth medium (Ahmad et al. 2002). Later on, *Aspergillus fumigatus* was used to produce extracellular Ag NPs in the size of 5–25 nm, which was longer as compared to what was reported in the case of *F. oxysporum*. Nevertheless, *A. fumigatus* exhibited very attractive prospect as the organism took only 10 min to reduce silver ions into NPs after exposure (Bhainsa and D'Souza 2006). *Trichoderma reesei* has also been found to synthesize extracellular Ag NPs within

Table 6.1 Microbe-mediated NPs' synthesis

Microorganisms	NPs	Size	Shape	References
Bacillus krulwichiae M2.5 and *Bacillus cellulosilyticus* M4.1	Silver	25.88 ± 10.49 nm	Spherical, irregular	Galvez et al. (2019)
Bacillus licheniformis strain GPI-2	Gold	20–45 nm	Triangular, irregular, hexagonal	Thakur et al. (2018)
Pleurotus ostreatus	Silver	30.5 ± 4.0 nm	Spherical	Al-Bahrani et al. (2017)
Trichoderma viride (MTCC 5661)	Silver	2–5 nm	Spherical	Kumari et al. (2017)
		40–65 nm	Rectangular	
		50–100 nm	Pentagonal and hexagonal	
Penicillium nalgiovense AJ12	Silver	25 ± 2.8 nm	Spherical	Maliszewska et al. (2014)
Macrophomina phaseolina	Silver	5–40 nm	Spherical	Chowdhury et al. (2014)
Humicola sp.	Silver	5–25 nm	Spherical	Syed et al. (2013)
Aspergillus terreus	Silver	1–20 nm	Spherical	Li et al. (2012)
Trichoderma Reesei	Silver	5–50 nm	–	Vahabi et al. (2011)
Penicillium brevicompactum	Silver	58.35 ± 17.88 nm	–	Shaligram et al. (2009)
Penicillium fellutanum	Silver	5–25 nm	Spherical	Kathiresan et al. (2009)
Phoma glomerata	Silver	60–80 nm	Spherical	Birla et al. (2009)
Cladosporium cladosporioides	Silver	35 nm	Polydisperse and spherical	Balaji et al. (2009)
Coriolus versicolor	Silver	25–75, 444–491 nm	Spherical	Sanghi and Verma (2009)
Aspergillus flavus	Silver	8.92 nm	Spherical	Vigneshwaran et al. (2007)
Pleurotus sajor-caju	Silver	30.5 ± 4.0 nm	Spherical	Vigneshwaran et al. (2007)
Pediococcus pentosaceus	Silver	28.2–122 nm	–	Shahverdi et al. (2007)
Phanerochaete chrysosporium	Silver	5–200 nm	Pyramidal	Vigneshwaran et al. (2006)
Aspergillus fumigatus	Silver	5–25 nm	Spherical, triangular	Bhainsa and D'Souza (2006)
Fusarium oxysporum	Silver	20–50 nm	Spherical	Durán et al. (2005)

72 h (Vahabi et al. 2011). Despite slower process of *T. reesei* in terms of NPs' synthesis, it is still used due to various facts. This fungus can be manipulated to produce high levels of enzymes, which may be up to 100 g/L, that helps in enhancing NPs' production.

6.2.3.3 Biosynthesis Using Plant or Plant Extracts

Different parts of plant, such as root, leaves, stem, shoot, flower, bark, and seeds, and their metabolites have been used successfully in biosynthesis of NPs. Production of simple silver NPs from silver trioxonitrate salts by using *Jatropha curcas* extracts was fairly homogenous (10–20 nm), and within 4 h the silver NPs were produced (Bar et al. 2009). Silver and gold NPs were produced using phyllanthin, which was extracted from the plant *Phyllanthus amarus*. This is a novel study as the metallic NPs were synthesized using only a single constituent of a plant extract, which is contrary to other studies described earlier in which whole plants or extracts were used. Phyllanthin concentrations played a vital role in the development of shape and size of the NPs because low concentrations produce triangular and hexagonal gold NPs while higher concentrations of phyllanthin develop greater levels of spherical NPs (Kasthuri et al. 2008). The synthesis of silver NPs with plant extracts was also initiated with leaf extracts from *Acalypha indica*. Silver NPs with 30 nm particle size, which was significantly homogenous, were produced. The antimicrobial properties of the Ag NP produced were tested on *Escherichia coli* and *Vibrio cholerae*; there was maximum growth inhibition at 10 μg/mL concentrations. Extract of *Artemisia nilagirica* (Asteraceae) was used to synthesize silver NPs of 70 nm with silver nitrate and hydrazine hydrate used as metal precursor and reducing agent, respectively (Vijayakumar et al. 2013). Extracts of various plants like *Ocimum tenuiflorum*, *Solanum trilobatum*, *Syzygium cumini*, *Centella asiatica*, and *Citrus sinensis* were also used in the production of silver NPs (Ag NPs) from silver nitrate solution. *Ocimum tenuiflorum* extracts were used in the reduction and stabilization of Ag NPs of the range 25–40 nm (Patil et al. 2012). An extensive research pertaining to the synthesis of NPs from various plants' leaf extracts, such as Pine, Persimmon, Ginkgo, Magnolia, and Platanus, was conducted. Out of all, Magnolia leaf broth exhibited to be the best silver reducer due to short time taken, that is, 11 min, for the reduction of 90% of the Ag+. During the process, the temperature was constantly monitored as it affects the shape, rate, and size of the NPs (Song and Kim 2009). Moreover, Parker et al. (2014) provided a method related to synthesis of NPs with the help of *Arabidopsis thaliana*. In an important example, the medium of *Arabidopsis* was replaced with potassium tetrachloropalladate (K_2PdCl) and allowed for the incubation for 24 h in salt sole solution. TEM analysis revealed palladium of 2–4 nm range, which were later on used in Suzuki-Miyaura coupling reactions. The results showed the higher catalytic activity as compared to commercial palladium NPs (Parker et al. 2014). Niraimathi et al. (2013) reported the microorganism-mediated synthesis of silver NPs from aqueous extract of *Alternanthera sessilis* and reported that leaf extract contains various important organic compounds, viz., carbohydrates, tannins, alkaloids, ascorbic acid, and proteins. These compounds play important role in the synthesis of NPs where they act as capping and reducing agents as well. Tippayawat et al. (2016) provided the report on the synthesis of Ag NPs from *Aloe vera* plant extracts that were characterized by various techniques such as UV-vis, SEM, TEM, and XRD. In brief, Ag NPs were confirmed on the basis of the appearance of a sharp peak at 420 nm in UV-vis region of the spectrum. In addition, they illustrated that reaction time and temperature played very important role in the synthesis of Ag NPs. Results revealed that the synthesized NPs were spherical in shape, which ranged from 70.70 ± 22 to 192.02 ± 53 nm. Aqueous leaf extract of *Trichodesma indicum* served as reducing agent in preparation of Ag NPs at 60 °C for 24 h. The NPs produced were spherical with cubic centered face structure and particle size of 50 nm; they exhibited larvicidal activity against *Mythimna separata* larvae, an army worm of many cereal crops at LC_{50} of 500 ppm (Buhroo et al. 2017). Nano-emulsion with acaricidal properties was synthesized from rosemary essential oil through sonication. The droplet size was 139.9 nm, with high acaricidal activity against adult spotted mites of *Tetranychus urticae*; 54.14 and 52.69% toxicity were established for immature and adult mites, respectively (Abdel-Tawab et al. 2019). The entomotoxic potential of zinc and silica NPs prepared through solvo-thermal method was assayed against *Sitophilus granarius* (Coleoptera: Curculionidae). Silica NPs were highly effective against *S. granarius* resulting in 100% mortality after 2 weeks of exposure; however, moderate activity was exhibited by Zn NPs (Mohammad et al. 2019). Latex from plants is used in the rapid preparation of NPs; latex of *Achras sapota* L. was used as medium in the synthesis of mono- and bimetallic silver and copper NPs (Thakore et al. 2015).

6.3 Application of NPs in Management of Plant-Parasitic Nematodes

Application of NPs in the management of various diseases of plant including plant nematodes is a new and holistic approach (Ladner et al. 2008). There is a potential prospect for the use of nanotechnology in plant disease management in various ways including in nematode infested plants. Direct application of NPs in the soil on seeds or foliage to stop the entry or multiplication of the pathogens is the simplest way. Pathogens may be eliminated in the way similar to pesticide application. However, direct soil application of NPs also affects the non-target beneficial microorganisms that provide many important ecological services. Nanomaterials such as carbon tubes, and cups might be used as carriers of some fancy chemicals such as pheromones, systemic acquired resistance (SAR)-inducing chemicals, polyamine synthesis inhibitors, or even concentrated active ingredients of pesticides for their controlled release especially under flooded conditions (Khan et al. 2014). Therefore, scope and application of NPs in diseases management can be managed through two major points, that is, (a) direct effects of NPs on pathogens, and (b) nanomaterials in nanopesticide formulations. In addition, NPs may also be helpful in the plant disease diagnosis and pesticide residue analysis. It can also be used in killing or paralyzing the nematodes and therefore could be good option in the management of nematode. The chemical and physical properties of NPs play very crucial role in controlling the nematodes. NPs of ultrasmaller size provide better results against plant-parasitic nematodes (PPNs; Gatoo et al. 2014). NPs like gold and silver may participate in the management of nematode disease and, therefore, it may provide alternatives to killer synthetic nematicides or consistent use of biological control agents. Combination of silver NPs with other agricultural practices such as irrigation system especially fertigation or tank-mixture with suitable chemicals, which may replenish the Ag NPs, may increase applicability of silver NPs in the management of PPNs (Thakur and Shirkot 2017). Plant extract-mediated NPs have been reported to be active against pests in agriculture; the small size allows easy penetration into the cell walls of bacteria, fungi, and plant-parasitic nematodes, which leads to cell deaths of the microorganisms (Sondi and Salopek-Sondi 2004). Maggie et al. (2016) reported successful control of *Meloidogyne incognita* using silver NPs. Atef and Nassar (2016) illustrated the effectiveness of *Urtica urens* mediated silver NPs against *Meloidogyne incognita*. Agricultural wastes such as corncobs, guinea corn chaff, and rice husk were used as stabilizers for the preparation of silver NPs. The waste-mediated silver NPs were applied on rice plants infested with *Heterodera sacchari* (rice cyst nematode) in screenhouse trials. Treated rice plants were robust with the 75 and 50 g treatment, which resulted in a considerable reduction in numbers of cysts at harvest. Silver NPs were also evaluated in the control of *Belonolaimus* spp. (sting nematodes) and the *Meloidogyne* spp. infecting Bermuda grass putting, where there was improvement in the quality of the turf grass. In laboratory studies, silver NPs consistently decreased the activity of *Meloidogyne incognita* juveniles. The effectiveness of silver nanoparticles mediated with extracts of *Ficus mucoso* under field studies revealed significant reduction in populations of *M. incognita* infesting groundnuts. Consistent improvement was observed in *M. incognita* infested groundnut plants treated with silver NP solution in a dose-dependent manner (Young-Ki et al. 2013; Cromwell et al. 2014; Fabiyi and Olatunji 2018; Fabiyi et al. 2018). Silver boron nanoparticles (AgB NPs) synthesized with gamma rays and polyvinylpyrrolidone (PVP) polymer as a cost-effective and eco-friendly method were found to possess in vitro and in vivo nematicidal activity. AgB NPs displayed 74.20% mortality of *M. incognita* juveniles after 96 h of exposure. There was a significant decrease in root gall formation and number of egg masses on tomato roots. The AgB NPs were characterized with energy-dispersive X-ray (EDX) mapping analysis, high-resolution topographic mapping (HRTM), DLS, XRD, FTIR, and UV-vis, to determine the morphology, crystallinity, purity, distribution, and the average particle size. Cu NPs prepared through chemical reduction method were established to be toxic to second-stage juveniles of *M. incognita*; mortality was found to be dose dependent; 100% mortality of juveniles was recorded at 0.2 g/L concentration of Cu NPs (Mohamed et al. 2019). The shape of the AgB NPs was spherical and was 2.55 nm in size (El-Batal et al. 2019). The growth and reproduction of *M. incognita* on tomato were truncated by the application of silver NPs synthesized with latex from *Euphorbia tirucalli*, which acted as the reducing agent in the formulation of Ag NPs. The NPs were lethal to second-stage juveniles of *M. incognita* and there was a high percentage egg hatch inhibition. Infestation of tomato roots by nematodes was significantly reduced when Ag NPs were used as root dip on tomato plants before transplanting. Treated tomato plants had healthier growth and reduction in overall gall formation (Kalaiselvi et al. 2019). Graphene oxide (GO) and zinc oxide (ZnO) nanoparticles (NPs) were examined at two concentrations, 0.05 and 0.10 mg/mL, on *M. incognita* infecting carrot. The application of GO and ZnO NPs significantly increased the vegetative growth of carrot plants. Carotenoids and proline contents were higher in carrot plants treated with ZnO NPs. In general, galling and nematode reproduction were significantly reduced by the application of the two NPs (Siddiqui et al. 2019). Silicon carbide (SiC) nanoparticles, though widely used in industrial applications, were evaluated against the proliferation of *M. incognita*. The synthesized SiC NPs with semi-crystalline shape and particle size of 50 nm affected the survival of first-stage larvae of *Caenorhabditis elegans*; the uptake of SiC NPs was feasible in the intestine of dead nematodes displaying a color change to black (Al Banna et al. 2018).

6.4 Conclusions and Future Prospects

Nanoscience is one of the most fascinating and emerging research areas that promises to deliver more pest management options. The use of NPs in nematode management is a good option for the farmers and the research in this direction is still on. It has been seen that synthesis of NPs is not an easy task, especially where facilities are inadequate. Out of all methods of synthesis, biological methods sound to be very good, environmentally safe, cheap, and easy to deliver to the farmers. Therefore, in the upcoming era these NPs may be used in the bio-nano formulation of various pesticides, including nematicides, which may be proven to be a reliable method of green technology. Although some lacunas are yet to be sorted out, this might be condoned in the future. Application of NPs in the intensification of agricultural commodities could also open a new road map to the researchers. Judicious utility of NPs in various plant-based cropping systems may improve our current understanding of sustainable agriculture.

References

Abd El-Rahman AF, Mohammad TGM (2014) Green synthesis of silver nanoparticle using *Eucalyptus globulus* leaf extract and its antibacterial activity. J Appl Sci Res 9(10):6437–6440

Abd FG, Al-Kawaz AJAH, Al-Dahmoshi HOM (2013) Phenotypic and genotypic investigation on silver nanoparticles of *Morganella morganii* recovered from (cauti), Iraq. Int J Med Pharm Sci 3:29–38

Abdel-Tawab HM, Sahar IA, Samia MMM, Badawi AA (2019) Rosemary essential oil nanoemulsion, formulation, characterization and acaricidal activity against the two-spotted spider mite *Tetranychus urticae* Koch (Acari: Tetranychidae). J Plant Protect Res 59 (1):102–112

Ahmad A, Mukherjee P, Mandal D, Senapati S, Khan MI (2002) Enzyme mediated extracellular synthesis of CdS nanoparticles by the fungus, *Fusarium oxysporum*. J Am Chem Soc 124:12,108–12,109

Al Banna L, Salem N, Ghrair AM, Habash SS (2018) Impact of silicon carbide nanoparticles on hatching and survival of soil nematodes *Caenorhabditis elegans* and *Meloidogyne incognita*. Appl Ecol Environ Res 16(3):2661–2662

Al-Bahrani R, Raman J, Lakshmanan H, Hassan AA, Sabaratnam V (2017) Green synthesis of silver nanoparticles using tree oyster mushroom *Pleurotus ostreatus* and its inhibitory activity against pathogenic bacteria. Mater Lett 186:21–25

Atef MK, Nassar B (2016) Effectiveness of silver nano-particles of extracts of *Urtica urens* (Urticaceae) against root-knot nematode *Meloidogyne incognita*. Asian J Nematol 5:14–19

Azizi S, Ahmad M, Mahdavi M, Abdolmohammadi S (2013) Preparation, characterization, and antimicrobial activities of ZnO nanoparticles/cellulose nanocrystal nanocomposites. Bioresources 8:1841–1851

Baco-Carles V, Datas L, Tailhades P (2011) Copper nanoparticles prepared from oxalic precursors. Nanotechnology 2011:1–7; ISRN

Balaji DS, Basavaraja S, Deshpande R, Mahesh DB, Prabhakar BK, Venkataraman A (2009) Extracellular biosynthesis of functionalized silver nanoparticles by strains of *Cladosporium cladosporioides* fungus. Colloids Surf B Biointerfaces 68:88–92

Bar H, Bhui DK, Sahoo GP, Sarkar P, De SP (2009) Green synthesis of silver nanoparticles using latex of Jatropha curcas. Colloids Surf A Physicochem Eng Asp 339:134–139

Bhainsa KC, D'Souza SF (2006) Extracellular biosynthesis of silver nanoparticles using the fungus *Aspergillus fumigatus*. Colloids Surf B Biointerfaces 47:160–164

Birla SS, Tiwari VV, Gade AK, Ingle AP, Yadav AP, Rai MK (2009) Fabrication of silver nanoparticles by Phoma glomerata and its combined effect against *Escherichia coli*, *Pseudomonas aeruginosa* and *Staphylococcus aureus*. Lett Appl Microbiol 43:173–179

Buffat P, Borel JP (1976) Size effect on the melting temperature of gold particles. Phys Rev A 13:2287. https://doi.org/10.1103/PhysRevA.13.2287

Buhroo AA, Nisa G, Asrafuzzaman S, Prasad R, Rasheed R, Bhattacharyya A (2017) Biogenic silver nanoparticles from *Trichodesma indicum* aqueous leaf extract against *Mythimna separata* and evaluation of its larvicidal efficacy. J Plant Protect Res 57(2):194–200

Chowdhury S, Basu A, Kundu S (2014) Green synthesis of protein capped silver nanoparticles from phytopathogenic fungus *Macrophomina phaseolina* (Tassi) Goid with antimicrobial properties against multidrug-resistant bacteria. Nano Res Lett 9:365

Cromwell WA, Yang J, Starr JL, Young-Ki J (2014) Nematicidal effects of silver nano particles on root-knot nematode in Bermudagrass. J Nematol 46(3):261–266

Deplanche K, Caldelari I, Mikheenko IP, Sargent F, Macaskie LE (2010) Involvement of hydrogenases in the formation of highly catalytic Pd(0) nanoparticles by bioreduction of Pd(II) using *Escherichia coli* mutant strains. Microbiology 156:2630–2640

Dhandapani P, Maruthamuthu S, Rajagopal G (2012) Bio-mediated synthesis of TiO2 nanoparticles and its photocatalytic effect on aquatic biofilm. J Photochem Photobiol B Biol 110:43–49

Dhuper S, Panda D, Nayak PL (2012) Green synthesis and characterization of zero valent iron nanoparticles from the leaf extract of *Mangifera indica*. Nano Trends J Nanotechnol Appl 13:16–22

Di Gregorio S, Lampis S, Vallini G (2005) Selenite precipitation by a rhizospheric strain of *Stenotrophomonas* sp. isolated from the root system of *Astragalus bisulcatus*: a biotechnological perspective. Environ Int 31:233–241

Dubchak S, Ogar A, Mietelski JW, Turnau K (2010) Influence of silver and titanium nanoparticles on arbuscular mycorrhiza colonization and accumulation of radiocaesium in *Helianthus annuus*. Span J Agric Res 8:103–108

Durán N, Priscyla D, Marcato PD, Alves O, De Souza G, Esposito E (2005) Mechanistic aspects of biosynthesis of silver nanoparticles by several *Fusarium oxysporum* strains. J Nanobiotechnol 3:1–7

Egorova EM, Revina AA (2000) Synthesis of metallic nanoparticles in reverse micelles in the presence of quercetin. Colloids Surf A 168:87–96

El-Batal AI, Attia MS, Mohamed M, Nofel MM, El-Sayyad GS (2019) Potential nematicidal properties of silver boron nanoparticles: synthesis, characterization, in vitro and in vivo root knot nematode (*Meloidogyne incognita*) treatments. J Clust Sci 30:687–705

Eppler AS, Rupprechter G, Anderson EA, Somorjai GA (2000) Thermal and chemical stability and adhesion strength of Pt nanoparticle arrays supported on silica studied by transmission electron microscopy and atomic force microscopy. J Phys Chem B 104:7286–7292

Fabiyi OA, Olatunji GA (2018) Application of green synthesis in nano particles preparation: *Ficus mucoso* extracts in the management of *Meloidogyne incognita* parasitizing groundnut *Arachis hypogea*. Indian J Nematol 48(1):13–17

Fabiyi OA, Olatunji GA, Saadu AO (2018) Suppression of *Heterodera sacchari* in rice with agricultural waste-silver nano particles. J Solid Waste Tech Manag 44(2):87–91

Feldheim DL, Foss CA (2002) Metal nanoparticles: synthesis, characterization, and applications. CRC, Boca Raton

Gade AK, Bonde P, Ingle AP, Marcato PD, Duran N, Rai MK (2008) Exploitation of *Aspergillus niger* for synthesis of silver nanoparticles. J Biobased Mater Bioenergy 2:243–247

Galvez AM, Ramos KM, Alexis JT, Baculi R (2019) Bacterial exopolysaccharide-mediated synthesis of silver nanoparticles and their application on bacterial biofilms. J Microbiol Biotechnol Food Sci 8(4):970

Gatoo MA, Naseem S, Arfat MY, Dar AM, Qasim K (2014) Physicochemical properties of nanomaterials: implication in associated toxic manifestations. BioMed Res Int 2014:1–8

Govindaraju K, Tamilselvan S, Kiruthiga V, Singaravelu G (2010) Biogenic silver nanoparticles by *Solanum torvum* and their promising antimicrobial activity. J Biopest 3:349–399

Gunalan S, Sivaraj R, Venckatesh R (2011) Green synthesis of zinc oxide nanoparticles by *Aloe barbadensis* miller leaf extract: structure and optical properties. Mater Res Bull 46:2560–2566

Gurunathan S, Kalishwaralal K, Vaidyanathan R, Venkataraman D, Pandian SRK (2009) Biosynthesis, purification and characterization of silver nanoparticles using *Escherichia coli*. Colloids Surf B Biointerfaces 74:328–335

He SY, Guo ZR, Zhang Y, Zhang S, Wang J, Gu N (2007) Biosynthesis of gold nanoparticles using the bacteria *Rhodopseudomonas capsulata*. Mater Lett 61:3984–3987

Hurst SJ, Lytton-Jean AKR, Mirkin CA (2006) Maximizing DNA loading on a range of gold nanoparticle size. Ann Chem 78:8313–8318

Jiang J, Oberdörster G, Biswas P (2009) Characterization of size, surface charge, and agglomeration state of nanoparticle dispersions for toxicological studies. J Nanopart Res 11:77–89

Johnston CW, Wyatt MA, Li X, Ibrahim A, Shuster J (2013) Gold biomineralization by a metallophore from a gold-associated microbe. Nat Chem Biol 9:241–243

Kalaiselvi D, Mohankumar A, Shanmugam G, Nivitha S, Sundararaj P (2019) Green synthesis of silver nanoparticles using latex extract of *Euphorbiatirucalli*: a novelapproach for the management of root knot nematode, *Meloidogyne incognita*. Crop Prot 117:108–114

Kalimuthu K, Suresh BR, Venkataraman D, Bilal M, Gurunathan S (2008) Biosynthesis of silver nanocrystals by *Bacillus licheniformis*. Colloids Surf B Biointerfaces 65:150–153

Kasthuri J, Kathiravan K, Rajendiran N (2008) Phyllanthin-assisted biosynthesis of silver and gold nanoparticles: a novel biological approach. J Nanopart Res 11:1075–1085

Kathiresan K, Manivannan S, Nabeal MA, Dhivya B (2009) Studies on silver nanoparticles synthesized by a marine fungus, *Pencillium fellutanum* isolated from coastal mangrove sediment. Colloids Surface B Biointerfaces 71:133–137

Khan MR, Rizvi TF (2014) Nano technology scope and application in plant disease management. Plant Pathol J 13:214–231

Khan MR, Altaf S, Mohidin FA, Khan U, Anwer A (2009) Biological control of plant nematodes with phosphate solubilizing microorganisms. In: Khan MS, Zaidi A (eds) Phosphate solubilizing microbes for crop improvement. Nova Science Publisher Inc., New York, pp 395–342

Khan AK, Rashid R, Martaza G, Zahra A (2014) Gold nanoparticles: synthesis and applications in drug delivery. Trop J Pharm Res 13(7):1169–1177. https://doi.org/10.4314/tjpr.1317.23

Khandel P, Yadaw RK, Soni DK (2018) Biogenesis of metal nanoparticles and their pharmacological applications: present status and application prospects. J Nanostruct Chem 8:217. https://doi.org/10.1007/s40097-018-0267-4

Kumari M, Pandey S, Giri VP, Bhattacharya A, Shukla R, Mishra A, Nautiyal CS (2017) Tailoring shape and size of biogenic silver nanoparticles to enhance antimicrobial efficacy against MDR bacteria. Microb Pathol 105:346–355

Ladner DC, Tchounwou PB, Lawrence GW (2008) Evaluation of the effect of ecologic on root knot nematode, *Meloidogyne incognita*, and tomato plant, *Lycopersicon esculenum*. Int J Environ Res Public Health 5:104–110

Law N, Ansari S, Livens FR, Renshaw JC, Lloyd JR (2008) Formation of nanoscale elemental silver particles via enzymatic reduction by *Geobacter sulfurreducens*. Appl Environ Microbiol 74:7090–7093

Li G, He D, Qian Y, Guan B, Gao S, Cui Y, Yokoyama K, Wang L (2012) Fungus-mediated green synthesis of silver nanoparticles using *Aspergillus terreus*. Int J Mol Sci 13:466–476

Maggie EMH, Hanaa S, Zawam SE, El-Nahas M, Abeer FD (2016) Comparison study between silver nanoparticles and two Nematicides against *Meloidogyne incognita* on tomato seedlings. Plant Pathol J 15:144–151

Maliszewska I, Juraszek A, Bielska K (2014) Green synthesis and characterization of silver nanoparticles using ascomycota fungi *Penicillium nalgiovense* AJ12. J Clust Sci 25:989–1004

Manonmani V and Juliet V (2011) Biosynthesis of Ag nanoparticles for the detection of pathogenic bacteria. In: Proceedings of the 2nd international conference on innovation, management and service, 16–18 Sep, Singapore, pp 307–311

Mohamed EA, Elsharabasy SF, Abdulsamad D (2019) Evaluation of *in vitro* nematicidal efficiency of copper nanoparticles against root-knot nematode *Meloidogyne incognita*. South Asian J Parasitol 2(1):1–6

Mohammad R, Mohammad AS, Mehdi Z, Khalil B, Mohammad G, Mohammad RA (2019) Synthesis and entomotoxicity assay of zinc and silica nanoparticles against *Sitophilus granarius* (Coleoptera: Curculionidae). J Plant Protect Res 59(1):26–31

Mourato A, Gadanho M, Lino AR, Tenreiro R (2011) Biosynthesis of crystalline silver and gold nanoparticles by extremophilic yeasts. Bioinorg Chem Appl 2011:546074. https://doi.org/10.1155/2011/546074

Mukherjee P, Ahmad A, Mandal D, Senapati S, Sainkar S (2001) Fungus-mediated synthesis of silver nanoparticles and their immobilization in the mycelial matrix—a novel biological approach to nanoparticle synthesis. Nano Lett 1:515–519

Mukunthan KS, Balaji S (2012) Cashew apple juice (Anacardium occidentale L.) speeds up the synthesis of silver nanoparticles. Int J Nanotechnol 4:71–79

Niraimathi KL, Sudha V, Lavanya R, Brindha P (2013) Biosynthesis of silver nanoparticles using *Alternanthera sessilis* (Linn.) extract and their antimicrobial, antioxidant activities. Colloids Surf B Biointerfaces 102:288–291

Pal S, Tak YK, Song JM (2007) Does the antibacterial activity of silver nanoparticles depend on the shape of the nanoparticle? A study of the gram-negative bacterium *Escherichia coli*. Appl Environ Microbiol 73:1712–1720

Park Y, Hong YN, Weyers A, Kim YS, Linhardt RJ (2011) Polysaccharides and phytochemicals: a natural reservoir for the green synthesis of gold and silver nanoparticles. IET Nanobiotechnol 5:69–78

Parker HL, Rylott EL, Hunt AJ, Dodson JR, Taylor AF (2014) Supported palladium nanoparticles synthesized by living plants as a catalyst for Suzuki-Miyaura reactions. PLoS One 9:e87192

Patel P, Agarwal P, Kanawaria S, Kachhwaha S, Kothari SL (2015) Plant-based synthesis of silver nanoparticles and their characterization. In: Siddiqui M, Al-Whaibi M, Mohammad F (eds) Nanotechnology and plant sciences. Springer, Cham

Patil RS, Kokate MR, Kolekar SS (2012) Bioinspired synthesis of highly stabilized silver nanoparticles using *Ocimum tenuiflorum* leaf extract and their antibacterial activity. Spectrochim Acta A Mol Biomol Spectros 91:234–238

Prasad K, Jha AK, Kulkarni AR (2007) *Lactobacillus* assisted synthesis of titanium nanoparticles. Nanoscale Res Lett 2:248–250

Pugazhenthiran N, Anandan S, Kathiravan G, Udaya Prakash NK, Crawford S (2009) Microbial synthesis of silver nanoparticles by *Bacillus* sp. J Nanopart Res 11:1811–1815

Ramanathan R, Field MR, O'Mullane AP, Smooker PM, Bhargava SK (2013) Aqueous phase synthesis of copper nanoparticles: a link between heavy metal resistance and nanoparticle synthesis ability in bacterial systems. Nanoscale 5:2300–2306

Sanghi R, Verma P (2009) Biomimetic synthesis and characterisation of protein capped silver nanoparticles. Bioresour Technol 100:501–504

Sankar R, Rizwana K, Shivashangari KS, Ravikumar V (2014) Ultra-rapid photocatalytic activity of Azadirachta indica engineered colloidal titanium dioxide nanoparticles. Appl Nanosci 5:731–736

Shahverdi AR, Minaeia S, Shahverdi HR, Jamalifar H, Nohi AA (2007) Rapid synthesis of silver nanoparticles using culture supernatants of Enterobacteria: a novel biological approach. Process Biochem 42:919–923

Shaligram NS, Bule A, Bhambure R, Singhal RS, Singh SK, Szakacs G, Pandey A (2009) Biosynthesis of silver nanoparticles using aqueous extract from the compactin producing fungal strain. Process Biochem 44:939–943

Shankar SS, Rai A, Ahmad A, Sastry M (2004) Rapid synthesis of au, Ag, and bimetallic au core–Ag shell nanoparticles using Neem (*Azadirachta indica*) leaf broth. J Colloid Interface Sci 275:496–502

Siddiqui ZA, Parveen A, Lukman Ahmad L, Hashem A (2019) Effects of graphene oxide and zinc oxide nanoparticles on growth, chlorophyll, carotenoids, proline contents and diseases of carrot. Sci Hortic 249:374–382

Sintubin L, De Windt W, Dick J, Mast J, van der Ha D (2009) Lactic acid bacteria as reducing and capping agent for the fast and efficient production of silver nanoparticles. Appl Microbiol Biotechnol 84:741–749

Sofi W, Gowri M, Shruthilaya M, Rayala S, Venkatraman G (2012) Silver nanoparticles as an antibacterial agent for endodontic infections. BMC Infect Dis 12:P60. https://doi.org/10.1186/1471-2334-12-S1-P60

Sondi I, Salopek-Sondi B (2004) Silver nanoparticles as antimicrobial agent: a case study on *E. coli* as a model for gram-negative bacteria. J Colloid Interface Sci 275:177–182. https://doi.org/10.1016/j.jcis.2004.02.012

Song JY, Kim BS (2009) Rapid biological synthesis of silver nanoparticles using plant leaf extracts. Bioprocess Biosyst Eng 32:79–84

Strasser P, Koh S, Anniyev T, Greeley J, More K, Yu C (2010) Lattice-strain control of the activity in dealloyed core–shell fuel cell catalysts. Nat Chem 2:454–460

Sun S, Murray CB, Weller D, Folks L, Moser A (2000) Monodisperse FePt nanoparticles and ferromagnetic FePt nanocrystal super lattices. Science 287:1989–1992

Syed A, Saraswati S, Kundu GC, Ahmad A (2013) Biological synthesis of silver nanoparticles using the fungus Humicola sp. and evaluation of their cytoxicity using normal and cancer cell lines. Spectrochim Acta A Mol Biomol Spectros 114:144–147

Thakore SI, Nagar PS, Jadeja RN, Thounaojam M, Devkar RV, Rathore PS (2015) Sapota fruit latex mediated synthesis of Ag, Cu mono and bimetallic nanoparticles and their in vitro toxicity studies. Arab J Chem 12:694–700. https://doi.org/10.1016/j.arabjc12.042

Thakur RK, Shirkot P (2017) Potential of biogold nanoparticles to control plant pathogenic nematodes. J Bioanal Biomed 9:220–222. https://doi.org/10.4172/1948-593X.1000182

Thakur RK, Shirkot P, Dhirta B (2018) Studies on effect of gold nanoparticles on Meloidogyne incognita and tomato plants growth and development. BioRxiv:428144. https://doi.org/10.1101/428144

Tippayawat P, Phromviyo N, Boueroy P, Chompoosor A (2016) Green synthesis of silver nanoparticles in aloe vera plant extract prepared by a hydrothermal method and their synergistic antibacterial activity. Peer J 4:e2589

Vahabi K, Mansoori GA, Karimi S (2011) Biosynthesis of silver nanoparticles by fungus *Trichoderma reesei* (a route for large-scale production of AgNPs). Inter Sci J 1:65–79

Vigneshwaran N, Kathe AA, Varadarajan PV, Nachane RP, Balasubramanya RH (2006) Biomimetics of silver nanoparticles by white rot fungus, *Phaenerochaete chrysosporium*. Colloids Surf B Interfaces 53:55–59

Vigneshwaran N, Ashtaputre NM, Varadarajan PV, Nachane RP, Paralikar KM, Balasubramanya RH (2007) Biological synthesis of silver nanoparticles using the fungus *Aspergillus flavus*. Mater Lett 61:1413–1418

Vijayakumar M, Priya K, Nancy FT, Noorlidah A, Ahmed ABA (2013) Biosynthesis, characterisation and anti-bacterial effect of plant-mediated silver nanoparticles using *Artemisia nilagirica*. Ind Crop Prod 41:235–240

Wang S, Lawson R, Ray PC, Yu H (2011) Toxic effects of gold nanoparticles on *Salmonella typhimurium* bacteria. Toxicol Ind Health 27:547–554

Yadav V, Sharma N, Prakash R, Raina KK, Bharadwaj LM, Tejo N (2008) Generation of selenium containing nano-structures by soil bacterium *Psudomonas aeruginosa*. Biotechnology 7:299–304. https://doi.org/10.3923/biotech.2008.299.304

Yong P, Rowson AN, Farr JPG, Harris IR, Mcaskie LE (2002) Bioaccumulation of palladium by *Desulfovibriode sulfuricans*. J Chem Technol Biotechnol 55:593–601

Young-Ki J, Starr JL, Deng Y (2013) Use of silver nano particles for nematode control on the Bermuda grass putting green turf grass. Environ Res 12(2):22–24

Chapter 7
Integrated Management of Phytopathogenic Nematodes Infesting Mushroom

Nishi Keshari and K. V. V. S. K. Kranti

Abstract Mushroom cultivation, being one of the fastest-growing, young, progressive industries worldwide, is attracting small-scale industrialists, rich farmers along with poor and marginal farmers and rural women. It does not need a defined land for its production. So, anyone can grow it. It is very nutritious, contains protein, minerals, fibres and have sufficient moisture. It also contains some medicinal properties. The common edible mushrooms that are grown commercially are white button mushroom, oyster mushroom, milky mushroom and paddy straw mushroom. Mushroom cultivation is very much prone to biotic and abiotic stresses. Amongst the biotic stresses, nematodes are the major pests in mushroom production. Myceliophagous nematodes (like *Aphelenchoides* spp., *Aphelenchus* spp. and *Ditylenchus myceliophagus*) and saprophagous nematodes both have great role in reducing the yield of mushroom sporophores. Myceliophagous nematodes feed the mycelium of the mushroom during spawn run. Hence, the fruiting bodies cannot develop properly. In severe infestation, the loss may go up to 100%. The saprophagous nematodes indirectly harm the production of mushroom by secreting some enzymes and toxins. The mushrooms are harvested and consumed fresh. So, application of nematicides is not advisable because of residual problems. Thus, management through integrated approach is the best method for getting better yield and less nematode population. Maintenance of hygienic conditions during the crop duration, i.e. from compost preparation to harvesting, is the most fruitful solution to avoid the nematode infestation in the crop. Application of heat for the sterilization of compost and implements also helps in reducing the contamination. Botanicals like neem seed kernel water extract (NSKWE), leaf incorporation of neem, pongamia, karanj, castor and eucalyptus in compost are useful in decreasing the nematode population. The predatory nematodes like *Seinura* spp. and *Fictor* spp., which are commonly found in compost, are very potential biocontrol agents and can be successfully used for the

N. Keshari (✉)
Department of Nematology, Faculty of Agriculture, RPCAU, Samastipur, Bihar, India

K. V. V. S. K. Kranti
AICRP (Nematodes on Agriculture), LBS Building, ICAR-Indian Agricultural Research Institute, New Delhi, India

R. A. Ansari et al. (eds.), *Management of Phytonematodes: Recent Advances and Future Challenges*, https://doi.org/10.1007/978-981-15-4087-5_7

management of myceliophagous nematodes and saprophagous nematodes. The major source of contamination of myceliophagous and saprophagous nematodes are through the unpasteurized or partially pasteurized compost. Since most of the mushroom growers are poor farmers who cannot afford a pasteurization chamber, they suffer the loss in yield and do not get a good return for their crop. Thus, a common pasteurization chamber facility should be provided by the Government and non-governmental organizations (NGOs). The researches on new botanicals and predatory nematodes and predatory fungi may fetch high return to the mushroom growers by managing the nematode population to a low level.

Keywords Mushroom · *Aphelenchoides* spp. · *Aphelenchus* spp. · *Ditylenchus myceliophagus* · Management

7.1 Introduction

Mushroom production is one of the fastest-growing small-scale industries, especially in rural areas, because it does not need any defined land. It is a young progressive industry all over the world today. Its flavour, texture, nutritional value, high productivity and other such qualities make it one of the most nutritive food. Mushroom production is a new potential tool for increasing the protein-rich food production to meet the future challenges in global agriculture and to ensure food security for the increasing population. Over 200 species of mushroom have long been cultivated for functional foods around the world (Kalac 2013) but only about 35 species have been commercially cultivated (Aida et al. 2009; Xu et al. 2011). It is full of vitamins, proteins, fibres, moisture and minerals and, in addition, it is having some medicinal values. It also contains some bioactive constituents, such as phenolic compounds, terpenes, steroids, polysaccharides (Kim et al. 2015; Royse 2014; Shang et al. 2015) and vitamins B, C and D (Panjikkaran and Mathew 2013). Mushroom contains 20–35% protein (dry wt.), are low in lipids and contain all the nine essential amino acids (Kalac 2009). They have many bioactive constituents (Sheu et al. 2007; Mariga et al. 2014). The biomolecules like phenols, terpenes, steroids and polysaccharides have various biological activities (Shang et al. 2015). They have antigenotoxic (Wang et al. 2005), antioxidative (Roupas et al. 2012), antiproliferative (Zhou et al. (2013), tumorigenic (Kim et al. 2015), antihypertensive (Vaz et al. 2011), hypocholesterolaemic (Han et al. 2011) and stress-reducing properties are good for diabetic patients (Akata et al. 2012) and can act as an appetite suppressant (Kim et al. 2011). Mushroom production can be an appropriate technology as it is capable of producing the highest quality of protein per unit area and time from agro-wastes that are available to the tune of more than 355 million tonnes per annum in India. If even 1% of these agro-wastes is used to produce mushrooms, India will become a major mushroom-producing country in the world, as India is blessed with varied agro-climates making it suitable for the cultivation of different types of mushrooms. Besides this, it is an employment- and income-generating venture. The commercial cultivation of mushroom in India is not very new. Its cultivation is growing very fast in Punjab, Haryana, Himachal Pradesh, Jammu and Kashmir and hilly areas. Its cultivation is also spreading in other states like Rajasthan, Bihar and Uttar Pradesh. The common edible mushrooms are of four types, which are grown for the commercial purpose:

1. White button mushroom, *Agaricus bisporus*
2. Oyster mushroom, *Pleurotus* spp.
3. Milky mushroom, *Calocybe indica*
4. Paddy straw mushroom, *Volvariella volvacea*

About 40% of the world production is white button mushroom, *Agaricus bisporus*, followed by *Pleurotus* spp. (25%), paddy straw mushroom, *Volvariella volvacea* (16%) and forest mushroom or shiitake mushroom, *Lentinula edodus* (19%) (Banik 2010).

7.1.1 White Button Mushroom, **Agaricus bisporus**

Although there are many species that are edible and people cultivate for their earnings but the button mushroom, *Agaricus bisporus* (Lange.) Singer, contributes more than 80% of the total mushroom production (NRCM 2007). This mushroom requires an indoor temperature ranging between 15 and 25 °C (22–25 °C for vegetative growth and 14–18 °C for fruiting). It is grown on a specially prepared substrate called compost, which is prepared by mixing different raw materials in specific proportions either by long or short method of composting. In long method of composting (LMC), the compost is prepared on the cemented floor, which requires almost 1 month of time and also more water for preparation. While in short method of composting (SMC), it requires only 14–15 days. In LMC, seven turnings of the compost are required while in SMC, only three turnings are required. In SMC, the compost is prepared in sterilization chamber or tunnels at a temperature ranging 56–58 °C after three turnings. So, this makes the compost completely sterilized, which is very helpful in reducing the initial nematode population. When the compost becomes ready for filling the trays or bags, spawning is done in which mushroom spawns are mixed with the compost at the rate of 7.5 g/kg of compost and after 21 days, casing soil is spread on the spawn run to give the strength to fruiting body.

7.1.2 Oyster Mushroom, **Pleurotus** *spp.*

It is also called as 'Dhingri' locally in India. The sporophore of this mushroom looks like a shell or spatula shaped. It is a cellulose loving fungus and grows naturally in the temperate and tropical regions. Its colour may be white, cream, off white, pink, yellow or light brown. It is also called white rot fungi because its mycelium looks like white rot. It can be cultivated on non-composted lignocellulosic substances (Curvetto et al. 2002). *Pleurotus* has many species, but *Pleurotus sajor caju* is the most cultivated species. It can be cultivated in bags, wooden boxes, trays etc. Paddy straw or wheat straw is used as a substrate after soaking overnight and proper sterilization is done by boiling in water or by using chemical like formalin (5% concentration). This straw is used for filling the bags or trays with spawn. After spawn run, the bags are removed and are irrigated with sterilized water daily. After a few days, the fruiting bodies develop and harvesting is done when these attain the proper size. It requires very little growth time compared to other mushrooms (Bellettini et al. 2017). Generally, it has been seen that this mushroom is resistant to the nematodes. One of the reasons is that the cultivation of this mushroom does not need compost preparation. It is more popular because it is easy to grow and needs low inputs. Also, it is nearly resistant to pests and diseases compared to button mushroom.

7.1.3 Milky Mushroom, Calocybe indica

It is newer edible mushroom. Its size is robust; colour is milky white, so is called milky mushroom. It is very delicious to taste and has a long shelf life. This mushroom variety was first identified in the eastern indian state of West Bengal and it can be cultivated on a wide variety of substrates, at a high temperature range (30–38 °C). However, no commercial cultivation was made until 1998. This edible mushroom has a long shelf life (5–7 days) compared to other mushrooms (Subbiah and Balan 2015). The nutritional value is at par with the white button mushroom but the production cost is less than the white button mushroom. It can be cultivated at a temperature range of 25–35 °C. Chopped straw is used for the cultivation, which are soaked in water for 8–16 h and then sterilized in special chamber. Spawning is done by mixing the spawns in layering or by broadcasting method. After 15–20 days of spawn run, casing is done. In 3–5 days, fruiting body comes out, which matures in a week. It has an excellent shelf life and an ability to grow at 30 °C and above (Bhardwaj 2004). It can be cultivated on a wide range of cellulosic substances like paddy straw, maize and sorghum stalks, palm rosa and vetiver grass, sugarcane bagasse, soybean hay and groundnut haulms. As it grows in hot and humid climate (temperature 30–35 °C, relative humidity 90%), the mushroom is highly suitable for cultivation in plains of India. A mycophagous nematode, *Paraphelenchus* sp., was found associated with this mushroom (Vats et al. 2006).

7.1.4 *Paddy Straw Mushroom,* Volvariella volvacea

It is also called straw mushroom and it is ranked third in consumption worldwide. They are prepared on rice straw beds and are harvested when immature. It is commonly grown in tropics and subtropics. It is a very fast-growing and easily cultivable mushroom. The best method of cultivation is the bed method and the best substrate is rice straw for paddy straw mushroom (Biswas 2014). It can be grown at a high temperature and so are available round the year. At first, paddy straw is soaked in water for 18 h and 70–75% moisture should be maintained during spawning. These straw bundles are kept in layers and spawning is done. Total crop cycle completes within 4–5 weeks. It can be grown at a high temperature range of 35–40 °C and at high relative humidity (85% and above).

7.2 Nematodes Infesting the Mushroom

Although the cultivation of mushroom is easy, it needs utmost care regarding hygiene and temperature and moisture conditions to which they are very sensitive. The mushroom cultivation is done mainly by the landless farmers, marginal farmers and housewives, where the hygienic cultivation of mushroom is still lacking in most of the places, which is a major cause for the limitation in the productivity of mushroom. This unhygienic condition invites many pests, like insects, nematodes, fungi, bacteria and other such biotic and abiotic stresses, that reduce the quality and quantity of the mushroom (Bellettini et al. 2015, 2016). Amongst the biotic stresses, nematodes present in the mushroom compost cause a great loss in the production. They are the most dreaded ones and can cause a complete crop failure. These nematodes are of many types. Some are myceliophagous, some are saprophagous and some are predatory. The myceliophagous and saprophagous nematodes are important economically because they are highly pathogenic to the very commonly cultivated mushroom, button mushroom, *Agaricus* spp. The commonly found myceliophagous nematode pests in mushroom houses are *Aphelenchus avenae*, *Aphelenchoides* spp. and *Ditylenchus myceliophagus* (Hesling 1979; Agdaci et al. 1990; Sharma 1995; Khanna and Chandran 2002; Bajaj and Kanwar 2011). These nematodes easily swim on the surface of water in compost and casing. The mycelium of mushroom is the favourable source of food for many species of nematodes. These nematodes may sometimes cause complete failure of the crop. The mycophagous nematodes are those noxious pests that once introduced in the mushroom compost beds, are very difficult to eliminate.

7.2.1 Myceliophagous Nematodes

The fungus-feeding nematodes present in the compost belong to the order Aphelenchida and Tylenchida. There are four genera belonging to the order Aphelenchida: *Aphelenchus* Bastian, *Aphelenchoides* Fischer, *Paraphelenchus* (Micol.) Micol. and *Seinura*. On the other hand, only one species, *Ditylenchus myceliophagus*, in Tylenchida group, was found. These nematodes are common in mushroom beds (Gitanjali and Nandal 2001a; Bajaj and Kanwar 2011). There are many species of *Aphelenchoides* and a few species of *Aphelenchus* and *Paraphelenchus* found in the mushroom beds. Devi (1999) reported that all the 11 strains of *A. bisporus*, viz. 31, 39, 44, 53, 56, U3, 100, 101, 102, 649 and 1927, tested under laboratory and field conditions were susceptible to *Aphelenchoides composticola*. In addition to this, some other myceliophagous nematodes like *Hexatylus* Goodey and *Deladinus* Thorne of Tylenchida order have also been encountered. About 21 species of the myceliophagous nematodes are reported to be associated with mushroom cultivation (Seth and Sharma 1986; Bajaj and Walia 1999; Nagesh and Reddy 2000; Kumar et al. 2007). All these nematodes have very short life cycle (8–10 days), high fecundity, wide fungal host range and ability to survive under adverse environmental conditions. They feed and kill a large number of fungal cells ultimately causing complete mycelia destruction leading to non-production of sporophores. The sporophores of *Calocybe indica* were found to harbour more nematodes as compared to that of *Agaricus bisporus* (Khanna et al. 2006). When these pests infect the mushroom crop in the very initial phase, complete crop failure may occur (Seth 1984; Khanna and Sharma 2001; Deepthi et al. 2004). In *Pleurotus ostreatus*, nematodes form knots on the gills. These nematodes live in the vegetative hyphae of this fungus (Tsuda et al. 1996). The fungus gnat, an insect, *Rhymosia domestica* is the vector of these nematodes (Thorn and Tsuneda 1993).

The common characteristics of the fungus-feeding nematodes of mushrooms are:

Wide Host Range

These nematodes feed on a large variety of fungi although the feeding rate differs with the different fungal genera, for example, *Agaricus bisporus*, *Gliocladium deliquescens*, *Penicillium* spp., *Fusarium solani*, *Mucor* sp., *Trichoderma viridae*, *Chaetomium globosum,* etc.

Short Life Cycle

These nematodes complete their life cycle in a very short duration under a wide range of temperature. Generally, their life cycle is of one week at an optimum temperature of 25 °C. For example, *Aphelenchoides agarici* and *A. composticola* have life cycle of 8 days, *Aphelenchoides sacchari* of 12 days and *Aphelenchus avenae* of 8 days.

High Fecundity

These nematodes have very high fecundity. On an average, one female lays up to three eggs per hour, for example, *Aphelenchoides* spp. and *Aphelenchus* spp. Due to this characteristic, the population of fungus-feeding nematodes increases manifold in mushroom beds (Table 7.1).

Table 7.1 Myceliophagous nematodes associated with the mushroom cultivation

S. No.	Nematodes	Mushroom	Reference
1.	*Aphelenchoides agarici*	*Agaricus bisporus*	Seth and Sharma (1986), Bajaj and Walia (1999), Seth (1984)
2.	*Aphelenchoides asterocaudatus*	*Agaricus bisporus*	Bajaj and Walia (1999)
3.	*Aphelenchoides bicaudatus*	*Agaricus bisporus*	Anonymous (2003)
4.	*Aphelenchoides brushimucronatus*	*Agaricus bisporus*	Bajaj and Walia (1999)
5.	*Aphelenchoides composticola*	*Agaricus bisporus*	Bajaj and Walia (1999), Khanna and Sharma (1988a, b), Gitanjali and Nandal (2001a, b)
6.	*Aphelenchoides delhiensis*	*Agaricus bisporus*	Bajaj and Walia (1999)
7.	*Aphelenchoides indicus*	*Agaricus bisporus*	Bajaj and Walia (1999)
8.	*Aphelenchoides minoris*	*Agaricus bisporus*	Bajaj and Walia (1999), Seth and Sharma (1986)
9.	*Aphelenchoides myceliophagus*	*Agaricus bisporus*	Anonymous (2003), Seth and Sharma (1986)
10.	*Aphelenchoides neocomposticola*	*Agaricus bisporus*	Seth and Sharma (1986)
11.	*Aphelenchoides sacchari*	*Agaricus bisporus*	Sharma et al. (1981), Thapa et al. (1981)
12.	*Aphelenchoides spinohamatus*	*Agaricus bisporus*	Bajaj and Walia (1999)
13.	*Aphelenchoides swarupi*	*Agaricus bisporus*	Anonymous (2003), Seth (1984), Seth and Sharma (1986)
14.	*Aphelenchus avenae*	*Agaricus bisporus*	Anonymous (2003)
15.	*Aphelenchus radicicolus*	*Agaricus bisporus*	Anonymous (2003)
16.	*Paraphelenchus* spp.	*Agaricus bisporus*, *Calocybe indica*	Vats et al. (2006)
17.	*Hexatylus viviparous*	*Agaricus bisporus*	Anonymous (2003)
18.	*Ditylenchus myceliophagus*	*Agaricus bisporus* and *Pleurotus sajor caju*	Sharma et al. (1981), Thapa et al. (1981), Anonymous (2003)

7.2.1.1 *Aphelenchoides* Species

These nematodes have a wide host range and feed upon a variety of fungi (Khanna and Sharma 1989). They are very active and voracious feeders. Their feeding rate is 6.22 per second on healthy, growing hyphae. They have stylet that helps in puncturing the hyphal cells. They puncture the hyphal cell by one to five rapid stylet thrusts. Thereafter, the median bulb pulsates at the rate of 1.86–2.44 pulsation per second. This results in the ingestion of cytoplasm of hyphal cells into nematode body. Due to rapid and continuous ingestion, the hyphal cells shrink. The following are some important species of *Aphelenchoides*.

Aphelenchoides swarupi

It was first found in a mushroom farm at Ambala (Haryana). The symptom was brown wet pinheads but no fruiting bodies. The population of this nematode was found as 10,000 individuals per 100 g of compost (Seth 1984). The crop produces no yield. They are amphimictic. They can survive through slow drying and starvation for at least 2 years. During dormant condition, they become spiral, coiled or form spring-like postures, which revive after getting moisture.

Aphelenchoides agarici

It is the most pathogenic species reported from Himachal Pradesh. This nematode has a life cycle of 8 days at 28 ± 1 °C and many generations are repeated during a single season (Seth 1984). The reproduction is by amphimixis and ratio between female and male is 1.1 to 1 (Khanna and Sharma 1992). Multiplication rate is more than 800 in 25 days at 28 °C (Khanna and Sharma 1989). *A. bisporus* is its main host but it can multiply on other fungi like *Fusarium*, *Trichoderma*, *Trichothecium* and *Gillaminello* (Khanna and Sharma 1989).

Aphelenchoides composticola

This is a common mushroom nematode found nearly in all the mushroom-growing regions of the world, and in India, in almost all the mushroom-growing states. Its life cycle is of 8 days at 23 °C, 10 days at 18 °C and 18 days at 13 °C. Hatching occurs 2–4 days after egg laying and mature adults are seen on the eighth day. It is the most predominant species (Gitanjali and Nandal 2001a, b). It is a severe limiting factor in the commercial cultivation of white button mushroom (*Agaricus bisporus*) in India

(Chhabra and Kaul 1982; Rao et al. 1992). The pathogenic potential of *Aphelenchoides composticola* to *Agaricus bisporus* was found to be dependent on the initial inoculum level and time of infestation of the mushroom beds (Gitanjali and Nandal 2005). Swarming and aggregation are common phenomena observed in this species. An initial inoculum of 10 nematodes can deplete mycelium completely in 40 days and multiply over 1000 times.

Aphelenchoides myceliophagus

This nematode is also very pathogenic and highly destructive pest for button mushroom. An inoculum of 10 individuals was able to destroy the mycelium completely within 40 days (Khanna and Sharma 1988a, b). In India, this nematode was reported from mushroom farm in Solan, Himachal Pradesh. This is also capable of reproducing by amphimixis.

Aphelenchoides sacchari

It is also one of the highly pathogenic nematodes responsible for damaging the mushroom cultivation and can reduce the sporophore yield by 94.5%. It can cause yield loss of 50–100% depending upon the initial nematode population (Sharma et al. 1984). Its life cycle completes within 12 days (Sharma and Seth 1986). In India this species was first recorded from white button mushroom (Sharma et al. 1981).

Aphelenchoides neocomposticola

This nematode is less destructive than the highly pathogenic species like *A. agarici*, *A. composticola* and *A. myceliophagus*. In India it was first recorded from Shimla, Himachal Pradesh. The range of temperature on which this nematode can multiply is 12–25 °C. The multiplication rates are 12.33, 677.3 and 942 at 15, 30 and 40 days, respectively (Khanna and Sharma 1988a, b).

Aphelenchoides minor

This is also of less economic importance. It was found first from the samples of button mushroom compost from Srinagar in Jammu and Kashmir (Seth and Sharma 1986).

Aphelenchoides aesterocaudatus

It was first reported in India from cropping beds of button mushroom by Bahl and Prasad (1985). In this species, males are generally absent. They can also survive slow drying and starvation for more than 2 years.

7.2.1.2 *Aphelenchus* Species

These nematodes are found in large number in mushroom beds during mid-season. Two species of this genus, *Aphelenchus avenae* Bastian and *Aphelenchus radicicola* (Cobb) Steiner are generally found. Their population declines drastically during the end of the crop. They are also very active and voracious feeder of mushroom mycelia like species of *Aphelenchoides*. *Aphelenchus* spp. have stylet without the stylet knobs. With the stylet, these nematodes penetrate the hyphal cells within 6–7 s. After that the median bulb pulsates for 8 s. This penetration occurs repeatedly, which results in the collapsing of hyphal cells (Juhl 1966; Fisher and Evans 1967). Although males are common in this nematode population, they reproduce by parthenogenesis. The fecundity is of 3 eggs per hour and a female can lay about 225 eggs in its life cycle. The generation time varies from 6 to 12 days depending on the temperature (Hechler 1962; Jairajpuri 1968). These nematodes also can survive through cryptobiosis during adverse conditions like food scarcity.

7.2.1.3 *Ditylenchus myceliophagus*

It is present in every crop-growing bed of button mushroom worldwide. It is mostly found in temperate regions (D'Errico et al. 1996). Initially it was identified as *Ditylenchus destructor* but later described as *D. myceliophagus* by Goodey. It is a very highly destructive pest for button mushroom. A population of 10–1000 individuals could cause 15–70% mycelial depletion within 60 days. The life cycle completes within 40 days at 13 °C, while the life cycle duration changes at different temperature, like 26 days at 18 °C and 11 days at 23 °C (Cayrol 1962; Evans and Fisher 1969). The embryonic development is completed within 75, 96 and 275 h at 25, 20 and 13 + 1 °C, respectively (Cayrol 1970). After first moulting, the second-stage juvenile starts thrusting its stylet inside the egg cell and hatching completes when the stylet punctures at the weak point of the egg cell. The nematode has a specific survival strategy. It survives by forming a cryptobiotic stage during off season. It was found that it can survive for 3½ years when dried slowly (Cayrol 1970). At the end of mushroom crop season, these nematodes form 'Nema wool'. The 'Nema wool' is the swarming dry form of this nematode population that breaks due to weight and gets attached with the mushroom beds, trays or other such things, and they thrive on them. When the next season comes, they again revive on the mushroom mycelium and infest the fresh crop. This 'nema wool' contains 40,000–45,000 nematodes in cryptobiotic stage. This species can multiply on a number of fungi, like *Alternaria alternata*, *Fusarium solani*, *Agaricus bisporus*, *Pleurotus sajor caju*, *Mucor* sp., *Rhizoctonia solani*, *Rhizoctonia bataticola*, *Trichoderma harzianum* and *Sclerotium species*.

The feeding of *Ditylenchus myceliophagus* on button mushroom, *Agaricus bisporus* (Lange) Singer, consists of exploration, penetration of stylet, injection of dorsal oesophageal gland secretions, ingestion of hyphal contents and stylet withdrawal phases. This nematode prefers mycelium and generally feeds near septum. Soon after stylet penetration, dorsal oesophageal gland nucleus becomes active and produces dorsal oesophageal gland secretions that accumulates at the ampulla. Pulsation of medium bulb, movement of basal bulb, anterior region of intestine and tail during feeding assist in ingestion of dorsal oesophageal gland secretions into the host cell as ingestion of cell contents. Several cycles of extracorporeal digestions and ingestion occur during each feed. Before cessation of feeding, the medium bulb pulsates rapidly for 4–5 min and nematodes withdraw their stylet with slight twitching of lip region. Feeding in a particular cell might continue for 2–5½ h. There is no shrinkage of the host cell (Anderson 1964; Sutherland and Fortin 1968; Siddiqui and Taylor 1969; Das and Bajaj 2008).

7.2.1.4 *Paraphelenchus* sp.

This nematode multiplies on all the fungi at both the temperatures (20 and 30 °C) except for *Pleurotus sajor caju* (Vats et al. 2006). A species of *Paraphelenchus* sp. was found associated with milky mushroom, *Calocybe indica*, where it caused complete destruction of mycelium in 30 days (Vats et al. 2006).

7.2.1.5 Saprophagous Nematodes

Many saprophagous nematodes are found in the mushroom compost beds. These nematodes belong to the order Rhabditida, Cephalobida and Diplogasterida. Their effects depend on the quality of pasteurization of compost, type of bacteria present in the compost/casing material and time of introduction of nematodes and bacteria (Grewal 1991). The important genera are *Rhabditis*, *Panagrolaimus*, *Bursilla*, *Cephalobus*, *Diplogaster* and *Acrobeloides*. Saprophagous nematodes are usually predominant and outnumber the myceliophagous forms. These nematodes lack the stylet. Buccal cavity is short and wide. These are generally clumsy looking worms with round caudal or clavate ends. Large populations of saprophytic nematodes (*Rhabditis* sp.) may produce by-products that adversely affect mushroom spawn growth. These nematodes multiply very fast. Chandel (1982) reported the pathogenic effects of *Rhabditis* species on *A. bisporus*. This nematode multiplied 500 times on *A. bisporus* and depleted mycelium by 50% within 30 days. Saprophagous nematodes also cause damage to the mushroom indirectly by the bacteria *Bacillus cereus*, *Bacillus* sp. and *Enterobacter amnigenus* (Grewal and Hand 1992). Theses bacteria cause significant reduction in mycelial growth of three strains: C43,

C54 and U3. In completely pasteurized compost, the saprophagous nematodes are found rarely because they could not find adequate moisture for their multiplication; also, the mushroom mycelia create nematostatic environment by producing volatile substances with antibiotic activity that restricts their development; the temperature rise during spawn run are lethal to many species of saprophagous nematodes. While in partially pasteurized compost, these nematodes survive because of the presence of suitable bacteria on which they can feed and multiply.

7.2.1.6 Predaceous Nematodes

Some predatory nematodes, like *Seinura* Fuchs and *Fictor composticola*, are found commonly in button mushroom compost along with other myceliophagous nematodes (Gitanjali and Nandal 2001a, b). They belong to the order Aphelenchida and Diplogasterida respectively. There are two species of *Seinura*, *Seinura paratenuicaudata* Geraert and *Seinura oxura* Goodey, found in mushroom compost in high numbers near the end of crop season (Vats et al. 2001, 2004). The species *S. paratenuicaudata* has a life cycle duration of 4–5 days. During feeding, this nematode inserts its stylet into the nematode body followed by injection of the contents of the dorsal oeshophageal glands, which paralyse the prey. In this way, the body contents of the nematode prey are sucked by the predator. It can feed on many fungivorous nematodes, like *Aphelenchoides bicaudatus*, *A. composticola*, *Aphelenchus avenae*, *A. radicicolus* and *D. myceliophagus*. They can feed at an optimum temperature of 20–35 °C. *Seinura* species do not feed on the mycelium. *F. composticola*, a predatory nematode of Diplogasterida order (Khan et al. 2008), is frequently found in the mushroom compost. It is a voracious feeder of myceliophagous nematodes as well as other nematodes. It has a very short life cycle and high fecundity. It also survives and multiplies on bacteria. These qualities make them a potential biological control agent (Bajaj and Kanwar 2015; Keshari 2016).

7.3 Nature of Damage

7.3.1 Myceliophagous Nematodes

Myceliophagous nematodes have needle-like structure in their mouth parts. The stylet is hollow inside and can be moved forward and backward by the contraction and relaxation of the muscles. They suck the cell sap resulting in cell death. After feeding from one cell, they shift to another cell and so on (Khanna 1994). The nematodes secrete variety of enzymes. These enzymes act immediately after ejection and help in penetration of stipe and to convert the cell contents in assimilable forms. The nematodes have a very fast rate of multiplication (50–100 fold per week). Rate of multiplication is faster during spawn run period (22–28 °C) (Thapa and Sharma

1987) than in the cropping period (14–18 °C). Beyond 30 °C, these do not reproduce. It has been found that initial infestation with three nematodes of *D. myceliophagus*/100 g of compost can entirely destroy the mycelium within a period of 70 days. These nematodes survive in a state of anabiosis (restoring to life from a death-like condition) for up to 2 years, if the compost is dried gradually, but they die if the compost is dried rapidly.

7.3.2 Saprophagous Nematodes

These nematodes invade the mushroom only after the invasion of myceliophagous nematodes. They secrete some enzymes and toxins, which increase the pH of compost and hampers spawn run. These are also known to create unhygienic conditions or sometimes they may carry harmful bacteria on their body surface. It was also found that saprophytic nematodes, *Caenorhabditis elegans*, can damage the sporophore of *A. bisporus*, if the population reaches 300–500 nematodes/g of casing mixture. Due to *Panagrolaimus fuchsia* infestation, there was yield reduction, changed flush pattern, shortened crop duration and qualitative reduction in sporophores (Chandran 2000). It was also reported that saprophagous nematodes at extremely high populations disturb the flush pattern and flush gap period in *A. bisporus* (Grewal and Richardson 1991; Khanna et al. 2005; Khanna and Kumar 2005). Some of the typical symptoms produced by these nematodes are:

1. Distorted, notched and kidney-shaped mushroom
2. Violet colour of gills
3. Browning of sporophores

These symptoms are produced because:

1. Many nematodes carry the viable bacteria in their digestive system, spreading them in the mushroom beds
2. There is production and accumulation of nematode by-products that ultimately inhibit the mycelium growth

Many of the hybrid strains are sensitive to saprophagous nematodes. *Caenorhabditis elegans* (Osche) Dougherty, when introduced along with casing material, rapidly colonize the casing material, which can be seen on the casing layer. This results in significant reduction in mushroom yield, disturbed flushing pattern and distorted buttons, and gills become violet (Grewal and Richardson 1991). These nematodes are found swarming on the sporophores. This happens because of the presence of sufficient moisture on the casing layer.

7.4 Sources of Contamination

Nematodes reach the mushroom house through several means. Various components of compost that support any fungal growth (wheat straw, chicken manure, horse dung, water etc.), casing soil (farm yard manure [FYM], soil, spent compost used as casing soil), platform soil, water, used infested trays/racks, handling implements, flies etc. are the possible source of nematode contamination. Sometimes the improper sterilization in short method of composting and lack of awareness about cultivation practices are responsible for the easy access of nematodes in cropping beds. In early days, wooden trays because of their repeated use and inadequate sterilization used to become an important source of nematode infestation. The most potential sources of nematode dissemination in *Agaricus bisporus* were wheat straw, chicken manure and spent compost often used as casing material. Besides, FYM, loam and platform soil also act as source of nematode inoculum (Khanna et al. 2006). Since chicken manure is not used in *Calocybe indica*, the source of contamination in this mushroom is wheat straw and spent compost. Animal wastes are the major source of contamination (Khanna et al. 2006). Compost prepared by long method of composting (LMC) may get contaminated with nematodes during the process of turning, besides at spawning and casing. In India, 90% of commercial mushroom production comes from marginal farmers who grow this crop in their houses under unhygienic conditions, where nematode incidence is extremely high. Dipteran flies hovering in the farms also disseminate nematodes from one bed/room to other. Many Aphelenchids are known to survive through anhydrobiosis and can be carried in dry form through wind and dirt to the composting area. These nematodes have many host ranges, so they survive on other fungi in the compost and build their population. Milky mushrooms are more susceptible to the myceliophagous nematodes than white button mushroom (Khanna et al. 2006). The sporophores of *Calocybe indica* were found to harbour more nematodes as compared to that of *Agaricus bisporus*. Spent compost and FYM used as casing are the most potential source of nematode inoculum in white button mushroom. Samples of irrigation water, reused trays and handling equipment contain the nematode inoculum. The pathogenic potential of *Aphelenchoides composticola* to *Agaricus bisporus* was found to be dependent on the initial inoculum level and time of infestation of the mushroom beds. Yield reduced at and above ten nematodes per kilogram of compost. Maximum nematode population was recorded when the nematodes were inoculated at the time of composting followed by spawning, and minimum at casing (Gitanjali and Nandal 2005). Infection at spawning time can have an effect on mycelial growth, thus making cultivation unprofitable from an economic point of view. Infection occurring later causes only minor crop losses and goes unnoticed by the mushroom grower.

7.5 Symptoms and Losses

The injury to mushroom by a fungal nematode (*Ditylenchus* sp.) was first recorded by Lambert et al. (1949). The myceliophagous and saprophagous nematodes produce many symptoms as a result of their feeding on the mycelium. These symptoms are not specific and depend upon the time of nematode access to mushroom beds. The intensity of feeding depends on the initial population of the nematodes. These nematodes feed the mushroom mycelia by piercing the mycelial wall with their hollow stylet and suck the contents of the pierced cells. The nematodes feed continuously from cell to cell, and in this process destroy the entire mycelium. They may attack the mycelium at any time, from spawning to fruiting. If the initial population is very high, the mushroom production shows a steady decline and 100% loss may result. There may be patchy or no growth of mycelium, sinking of mushroom beds and soggy and foul smell of the compost, which leads to severe reduction in the sporophore yield (Kumar et al. 2008a, b). The foul smell is due to production of anaerobic bacteria. The other symptoms are sparse mycelial growth, delayed and poor mushroom flushes, decline in mushroom yield and no mushroom production. Whiteness of spawn run slowly changes to brown. There can also be alternate high and poor yield in successive flushes, browning of pinheads and white fungal growth over casing in patch areas. However, in case the nematodes get introduced at the time of casing, normal first and second flushes are followed by poor or no successive flushes as the nematode population rapidly builds up. Since the symptoms are non-specific, the compost samples should be examined by the nematologists for the presence of nematodes as these nematodes are microscopic and cannot be seen through naked eye. Myceliophagous nematodes are highly destructive and are known to cause damage ranging from 41 to 100% crop loss in button mushrooms, depending on the nematode species involved, its population density and the cropping stage at the time of infestation (Sharma et al. 1984; Khanna 1991, 1993; Cairns and Thomas 1950; Khanna and Kumar 2005). The damage potential of *Aphelenchoides swarupi* and *Aphelenchus avenae* was high when inoculated at spawning time and restricted the mycelial growth significantly. Mycelial growth was restricted significantly more by *A. swarupi* than *A. avenae* at corresponding spawning and casing time inoculations. Despite the lower multiplication potential of *A. avenae*, it incurred significant losses to *A. bisporus*, indicating its high pathogenic potential (Kumar et al. 2007). *Aphelenchoides composticola* acts as a severe limiting factor in the commercial cultivation of white button mushroom, *Agaricus bisporus*, in India (Bhardwaj et al. 1973; Chhabra and Kaul 1982; Rao et al. 1992). Mycelial growth was restricted and the sporophore yield was reduced significantly by *Aphelenchoides swarupi* and *Aphelenchus avenae* (Kumar et al. 2008a, b). Sporophore yield loss in white button by myceliophagous nematodes like *A. sacchari*, *A. composticola* and *D. myceliophagus* was recorded to be 59.5–100.0%, 95.0% and 89.0–100.0% respectively, when inoculated at spawning time (Janowicz 1978; Sharma et al. 1984; Khanna 1991, 1993). Khanna and Jandaik (2002) reported 6.8–26.5% losses due to *A. avenae* at different spawning times. The parasitic nematodes *Ditylenchus myceliophagus*, *Aphelenchoides composticola* and

Paraphelenchus myceliophthorus severely damage mushroom mycelium and, in great numbers, reduce or even prevent cropping (Goodey 1960). *Aphelenchoides agarici* and *A. composticola* depleted more mycelium when inoculated at spawning rather than casing (Khanna and Kumar 2005). In *Agaricus bitorquis*, the population of *A. swarupi* and *A. avenae* restricted the mycelium significantly when inoculated at spawning time. Also, the yield losses were 72% and 53% respectively by *A. swarupi* and *A. avenae*. The nematodes also disrupted the flush pattern and cropping period (Kumar et al. 2008b). Peak populations of *D. myceliophagus* and *A. composticola* were reached at 7 and 13 weeks after spawning, respectively, when inoculated at spawning (Arnold and Blake 1968). High yield and number of fruiting bodies were recorded when nematodes were inoculated at the time of casing followed by spawning and lowest at composting (Gitanjali and Nandal 2005). The fungus-feeding nematodes are more pathogenic when introduced in large-sized mushroom beds than in the polythene bags due to different environmental conditions. It was also seen that *D. myceliophagus* is more pathogenic in European countries if present at the time of spawning and can cause 100% loss in the yield (Arnold and Blake 1968; Khanna 1993; Vats et al. 2002). The saprophagous nematodes feed on organic matter and bacteria. These nematodes indirectly harm the mushroom production by acting as carriers of many disease-causing organisms. Cairns and Thomas in 1950 reported that the saprophagous nematode, *Rhabditis lambdiensis*, was a carrier of pathogenic bacteria, *Pseudomonas tolaasii*, which causes 'bacterial pit' and disfiguration of the mushroom caps. Bacterial and fungal spores are either transported on nematode cuticle or may be ingested and passed unchanged. The abundance of saprophagous nematodes may encourage spread of bacterial disease such as 'blotch' or 'pit'. *Rhabditis* sp. caused mycelial destruction in 7 days at a very high level of population as reported by Haan et al. (1974). This mainly happened due to excretory products of this saprophagous nematodes. Again, Chandel in 1982 observed some pathogenic effects on *Agaricus bisporus* mycelium grown on malt extract agar medium. Infestation of mushroom nematodes at the time of composting results in poor spread of mycelium and if it is at the time of spawning, there is a very poor spawn run. Devi (2005) observed effect of *Aphelenchoides composticola* on white button mushroom. The spawn run was poor at the inoculum level of 500 and 1000 nematodes per kg compost and irrespective of the time of inoculation, the fruiting bodies and yield were significantly less at and above the inoculum level of 100 nematodes/kg compost compared with the uninoculated control. With the increase in inoculum level of *Aphelenchodes swarupi* and *Aphelenchus avenae* from 10 to 1000, there was significant decrease in mycelial growth of *Agaricus bisporus* and *Agaricus bitorquis* (Kumar et al 2009). Nematode introduction at initial stages results in the quick build-up of massive nematode populations that inflict heavy damage, while introduction at very later stages, like at casing, causes less damage due to smaller nematode population build-up during shorter duration and better establishment of spawn. With regard to *Aphelenchoides swarupi*, when screened against different strains of *Agaricus bisporus*, all strains except two were found susceptible to this nematode, and mycelial damage resulted. The nematode population and reproduction factor were higher after 30 days. But, the strains of *Pleurotus eous* were found resistant (Madhuri and Kanwar 2016).

Although the nematode population in the mushroom beds are higher during the cropping time, rather it is maximum during early cropping phase, it declines suddenly at the crop cessation, like in *A. agarici* and *A. composticola* infesting *A. bisporus* (Khanna and Sharma 1988a, b; Khanna 1991).

7.6 Management of Mushroom Nematodes

7.6.1 Prophylactic Measures

Prophylactic measures are the best method to avoid the diseases and other pests. The following measures could be taken under this method:

- The mushroom rooms should be properly ventilated
- The water used for irrigation should be pest-free
- While preparing compost through manual methods, the floor should be concrete and it should be well irrigated with disinfecting agents like formalin (5% concentration) before compost preparation
- Hygienic conditions and cleanliness should be maintained for the total crop duration (Bruno et al. 2013)
- The workers and visitors should use disinfectants (formalin, 5% concentration) before entering and after coming from the mushroom room
- The compost that is going to be used for spawning should have 70% moisture and pH 7–7.2
- The mushroom compost should be pasteurized in a proper pasteurization chamber
- The implements and other equipment used in the mushroom cultivation should be sterilized before and after use; for this, hot water can be used
- Manage insect pests inside the mushroom beds to avoid the nematode infestation
- The spent compost should be used properly; it should not be reused in the mushroom beds
- The platform used for composting must be cemented and slightly raised so that compost does not come in contact with soil
- Limited and recommended use of fungicides/insecticides should be ensured
- Steaming compost for 2 h at 60 °C should be done prior to spawning
- The casing material should be well sterilized before use; formalin at 5% concentration can be used for this purpose; use of spent compost as casing soil should be avoided to maximum extent
- Proper steam cookout of mushroom house at 70 °C for 5–6 h is necessary
- Proper disposal of spent compost far away from the mushroom farm is necessary after the crop is over
- Washing of mushroom house with some disinfectant and proper drying before going for the next crop should be done
- The temperature should be maintained accordingly in the mushroom chamber
- The casing ingredients should be stored and mixed in clean area and it should be pasteurized properly

- All spent compost and casing soil should be removed from the mushroom house after the harvest of the crop
- When steaming could not be done, the compost should be fumigated with methyl bromide
- In mushroom bed, the equipment and tools should be disinfected by chlorine
- The interior of the room should be made from non-biodegradable substances
- All surfaces should be smooth and easy to clean
- The shelves should be made of galvanized iron or formica (Oei and Nieuwenhuijzen 2005)
- Disinfection may be done by cleaning with a 10% Clorox solution or 70% ethyl alcohol
- The floor should be cemented to avoid the direct contact of compost with the infested soil; it should be slightly inclined to allow excess water to drain
- The drainage system of the different rooms should not be connected to prevent passing of a disease from one room to another (Oei and Nieuwenhuijzen 2005)
- The irrigation water should be cleaned and a proper sewage disposal system should be there
- The grain and saw dust used should be properly sterilized as they may contain thousands of bacteria, fungi and actinomycetes
- Picking should be done from new crops to older ones
- The waste from various operations should be collected and disposed off daily from the working areas immediately (Bellettini and Fiorda 2016)
- The growing rooms should be properly ventilated, having doors and windows with wire net of 14–16 mesh/cm to avoid the entry of insect pests (Singh and Sharma 2016)

7.6.2 Physical Control

In this method, heat is used to control the nematodes as well as to restrict the development and infestation of nematodes. So, from the beginning, that is, from the compost preparation, maintenance of heat at a certain temperature is necessary for avoiding the nematode introduction. In the pasteurization chamber, the bed temperature should be maintained at 60 °C at least for 2 h and the incubation period of the chamber should be 70 °C for 5–6 h or 80 °C for 30–60 min (Thapa and Sharma 1981). The trays and other handling tools and equipment should be used after disinfection through the use of boiling water for 1–2 min or by using formalin (5% concentration) or cresylic acid (Seth 1984; Seth and Sharma 1986). This would be sufficient for the complete destruction of nematodes. In case of *D. myceliophagus* nematode infestation, the disinfection becomes very important because this nematode has a very high survival capacity and it can survive in dry conditions for more than 3 years. In this case the disposal of spent compost also becomes very important. The room temperature at the time of spawning and fruiting should be maintained according to the requirement. It is very important to maintain the temperature of the mushroom chamber below 14 °C for proper fruiting. If the compost dries during warm-up to pasteurization, then some nematodes may go into a resistant stage and infest the crop at spawning. Once the infection site is identified, it is quite important to maintain vigilant control of flies. Covering the infected area with plastic will minimize movement by personnel, insects or irrigation. For the management of fungus gnat, *Rhymosia domestica*, the logs used to grow oyster mushrooms should be covered with a 1 mm mesh screen net (Kaneko 1983). Steam sterilization is beneficial for the disinfection of the room after the crop cycle (cooking out) by applying steam vapour at 70 °C for 12 h (Fletcher and Gaze 2008). Casing soil is chemically pasteurized by formaldehyde (Garcha and Sekhon 1981) or with steam treatment (Haynes and Shandilya 1977). Solar energy can also be used for the sterilization (Grewal and Grewal 1988).

7.6.3 Chemical Control

Mushrooms are the crops that are eaten fresh and have little space for chemical usage. Being a very short cropping duration, harmful effect of pesticides on the mycelium and residue problems limit the use of chemicals. It is also not advisable because resistance develops amongst the pests against the pesticides (Gea et al. 2003). However, some of the chemicals can be used during composting itself, which are effective in checking nematode population, especially in long method of composting. Application of chemicals for the control of nematodes has been reported by several authors (Sharma et al. 1984, Shandilya et al. 1975, Sharma et al. 1981, Chandel 1982, Chhabra and Kaul 1982, Seth 1984, Thapa and Sharma 1987, Kaul and Chhabra 1992, Rao and Pandey 1991; Kaur et al. 1987). The first group of chemicals used for the control of mushroom pests was the organochlorine pesticides. The use of chemicals like lindane, acephate, chlorpyriphos, deltamethrin, diazinon, dimethoate, ethoprop, fenitrothion, methepreme, diflubenzuron, dichlorvos and malathion as prophylactic treatment, by spraying on the compost walls of the mushroom house surroundings, and their incorporation in the compost have been found effective in controlling the flies (Read 1968). Organophosphates are also used as nematicides. Thionazin at 80 ppm when used in the compost gave an excellent control of *D. myceliophagus* and *A. composticola*. Dazomet is a slow-releasing fumigant and its nematicidal properties are due to the ability of isothiocynates to react with nucleophilic centres, that is, thiol group of vital enzymes in nematodes. Dichlorvos (0.04%) under polythene cover for 3–4 days was found to be most effective for control of *A. composticola* and *Rhabditis* sp. Fenamiphos EC at 20 mg/kg incorporated in compost is a practical preventive measure (Grewal and Sohi 1987). Thionazin at 80 ppm (0.008%) in compost and its spray on the bed surface during spawn run effectively controls the nematodes without any detectable residue on mushrooms (Singh and Sharma 2016). Phenamiphos (E.C.) at 20 mg a.i./kg incorporated in compost is a practical preventive measure for control of *Aphelenchoides composticola* (Mcleod and Khair 2008). It was also reported that diflubenzuron (Gahukar 2014) and the fungicides, benomyl and thiabendiazole (Mcleod and Khair 1978) may reduce population growth of *Aphelenchoides* spp.

Carbofuran has been successfully used for the control of myceliophagous and saprophagous nematodes (Sharma et al. 1981). But many chemicals, like dichlorovos and phenamiphos, are reported to leave residues in the mushroom crop (Bahl and Agnihotri 1987, Kaur et al. 1987; Grewal and Sohi 1987) and some of them affect the mushroom mycelium adversely (Grewal and Sohi 1987).

7.6.4 Biocontrol

Mushroom crop continues for 6–8 weeks, and the mushrooms are harvested and consumed soon after they appear. Therefore, during the cropping season, using nematicide is not safe for health because of toxicity and residual problems. Mushroom is very sensitive to chemicals and consuming fresh mushroom after spraying nematicide is not advisable for health reasons. Thus, the role of biocontrol agents become very important. For this, many microorganisms, predatory nematodes, entomopathogenic nematodes, fungi, bacteria and mites are used. Biological control agents can be used for the eco-friendly management of nematodes (Ansari and Khan 2012a, b; Ansari et al. 2017a, b, 2019; Ansari and Mahmood 2017a, b, 2019a, b). These are found in the mushroom compost itself, which can be isolated and multiplied and can be used in required population. The nematode-trapping fungi such as *Arthrobotrys irregularis* and *Candelalretta musiformis* were found as highly effective in reducing the nematode population in mushroom cultivation (Khanna and Sharma 1990; Grewal and Sohi 1988; Khanna and Sharma 1988a, b). These fungi trap the nematodes through their different trapping devices and the population of nematodes reduce and mushroom yield increases. *Arthrobotrys robusta* has been recommended against mycophagous nematodes. This fungus is formulated as 'Royal 300' strain in Antipolis in France. *Candelalretta musiformis* isolated from spent compost has been found highly effective in checking nematode multiplication. Another fungus, *A. irregularis*, is highly effective against *A. composticola* (Grewal and Sohi 1988; Khanna and Sharma 1988a, b). Presence of naturally occurring parasites and predators in the mushroom ecosystem can be exploited for the management of these nematodes. The biocontrol agents are cheaper, non-toxic and provide pollution-free control of pests. The bioagents may be preventive as well as curative because they can help in invading the pest or disease, but if the disease is already set, it can be corrected by reducing the population densities of the pest or pathogen. Amongst the biocontrol agents, predacious nematodes can play a vital role in nematode management if given equal importance and opportunity. The predatory nematodes present in the mushroom compost are *Seinura* sp. and *Fictor composticola* (Khan et al. 2008). *Seinura* sp. belongs to the order Aphelenchida and *Fictor composticola* belongs to the order Diplogasterida. Both these nematodes predate the myceliophagous nematodes. They are voracious feeder although they have different feeding habits. *Seinura* sp. feeds the preys by paralysing them by secreting some chemicals while *Fictor composticola* pierces and sucks the body contents of the prey body with the teeth (Bajaj and Kanwar 2015). *F. composticola* is generally found abundantly in decomposing organic manure. Their life cycle is short and they can be easily cultured and maintained on simple nutrient media containing bacteria (Yeates 1969). Thus, they are promising biocontrol agents. They have very short life cycle, high rate of predation, easily cultured and high fecundity (Siddiqi et al. 2004; Bilgrami et al. 2005). They have a very special quality, which is, their specificity to the prey. This is a very good quality of the biocontrol agent (Chitamber and Noffsinger 1989). *Fictor composticola* was found prevalent in compost used for cultivating button mushroom in Haryana and Bihar states of India (Khan et al. 2008). The entomopathogenic nematodes like *Steinernema* species can be used for the management of insects, which help in spread of nematode pests. Application of entomopathogenic nematodes *Steinernema feltiae* and *Heterorhabditis heliothis* before casing gives up to 97% control of flies that may act as vector of nematodes. *Pleurotus ostreatus*, *Pleurotus florida* and *Pleurotus citrinopileatus* have nematicidal abilities against phytonematodes, *Pratylenchus*, *Xiphinema*, *Tylenchorhynchus*, *Tylenchus*, *Helicotylenchus*, *Ditylenchus*, *Psilenchus*, *Aphelenchus*, *Hoplolaimus*, *Longidorus*, *Aphelenchoides* and *Paralongidorus* spp. They are found effective for killing the nematodes. Amongst the three, *P. citrinopileatus* was found more effective than other species and it killed 100% nematodes after 24 h followed by fruiting body extracts of *P. florida*, 99% and waste of *P. ostreatus*, 77% (Khan et al. 2014).

7.6.5 Botanicals

Botanical extracts from many plants are used for the management of these nematodes (Nath et al. 1982, Khanna et al. 1988; Grewal 1989a, b). They can be used by mixing in the compost. The dried leaves of Neem, *Azadirachta indica*, *Cannabis sativa*, *Eucalyptus tereticornis* and *Ricinus communis* mixed at the rate of 3 kg/100 kg of dry wheat straw help in reducing the nematode population; also, it can enhance the population of thermophilic fungi and mesophilic antibiotic-producing fungi, but can reduce the number of mesophilic competitor/pathogenic moulds. Populations of the mycophagous nematode, *Aphelenchoides composticola* were reduced below economic injury level in dried leaf treated composts. Composts obtained from *C. sativa* and *R. communis* treatments were more rapidly colonized by the mushroom mycelium (*Agaricus bisporus*) than composts from other treatments. Compared with carbofuran, the composts treated with *R. communis*, *C. sativa* and *A. indica* significantly increased mushroom yield by 19.4, 8.1 and 6.5% and by 108.5, 88.7 and 85.9% (Grewal 1989a). Karanj leaves, when mixed with the compost, also gives good result in reducing nematode population. Neem leaf powder at the rate of 2% when added on w/w basis to the compost at spawning reduces the multiplication of *A. composticola*. Use of oil cakes like neem cake (*Azadirachta indica*), karanj (*Pongamia pinnata*), coconut (*Cocos nucifera*), castor (*Ricinus communis*) and groundnut (*Arachis hypogea*) in compost before spawning has been found to reduce nematode multiplication (Rao et al. 1991). Two per cent neem-based seed kernel water extract and 800 ppm of Achook, a neem-based pesticide, if applied at the time of spawning help in reducing the nematode damage

(Vats et al. 2003). Mixing Nimbecidine or neem seed kernel water extract (NSKWE) at the time of last turning of compost along with the synthetic insecticides, that is, endosulfan or chlorpyriphos, has been found effective (Reddy et al. 1997). Neem products were equal in bioefficacy to diflubenzuron 20 wettable powder (WP; 0.025% w/v) against *Aphelenchoides composticola* infesting button mushroom (Baba 1990). In button mushroom, in vitro application of Neemazal EC (1% v/v) or Nimbecidin EC (0.03% v/v) applied at 4–6 mL/L and in vivo application of neem seed kernel powder at 4 g/kg compost resulted in >50% mortality in *Aphelenchoides composticola* population and increased mushroom yield by 26–33% (Katyal et al. 2007). Neem leaf (oven dried) at the rate of 16 g/2 kg compost, neem cake at the rate of 20 g/2 kg compost and dazomet at the rate of 1.2 g/2 kg compost effectively increased number of fruiting bodies as well as yield of *Agaricus bisporus* by reducing the population of *Aphelenchoides composticola* (Gitanjali and Nandal 2001b). Neem cake decomposes in the presence of adequate moisture and releases many nematicidal compounds like ammonia, phenols, aldehydes, amino acids and fatty acids, which are highly deleterious to nematodes (Reddy et al. 1997). Also, some chemicals including metabolites produced by microorganisms, which multiply and become active during neem cake decomposition, also contribute to nematoxicity. Besides direct toxicity, neem cake contains 5–7% nitrogen and its incorporation results in increase in pH of the finished product leading to rapid colonization of mushroom mycelium (Mojumder 1997). Neem leaves when decompose increase the compost temperature, which might enhance the growth of thermophilic microorganisms and actinomycetes that help in rapid colonization of mushroom mycelium, thereby increasing the yield. Neem leaves also release allelochemicals like limonoids, nimbidic acid and meliacin that are directly toxic to nematodes (Grewal 1989a, b). A neem-based formulation, Achook, containing Azadirachtin as an active principle, is a strong antifeedant that affects the physiological process of nematodes (Parmar 1997). Rao and Pandey (1992) reported that karanj leaf (5%) treatment and carbofuran were at par with each other in reducing the population of *A. composticola* and increasing the yield of *Agaricus bisporus*.

7.6.6 Resistant Strains

To date, there is no strain of *Agaricus bisporus* that is found resistant against the myceliophagous nematodes. *Pleurotus sajor caju* is also resistant to several species of *Aphelenchoides* and *Ditylenchus myceliophagus* (Khanna and Sharma 1989; Thapa et al. 1983). *Pleurotus sajor caju* is resistant to *Paraphelenchus* spp., *Aphelenchoides* spp. and *Ditylenchus myceliophagus* (Khanna and Sharma 1989; Thapa et al. 1983). The resistance in mushrooms is characterized by a lower rate of nematode multiplication and less mycelial damage. This is due to biochemical changes in mushroom spawns (Sharma and Seth 1993).

7.7 Conclusions and Future Prospects

Mushroom cultivation is a very rapidly growing small-scale industry in the world, and especially prevalent amongst the poor and marginal farmers along with the rich farmers and the rural women who can earn at their home. The nematodes can decrease the yield of mushroom. The mycophagous and saprophagous nematodes, once established in the crop, are very difficult to eliminate. They can only be eradicated after destroying the crops. They can be managed only if proper sanitary conditions are followed from the very beginning, from compost preparation to fruiting stage. For the seasonal growers, the maintenance of hygienic conditions is not feasible. The pasteurization chamber is not affordable for every grower. So, the government and NGOs or cooperatives may help in setting up a common pasteurization facility for composting. The resistant strains may be developed through researches using biotechnological tools. Chemicals are not advisable in mushroom cultivation, so botanicals should be tried for safe protection. Neem seed kernel water extract, Achook and some other botanicals are already found to be promising controlling agents. Researches on identification of new plant products and their formulations should be done for managing the pests and pathogens along with the nematodes. The mushroom compost is also a complex ecosystem containing many biocontrol agents like predacious fungi, predacious nematodes and other parasites and predators, which should be exploited. Predatory nematodes *Fictor composticola* and *Seinura* spp. can be the potential biocontrol agents (Kanwar et al. 2009; Bajaj and Kanwar 2015; Keshari 2016). They can be multiplied in the compost and can be formulated as ready to use. They can be made such that they can be transported to other places. The efficiency of *Seinura* spp. to paralyse the prey nematodes can be analysed biochemically and if the chemical composition is found, it can be used in preparation of effective nematicides for the management of nematodes in mushroom production.

References

Agdaci M, Isik SE, Erkel I (1990) Studies on nematode damaging cultivated mushroom in Marmara region. J Yalova-Ataturk-Central Hortic Res Inst 19:11–16

Aida FMNA, Shuhaimi M, Yazid M, Maaruf AG (2009) Mushroom as a potential source of prebiotics: a review. Trends Food Sci Technol 20:567–575

Akata I, Ergonul B, Kalyoncu F (2012) Chemical compositions and antioxidant activities of 16 wild edible mushroom species grown in Anatolia. Int J Pharmacol 8:134–138

Anderson RV (1964) Feeding of *Ditylenchus destructor*. Phytopathology 54:1121–1126

Anonymous (2003) Nematode problems of mushroom in Haryana and their management. Final Technical Report. Department of Nematology, CCS, HAU, Hisar, India. pp 35

Ansari RA, Khan TA (2012a) Parasitic association of root-knot nematode, *Meloidogyne incognita* on guava. e-J Sci Technol 5:65–67

Ansari RA, Khan TA (2012b) Diversity and community structure of phytonematodes associated with guava in and around Aligarh, Uttar Pradesh, India. Trends Biosci 5(3):202–204

Ansari RA, Mahmood I (2017a) Optimization of organic and bio-organic fertilizers on soil properties and growth of pigeon pea. Sci Hortic 226:1–9

Ansari RA, Mahmood I (2017b) Determination of disease incidence caused by *Meloidogyne* spp. and or *Fusarium udum* on pigeonpea in Aligarh district: a survey. Trends Biosci 10 (24):5239–5243

Ansari RA, Mahmood I (2019a) Plant health under biotic stress: volume 2: microbial interactions. Springer, Singapore. https://doi.org/10.1007/978-981-13-6040-4

Ansari RA, Mahmood I (2019b) Plant health under biotic stress: volume 1: organic strategies. Springer, Singapore. https://doi.org/10.1007/978-981-13-6043-5

Ansari RA, Rizvi R, Sumbul A, Mahmood I (2017a) PGPR: current vogue in sustainable crop production. In: Kumar V, Kumar M, Sharma S, Prasad R (eds) Probiotics and plant health. Springer, Singapore, pp 455–472

Ansari RA, Mahmood I, Rizvi R, Sumbul A (2017b) Siderophores: augmentation of soil health and crop productivity. In: Kumar V, Kumar M, Sharma S, Prasad R (eds) Probiotics in agroecosystem. Springer, Singapore, pp 291–312

Ansari RA, Sumbul A, Rizvi R, Mahmood I (2019) Organic soil amendments: potential tool for soil and plant health management. In: Ansari RA, Mahmood I (eds) Plant health under biotic stress. Springer, Singapore, pp 1–35

Arnold NP, Blake CD (1968) Some effects of nematode *Ditylenchus myceliophagus* and *Aphelenchoides composticola* on the yield of the cultivated mushroom. Ann Appl Biol 61:161–166

Baba ZA (1990) Studies on diflubenzuron and *Azadirachta indica* seed on the development and control of *Bradisia tritici* (coq.) and *Ditylenchus myceliophagus* (Goodey) infesting button mushrooms. Ph.D. Dissertation, Plant Pathology Department, Punjab Agricultural University, Ludhiana, Punjab, India

Bahl N, Prasad D (1985) Nematodes associated with white button mushroom. Indian Bot Reptr 4:74

Bahl N, Agnihotri VP (1987) Translocation of nematode residue in mushroom following their application in composting. Mushroom Sci 12:3

Bajaj HK, Kanwar RS (2011) Nematode infestations in mushrooms. In: Khan MR, Jairajpuri MS (eds) Nematode infestations part III: horticultural crops. NAS, Allahabad, India, pp 357–382

Bajaj HK, Kanwar RS (2015) Biology and predatory attributes of a diplogasterid nematode, *Fictor composticola* Khan et al., 2008. Helminthologia 52:50–57

Bajaj HK, Walia KK (1999) Three new species of *Aphelenchoides* Fisher 1894 (Aphelenchina: Nematoda) from India along with a compendium of Indian species. Indian J Nematol 29:154–170

Banik S (2010) Mushrooms: the magic store of health benefits. Everyman's Sci 64:360–365

Bellettini MB, Fiorda FA (2016) Production of pests and diseases in mushroom, *Pleurotus* spp. crops. Apprehendere, Guarapuava, p 152

Bellettini MB, Fiorda FA, Bellettini S (2015) Aspectos gerais do cultivode cogumelo Pleurotus ostreatus e djamor pela técnica Jun-Cao, Guarapuava, Apprehendere, p 92

Bellettini MB, Fiorda FA, Maieves HA, Teixeira GL, Ávila S, Hornung PS, Maccari-Júnior A, Ribani RH (2016) Factors affecting mushroom *Pleurotus* spp. Saudi J Biol Sci 12

Bellettini MB, Bellettini S, Vítola FMD, Fiorda FA, Maccari JA, Soccol CR (2017) Residual compost from the production of *Bactrisgasipaes* Kunth and *Pleurotus ostreatus* as soil conditioners for *Lactuca sativa* 'Veronica'. Semin-Cienc Agrar 8:581–594

Bhardwaj G (2004) Studies of *Calocybe indica* (P & C) the milky mushroom. Ph.D. thesis submitted to Dr. Y. S. Parmar University of Horticulture and Forestry, Nauni, Solan, H.P, pp 66

Bhardwaj HK, Sharma PL, Thakur JR (1973) Mushroom nematodes. Current Sci 42:257–258

Bilgrami AL, Gaugler R, Brey C (2005) Prey preference and feeding behaviour of the diplogasterid predator *Mononchoides gaugleri* (nematode: Diplogasterida). Nematology 7:333–342

Biswas MK (2014) Cultivation of paddy straw mushrooms (*Volvariella volvacea*) in the lateritic zone of West Bengal-a healthy food for rural people. Int J Econ Plants 1:43–47

Bruno GL, Rana GL, Sermani S, Scarola L, Cariddi C (2013) Control of bacterial yellowing of cardoncello mushroom *Pleurotus eryngii* using acetic or hydrochloric acid solutions. Crop Prot 50:24–29

Cairns J, Thomas CA (1950) Proceeding of the first international conference on scientific aspects of mushroom growing. Mushroom Sci 1:89–91

Cayrol JC (1962) Etude du cycle evolutif d' *Aphelenchoides composticola*. Nematologica 10:361–368

Cayrol JC (1970) Phenomene de survie du nematode mycophage *Ditylenchus myceliophagus*: anabiose. Bull Feder Natn Synds Agric Cult Champ 189:2329–2333

Chandel YS (1982) M.Sc. thesis, Department of Entomology and Apiculture, HPKVV, College of Agriculture, Nauni, Solan (HP), India

Chandran R (2000) Studies on saprophagous nematodes associated with white button mushroom. M.Sc. thesis, Deptt. of Entomology and Apiculture, COH, UHF, Nauni, Solan, pp 77

Chhabra HK, Kaul VK (1982) Occurrence and control of *Aphelenchoides composticola* on mushrooms. Indian J Nematol 12:191–192

Chitamber JJ, Noffsinger M (1989) Predacious behaviour and life history of *Odontopharynx longicaudata* (Diplogasterida). J Nematol 21:284–291

Curvetto NR, Figlas D, Devalis R, Delmastro S (2002) Growth and productivity of different *Pleurotus ostreatus* strains on sunflower seed hulls supplemented with N-NH_4 and/or Mn (II). Bioresour Technol 84:171–176

D'Errico FP, Tacconi R, Abdel Ghany MM, Caprio E, Langellotti RR (1996) Presence of *Ditylenchus myceliophagus* on mushrooms in Italy. Informatore Fitopatologico 46:17–19

Das D, Bajaj HK (2008) Feeding behaviour of myceliophagous nematode, *Ditylenchus myceliophagus* Goodey on button mushroom, *Agaricus bisporus* (Lange.) Singer. Indian J Nematol 38:232–235

Deepthi S, Suharban M, Geetha D, Sudharma K (2004) Pests infesting oyster mushrooms in Kerala and the seasonality of their occurrence. Mushroom Res 13:76–81

Devi G (1999) Studies on nematodes damaging cultivated mushroom, *Agaricus bisporus* (Lange) singer and their management using eco-friendly measures. Ph.D. thesis, CCS Haryana Agricultural University, Hisar, 94 pp

Devi G (2005) Damage potential of *Aphelenchoides composticola* on white button mushroom. Indian J Plant Prot 33:307–308

Evans AAF, Fisher JF (1969) Development and structure of population of *Ditylenchus myceliophagus* as affected by temperature. Nematologica 15:395–402

Fisher JF, Evans AAF (1967) Penetration and feeding by *Aphelenchus avenae*. Nematologica 13:425–428

Fletcher JT, Gaze RH (2008) Mushroom pest and disease control. CRC Press, Boca Ratón, p 192

Gahukar RT (2014) Mushroom pest and disease management using plant derived products in the tropics: a review. Int J Veg Sci 20:78–88

Garcha HS, Sekhon A (1981) Evaluation of casing material for mushroom culture in Punjab (India). Mushroom Sci 11:411–417

Gea FJ, Tello JC, Navarro MJ (2003) Occurrence of *Verticillium fungicola* var. *fungicola* on *Agaricus bitorquis* mushroom crops in Spain. Mol Plant Pathol 151:98–100

Gitanjali, Nandal SN (2001a) Nematodes associated with white button mushroom (*Agaricus bisporus*) in Sonipat district of Haryana. Indian J Nematol 31:79–98

Gitanjali, Nandal SN (2001b) Effect of neem products and dazomet for the management of *Aphelenchoides composticola* on white button mushroom (*Agaricus bisporus*) under semi-commercial conditions. Indian J Nematol 31:52–57

Gitanjali, Nandal SN (2005) Pathogenicity of *Aphelenchoides composticola* on white button mushroom (*Agaricus bisporus*). Indian J Nematol 35:120–122

Goodey JB (1960) Observations on the effects of the parasitic nematodes, *Ditylenchus myceliophagus, Aphelenchoides composticola and Paraphelenchus myceliophthorus* on the growth and cropping of mushrooms. Ann Appl Biol 48:655–664

Grewal PS (1989a) Nematicidal effects of some plant extracts to *Aphelenchoides composticola* (Nematoda) infesting mushroom, *Agaricus bisporus*. Rev Nematol 12:317–322

Grewal PS (1989b) Effects of leaf matter incorporation on *Aphelenchoides composticola* (Nematoda), mycoflora composition, mushroom compost quality and yield of *Agaricus bisporus*. Annu Appl Biol 115:299–312

Grewal PS (1991) Effect of *Caenorhabditis elegans* (Nematoda: Rhabditidae) on the spread of the bacterium, *Pseudomonas tolasii* in mushrooms (*Agaricus bisporus*). Annu Appl Biol 115:299–312

Grewal PS, Sohi HS (1987) Studies on the different pesticides on the growth of *Agaricus bisporus* (Lange) Singer and *Pleurotus sajor-caju* (Fr.) Singer. Mush J Tropics 7:25–29

Grewal PS, Grewal SK (1988) Selective fungicidal properties of some plant extracts in mushroom weed moulds. Phytopathol Mediterr 27:112–114

Grewal PS, Hand P (1992) Effects of bacteria isolated from a saprophagous rhabditid nematode, *Caenorahbditis elegans* on the mycelial growth of *Agaricus bisporus*. Annu Appl Biol 72:173–179

Grewal PS, Richardson PN (1991) Effects of *Caenorhabditis elegans* (Nematoda: Rhabditidae) on the yield and quality of cultivated mushroom, *Agaricus bisporus*. Annu Appl Biol 118:581–594

Grewal PS, Sohi HS (1988) A new and cheaper technique for rapid multiplication of *Arthrobotrys conoides* and its potential as a bio-nematicide in mushroom culture. Curr Sci 57:44–46

Haan YS, Park JS, Kim DB, Shim KC (1974) Research report of the office of rural development, Fertilizer. Plant Protect Microb 16:35–44

Han EH, Hwang YP, Kim HG, Choi JH, Im Yang JH (2011) Inhibitory effect of *Pleurotus eryngii* extracts on the activities of allergic mediators in antigen-stimulated mast cells. Food Chem Toxicol 49:1416–1425

Haynes WA, Shandilya TR (1977) Casing soil and compost substrates used in the artificial culture of *Agaricus bisporus* (Lange) Singer - the cultivated mushroom. Indian J Mycol Plant Pathol 7:5–10

Hechler HC (1962) The development of *Aphelenchus avenae* Bastian, 1865 in fungus culture. Proc Helminthol Soc Washington 29:162–167

Hesling JJ (1979) The role of eelworms in culture of mushrooms in UK. Mushroom J 83:423–429

Jairajpuri MS (1968) Some studies on the morphology and biology of *Aphelenchus avenae* Bastian 1965 (Nematoda: Aphelenchoidea). Lab Dev J Sci Technol 6:27–34

Janowicz K (1978) Feeding and destructiveness of *Aphelenchoides sacchari* Hooper for mycelium of mushrooms. Rozniki Nauk Rolniczych E 7:209–220

Juhl H (1966) Nematode I Champignonkulturer. Tidsskr PIAvl 69:511–531

Kalac P (2009) Chemical composition and nutritional value of European species of wild growing mushrooms: a review. Food Chem 113:9–16

Kalac P (2013) A review of chemical composition and nutritional value of wild-growing and cultivated mushrooms. J Sci Food Agric 93:209–218

Kaneko S (1983) The wart disease of the oyster mushroom and its control. For Pests 32:201–203

Kanwar RS, Bajaj HK, Nandal SN (2009) Potential of a predatory nematode, *Fictor composticola* as biocontrol agent of nematode pest of mushroom. Abstract of the paper presented in National Symposium, Y.S. Parmar University of Horticulture and Forestry, Solan. 8–9 Jul 2009

Katyal P, Khanna PK, Kapoor S, Kaul VK (2007) Evaluation of azadirachtin formulations for the management of nematode, *Aphelenchoides composticola* infesting white button mushroom. Pestology 31:31–35

Kaul VK, Chhabra HK (1992) Effect of diflubenzuron on *Aphelenchoides composticola* infesting white button mushroom. Plant Dis Res 7:147

Kaur DJ, Thapa CD, Nath A, Sharma NK (1987) Evaluation of some nematicides against *Aphelenchoides sacchari*. Indian J Agric Sci 57:867–868

Keshari, N. (2016) Predatory behaviour of *Fictor composticola* Khan et al and its potential for the management of nematode pests of button mushroom. Ph.D. thesis, pp 81

Khan R, Bajaj HK, Sultana R, Tahseen Q (2008) Description of *Diplogastrellus gracilis* (Butschli 1876) Paramonov, 1952, *sikorai* sp. n. and *Fictor composticola* sp. n. (Nematoda: Diplogasterida) from India. Nematology 10:153–166

Khan A, Saifullah IM, Hussain S (2014) Organic control of phytonematodes with *Pleurotus* species. Pak J Nematol 32:155–161

Khanna AS (1991) Yield losses in white button mushrooms due to nematode infestation. Nematol Mediterr 19:35–36

Khanna AS (1993) Estimation of yield losses in white button mushrooms due to nemic infestation. Curr Nematol 4:173–176

Khanna AS (1994) Nematode problems in mushrooms : management. In: Nair MC, Gokulapalan C, Das L (eds) Advances in mushroom biotechnology. Scientific Publishers, Jodhpur, pp 157–159. pp:193

Khanna AS, Chandran R (2002) Studies on sources of nematode inoculum/dissemination in white button mushrooms. National Symposium on biodiversity and Management of Nematodes in cropping systems of sustainable agriculture held in agriculture research station, Durgapura, Jaipur during 11–13 Nov 2002, pp 32–33

Khanna AS, Jandaik S (2002) Role of *Aphelenchoides avenae* Bastian 1865 in cultivation of white button mushroom. Indian J Mushrooms 20:7–8

Khanna AS, Kumar S (2005) Effect on myceliophagous nematodes on flush pattern and yield of *Agaricus bisporus* (Lange) Imbach. Indian J Mushrooms 23:34–36

Khanna AS, Sharma NK (1988a) Pathogenicity of various *Aphelenchoides* species on *Agaricus bisporus*. Indian Phytopathol 41:472–473

Khanna AS, Sharma NK (1988b) Effect of population levels and time of infestation of *Aphelenchoides agarici* on mycelial growth of *Agaricus bisporus*. Nematol Mediterr 16:125–127

Khanna AS, Sharma NK (1989) Resistance of the cultivated mushroom, *Pleurotus sajor caju* to *Aphelenchoides agarici*. Nematol Mediterr 17:61

Khanna AS, Sharma NK (1990) Efficacy of nematophagous fungi against myceliophagous nematodes *in vitro*. In: Proceedings of national seminar on bio-agents in nematode management, New Delhi, IARI, pp 83–84

Khanna AS, Sharma NK (1992) Significance of males in the reproduction of mushroom nematode, *Aphelenchoides agarici*. Curr Nematol 3:101–102

Khanna AS, Sharma NK (2001) Incidence and frequency of distribution of mushroom nematodes in northern India. Indian J Mushrooms 19:56–59

Khanna AS, Grewal PS, Sharma NK, Sohi HS (1988) Screening of plant extracts for nematicidal action against *Aphelenchoides composticola* infesting mushroom beds. Indian J Hill Farm 1:41–44

Khanna AS, Chandran RA, Kumar SG. (2005) Effect of *Panagrolaimus fuchsia* Ruhm on mycelial growth, flush pattern and sporophore yield of *Agaricus bisporus* (Lange.) Imbach and nematode multiplication. In: National symposium on recent advances and research priorities in indian nematology, 9–10 Dec 2005, IARI, New Delhi, India, p 8

Khanna AS, Chandran RA, Kumar S (2006) Sources of nematode inoculum in white button and milky mushroom. Indian J Nematol 36:267–269

Kim K, Choi B, Lee I, Lee H, Kwon S, Oh K, Kim AY (2011) Bioproduction of mushroom mycelium of *Agaricus bisporus* by commercial submerged fermentation for the production of meat. J Sci Food Agric 91:1561–1568

Kim SH, Jakhar R, Kang SC (2015) Apoptotic properties of polysaccharide isolated from fruiting bodies of medicinal mushroom *Fomes fomentarius* in human lung carcinoma cell line. Saudi J Biol Sci 22:484–490

Kumar S, Khanna AS, Chandel YS (2007) Effect of population levels of *Aphelenchoides swarupi* and *Aphelenchus avenae* inoculated at spawning on mycelial growth of mushrooms and nematode multiplication. Nematol Mediterr 35:155–163

Kumar S, Khanna AS, Chandel YS (2008a) Damage potential of *Aphelenchoides swarupi* and *Aphelenchus avenae* in white button mushroom, *Agaricus bisporus*. Indian J Nematol 38:146–153

Kumar S, Khanna AS, Verma AK, Chandel YS (2008b) Effects of inoculation time on damage potential of *Aphelenchoides swarupi* and *Aphelenchus avenae* to Rodman's agaricus. *Agaricus bitorquis* 36:171–178

Kumar S, Khanna AS, Verma AK, Chandel YS, Jandaik S (2009) Effect of population levels of *Aphelenchoides swarupi* and *Aphelenchus avenae* on mycelial growth of mushrooms and nematode population. Indian J Nematol 39:14–24

Lambert EB, Steiner G, Drechler C (1949) The "Cephalothecium disease" of cultivated mushrooms caused by a nematode (*Ditylenchus spp.*) evidenced by surface development of predacious fungi. Plant Dis Rep 33:252–253

Madhuri, Kanwar RS (2016) Relative susceptibility of some mushroom strains to *Aphelenchoides swarupi* and nematode reproduction on them. Int Res J Nat Appl Sci 3:208–219

Mariga AM, Yang WJ, Mugambi DK, Pei F, Zhao L, Shao Y (2014) Antiproliferative and immunostimulatory activity of a protein from *Pleurotus eryngii*. J Sci Food Agric 94:3152–3162

McLeod RW, Khair GT (1978) Control of *Aphelenchoides composticola* in mushroom compost with nematicides. Ann Appl Biol 88:81–88

Mcleod RW, Khair GT (2008) Control of *Aphelenchoides composticola* in mushroom compost with nematicides. Ann Appl Biol 88:81–88

Mojumder V (1997) Nematode management. In: Narwal SS, Tauro P, Bisla SS (eds) Neem in sustainable agriculture. Scientific Publishers, Jodhpur, India, pp 163–177

Nagesh M, Reddy PP (2000) Status of mushroom nematodes and their management in India. Integr Pest Manag Rev 5:213–224

Nath A, Sharma NK, Bhardwaj S, Thapa CD (1982) Nematicidal properties of garlic. Nematologica 28:253–255

NRCM (2007) Perspective Plan Vision-2025, 48 pp. http://nrcmushroom.com

Oei P, Nieuwenhuijzen BV (2005) Small scale mushroom cultivation: oyster, shiitake and wood ear mushrooms. Agromisa Foundation and CTA, The Netherlands, p 86

Panjikkaran ST, Mathew D (2013) An environmentally friendly and cost effective technique for the commercial cultivation of oyster mushroom *Pleurotus florida*. J Sci Food Agric 93:973–976

Parmar BS (1997) Industrial uses of neem and future commercialization. In: Narwal SS, Tauro P, Bisla SS (eds) Neem in sustainable agriculture. Scientific Publishers, Jodhpur, India, pp 251–260

Rao MS, Pandey M (1991) Comparative efficacy of karanj leaf and carbofuran on *Aphelenchoides composticola*. Indian J Nematol 21:158–159

Rao MS, Pandey M (1992) Integrated management of mushroom nematodes. Mush Res 1:137–138

Rao MS, Parvatha Reddy P, Tewari RP (1991) Comparative efficacy of certain oil cakes against mushroom nematode, *Aphelenchoides sacchari* and their effect on yield of *Agaricus bisporus*. Indian J Nematol 21:101–106

Rao MS, Tewari RP, Reddy PP (1992) Occurrence of mushroom nematodes in South India. Mushroom Res 1:135–136

Read WM (1968) Pesticides for mushroom. M G A Bull 194:77–82

Reddy PP, Rao MS, Nagesh M (1997) Nematode management in horticultural crops. In: Grewal SS, Tauro P, Bisla SS (eds) Neem in sustainable agriculture. Scientific Publishers, Jodhpur, India, pp 229–246

Roupas P, Keogh J, Noakes M, Margetts C, Taylor P (2012) The role of edible mushrooms in health: evaluation of the evidence. J Funct Foods 4:687–709

Royse DJ (2014) A global perspective on the high five: *Agaricus, Pleurotus, Lentinula Auricularia & Flammulina*. In: Proceedings of the 8th international conference on mushroom biology and mushroom products, pp 1–6

Seth A (1984) Studies on the incidence, pathogenicity, biology and control of *Aphelenchoides* species associated with mushroom. Ph. D. thesis, HPKV, Solan (HP), India

Seth A, Sharma NK (1986) Five new species of *Aphelenchoides* (Nematoda: Aphelenchida) infesting mushroom in North India. Indian J Nematol 16:205–215

Shandilya TR, Munjal RL, Seth PK (1975) Combating insect pests of mushroom, *Agaricus bisporus* Sing. Indian J Mushroom 1:13–15

Shang HM, Song H, Xing YL, Niu S, Ding GD, Jiang YY, Liang F (2015) Effects of dietary fermentation concentrate of *Hericium caputmedusae* (bull.:Fr.) Pers. on growth performance, digestibility and intestinal microbiology and morphology in broiler chickens. J Sci Food Agric 96:215–222

Sharma NK (1995) Management of mushroom nematodes. In: Chada KL, Sharma SR (eds) Advances in Horticulture, vol 13. Malhotra Publishing House, New Delhi, pp 261–275

Sharma NK, Seth A (1986) Nematode problems in mushrooms: losses, symptomatology and management. In: Swarup G, Dasgupta DR (eds) Plant parasitic nematodes in india: problems and progress. ICAR, New Delhi, pp 384–394; pp 497

Sharma NK, Seth A (1993) Nematode problems of mushroom. In: Sitaramaih K, Singh RS (eds) Handbook of nematology. Cosmo Publications, New Delhi, pp 210–219. pp:336

Sharma NK, Thapa CD, Nath A (1981) Pathogenicity and identity of *Ditylenchus myceliophagus* nematodes infesting *Agaricus bisporus* (Lange.) singer in H. P. (India). Indian J Nematol 11:230–231

Sharma NK, Thapa CD, Damanjeet (1984) Estimation of losses in mushroom yield due to myceliophagous nematode, *Aphelenchoides sacchari*. Indian J Nematol 14:121–124

Sheu F, Chien PJ, Wang HK, Chang HH, Shyu YT (2007) New protein PCiP from edible golden oyster mushroom (*Pleurotus citrinopileatus*) activating murine macrophages and splenocytes. J Sci Food Agric 87:1550–1558

Siddiqi MR, Bilgrami AL, Tabassum KA (2004) Description and biology of *Mononchoides gaugleri* sp. N. (Nematoda: Diplogastrida) from New Jersey, USA. Int J Nematol 14:129–134

Siddiqui IA, Taylor DP (1969) Feeding mechanisms of *Aphelenchoides bicaudatus* on three fungi and an alga. Nematologica 15:503–509

Singh UA, Sharma K (2016) Pests of mushroom. Adv Crop Sci Technol 4:1–6

Subbiah KA, Balan V (2015) A comprehensive review of tropical milky white mushroom (*Calocybe indica* P & C). Mycobiology 43:184–194

Sutherland JR, Fortin JA (1968) Effect of the nematode, *Aphelenchus avenae* on some ectoparasitic mycorrhizal fungi and on a red pine mycorrhizal relationship. Histopathology 58:519–523

Thapa CD, Sharma NK (1981) Efficient peak heating of compost for eradication of mushroom nematodes. Indian J Mushroom 7:56–59

Thapa CD, Sharma NK (1987) Effect of temperature on pathogenicity and multiplication of two myceliophagous nematodes. Indian Mushroom Sci 2:169–172

Thapa CD, Sharma NK, Nath A (1981) Mushroom nematodes and their sources of contamination in mushroom beds in India. Mushroom Sci 11:443–448

Thapa CD, Jandaik CL, Sharma NK (1983) Comparative resistance in some cultivated edible fungi against two mushroom nematodes, *Aphelenchoides sacchari* and *Ditylenchus myceliophagus* occurring in India. In: Symposium of science and cultivation technology of edible fungi, Regional Research Laboratory, Jammu and Kashmir, India

Thorn G, Tsuneda A (1993) Interactions between *Pleurotus* species, nematodes and bacteria on agar and in wood. Trans Mycol Soc Jap 34:449–464

Tousson TA, Weinhold RC, Lindermann PZA (1968) Nature of phytotoxic substances produced during plant residue decomposition in soil. Phytopathology 58:41–45

Tsuda K, Kosaka H, Futai K (1996) The tripartite relationship in gill-knot disease of the oyster mushroom. *Pleurotus ostreatus* (Jacq.: Fr.) Kummer. Can J Zool 74:1402–1408

Vats R, Kanwar RS, Bajaj HK (2001) Biology of *Seinura paratenuicaudata* on *Aphelenchus avenae*. In: National congress on centenary of nematology in India: appraisal and future plans, 5–7 Dec 2001, IARI, New Delhi, pp 90

Vats R, Bajaj HK, Mann SS (2002) Assessment of losses in button mushroom due to myceliophagous nematodes. National symposium on biodiversity and management in nematodes in cropping systems for sustainable agriculture, 11–13 Nov 2002. Agriculture Research Station, Durgapura, Jaipur, p 38

Vats R, Bajaj HK, Rana RS (2003) Management of a myceliophagous nematode, *Aphelenchoides composticola* Franklin in *Agaricus bisporus* (Lange) Singer using non-chemical methods. Upadhyay RC, Singh SK, Rai RD (eds). In: Current vistas in mushroom biology and production, Mushroom Society of India, India pp 210–213

Vats R, Kanwar RS, Bajaj HK (2004) Biology of *Seinura paratenuicaudata Geraert*. Nematol Mediterr 32:117–121

Vats R, Singh S, Bajaj HK (2006) Pathogenicity of *Paraphelenchus* species of milky mushroom (*Calocybe indica*) and its multiplication on other fungi. Indian J Nematol 36:146–147

Vaz JA, Barros L, Martins A, Santos-Buelga C, Vasconcelos MH, Ferreira ICFR (2011) Chemical composition of wild edible mushrooms and antioxidant properties of their water soluble polysaccharidic and ethanolic fractions. Food Chem 126:610–616

Wang JC, Hu SH, Liang ZC, Lee MY (2005) Antigenotoxicity of extracts from *Pleurotus citrinopileatus*. J Sci Food Agric 85:770–778

Xu X, Yan H, Chen J, Zhang X (2011) Bioactive proteins from mushrooms. Biotechnol Adv 29:667–674

Yeates GW (1969) Predation by *Mononchoides potohikus* (Nematoda: Diplogasteridae) in laboratory culture. Nematologica 15:1–9

Zhou J, Chen Y, Xin M, Luo Q, Gu J, Zhao M (2013) Structure analysis and antimutagenic activity of a novel salt-soluble polysaccharide from *Auricularia polytricha*. J Sci Food Agric 93:3225–3230

Chapter 8
Plant-Parasitic Nematodes and Their Biocontrol Agents: Current Status and Future Vistas

Mahfouz M. M. Abd-Elgawad

Abstract Plant-parasitic nematodes (PPNs) pose so considerable threats to growing crops in size and quality that the figures reflecting averages of worldwide crop losses annually are staggering. Biological control agents (BCAs) rank high among other PPN management options, given mounting care to lessen application of chemical nematicides with a clear aim at the avoidance of human health hazards and attaining pollution-free environment. Nevertheless, BCAs are frequently slower acting, less effective, and more inconsistent than control normally achieved with chemicals. Therefore, the different groups of BCAs were reviewed herein to identify conditions and practices that affected their use for nematode management and alternatives to maximize their useful applications against PPNs. Various approaches to minimize costs, facilitate availability, optimize application, and enhance efficacy of these BCAs have been discussed. Researchers should further grasp the complex network of interactions among biotic and abiotic factors in intimate contact with these BCAs to maximize their gains via safe and skillful application and advanced technology. Integrated pest management programs in ways that make BCAs complementary or superior to chemical nematicides are highlighted.

Keywords Nematodes · Biological control agents · Microorganisms · Integrated management · Pesticide market

8.1 Introduction

In the whole world, there is a tendency to pay more attention to human health nowadays. This implies growing dissatisfaction with traditional farming patterns that rely on the use of the means of pest control, including chemically produced pesticides as well as other chemicals that cause health hazards; even if not certain, are sometimes expected. In one sense, with the increase in the human population

M. M. M. Abd-Elgawad (✉)
Plant Pathology Department, National Research Centre, Giza, Egypt

R. A. Ansari et al. (eds.), *Management of Phytonematodes: Recent Advances and Future Challenges*, https://doi.org/10.1007/978-981-15-4087-5_8

density, the need to enhance crop production in terms of quantity and quality by using nonconventional methods in combating harmful pests and pathogens is one of our present pressing problems so as to rid man and his environment of the harmful effects of such synthetic chemicals. Therefore, the current approaches of pest and pathogen management seek to furnish an integrated types of biological control that do not exclude the integration of the safe traditional control methods and maximize production with the effectiveness of nontraditional biocontrol methods to achieve the major objectives: access to a crop as safe as possible while doubling its productivity to meet the needs of the growing population density. Such current approaches should be applied to many agricultural crops that are quite susceptible to various types of pests, parasites, and pathogens. Admittedly, plant-parasitic nematodes (PPNs) rank high among these organisms that can cause considerable yield losses. However, because of the subterranean parasitic nature of most PPNs, their damage, especially to plant roots, can occur invisibly, which may wrongly minimize the importance of these nematodes. Typically, this improper estimation occurs from farmers and growers with little or insufficient knowledge of these nematodes. Moreover, PPNs are as ubiquitous as many other pathogens. Therefore, many economically important crops are plagued by different nematode pests worldwide. Eventually, PPNs are linked to almost every important crop globally and hence exert considerable restrictions on food security worldwide (Ansari and Khan 2012a, b; Abd-Elgawad 2014; Abd-Elgawad and Askary 2015). On the other hand, it is well-known that some PPN species may be endemic or indigenous to specific regions. So, an adequate biological control agent(s) (BCAs) should be searched to match such, or any other, nematode species. Thus, grasping dispersal methods and genetic relationships among PPN species/pathotypes are essential to develop sound quarantine regulations and adequate control measures (Blacket et al. 2019). Likewise, a biocontrol agent that works for a particular environment (e.g., hot climate) does not necessarily actively correspond to another (cold) environment. Therefore, factors such as moisture, temperature, soil characteristics, and different flora and fauna in the environment should be carefully sought in relation to other biological, chemical, and physical impact on the PPNs and their applied management measure(s). Factually, because of differences in such biological, chemical, edaphic, and physical factors, the presence of specific nematode pest(s) requires a biocontrol agent-nematode host matching, which should be virulent, and ecologically suitable to the relevant nematode species. Therefore, it is necessary to carry out experiments under different ecosystems and apply various biocontrol agent tactics such as nematodes, fungi, and bacteria to detect the most suitable and effective one in the laboratory and then in the greenhouse before field application under natural conditions (Moosavi and Zare 2015; Ansari and Mahmood 2017a, b, 2019a, b). Ultimately, the objectives are to obtain a safe and clean crop yield without hazards to man and environment, in the first place, and to maximize economic gains, if possible, in the second place.

New directions of research on biocontrol agents are in progress, given the PPN losses and new molecular and technological tools for harnessing them (Blacket et al. 2019; Blyuss et al. 2019). One of them is based on the thought of finding such an agent that can adequately manage one type of harmful pathogens, such as PPNs, on one hand, and also directly enhance plant production, on the other hand. Another direction aims at controlling more than one harmful nematode species and/or a set of pests using a common biocontrol agent. Other trends include products that contain bacteria or fungi that are also sold as plant strengtheners, plant growth promoters, or soil conditioners in order to promote plants' ability to withstand PPN attack. So far, these directions are being applied via numerous marketed products, while others are still under experimentation globally as it is required to do much of the laboratory experiments in the light of the multiple hypotheses to test the validity of each of them. Hence, functional sampling aimed at timely centering on isolating biocontrol agents of certain pests should be expanded in parallel to large-scale surveys for coordination in search of such reliable agents. Such functional sampling is based on adequate sampling method, time and site targeted, and extraction technique used for effective strain isolation via location of appropriate environment. This technique may avoid hurdles and costs of strain improvement, which means to modify genetics of the biocontrol agent in order to carry out a process more effectively (Abd-Elgawad 2020). In this chapter, the chief emphasis has been given over the losses, and their assessments, caused by PPNs on agricultural crops. Consequently, the importance of managing these nematodes especially via inclusion of BCAs along with their merits and demerits is shown. Then, general biological control strategies are outlined followed by a concise review that briefly classifies these BCAs with the advantages and disadvantages of each group. Current status of some issues and limitations related to BCA applications against PPN is given. Finally, the prospects of adequate use of these BCAs in integrated pest management (IPM) systems that have BCAs as main components to enhance their impact and expand their application and effectiveness are reviewed with conclusions for future prospects (Stirling 2011, 2014; Abd-Elgawad 2016a, b).

8.2 Assessment of Losses Caused by Plant-Parasitic Nematodes

Although plant nematodes have existed since time immemorial, their incurred losses have only been identified and estimated in recent decades. The successive estimates of these losses caused by PPNs have shown a steady increase in the figures resulting from these assessments (Abd-Elgawad and Askary 2015). Most of these estimates were made and published in the USA due to the remarkable scientific progress. For example, Poinar Jr. (1983) reported that PPNs can spoil about 15% of each year's agricultural yields related to different nematode-susceptible crop varieties in the USA. Later, a notable record on US agricultural crops and estimated nematode-losses became available (McSorley et al. 1987). The outcome of their bibliography was based on both scientists' opinions of the 50 American states and documented reports mostly including survey data. Sasser and Freckman (1987) offered a relevant and global estimate for specific crops. Their assessments relied on data from PPN surveys intentionally gathered for such estimations. Thus, they found an average loss of annual yield on the main crops to be 12.3%. The loss was 10.7% when the estimate extended to include the 20 life-sustaining crops, which were previously presented by Wittwer (1981). These crops were maize, wheat, barley, sorghum, field bean, banana, chickpea, soybean, sugarcane, sugar beet, coconut, cassava, sweet potato, millet, potato, pigeon pea, oat, rice, rye, and groundnut. Those authors did not exclude crops that are valuable for food or export, such as grape, coffee, guava, cacao, okra, papaya, citrus, melons, cotton, cowpea, pepper, yam, aubergine, tobacco, forages, tea, tomato, pineapple, and ornamentals, from estimating their PPN-related losses. They were subjected to such an assessment too. They also suffered a yield loss of 14% annually due to PPNs. Jointly, a total 12.3% loss is the overall average. Admittedly, nematode losses in developing countries are higher than those in developed countries because of the progress of the PPN control methods in the latter states. Hence, these losses averaged 8.8% in developed countries, but reached 14.6% in developing countries. On the other hand, when the most economically important 21 crops, including 15 life-sustaining plant species, were summed together, their nematode losses were found to be US$77 billion annually. The US share in it was US$5.8 billion. Many scientists have sensed that total losses due to the nematodes are higher than these figures at least because other crops have not been estimated. Others, like McCarter (2008), stated that such assessments are outcomes of informed opinions but not necessarily actual data, and so dependability of these values remains difficult to rate. Moreover, since more than three decades have passed since the publication of Sasser and Freckman's (1987) report, updated assessments are reviewed herein. In this vein, Abd-Elgawad and Askary (2015) assessed damages caused by PPNs globally as US$358.24 billion annually. Such an estimate was based on the weighted mean of losses and included 371 responses from the International *Meloidogyne* Project for 20 life-sustaining and 20 economically important crops. An annual yield loss of PPNs averaged 12.6% (or US$215.77 billion) for 20 life-sustaining crops but was 14.45% (or US $142.47 billion) for the other food or export crops; both groups averaged 13.5% (Abd-Elgawad and Askary 2015).

8.3 Phytonematodes Management through BCAs

Clearly, these figures of losses caused by PPNs reflect serious situation for the world's food security especially in less developed countries with overpopulations. In fact, the economic importance of PPNs is sentenced by its geographical distribution, pathogenic or parasitic potential, interaction with other pests and pathogens, and the specific crop value. This importance is extended to include farmers' awareness of these nematodes, and thus the available facilities and materials as well as the appropriate and effective means for their management in integrated pest management systems (Abd-Elgawad 2014; Abd-Elgawad and Askary 2015; Ansari et al. 2017a, b). On the other hand, accelerated public concern over chemical nematicides has created markets for biopesticide products worldwide. Although the long-standing concerns were related to health hazards and environmental pollutions, there are recent findings that call for other types of troubles and pose additional restrictions on using chemical nematicides for the management of plant-parasitic nematodes. For example, nematode-resistant cultivars and varieties have been implemented as an effective strategy for nematode management. Nevertheless, the sophisticated relationship between the parasitic nematode and plant has led up to an "evolutionary arms race" where resistance-breaking pathotypes of plant-parasitic nematode species may develop. In this respect, PPNs have sometimes evolved strategies to inhibit plant immune responses for forming feeding sites necessary for nematode feeding and development within plant roots. Various agricultural practices are also employed for nematode control, but extensive yield losses in crop quantity and quality demonstrate a desperate need for novel, environmentally friendly approaches to promote the present systems of nematode management. Utilizing or manipulating PPN-antagonistic microbes seems a promising scope to address this deficiency. Usually, these nematodes live in soil with many other groups of organisms and are subjected to suppression by many natural antagonists such as predatory nematodes (PNs), fungi, viruses, insects, bacteria, mites, and protozoa (Stirling 2011). However, even various species within each group may react differently with the nematodes. Admittedly, there are various ways for the antagonistic activities of BCAs. Thus, BCAs have been hitherto known to effectively suppress nematode diseases via competition, antibiosis, plant growth promotion, mycoparasitism, induced plant resistance, cell wall degradation, and rhizosphere colonization capability. Therefore, the relative impact on PPNs may differ from one BCA to another given also the various biotic and abiotic factors that affect BCAs' ability to suppress PPN populations. Eventually, PPN can reduce and/or degrade the value of various crops used for human food, clothing, and housing, as well as for feeding his cattle and domestic animals. They can disrupt the right balance afforded by the normal ecological modulation of organisms. So, given the aforementioned options of PPN management, it seems a sound approach to use natural enemies in controlling PPN populations so that their negative effects on economically important crops and environmental practices could be minimized. Such a strategic approach is called the biological control of PPNs. Almost all relevant researchers believe that the biological control of PPNs as an alternative to chemical nematicides and poisons can imply specific merits and demerits. Generally, merits and demerits of biological control can be summarized as follows (Askary and Martinelli 2015; van Bruggen and Finckh 2016; Barratt et al. 2018).

8.3.1 Merits of BCAs

1. Biocontrol is often intended to control a nematode species or specific group of PPNs (PPN community found in plant rhizosphere). Admittedly, biocontrol agents are usually introduced into cropping systems plagued with high and

consequently damaging levels of PPN populations to control them and bring the nematode population(s), they are meant to target, into economically harmless levels. Thus, they change the existing plant-damaging nematode densities to a better investing level(s) of crop profitability. Therefore, BCAs are regarded as green alternatives to mechanical or chemical control processes.

2. Natural enemies or antagonistics of phytonematodes introduced to the environment are capable of upholding themselves, usually by decreasing whatever nematode population(s) they are thought to manage. Conceivably, after their initial introduction, they become in intimate contact with their new but sustainable environment given enough PPN population(s) to feed on. Hence, BCAs usually face very little stress in their new ecosystem to keep the medium operating naturally. Interestingly, BCAs can be kept in such new ecosystem for a much longer time than other ways of nematode control.
3. This latter statement implies that BCAs can be cost-effective on the long run. Initially, it may be expensive to introduce BCAs to the system that has nematodes to control. Nevertheless, the persistence of these biocontrol agents as living organisms capable of living and reproducing (self-continuing nature) gives it the merit of superiority to other nematicides as well as their being harmless and environmentally friendly. Sometimes, BCAs may need to be applied only once when nematode-suppressive soil is obtained.
4. It is an effective method for nematode management once we are assured that the introduced BCA will be naturally inclined to target the damaging PPNs; mostly the harmful nematode population will decline.

8.3.2 Demerits of BCAs

1. Biological control's predicted results can be changeful. Ultimately, there are many unmeasured, hidden, physical, and biotic variables affecting whatever natural enemy you set loose in an environment. While it is targeted to manage a definite nematode species, there is always the possibility that this BCA will switch to a different process or target, even a different but beneficial nematode species. An outstanding example would be introducing predatory nematodes (PNs), as BCAs, of course, to control PPNs into soil containing entomopathogenic nematodes (EPNs) too. Indiscrimination may lead PNs to prey the beneficial EPNs. There is also a risk of disrupting the natural food chain via introducing a new species or BCA to an ecosystem.
2. Contrary to chemical nematicides that provide immediate results, BCA undergoes a relatively slow biological process. It may consume a considerable time to reach the required balance of suppressing the targeted nematode population(s). Upon its success, however, BCA provides long-term effect biological control.
3. It is not intended to completely eradicate PPNs in the targeted agro-ecosystem. Biological control agent of PPNs can survive only if there is something to feed on. On the contrary, eradicating their food population(s) would risk their own

safety. Therefore, they can only keep the biological balance adequate for them via decreasing, but not exhausting, all the PPN population(s).

4. Establishing a biological control system is usually an expensive process in order to get a reliable efficacy. These costs should cover all stages associated with development of biocontrol agents from the surveys to explore a potential BCA and goes through its tests of efficacy under different laboratory, greenhouse, and field conditions and ending with mass-production methods and appropriate formulation and packaging of this BCA to match the targeted nematode pest (Btryon 2019).

8.4 Biological Control Strategies

Nematode management should be carefully thought of even before seed sowing or transplantation of seedlings especially if preplant sampling revealed that PPN population density is above the economic threshold level. In this vein, a recent computation for improved nematode-sample size (Abd-Elgawad 2016c) and examples for calculating nematode management costs following both sampling and nematicidal application were reported (Abd-Elgawad et al. 2016). On the other hand, a slow management practice using BCA(s) that consumes considerable time to reduce the pest density is not a sound technique. Admittedly, IPM should always be applied either during a definite stage of plant growth or during all over the growing season. Thus, BCA usages combined with soil amendment, phytochemical/plant product, compatible chemical nematicide, resistant variety/cultivar, and/or include them, partially or in full, in crop rotation should be examined for best pest management systems. Moreover, a superior biocontrol agent should exercise multiple mechanisms in antagonizing the nematode pests. For example, *Pseudomonas* can exercise both induction of host resistance and antibiosis to suppress the disease-causing nematodes. It is substantial to examine compatibility of the BCA with such other IPM components to affirm that none of the materials included in the integrated application are mutually suppressed. In other words, BCA in IPM should be adequately handled, wisely exploited, and economically commercialized to manage PPN as synergistic or complementary component. Generally, there are three different strategies in biological control of pests as follows: classical, augmentation, and conservation biological control. A brief description of each strategy is given herein to point out its concept as a prerequisite to its proper but exclusive use in integrated pest control programs.

8.4.1 Classical Biological Control

Classical biological control is the introduction of the biocontrol agent into a new area where they do not occur naturally. In this case natural enemies are exotic and not indigenous to this new locale. The goal of classical biocontrol is to detect beneficial BCAs and transfer them into another place where the target pest is found. After such an introduction, it is expected that the BCA(s) will be permanently established to furnish continuous pest suppression without further human intervention. The process of classical biological control usually goes through the following stages: deciding the origin of the introduced pest, gathering an adequate amount of natural enemies accompanying the pest or closely related species, and then examining the chosen natural enemies thoroughly to ensure their efficacy, reliability, and adequacy for quarantine regulations. These tests are followed by mass production, formulation, packaging, and release of these natural enemies. After their release, investigations should confirm whether the newly introduced BCA(s) have been successfully established at the intended area, and to determine the long-term service of its existence (Sanda and Sunusi 2014).

8.4.2 Augmentation Biological Control

Augmentation is delivering a natural enemy that is not found naturally in enough numbers to control a pest below the economic threshold level. It is assumed that this biocontrol agent is found but in too few numbers to control a pest. So, its addition should improve the pest control. The augmentation approach is divided into two branches of its releasing strategy: inundative and inoculative. An early release of BCA at intervals starting with small population densities of both organisms, BCA and the pest, is inoculation biological control. Hence, this strategy expects that the BCA will manage the targeted pest after its reproduction and usually can persist in the pest ecosystem for a long time. On the contrary, inundative technique depends on releasing huge numbers of BCAs for immediate knockdown or significant suppression of pest population. So, it is usually used when the pest population is near or reaching its damaging level (Sanda and Sunusi 2014).

8.4.3 Conservation Biological Control

It is also aptly named as it is aimed at manipulating the pest habitat so that it conserves the activities of BCA on the targeted pest(s). This usually requires modification of the existing practice/eco-system so that it complies with such activities. So, this conservation technique is centered on methods that can be employed to enhance, maintain, or restore the natural PPN suppressive status; that is, PPNs are constrained by the bio-buffering mechanisms that restrain the nematodes to nondamaging level (Sanda and Sunusi 2014; Btryon 2019). Given the three aforementioned strategies for biological pest control generally, biocontrol of PPNs can often be done via (1) introducing antagonistic organisms, (2) conservation and increasing indigenous antagonists, or (3) a combination of both strategies, that is, conservation and increasing indigenous and applied antagonists (Timper 2011). The application of such biocontrol methods for PPNs has sometimes faced erratic or inconsistent results concerning reducing nematode populations. Frankly, such results are frequently unsurprising, given the countless, undetermined, and thus hidden, biotic, and abiotic variables influencing both BCAs and their nematode victims. These BCAs may be impacted not only by other soil organisms but also by the host plant. Therefore, current research work should provide more insights of such relations between the candidate BCA and relevant factors especially those in intimate contact with it. In one sense, BCA efficacy under field conditions is as sound as it considers the interaction of BCA with such factors to realize real optimism concerning successful application and more commercialization of bionematicides. Needless to remind that the huge amount of research in laboratory assays and greenhouse pots demonstrating significant levels of PPN suppression with antagonistic organisms was conducted under unrealistic conditions of heat-treated or fumigated soil to eliminate resident, neither calculated nor considered, plant-parasitic nematodes and other biotic factors such as other plant pathogens. Yet, a few researchers have confidence only in studies conducted in natural soil under field applications. As a compromise, laboratory assays and greenhouse experiments may be regarded as a rationale only for experimentation under real field conditions.

8.5 Classification of BCAs with the Advantages and Disadvantages of Each Group

Because of the abundant research on the use of BCAs against PPNs (e.g., Blyuss et al. 2019; El-Eslamboly et al. 2019), various aspects of applying each but common BCA groups and applications are addressed as follows.

8.5.1 Bacteria

Nematophagous bacteria may be classified according to their mode of action into parasitic bacteria and nonparasitic rhizobacteria (Siddiqui and Mahmood 1999; Eissa and Abd-Elgawad 2015), but these bacteria may also be divided into six groups (Tian et al. 2007). These groups are obligate parasitic bacteria (*Pasteuria*), opportunistic parasitic bacteria, rhizobacteria, cry protein-forming bacteria, endophytic bacteria, and symbiotic bacteria. Details about each of these groups as well as their interaction with biotic and abiotic factors were recently reported (Eissa and Abd-Elgawad 2015). Moreover, excellent references and illustrations on BCAs of the whole bacterial group were reviewed by several authors in Davies and Spiegel (2011) and Askary and Martinelli (2015). As obligate parasites, *Pasteuria* spp. is an outstanding example of such biocontrol agents against plant-parasitic nematodes

(Chen and Dickson 1998; Timper 2011). Four *Pasteuria* species parasitize nematodes: *Pasteuria thornei* infects root lesion nematodes, (*Pratylenchus* spp.), *Pasteuria penetrans* primarily infects *Meloidogyne* spp., *Pasteuria nishizawae* infects *Heterodera* spp. and *Globodera* spp., and *Pasteuria usage* infects *Belonolaimus* spp. (Wilson and Jackson 2013). Such specificity may be a merit since the bacterium does not harm nontarget nematodes that could be useful such as predacious or entomopathogenic nematodes. The dried plant material containing its spore-filled females or its separated endospores was employed for its field introduction given its relatively cheap costs in addition to previous problematic of its in vitro mass production. Currently, methods for its in vitro culture have been developed and it is being commercialized as a biopesticide (Gerber and White 2005; Abd-Elgawad and Askary 2018). Its isolate "Candidatus *Pasteuria usgae*" achieved notable successes where a reduction in *Belonolaimus longicaudatus* population densities occurred 13 months after inoculation in field plots (Giblin-Davis et al. 2003) as well as in other field trials (Luc et al. 2010a, b). Also, it has good results with root–knot nematodes. Kokalis-Burelle (2015) stated three methods of its application: seed, transplant, and postplant treatments for management of *Meloidogyne incognita* on tomato and cucumber and *M. arenaria* on snapdragon. Also, 3 years after *P. penetrans*-infested dried roots transfer to another field site could bring peanut-root galling in the infested plots to as low level as those in plots fumigated with 1,3-dichloropropene (Kariuki and Dickson 2007). In contrast, Tzortzakakis et al. (1996) supposed that indigenous *P. penetrans* in some trials could not increase in spite of continuous cropping of a root-knot nematode (RKN)-host due to the nematode selection process for decreased attachment of the bacterial endospores. Thus, decreased levels of bacterial parasitism occurred. Bharali et al. (2019) reported that bacterial bioagents showed more efficacy than fungal ones for managing RKN infecting black gram using seed treatment under pot condition.

As mentioned previously, additional bacterial groups and species, other than *Pasteuria* spp., have been applied for control of PPNs. Some of the commonly used bacteria as BCAs are listed herein (Table 8.1). *Burkholderia cepacia* reduced the *M. incognita* eggs numbers by 60–69% on bell pepper roots (Meyer et al. 2001). Also, *Pseudomonas aeruginosa* decreased *Heterodera cajani* in sesame by up to 58% and increased yield (Kumar et al. 2009). Examples of the many bacterial products that are being marketed for control of the nematodes were given (Abd-Elgawad and Askary 2018). Among them, Wilson and Jackson (2013) have focused on *Pasteuria* spp. and *Bacillus firmus* as commercial products against PPNs. This latter was also effective on *B. longicaudatus* infecting bermudagrass (Crow 2014). Currently, many other bacterial products are being produced or tested with varying degrees of success for PPN management.

Table 8.1 Microbial agents used in the management of phytonematodes

S. No.	Microbial agents	Crop	PPN	Mode of action/ mechanisms	References
1.	*Bacillus subtilis*, *B. pumilus*, or *Pseudomonas fluorescens*	Cowpea	*Meloidogyne incognita*	Antibiosis, production of enzymes, and toxins	Abd-El-Khair et al. (2019)
2.	(*Trichoderma asperellum* M2RT4; *T. atroviride* F5S21 and *T. harzianum* F2L4, *Trichoderma* sp. (MK4), *P. lilacinum* (KLF2 and MR2)	Pineapple	*Meloidogyne javanica*	Antibiosis, increased nutrient uptake	Kiriga et al. (2018)
3.	*Pseudomonas fluorescens*, *Bacillus megaterium*, *Pochonia chlamydosporia*, and *Purpureocillium lilacinum*	Black gram (*Vigna mungo*)	*Meloidogyne incognita*	Antibiosis, nematicidal toxins, parasite of nematode eggs and adult females, nematode juvenile inside the egg is destroyed by the rapidly growing hyphae.	Bharali et al. (2019)
4.	*Bacillus pumilus*, *Paenibacillus castaneae*, *Mycobacterium immunogenum*	Tomato	*Meloidogyne incognita*	Growth-promoting rhizobacteria, Antagonize the nematodes	Cetintas et al. (2018)
5.	*Bacillus firmus*	Bermudagrass	*Belonolaimus longicaudatus*	Enzymatic action, degradation of root exudates, root protection, and the production of a phytohormone	Crow (2014)
6.	*Spirulina platensis*, *Amphora coffeaeformis*	Cucumber	*Meloidogyne incognita*	Nematicidal effects of algal compounds such as flavonoids, polyphenols, phenylalanine, And	El-Eslamboly et al. (2019)

(continued)

Table 8.1 (continued)

S. No.	Microbial agents	Crop	PPN	Mode of action/ mechanisms	References
				antioxidants as well as mineral salts that facilitate the rate of Penetration of algal harmful by-products	
7.	*Bacillus firmus*	Maize, cotton, Sorghum, soybean, Sugar bee, turfgrass	Species of *Heterodera Hoplolaimus*, *Criconema*, *Pratylenchus*, and *Meloidogyne*	Enzymatic action, degradation of root exudates, root protection, and the production of a phytohormone	Wilson and Jackson (2013)
8.	*Trichoderma harzianum*	Many different crops such as tomato, maize, and cotton	*Heterodera cajani*, *Meloidogyne* spp.	Lytic enzymes like chitinase, glucanases, and proteases that help parasitism of *Meloidogyne* and *Globodera* eggs. The hyphae of *T. harzianum* penetrate the eggs and juvenile cuticle, Proliferate within the organism, and produce toxic metabolites.	Abd-Elgawad and Askary (2018)
9.	*Pasteuria Penetrans*; *P. thornei*; *P. Nishizawae*; *Candidatus Pasteuria usgae*	Many different crops such as tomato, maize, soybean, turfgrass, and cotton	323 nematode Species of 116 genera	Parasitism	Eissa and Abd-Elgawad (2015)

(continued)

Table 8.1 (continued)

S. No.	Microbial agents	Crop	PPN	Mode of action/ mechanisms	References
10.	*Bacillus* (more than 15 species) And *Pseudomonas* (more than 11 species) Are two common genera of rhizobacteria	Tomato, faba bean, cowpea, maize, strawberry, and others	Species of *Heterodera Hoplolaimus*, *Criconema*, *Pratylenchus*, and *Meloidogyne*	Interfering with recognition, production of toxin, nutrient competition, plant-growth promotion; induction of Systemic resistance	Eissa and Abd-Elgawad (2015); Abd-Elgawad and Vagelas (2015)

8.5.2 Fungi

Nematode parasitic and antagonistic fungi have also two major categories: nematophagous and endophytic fungi. Nematophagous fungi could further be divided into nematode-trapping fungi, endoparasitic fungi, egg- and female-parasitic fungi, and toxin-producing fungi. Their distribution, taxonomy, biology, ecology, and their mode of action were reviewed (Askary and Martinelli 2015). Most microplot and field experimentations on fungi as BCAs of PPNs have recently focused on different stages of sedentary nematodes (*Heterodera*, *Globodera*, and *Meloidogyne* spp.), that is, eggs, developing juveniles, and females of cyst and root-knot nematodes. *Purpureocillium lilacinus* (synonym: *Paecilomyces lilacinus*), *P. chlamydosporia*, and *Trichoderma* spp. are all widely distributed and searched soil inhabitants and some strains are aggressive parasites of the aforementioned nematode stages (Timper 2011). Species of *Trichoderma* can produce toxic metabolites as well (Sharon et al. 2001; Abd-Elgawad and Kabeil 2012). Like other BCAs, a good knowledge of the biology and ecology of these fungi should be wisely employed in their application strategies. Most research work has focused on trapping (predaceous) and endoparasitic fungi as antagonists of PPN. Abd-Elgawad and Askary (2018) reported commercial products that have fungi such as *Purpureocillium lilacinus*, *Pochonia chlamydosporia*, *Trichoderma harzianum*, *Aspergillus niger*, and *Arthrobotrys oligospora* as active ingredients. Yet, such fungi are not exceptions; that is, as other BCAs, they are frequently less reliable, more inconsistent, and slower acting than control usually obtained by chemical nematicides. In this vein, *P. lilacinus* is an excellent colonizer (Cabanillas et al. 1988) on young RKN females and their egg masses embedded in plant roots. Nevertheless, *P. lilacinus* lacks a mechanism of aggressive trapping or attachment device leading to limitation in controlling mobile nematodes (Esser and El-Gholl 1993). Moreover, different isolates of these fungi vary in their potential as BCAs under field conditions due to their interaction with various biotic and abiotic edaphic factors. Therefore, when applied, such fungi should be integrated with other PPN management tactics such as a compatible chemical nematicide, crop rotation, plant product (i.e., its active ingredient is phytochemical antagonistic toward PPNs), and/or resistant cultivar. Integrated pest management via seed treatment with compatible pesticide, BCA (fungus), and phytochemical antagonistic could be a reasonable, economical, and safe option under field conditions (Askary 2015). Hence, further compatibility tests should investigate the best combinations of fairly effective BCA with compatible nematicide/pesticide and/or botanicals in ways that make them superior or complementary to chemical nematode control method alone (Abd-Elgawad and Askary 2018). Clearly, none of the components contained in such treatments are mutually suppressed. In this respect, tomato growth parameters were better ($p \leq 0.05$) with *Pseudomonas fluorescens* GRP3 and organic manure together than either alone (Siddiqui et al. 2001). Factually, such an IPM concept along with recent technological developments and significant shifts in the relevant settings for wider BCA applications (Glare et al. 2012; Abd-Elgawad 2016a; van Bruggen and Finckh 2016; Barratt et al. 2018) should continue to be a major focus moving forward. In other words, IPM programs may further be properly utilized, cost-effectively exploited and environmentally friendly applied for PPN management. Such programs necessitate experimentations under different field conditions, and various modes of actions for both BCAs and chemical standards. Agricultural extensions should also follow up these IPM programs to examine their outputs especially for assessing the treated crops in their yield size and quality as well as population levels of PPNs, while guiding the growers/farmers via extension services. Consequently, wise extension services should also comply with the real field conditions: (1) consider the presence of antagonists of the beneficial fungi in plant rhizosphere and soil. Other microflora and microfauna in intimate contact with plant roots might affect the applied BCA in one way or the other, and (2) the frequent occurrence of high PPN population densities. In both cases, more suitable and consequently effective components of the IPM programs should be sought of. Therefore, permutations and combinations of such components for effective IPM programs in order to meet various ecosystems should be materialized via intensifying international coauthorship and increasing multilateral collaboration in relevant fields of biological control.

8.5.3 Mites

Because mites usually have aboveground life and activities, their biocontrol potential was also studied against aboveground PPNs. In this respect, the acarid mite *Rhizoglyphus echinopus* was recorded as an exclusive predator of *Aphelenchoides* spp., which infects plant leaves and buds (Sturhan and Hampel 1977). Nevertheless, nematophagous mites were reported to have low biocontrol potential whether alone or with other predators due to their short-term, noneconomic PPN management (Gerson 2015). Hence, multiple techniques via various combinations of different components including *R. echinopus* for effective IPM programs may be tried. These may include agrotechnical approaches, for example, solarization and soil aeration, adding soil amendments, including various green manures and composts, resistant and tolerant cultivars, crop sequence, and chemicals in terms of compatible pesticides. Eventually, a tactful utilization of BCAs should rely on informed decision with regard to relevant knowledge, that is, targeted PPN species, their biology, ecology, and natural enemies, and perfect comprehension of the other relevant biological and edaphic factors that may interact with each other and with the host plants (Kerry 1997; Abd-Elgawad 2016a, b). On the other hand, the distribution patterns of the nematodes and their natural enemies, as well as their compatibility with relevant chemicals, are fundamental to better grasping of BCAs persistence, effect on populations of other organisms, and development of predictive models for control programs (Jatala 1986; Abd-Elgawad 2016b). Stirling (2014) stressed the need to get more accurate assessment of predation rates in the soil so that a decision maker can adequately construct prediction model using any predator's influence on PPN population densities. Such an assessment should take into account competing organisms (e.g., fungi, bacteria, and other predators, including collembola, rotifers, and predatory nematodes) that can kill PPNs in the soil.

8.5.4 Plant Growth-Promoting Rhizobacteria and Fungi

Some researchers prefer to acknowledge them as plant growth promoters rather than bionematicides or BCAs (e.g., Wilson and Jackson 2013). As is aptly named, these rhizobacteria and fungi have the ability to enhance plant growth. Consequently, they can also contribute in raising plants' capability to tolerate/resist nematode infection and feeding. Also, Wani (2015) mentioned such bacteria as *Azotobacter*, *Bradyrhizobium*, *Bacillus*, *Burkholderia*, *Azospirillum*, *Pseudomonas*, *Rhizobium*, *Serratia*, and *Streptomyces* to play a role in the PPN management. They have different activities that may include colonizing plant root system and seed surface, uptaking nutrients, solubilizing minerals, producing siderophores, fixing atmospheric nitrogen, and releasing phytohormones (auxins, cytokinin, and gibberellins) and antibiotics. Therefore, plant growth-promoting rhizobacteria and fungi (PGPR and PGPF) may, in one way or the other, contribute in the biocontrol of PPN. However, their further experimentations under real field conditions are still needed given the reliance of their most experimental data on pot tests or under controlled conditions. Although commercial formulations and production of PGPR as biopesticides are available at niche markets, most PGPR products are sold as plant strengtheners for enhancing crop yield. In one sense, PGPR and PGPF utilization may be harnessed for substantial promotion of yield for specific crops via possible enhancement of plant growth and suppression of PPNs. Yet, further research is required to adjust the various factors interacting with both the growing plants and BCAs as well as to verify the validity of the results and find out the causes of disparities and differences in some of their applications before the practical application on large-scale treatments as well as before adopting them in IPM-based strategies. For example, two *Paenibacillus castaneae* isolates decreased the number of root galls and egg masses but had no influence on plant growth compared to the control. In another inconsistent results *Mycobacterium immunogenum* isolate could enhance both shoot fresh weight and plant height while another isolate of the same bacterium increased only plant height (Cetintas et al. 2018). Such examples demonstrate that we are in the critical phase concerning the use of microbe-based products in agriculture. The gains of such microbe-based products are known by various names that suit their mostly expected benefits such as bionematicides, biostimulants, biofertilizers, bioinoculants, bioinsecticides, biopesticides, and bioformulations. Yet, impediments related to their efficacy, quality, and performance reduce their demand in the markets. Because of these restrictions, bionematicides are not an exception and, therefore, are lacking quite beyond the synthetic chemicals. Thus, novel approaches and directions should be explored to solve these problems and consequently build good confidence in such products among the end users/farmers. Researchers have been manifesting the trends for developing formulations that will be more effective and coordinated (Glare et al. 2012; Arora and Mishra 2016). In this respect, metabolites combined with plant growth-promoting (PGP) microbes are demonstrating great promise, showing more reliability with multiple benefits and consequently they are a way forward in nematode management and crop protection (Arora and Mishra 2016).

8.5.5 Arbuscular Mycorrhizal Fungi

These root fungi are obligate symbionts. They were assessed to colonize more than 80% of plant species globally (Schouteden et al. 2015). They can enter the cortical cells of vascular plant roots to establish arbuscules. An arbuscule resembles a tuft of hairs or cilia or a branched treelike organ. While vascular plants help arbuscular mycorrhizal fungi (AMF) to develop, these fungi support plants to absorb nutrients such as nitrogen, sulfur, phosphorus, and other micronutrients from the soil. It is supposed that plants with good nutrient status can tolerate higher PPN population densities in their roots. Such a supposition was materialized in cotton fields infested with the reniform nematode *Rotylenchulus reniformis* (Pettigrew et al. 2005). Yet, no solid data are available to affirm that the AMF-promoted nutrient status is inducing a higher resistance against PPN. The employment of AMF to manage PPN on numerous crops is also an eco-friendly plausible avenue. Growing satisfaction in their use against PPN was reviewed by Sankaranarayanan (2015). They could also reduce losses induced by other soil-borne plant pathogens. As an outstanding group among PPN, sedentary nematodes have been the paradigm of many researchers to be tried for control by species/strains of AMF. They are also effective against the migratory nematodes. Recently, AMF have demonstrated capability to mitigate salt stress in plants. Moreover, two AMF strains mitigated stresses on the nutrient uptake of *Zelkova serrata* seedlings, photosynthesis, and consequently inhibition of growth under salt stress (Wang et al. 2019). Abd-Elgawad 2016b pointed out the importance of AMF culturing in the field using suitable plants, to promote the AMF inoculum in the soil and allow rapid and more densities of root colonization, prior to the next targeted crop. Adequate integration of BCAs such as PGPR, PGPF, nematophagous fungi, and AMF may enhance plant growth and control PPN. Schouteden et al. (2015) reviewed various mechanisms involved in biocontrol effect of AMF against PPN. These mechanisms comprise induced systemic resistance (ISR), direct competition for nutrients and space, enhanced plant tolerance, and altered rhizosphere interactions. Such mechanisms are fairly interdependent. Therefore, their efficacy as biocontrol agents probably results from a combination of several mechanisms (Cameron et al. 2013). The relative significance of a specific mechanism can differ according to the specific AMF–pathogen–plant interaction. The current advances in the disciplines of induced systemic resistance (Pieterse et al. 2014) and on the role of the root zone in biological control (Cameron et al. 2013) will possibly guide more reliable applications of AMF for PPN management.

On the other hand, because of their obligate nature, in vitro production of AMF is still difficult to obtain as the fungi rely on their propagation manner on plant roots. Admittedly, some plants can display a defense reaction to the fungal infection. Interestingly, such mycorrhiza-induced resistance (MIR) may provide systemic protection against a broad range of pathogens attacking the plant. This is similar to induced systemic resistance (ISR) following root colonization by nonpathogenic rhizobacteria and systemic acquired resistance (SAR) after pathogen infection. Yet, MIR reflects plant reactions to colonization of the fungi directly but indirectly implies plant responses to ISR-eliciting rhizobacteria (Cameron et al. 2013). Martínez-Medina et al. (2011) evaluated the interactions between fungus *T. harzianum* and four AMF (i.e., *Glomus claroideum*, *Glomus constrictum*, *Glomus mosseae*, and *Glomus intraradices*) for their integrated efficacy against Fusarium wilt on melon crop growth under conventional and field conditions. They found a synergistic effect on AMF root colonization due to the interaction between *G. intraradices* or *G. constrictum* and *T. harzianum* under a reduced fertilizer dosage. The AMF-inoculated plants were effective in controlling Fusarium wilt. Coinoculating *T. harzianum* and AMF showed a more effective control of Fusarium wilt than each AMF inoculated alone, but with similar efficacy to that of *T. harzianum*-inoculated plants. Coinoculation of *T. harzianum* and *G. mosseae* was more efficient than any other combination tested. Developed molecular tools are available to actually unravel the underlying mechanisms of BCAs, making this a timely opportunity for further progress and facing challenges ahead. For example, the use of a proteinase Prb1-transformed line (P-2) that had multiple copies of proteinase Prb1 gene could enhance biocontrol efficacy of *T. harzianum*, as BCA against the root-knot nematodes, compared with the nontransformed wild-type strain. While the wild *T. harzianum* was capable to colonize *Meloidogyne javanica*-separated eggs and second-stage juveniles (J2) in sterile in vitro assays, the engineered strain P-2 could further penetrate the egg masses (Sharon et al. 2001). Those authors concluded that enhanced proteolytic activity of the BCA might be significant for the biological control of the nematodes.

8.5.6 Predatory Nematodes

A solid background has been established concerning predation capacity, prey searching power, and other biological features of these predatory nematodes (PN). Nevertheless, more research is needed about their utilization as BCAs especially for effective introduction against PPN based on cost-effective mass production, longevity in soil, and long stability (storage). Their efficacy relies on prey-searching capability and the product of predation ability in terms of the real number of PPN preys captured and killed. The process of prey predation may be divided into the following phases: encountering the prey and assaulting response followed by feeding (composed of attack, extracorporeal digestion, and ingestion; Bilgrami 2008). Mononchid predators of the families Anatonchidae, lotonchidae, and Mononchidae meet prey nematodes via incidental contact. Consequently, they have a low chance of prey reaching but other PNs such as aphelenchids, diplogasterids, and dorylaimids make conscious contact with prey nematodes via chemosensory responses to kairomones sent out by the prey and delivered by the predator which promote the possibility of predator–prey meet (Kim 2015). On meeting their prey, *Seinura* species, as aphelenchid predators, inject toxic substances in the prey resulting in loss of its mobility which facilitates the process of predation. Contrary to dorylaimids and mononchids, aphelenchids and diplogasterids possess superior attributes as BCA due to their remarkable colonizing power, short life cycle, and elevated reproduction potential (Kim 2015). On the other hand, among these various groups of PN, diplogasterids are the best biocontrol agent of nematodes due to their resistance to adverse conditions, easy culture, chemotaxis sense, short life cycles, and prey specificity (Khan and Kim 2007). Thus, *Mononchoides* spp., related to diplogasterids, can be virtually introduced for PPN control by enhancing their population densities using soil amendments of compost and manure to encourage them to reproduce rapidly. Kim (2015) suggested a huge introduction of mononchids as a strong instant knockdown of high PPN population levels in crops growing in protected areas, for example, high plastic tunnels and glasshouses especially because of their vigorous and aggressive feeding manners.

Clearly, conscious cultural exercises should be utilized to maintain dorylaimids for the long-term control of chronic nematode diseases. Because of their feeding strategy on bacteria and nematodes and huge reproduction ratio, mass production is most reasonable for diplogastrids, contrary to mononchids that have critical strategy as predators. Aphelenchids and dorylaimids-cultured growth are fairly acceptable because of their feeding manner and modest culture in vitro. Kim (2015) pointed out that formulations utilized for EPNs can be applied in the PN commercialization to represent acceptable BCAs. Diplogasterids are prime candidates since they share a similar biological feature relative to the infective juveniles of EPN which can survive harsh conditions. In one sense, the PN-mass production with a long stability and storage (shelf-life) should be sought of in the biological control of PPN. Such a statement reflects merits for the attributes of diplogasterids, but demerits for mononchids and neutral features for both aphelenchids and dorylaimids. Yet, even more pronounced than other groups of BCAs, the general poorest attributes of PN for the biocontrol of plant nematodes are their difficult mass production and short shelf-life. These barriers should be compromised to enhance their biocontrol efficacy. On the other hand, the PN predation strategy is superior or equivalent to the antagonistic ability of other BCAs such as fungi and bacteria. These latter depend mostly on slow activities of antibiosis, competition, and parasitism contrary to PN, which usually have instant lethal predation (Kim 2015).

8.5.7 Nematode-Suppressive Soils

These soils are the best site for detecting potent natural BCAs (Timper 2011). Their suppression of soil-borne pathogens, including PPN, is due to having elevated antagonistic potential. In a suppressive soil, PPNs do not establish, persist, or reproduce rapidly and therefore do not significantly damage the growing plants (Ansari et al. 2019). Researchers have been studying suppression of, especially sedentary, nematodes in such soils intensively. In this vein, fungi as BCAs could induce suppression in soybean (Chen 2007), sugar beet (Westphal and Becker 1999, 2001), and cereal cyst nematodes (Kerry 1980), as well as RKN (Pyrowolakis et al. 2002; Adam et al. 2014). However, the exact mechanisms of the suppression are not adequately grasped. Admittedly, the role of biotic and abiotic factors in such suppression is obvious (Janvier et al. 2007), where microfauna and microflora have a significant contribution (Weller et al. 2002) as they maintain biological productivity (Garbeva et al. 2004). In other words, altered soil microbial diversity with the consequent soil's disease suppressiveness relies on the plant type, edaphic factors, agricultural practices (including PPN control options), and reciprocal action among soil microorganisms as well as their interaction with those other factors (Garbeva et al. 2004; Giné et al. 2016). The identification and determination of microbial and nematode population levels could be done by culture-independent methods, for example, denaturing gradient gel electrophoresis (Muyzer et al. 2004). This approach permitted detection of the total microbial structure in soil (Smalla and Heuer 2006). Recently, however, identifying natural enemies that potentially regulate the temporal abundance of naturally occurring nematodes were conducted via DNA extraction and measuring target organisms using real-time polymerase chain reaction (qPCR) (Campos-Herrera et al. 2019). The fluctuations of the RKN population levels in *M. javanica*-suppressive soil as well as in *M. javanica* and *M. incognita*-suppressive soil were identified by Giné et al. (2016). Both are suppressive greenhouse soils where vegetables had been organically grown. Comparing microbial profiles of suppressive and nonsuppressive soils, they noted that these fluctuations were affected by fungal egg parasites during the crop rotation. In addition to the contribution of biotic parameters specified in both sites to soil suppressiveness, a series of agricultural exercises, including the addition of organic amendments, crop rotation (with cover crop as green manure and RKN-resistant cultivars), and cultivation date, can suppress nematode reproduction. Nevertheless, the correlations between crop management practices and microbial community levels that cause soil suppressiveness should further be examined to plan strategies that can promote the antagonistic potential of soil. Eventually, it is expected that the recently used molecular tools (e.g., Campos-Herrera et al. 2019) will significantly facilitate analysis of soil ecosystems with a complex network of interactions among biotic and abiotic parameters to maximize the benefits and various applications of BCAs.

8.6 Have Biocontrol Agents of Plant-Parasitic Nematodes Come of Age (Glare et al. 2012)?

Many businessmen, scientists, and researchers have been trying to predict and react to the pesticides market analysis and demand. So, the aforementioned question has been raised but probably more extensively to include various types of pesticides with special interest in the safe biopesticides (e.g., Glare et al. 2012). Orders for pesticides are expected to increment at a positive compound annual growth rate (CAGR) of about 3.67% during the years 2019–2025 (https://industryarc.com/Report/18229/pesticides-market-research-report-analysis.html). Such an increase will reflect advancements that have obviously been made resulting in many new effective biopesticides on the market. This progress is mainly due to intensifying hard work for novel biopesticides in addition to relevant technological developments, new molecular tools, and gene transformation used. Admittedly, biocontrol agents of plant-parasitic nematodes are not exceptions in this general progressive status. Nevertheless, they have not yet attained their hoped prospects, even though all expectations advocate that general bionematicides, which may include plant extracts and other naturally sourced materials in addition to BCAs, will outperform other PPN control measures. In other words, their market share is predicted to expand continuously. Such a prediction may be materialized; the keys should be increased academic–industry partnerships and a change in mindset away from the further use of the conventional chemical pesticide model. Therefore, the following are several factors that may drive more market demand for such BCAs.

8.6.1 Enhancing Commercial Demand of Biocontrol Agents

The important operators for more commercial demands of a BCA can be attributed to intrinsic, ecological, technological, societal, regulatory, and commercial items (Moosavi and Zare 2015).

8.6.1.1 Intrinsic Parameters

These are observable attributes of a BCA that reflect its genetic constitution or result from the interaction of its genotype with the environment. Such many attributes as production of nematotoxic materials, proper establishment and performance in certain ecosystems, creation of robust survival structures, promoting plant growth, compatibility with other control methods, safe to the environment, tolerance to antibiotics and agrochemicals, inducing plant host defense mechanisms, and capability of establishing and reproducing in soil, and secretion of antibiotics are present or absent naturally in the BCAs. Of course, the degree of importance of each attribute depends on its contribution to reduce PPN population level(s) in specific environment. Generally, factors such as aggressiveness and persistence of a BCA are so worthy provided that they sustain reproducible levels of PPN control under field conditions. For instance, isolates of *Pochonia* species obtained from similar soils considerably differed in their parasitic ability (Moosavi et al. 2010). Also, an innate attribute to kill the PPN target is superior to changing behavior or sublethal effects on the nematode that can be recovered. Other such factors have been presented in detail (Moosavi and Zare 2015).

8.6.1.2 Ecological Parameters

There are many such factors that influence the degree of success or failure of a BCA to control PPN. They may be divided into abiotic factors (soil temperature, soil type, soil moisture, soil pH, soil nutritional status, concentration of heavy metals and interactions among soil abiotic factors) and biotic factors (soil organisms, host plants, nematode target and interactions among soil biotic factors). Various aspects of all important ecological factors should be considered at selection, development, and application of a BCA.

8.6.1.3 Technological Parameters

Successful commercialization of BCAs may require technological development in aspects related to scale-up production, formulation, stabilization, and delivery system. Otherwise, BCAs may fail to commercialize because of high cost for their production, unavailability of suitable and stable formulation, unpractical dosage advice, special conditions for their storage, inaccessibility to proper immobilizing materials, and inappropriate delivery systems (Moosavi and Zare 2015). Moreover, a few defects in experimentations and applications against PPNs should be considered for improving (Abd-Elgawad 2016a). For instance, estimating BCA efficacy, based on nematode-egg mass index (EI) is preferable to that based on gall index (GI) because the former index measures nematode fecundity. Also, El does not scale reproduction properly because it does not quantify the number of eggs. Egg numbers may be a better parameter of sedentary nematode multiplication than GI, EI, or other developmental stages. In addition, fragile links in a nematode's life cycle that can be targeted by BCA should be addressed more properly. The efficacy of BCAs not only on nematodes but also on crop yields as the crux of the matter should be examined. Such an efficacy is usually based on PPN development but targeting pathogenesis-related proteins and/or relevant compounds as fast and accurate biochemical markers or components of systemic or induced resistance in plants is a useful approach (Abd-Elgawad 2016a).

8.6.1.4 Societal Parameters

Accelerating public concern about overuse of chemical pesticides has sparked wide interest in the development of environmentally friendly biological alternatives such as BCAs against PPN. It is interesting to know that damage costs of chemical pesticides to the environment have evidently exceeded the total purchase value of all pesticides. Therefore, actual charges of using pesticides are greater than twofold of what is paid by the farmers and could be considered as society financial assistance to the chemical control of pests. Such charges increase yearly because of bulking of such chemical hazards and contamination (Moosavi and Zare 2015).

8.6.1.5 Regulatory Parameters

After isolation, screening, characterization, mass production, formulation, and packaging of a qualified BCA, there are other essential commercial, legal, and financial charges. There are a variety of regulations that may vary from one country to another for the handling and use of BCAs. These commercial products are usually subjected to the rigorous and costly registration processes in North America and Europe. On the other hand, other countries in Asia and Latin America possess numerous small industries using cheap labor to produce microbial products at low cost primarily for domestic markets, although some aim to export many of these products that may be efficacious (Wilson and Jackson 2013). The registration of the BCA product is the main obstacle for commercialization of microbial products. It is very complex, elaborate, time-consuming, and expensive. These issues have caused many products being brought to the market without meeting registration processes and are sold under other names such as biofertilizers, biological activators, or soil conditioners. The use of these wrongly named materials will usually be linked to an unestimated level of hazard, given that the product data on aspects such as ecological impact and toxicity are not measured. Yet, protection of intellectual property rights (product or technological idea) is vital for the future of the biopesticide companies via patent protection.

8.6.1.6 Commercial Parameters

A perfect detailed business plan (BP) based on costs and profits is sought of for each BCA product. So, precise and careful estimation of sales volume and an in-depth knowledge of markets and customers must be accurately predicted on the long run. A qualified manager must have proficiency in areas such as production, buying, selling, finance, and management of the personnel. Also, rapid profit-making is frequently not possible and so many businesses have withdrawn from these projects. Frequently, a BCA will be sold only if it could ensure an economic return.

8.6.2 Market Parameters

Restrictions in selling a BCA are related to market share and market size. Therefore, analysis of these factors is necessary to determine whether to process this trade or not. Usually, the real market size is much smaller than the potential market size. It is also necessary to assign a rational end-user charge for the product. So, one should be aware of prices of other products that they are trying to substitute the BCA and should compensate for all costs involved (Moosavi and Zare 2015).

8.6.3 Distribution Parameters

Large companies often have related distribution units that may also have contracts with small corporations and cottage industries. In some countries such as Egypt (Abd-Elgawad 2017), BCA producers, frequently of cottage industry, have adopted a cost-effective plan as distributors too where they cater mainly to the local markets. In this case, distributors are experienced in different aspects of BCA application and performance against PPN. They can recognize relevant issues of PPN control such as the viability of BCAs, favoring factors for BCA success and persistence. Also, being responsible for following up BCA performance will increase profit margins while ensuring rigorous quality control, self-regulation, and proper feedback. Eventually, costs, profit margins, product efficacy, grower acceptance, crop value, and size of market are determinative factors.

8.7 Current Issues and Future Prospects in the Biological Control of PPNs

Having presented the contemporary status of biological control agents, the current issues for their utilization against PPN are related to the following key factors (Askary 2015).

8.7.1 Isolation of a BCA

Pros and cons of an isolated BCA should be carefully assessed and decided for its potential as BCA on a case-by-case basis. For instance, *P. penetrans*, as an obligate parasite of active and sedentary nematodes with storage capability due to its endospores, is poorly spread in soil. Some of its isolates are very host specific and an isolate selection is difficult because spore burden is not always correlated with virulence. The problem of its in vitro mass production has been pending for a while until it has been resolved. Species of *Arthrobotrys* enjoy with broad host range and are readily in vitro mass produced, but they are frequently hard to formulate. Likewise, *P. lilacinus* is applied successfully worldwide although it has short duration survival and consequently needs multiple applications. Admittedly, many isolates of a definite species, employed as BCA, have great variations among them as mentioned previously.

8.7.2 Product Cost

Openly, costs and reliability are two major factors that impede the use of many BCAs. So, processes involved in their isolation, screening, characterization, mass production, formulation, and packaging should be cost-effective and well-qualified as best as we can (Abd-Elgawad and Vagelas 2015).

8.7.3 Ecological Factors

These factors have been mentioned previously as one of the main sources for the failure of a BCA. Thus, the interaction of BCAs with edaphic biotic and abiotic factors should be adequately considered.

8.7.4 Application Technology

An adequate delivery of BCA to the targeted point is essential for success. Rhizosphere bacteria have short-term control against PPNs if applied as seed treatment. Other BCAs are species-specific and therefore should not be applied to polyspecific nematode communities. The mass production and formulation methods should be suitable for and comply with application technique. Alginate pellets for spores formulations can efficiently store, distribute, and apply the fungal BCA. Yet, the pellets may be destroyed by mites and collembola found in the soil (Askary 2015; Moosavi and Zare 2015).

8.7.5 *Variability in Efficacy*

Usually BCAs are added to the soil that has biologically, physically, and chemically complex ecosystem. The interactions between BCA and all such variables in the rhizosphere are also complicated as mentioned previously, given that the nematode control with BCA is a slow process that depends upon the efficacy of the BCA as affected by these variables. That is why there is usually a BCA efficacy gap between laboratory and field results. A reliable utilization of egg- and female-parasitic fungi, *P. lilacinus*, relies on such factors as viability, virulence, method of application, inoculum concentration, and environmental conditions; its virulence varies under different conditions.

8.7.6 *Limitations of Expanding the Use of BCAs against PPN*

Several major factors are responsible for limiting the expansion of BCA application. Although they are general factors, but their degree of importance varies from one region to another. They may fluctuate from lack of know-how as well as relevant knowledge and experience of the growers, farmers and end-users especially in developing countries to the high product cost, low availability, and poor efficacy of the BCAs. Therefore, future vistas in maximizing their beneficial applications should be explored.

8.8 Future Vistas in Maximizing the Useful Applications of BCAs against PPNs

The factors that help to develop beneficial applications of BCAs against PPN are mainly based on addressing the BCAs deficiencies and other shortcomings mentioned in detail in this chapter. In addition, we should maximize the utilization of the BCA useful aspects. This could be attained through the following.

8.8.1 *Minimizing Costs*

Availability of inexpensive BCAs in sustainable agriculture requires optimizing their cost. Prices of BCAs products should be addressed locally to suit the conditions of agriculture, types of the cultivated crops, and the standard of living in a specific region. For example, most developing countries lack the capital and expertise to develop a biopesticide production industry, especially one based on costly and stainless steel fermenters and centrifuges, which require highly skilled workers.

Lacking capital or expertise for conventional fermentation, the default strategy is to optimize BCA in vivo culture whenever available. For instance, the dried plant material containing the required *Pasteuria* species via its spore-filled females or its separated endospores was employed for field application, given its relatively inexpensive costs. Furthermore, efforts should always be directed towards increasing the BCA production efficiency and/or decreasing the culture process time.

8.8.2 Easy Availability

The accessibility of BCA products can be made easier via enlightening growers and farmers by the biological control schemes and gains, and increasing the BCA production and its shelf-life via proper manipulation.

8.8.3 Simple and Skillful Application

Being aware of the complex network of interactions among biotic and abiotic factors in the treated ecosystem, the applicator should consider all factors necessary to maximize benefits of the used BCAs. For instance, the users/applicators should expect the best BCA rate/concentration to be applied, the interactions among BCAs, target nematode species, other soil microorganisms, and the plant. If an unformulated BCA is produced cheaply and locally in relatively large amounts, it can be mixed into the soil. On the contrary, under a developed agriculture system, crops are grown in monocultures over large areas using standard application machinery.

8.8.4 Enhancing Efficacy

An excellent BCA should possess the attributes of an ideal biological control agent. These could be summarized in its good initiation in soil, its self-continuing ability and rapid growth using the targeted nematodes, survival under adverse condition, and long-lasting virulence. Molecular tools should be harnessed for a better grasping of the biology and ecology and metabolism from whole organism to molecular scales and consequently effective BCAs could be manipulated. Researchers have developed some molecular techniques such as qPCR, restriction fragment length polymorphism (RFLP), and biological methods such as dilution plating on a selective medium, to monitor the occurrence, abundance and activity of the fungus in the soil, rhizosphere, and nematode egg masses (Atkins et al. 2003). Zhang et al. (2008) applied green fluorescent protein (GFP)-like proteins as living cell markers to study the natural mode of action of genetically engineered nematophagous fungi. Grasping the populations of BCAs and their action mechanisms against PPN at the molecular level will provide an informative tool for enhancing their pathogenic activities.

8.8.5 Integrated Pest Management Programs

Recent research has found that bionematicides can act additively or synergistically with other cultural inputs in IPM programs (Abd-Elgawad and Askary 2018). For example, tomato had better ($p \leq 0.05$) growth using *P. fluorescens* and organic manure together than either of them alone (Siddiqui et al. 2001). Moreover, combined utilization of BCAs and other pesticides may be explored for practices on wider scales. In this vein, there are many BCAs that are being in the production pipeline or will be swiftly available. Thus, priorities for harnessing bacterial and fungal nematicides in sustainable agriculture and grasping their biology, ecology, mode of action, and interaction with other agricultural inputs deserve further research. Such research should address developing specialized useful techniques and facilitate ways of enlightening growers by pros and cons of BCAs. The growers/farmers should be adequately taught to satisfy and optimize their needs for sustainable and environmentally friendly PPN management tactics. A combination of BCAs may result in more efficacies on nematode targets. Integrated application has merits such as ability to act over a wider range of edaphic factors, ability to kill the target PPN in more than one stage of its life cycle, and multiple modes of action against the PPN. Management should be done so wisely that mutually suppressive inputs or incompatibility between BCAs or other cultural applications are avoided.

8.9 Conclusions and Future Prospects

Biological control of plant-parasitic nematode (PPN) pests reflects a wise application of natural methods and most sustainable approach for nematode management with consequent crop yield increase in size and quality. So, it should be exploited as best as we can via safe and skillful application and advanced technology. These aspirations and caveats require considering an alternative paradigm of chemical nematicides. Biological control agents (BCAs) against PPN are slower acting, less effective, and more inconsistent than control normally achieved with chemicals. Moreover, instant efficacy gaps between BCAs and nematicides are frequently in favor of the chemicals, while these latter are sometimes less costly than BCAs. Nevertheless, contrary to chemical nematicides, these BCAs are harmless to human beings, maintain wildlife and unpolluted environment, and do not give rise to resistance-breaking nematode pathotypes. The beneficial activities of BCAs with their merits and limitations for PPN management, and their future prospects were reviewed herein to stimulate further research in this fascinating area and to provide a road map for practical application of BCAs for the integrated pest management

(IPM). Such an IPM concept along with recent technological developments and significant shifts in the relevant settings for wider BCA applications should continue to be a major focus moving forward. These IPM programs may further be properly utilized, cost-effectively exploited, and environmentally friendly applied for PPN management. Therefore, researchers should grasp the complex network of interactions among biotic and abiotic factors in intimate contact with the BCA(s) to maximize their benefits and expand applications. Molecular research on BCAs has entered an exciting era that provides the tools to actually unravel this complex network of interactions and make timely opportunity to advance biocontrol forward. In parallel, growers/farmers should be enlightened to the right use of these natural enemies to achieve economic and successful PPN control.

Acknowledgments This study was supported in part by STDF via the US-Egypt Project cycle 17 (No. 172) entitled "Preparing and Evaluating IPM Tactics for Increasing Strawberry and Citrus Production." This chapter is derived from the Subject Data funded in part by NAS and USAID, and that any opinions, findings, conclusions, or recommendations expressed in it are those of the authors alone, and do not necessarily reflect the views of USAID or NAS. The facilities and financial support offered by The National Research Centre via In-house project No. 12050105 entitled "Pesticide Alternatives Against Soil-Borne Pathogens and Pests Attacking Economically Important Solanaceous Crops" are appreciated.

References

Abd-Elgawad MMM (2014) Plant-parasitic nematode threats to global food security. J Nematol 46:130

Abd-Elgawad MMM (2016a) Comments on the use of biocontrol agents against plant-parasitic nematodes. Int J PharmTech Res 9(12):352–359

Abd-Elgawad MMM (2016b) Biological control agents of plant-parasitic nematodes: a review. Egypt J Biol Pest Control 26(2):423–429

Abd-Elgawad MMM (2016c) Use of Taylor's power law parameters in nematode sampling. Int J PharmTech Res 9(12):999–1004

Abd-Elgawad MMM (2017) Status of entomopathogenic nematodes in integrated pest management strategies in Egypt. In: Abd-Elgawad MMM, Askary TH, Coupland J (eds) Biocontrol agents: Entomopathogenic and slug parasitic nematodes. CAB International, Wallingford, pp 473–501

Abd-Elgawad MMM (2020) Can rational sampling maximise isolation and fix distribution measure of entomopathogenic nematodes? Nematology 22. https://doi.org/10.1163/15685411-00003350

Abd-Elgawad MMM, Askary TH (2015) Impact of phytonematodes on agriculture economy. In: Askary TH, Martinelli PRP (eds) Biocontrol agents of phytonematodes. CAB International, Wallingford, pp 3–49

Abd-Elgawad MMM, Askary TH (2018) Fungal and bacterial nematicides in integrated nematode management strategies. Egypt J Biol Pest Control 28:74. https://doi.org/10.1186/s41938-018-0080-x

Abd-Elgawad MMM, Kabeil SSA (2012) Biological control of *Meloidogyne incognita* by *Trichoderma harzianum* and *Serratia marcescens* and their related enzymatic changes in tomato roots. Afr J Biotechnol 11:16,247–16,252

Abd-Elgawad MMM, Vagelas IK (2015) Nematophagous bacteria: field application and commercialization. In: Askary TH, Martinelli PRP (eds) Biocontrol agents of phytonematodes. CAB International, Wallingford, pp 276–309

Abd-Elgawad MMM, Koura FFH, Montasser SA, Hammam MMA (2016) Distribution and losses of *Tylenchulus semipenetrans* in citrus orchards on reclaimed land in Egypt. Nematology 18:1141–1150

Abd-El-Khair H, El-Nagdi WMA, Youssef MA, Abd-Elgawad MMM, Dawood MG (2019) Protective effect of *Bacillus subtilis*, *B. pumilus*, and *Pseudomonas fluorescens* isolates against root knot nematode *Meloidogyne incognita* on cowpea. Bull Natl Res Centre 43:64. https://doi.org/10.1186/s42269-019-0108-8

Adam M, Westphal A, Hallmann J, Heuer H (2014) Specific microbial attachment to root knot nematodes in suppressive soil. Appl Environ Microbiol 80:2679–2686. https://doi.org/10.1128/AEM.03905-13

Ansari RA, Khan TA (2012a) Parasitic association of root-knot nematode, *Meloidogyne incognita* on guava. e-J Sci Technol 5:65–67

Ansari RA, Khan TA (2012b) Diversity and community structure of phytonematodes associated with guava in and around Aligarh, Uttar Pradesh, India. Trends Biosci 5(3):202–204

Ansari RA, Mahmood I (2017a) Optimization of organic and bio-organic fertilizers on soil properties and growth of pigeon pea. Sci Hortic 226:1–9

Ansari RA, Mahmood I (2017b) Determination of disease incidence caused by *Meloidogyne* spp. and or *Fusarium udum* on pigeonpea in Aligarh district: a survey. Trends Biosci 10 (24):5239–5243

Ansari RA, Mahmood I (2019a) Plant health under biotic stress: volume 2: microbial interactions. Springer, Singapore. https://doi.org/10.1007/978-981-13-6040-4

Ansari RA, Mahmood I (2019b) Plant health under biotic stress: volume 1: organic strategies. Springer, Singapore. https://doi.org/10.1007/978-981-13-6043-5

Ansari RA, Rizvi R, Sumbul A, Mahmood I (2017a) PGPR: current vogue in sustainable crop production. In: Kumar V, Kumar M, Sharma S, Prasad R (eds) Probiotics and plant health. Springer, Singapore, pp 455–472

Ansari RA, Mahmood I, Rizvi R, Sumbul A (2017b) Siderophores: augmentation of soil health and crop productivity. In: Kumar V, Kumar M, Sharma S, Prasad R (eds) Probiotics in agroecosystem. Springer, Singapore, pp 291–312

Ansari RA, Sumbul A, Rizvi R, Mahmood I (2019) Organic soil amendments: potential tool for soil and plant health management. In: Ansari RA, Mahmood I (eds) Plant health under biotic stress. Springer, Singapore, pp 1–35

Arora NK, Mishra J (2016) Prospecting the roles of metabolites and additives in future bioformulations for sustainable agriculture. Appl Soil Ecol 107:405–407

Askary TH (2015) Nematophagous fungi as biocontrol agents of phytonematodes. In: Askary TH, Martinelli PRP (eds) Biocontrol agents of phytonematodes. CAB International, Wallingford, pp 81–125

Askary TH, Martinelli PRP (eds) (2015) Biocontrol agents of phytonematodes. CAB International, Wallingford; 470 pp

Atkins SD, Hidalgo-Diaz L, Kalisz H, Mauchline TH, Hirsch PR, Kerry BR (2003) Development of a new management strategy for the control of root-knot nematodes (*Meloidogyne* spp.) in organic vegetable production. Pest Manag Sci 59:183–189

Barratt BIP, Moran VC, Bigler F, van Lenteren JC (2018) The status of biological control and recommendations for improving uptake for the future. BioControl 63:11–26

Bharali A, Bhagawati B, Uday K (2019) Bio-efficacy of native bioagents and biofertilizers for the management of root-knot nematode *Meloidogyne incognita* infecting black gram *Vigna mungo*. Int J Curr Microbiol App Sci 8(2):1484–1501

Bilgrami AL (2008) Biological control potentials of predatory nematodes. In: Ciancio A, Mukergi KG (eds) Integrated management and biocontrol of vegetable and grain crops nematodes. Springer, New York, pp 3–28

Blacket MJ, Agarwal A, Wainer J, Triska MD, Renton M, Edwards J (2019) Molecular assessment of the introduction and spread of potato cyst nematode, *Globodera rostochiensis*, in Victoria, Australia. Phytopathology 109(4):659–669

Blyuss KB, Fatehi F, Tsygankova VA, Biliavska LO, Iutynska GO, Yemets AI, Blume YB (2019) RNAi-based biocontrol of wheat nematodes using natural poly-component biostimulants. Front Plant Sci 2019:10. https://doi.org/10.3389/fpls.2019.00483

van Bruggen AHC, Finckh MR (2016) Management approaches in organic farming systems. Annu Rev Phytopathol 54:25–54

Btryon (2019). Advantages & disadvantages of biological control. https://owlcation.com/stem/Advantages-Disadvantages-of-Biological-Control. Accessed 26 April 2019

Cabanillas E, Barker KR, Daykin ME (1988) Histology of the interaction of *Paecilomyces lilacinus* with *Meloidogyne incognita* on tomato. J Nematol 20:362–365

Cameron DD, Neal AI, van Wees SCM, Ton J (2013) Mycorhiza induced resistance: more than the sum of its parts. Trends Plant Sci 18:539–545

Campos-Herrera R, Stuart RJ, Pathak E, El-Borai FE, Duncan LW (2019) Temporal patterns of entomopathogenic nematodes in Florida citrus orchards: evidence of natural regulation by microorganisms and nematode competitors. Soil Biol Biochem 128:193–204

Cetintas R, Kusek M, Fateh SA (2018) Effect of some plant growth-promoting rhizobacteria strains on root-knot nematode, *Meloidogyne incognita*, on tomatoes. Egypt J Biol Pest Control 28:7. https://doi.org/10.1186/s41938-017-0008-x

Chen S (2007) Suppression of Heterodera glycines in soils from fields with long-term soybean monoculture. Biocontrol Sci Tech 17:125–134

Chen ZX, Dickson DM (1998) Review of *Pasteuria penetrans*: biology, ecology, and biological control potential. J Nematol 30:313–340

Crow WT (2014) Effects of a commercial formulation of *Bacillus firmus* I-1582 on golf course bermudagrass infested with *Belonolaimus longicaudatus*. J Nematol 46(4):331–335

Davies K, Spiegel Y (eds) (2011) Biological control of plant-parasitic nematodes: 1 building coherence between microbial ecology and molecular mechanisms, progress in biological control. Springer Science+Business Media B.V, London

Eissa MFM, Abd-Elgawad MMM (2015) Nematophagous bacteria as biocontrol agents of phytonematodes. In: Askary TH, Martinelli PRP (eds) Biocontrol agents of phytonematodes. CAB International, Wallingford, pp 217–243

El-Eslamboly AASA, Abd El-Wanis MM, Amin AW (2019) Algal application as a biological control method of root-knot nematode *Meloidogyne incognita* on cucumber under protected culture conditions and its impact on yield and fruit quality. Egypt J Biol Pest Control 29:18. https://doi.org/10.1186/s41938-019-0122-z

Esser, R.P. & El-Gholl, N.E. (1993). *Paecilomyces lilacinus*, a fungus that parasitizes nematode eggs. Nematology Circular No. 203, Division of Plant Industry, Florida, USA, 3 pp

Garbeva P, van Veen JA, Van Elsas JD (2004) Microbialdiversity in soil: selection of microbial populations by plant and soil type and implications for disease suppressiveness. Annu Rev Phytopathol 42:243–270. https://doi.org/10.1146/annurev.phyto.42.012604.135455

Gerber JF & White JH (2005) Materials and methods for the efficient production of Pasteuria. U.S. Patent 6, 919, 197 B2

Gerson U (2015) Mites as biocontrol agents of phytonematodes. In: Askary TH, Martinelli PRP (eds) Biocontrol agents of phytonematodes. CAB International, Wallingford, pp 323–335

Giblin-Davis RM, Williams DS, Bekal S, Dickson DW, Becker JO, Preston JF (2003) 'Candidatus Pasteuria usgae' sp. nov., an obligate endoparasite of the phytoparasitic nematode, *Belonolaimus longicaudatus*. Int J Syst Evol Microbiol 53:197–200

Giné A, Carrasquilla M, Martínez-Alonso M, Gaju N, Sorribas FJ (2016) Characterization of soil suppressiveness to root-knot nematodes in organic horticulture in plastic greenhouse. Front Plant Sci 7:164. https://doi.org/10.3389/fpls.2016.00164

Glare TR, Caradus J, Gelernter W, Jackson T, Keyhani N, Kohl J, Marrone P, Morin L, Stewart A (2012) Have biopesticides come of age? Trends Biotechnol 30:250–258

Janvier C, Villeneuve F, Alabouvette C, Edel-Hermann V, Mateille T, Steinberg C (2007) Soil health through soil disease suppression: which strategy from descriptors to indicators ? Soil Biol Biochem 39:1–23. https://doi.org/10.1016/j.soilbio.2006.07.001

Jatala P (1986) Biological control of plant-parasitic nematodes. Annu Rev Phytopathol 24:453–489

Kariuki GM, Dickson DW (2007) Transfer and development of *Pasteuria penetrans*. J Nematol 39:55–61

Kerry B (1980) Biocontrol: fungal parasites of female cyst nematodes. J Nematol 12:253–259

Kerry B (1997) Biological control of nematodes: prospects and opportunities. In: Maqbool MA, Kerry B (eds) Plant nematode problems and their control in the near east region, vol 144. FAO Plant Production and Protection Paper, Pakistan, 326 pp

Khan Z, Kim YH (2007) A review on the role of predatory soil nematodes in the biological control of plant parasitic nematodes. Appl Soil Ecol 35:370–379

Kim YH (2015) Predatory nematodes as biocontrol agents of phytonematodes. In: Askary TH, Martinelli PRP (eds) Biocontrol agents of phytonematodes. CAB International, Wallingford, pp 393–420

Kiriga AW, Haukeland S, Kariuki GM, Coyne DL, Beek NV (2018) Effect of *Trichoderma* spp. and *Purpureocillium lilacinum* on *Meloidogyne javanica* in commercial pineapple production in Kenya. Biol Control 119:27–32

Kokalis-Burelle N (2015) *Pasteuria penetrans* for control of *Meloidogyne incognita* on tomato and cucumber, and *M. arenaria* on snapdragon. J Nematol 47(3):207–213

Kumar T, Wahla V, Pandey P, Dubey RC, Maheshwari DK (2009) Rhizosphere competent *Pseudomonas aeruginosa* in the management of *Heterodera cajani* on sesame. World J Microbiol Biotechnol 25:277–285

Luc JE, Crow WT, McSorley R, Giblin-Davis RM (2010a) Suppression of *Belonolaimus longicaudatus* with in vitro-produced *Pasteuria* sp. endospores. Nematropica 40(2):217–225

Luc JE, Wenjing P, Crow WT, Giblin-Davis RM (2010b) Effects of formulation and host nematode density on the ability of in vitro-produced *Pasteuria* endospores to control its host *Belonolaimus longicaudatus*. J Nematol 42(2):87–90

Martínez-Medina A, Roldán A, Pascual JA (2011) Interaction between arbuscular mycorrhizal fungi and Trichoderma harzianum under conventional and low input fertilization field condition in melon crops: growth response and Fusarium wilt biocontrol. Appl Soil Ecol 47(2):98–105

McCarter JP (2008) Molecular approaches toward resistance to plant-parasitic nematodes. In: Berg RH, Taylor CG (eds) Cell biology of plant nematode parasitism—plant cell monographs. Springer, Berlin, pp 239–267

McSorley R, Arnett JD, Bost SC, Carter WW, Hafez S, Johnson AW, Kirkpatrick T, Nyczepir AP, Radewald JD, Robinson AF, Schmitt DP (1987) Bibliography of estimated crop losses in the United States due to plant parasitic nematodes. Ann Appl Nematol 1:6–12

Meyer SLF, Roberts DP, Chitwood DJ, Carta LK, Lumsden RD, Mao W (2001) Application of *Burkholderia cepacia* and *Trichoderma virens*, alone and in combinations, against *Meloidogyne incognita* on bell pepper. Nematropica 31:75–86

Moosavi MR, Zare R (2015) Factors affecting commercial success of biocontrol agents of phytonematodes. In: Askary TH, Martinelli PRP (eds) Biocontrol agents of phytonematodes. CAB International, Wallingford, pp 423–445

Moosavi MR, Zare R, Zamanizadeh HR, Fatemy S (2010) Pathogenicity of *Pochonia* species on eggs of *Meloidogyne javanica*. J Invertebr Pathol 104:125–133

Muyzer G, Brinkhoff T, Nübel U, Santegoeds C, Schäfer H, Wawer C (2004) Denaturing gradient gel electrophoresis (DGGE) in microbial ecology. In: Kowalchuk GA, de Bruijn FJ, Head IM, Akkermans AD, van Elsas JD (eds) Molecular microbial ecology manual, vol 1–2, 2nd edn. Springer, Dordrecht, pp 743–769

Pettigrew WT, Meredith WR, Young LD (2005) Potassium fertilization effects on cotton lint yield, yield components, and reniform nematode populations. Agron J 97:1245–1251. https://doi.org/10.2134/agronj2004.0321

Pieterse CMJ, Zamioudis C, Berendsen RL, Weller DM, vanWees SCM, Bakker PAHM (2014) Induced systemic resistance by beneficial microbes. Annu Rev Phytopathol 52:347–375. https://doi.org/10.1146/annurev-phyto-082712-102340

Poinar GO Jr (1983) The natural history of nematodes. Prentice-Hall, Englewood Cliffs, NJ

Pyrowolakis A, Westphal A, Sikora RA, OleBecker J (2002) Identification of root-knot nematode suppressive soils. Appl Soil Ecol 19:51–56

Sanda NB, Sunusi M (2014) Fundamentals of biological control of pests. IJCBS 1(6):1–11

Sankaranarayanan C (2015) Arbuscular mycorrhizal fungi as biocontrol agents of phytonematodes. In: Askary TH, Martinelli PRP (eds) Biocontrol agents of phytonematodes. CAB International, Wallingford, pp 365–389

Sasser JN, Freckman DW (1987) A world perspective on nematology: the role of the society. In: Veech JA, Dickson DW (eds) Vistas on nematology. Society of Nematologists, Hyattsville, MD, pp 7–14

Schouteden N, De Waele D, Panis B, Vos CM (2015) Arbuscular mycorrhizal fungi for the biocontrol of plant-parasitic nematodes: a review of the mechanisms involved. Front Microbiol 6:1280. https://doi.org/10.3389/fmicb.2015.01280

Sharon E, Bar-Eyal M, Chet I, Herrera-Estrella A, Kleifeld O, Spiegel Y (2001) Biological control of the root-knot nematode *Meloidogyne javanica* by *Trichoderma harzianum*. Phytopathology 91:687–693

Siddiqui ZA, Mahmood I (1999) Role of bacteria in the management of plant parasitic nematodes: a review. Bioresour Technol 69:167–179

Siddiqui ZA, Iqbal A, Mahmood I (2001) Effects of *Pseudomonas fluorescens* and fertilizers on the reproduction of *Meloidogyne incognita* and growth of tomato. Appl Soil Ecol 16:179–185

Smalla K, Heuer H (2006) How to assess the abundance and diversity of mobile genetic elements in soil bacterial communities? In: Nannipieri P, Smalla K (eds) Nucleic acids and proteins in soil. Springer, Berlin, Heidelberg, pp 313–330

Stirling GR (2011) Biological control of plant-parasitic nematodes: an ecological perspective, a review of progress and opportunities for further research. In: Davies KG, Spiegel Y (eds) Biological control of plant-parasitic nematodes: building coherence between microbial ecology and molecular mechanisms, progress in biological control 11. Springer Science + Business Media, Dordrecht, the Netherlands, pp 1–38

Stirling GR (2014) Biological control of plant-parasitic nematodes, 2nd edn. CAB International, Wallingford

Sturhan D, Hampel G (1977) Plant-parasitic nematodes as a prey of the bulb mite *Rhizoglyphus echinopus* (Acarina, Tyroglyphidae). Anzeiger fűr Schadlingskde, Pflanzenschutz, Umweltschutz 50:115–118

Tian B, Yang J, Zhang KQ (2007) Bacteria used in the biological control of plant-parasitic nematodes: populations, mechanisms of action, and future prospects. FEMS Microbiol Ecol 61:197–213

Timper P (2011) Utilization of biological control for managing plant-parasitic nematodes. In: Davies K, Spiegel Y (eds) Biological control of plant-parasitic nematodes: 1 building coherence between microbial ecology and molecular mechanisms, progress in biological control. Springer Science+Business Media B.V, London, pp 259–290

Tzortzakakis EA, Gowen SR, Goumas DE (1996) Decreased ability of *Pasteuria penetrans* spores to attach to successive generations of *Meloidogyne javanica*. Fundam Appl Nematol 19:201–204

Wang J, Fu Z, Ren Q, Zhu L, Lin J, Zhang J, Cheng X, Ma J, Yue J (2019) Effects of arbuscular mycorrhizal fungi on growth, photosynthesis, and nutrient uptake of *Zelkova serrata* (Thunb.) Makino seedlings under salt stress. Forests 10:186. https://doi.org/10.3390/f10020186

Wani AM (2015) Plant growth-promoting rhizobacteria as biocontrol agents of phytonematodes. In: Askary TH, Martinelli PRP (eds) Biocontrol agents of phytonematodes. CAB International, Wallingford, pp 339–362

Weller DM, Raaijmakers JM, Gardener BBM, Thomashow LS (2002) Microbial populations responsible for specific soil suppressiveness to plant pathogens. Annu Rev Phytopathol 40:309–348. https://doi.org/10.1146/annurev.phyto.40.030402.110010

Westphal A, Becker JO (1999) Biological suppression and natural population decline of *Heterodera schachtii* in a California field. Phytopathology 89:434–440

Westphal A, Becker JO (2001) Components of soil suppressiveness against *Heterodera schachtii*. Soil Biol Biochem 33:9–16. https://doi.org/10.1016/S0038-0717(00)00108-5
Wilson MJ, Jackson TA (2013) Progress in the commercialization of bionematicides. BioControl 58:715–722
Wittwer SH (1981) The 20 crops that stand between man and starvation. Farm Chem 144:17–28
Zhang L, Yang J, Niu Q, Zhao X, Ye F, Liang L, Zhang KQ (2008) Investigation on the infection mechanism of the fungus *Clonostachys rosea* against nematodes using the green fluorescent protein. Appl Microbiol Biotechnol 78:983–989

Chapter 9
Importance of Biopesticides in the Sustainable Management of Plant-Parasitic Nematodes

K. P. Roopa and Anusha S. Gadag

Abstract The uncontrolled and ever-increasing population in the world has put up a major challenge in the agriculture sector for the production and supply of food. Various pests have been the major concern for food production. Few pests like nematodes are gaining more importance due to their attacking nature on wide range of hosts. The major crops affected by nematodes include legumes, cereals, citrus family, grasses, and horticulture crops, in which they cause complete crop losses, that is, 100%. The main plant-parasitic nematode species include those from genera *Meloidogyne*, *Heterodera*, *Pratylenchus*, and *Globodera*, accounting for an annual loss of $100 billion every year. These losses account 90% of the yield of cotton, wheat, tomato, beans, soybeans, and many horticultural crops. There are integrated approaches to manage nematodes, such as cultural practices, like sanitation and crop rotation, or chemical method—judicial use of nematicides are found to be most popular, economical, and quite effective. The major drawback of chemical management is health and environmental hazards. Alternative methods with minimal environmental pollution and less health hazards are being adopted in recent years. Biopesticides play a crucial role in this context. A bacterium that is prominently used for nematode control is of the genus *Bacillus* (*Bacillus subtilis*); other commercially available biopesticides for nematode control include Bio-Act® (*Paecilomyces lilacinus*), Botanigard® (*Beauveria bassiana*), Bioarc® (*Bacillus megaterium*), Bio Zeid® (*Trichoderma album*), and brown alga, Algaefol® (*Ascophyllum nodosum*). However, there is a need to increase the use and application of ecofriendly biopesticides for control of nematodes in economically important crop plants.

Keywords Agriculture · Crop pest · Plant-parasitic nematodes · Biocontrol · Biopesticides

---

K. P. Roopa (✉) · A. S. Gadag
Department of Biotechnology, University of Agricultural Sciences, Dharwad, Karnataka, India

R. A. Ansari et al. (eds.), *Management of Phytonematodes: Recent Advances and Future Challenges*, https://doi.org/10.1007/978-981-15-4087-5_9

9.1 Introduction

Pest is any organism that is detrimental to humans, destroys crop plants, imposes threats to animals and human health, and further reduces aesthetic and recreational values. The main pests include insects, mites, plant pathogens, fish, birds, weeds, mollusks, and mammals. Crop losses occur due to numerous pests, like plant pathogens (fungi, bacteria, virus, nematodes, etc.), insects, and weeds, that have been major issue in agriculture as they cause drastic reduction in yield and productivity (Salma and Jogen 2011). Plant-parasitic nematodes (PPNs) pose a serious threat to economically important crops across the world. The crop losses could be accounted to about 12.6% (around $215.77 billion) per annum due to attack by nematodes on 20 important crops of the world (Abd-Elgawad and Askary 2018). Crop losses caused by pests coupled with other problems like unpredicted weather conditions, limited technical knowledge of farmers, and poor soil fertility further affect global food security. About 40% reduction in the world's crop yield due to pests has been estimated (Oerke et al. 1994). There is a serious need for the management of the economically threatening pests to increase food security to meet the demands of fast-increasing human population. The pest management practices should be such that they should cause minimal or no damage to human health, animals, birds, and other living organisms. Further, the management strategies should be economically feasible and environment friendly (Chandler et al. 2011; Bastiaans et al. 2008).

9.2 Nematodes

The word "nematode" is derived from the Greek term "nema," meaning thread. Nematodes are long, slender, thread-like, cylindrical, non-segmented bodies tapering toward the head and tail. Some female parasitic nematodes have shapes of lemon, kidney, or pear. Nematodes survive in varied environments and habitats and are named as omnipresent inhabiting organism (Ansari and Khan 2012a, b; Ali et al. 2015; Ansari and Mahmood 2017b). They belong to the phylum of "Nematoda," which includes plant and animal parasites and free-living species. Plant-parasitic nematodes (PPNs) are considered to be obligate biotrophic parasites affecting huge damage and crop yield losses. They are causing significant losses in crop production. Infecting many crops and imposing constraint on quality and quantity of yield further affects global food security. About 7% of the plant-parasitic nematodes (PPNs) comes from Nematoda phylum, that is, they account to 197 genera and 4300 species, which infect diverse economically important crop plants like tomato, potato, wheat, soybean, maize, and sugar beets (Decraemer and Hunt 2006). Penetration of PPNs into the roots leads to blockage in translocation of water and nutrients. The established nematodes result in plant symptoms like chlorosis, growth, necrosis, wilting, stunting, and increase in plants' susceptibility to other pathogens and pests (Webster 1995). Symptoms on roots include root gall, root knots, abbreviations, lesions, and necrosis. The major damage is loss in yield. These worms act as vectors for viral disease transmission, hence causing abrupt loss economically and having huge impact socially also (Ali et al. 2015). It is of great importance to search for management strategies, in various economically important crops, against PPNs.

9.3 Plant-Parasitic Nematode Evolution

There is a dynamic association between nematodes and host plants that has led to what is called as plant parasitism, which has evolved faster and has benefited much for the growth and survival of nematodes (Kiontke and Fitch 2013; Maier et al. 2013). The evolution of these parasitic nematodes can be traced back to 400 million years before (Poinar 1983). The initial occurrence of plant-parasitic nematodes is seen in 235 BC (Noel 1992). The first plant-parasitic nematode was confirmed by observing galls on wheat (Needham 1742). Root-knot nematodes (RKNs) were identified in roots of cucumber (Berkeley 1855). The plant-nematode association has resulted in specific feeding structures and secretory products for nutrient absorption and infecting host. Specialized structures include stylet, and subventral and dorsal esophageal glands—the most significant adaptations involved in parasitism (Maier et al. 2013; Quentin et al. 2013). A needle-like, protrusible stylet is used by these nematodes to protrude plant tissue and later release proteinaceous secretions and ease the nematode entry into the plants (Davis et al. 2008). The plant-parasitic nematodes can be classified as ectoparasitic and endoparasitic nematodes. Ectoparasites will gain entry into the plant cells and reside internally and obtain the nutrients from the plants, whereas endoparasites reside outside the plant and obtain nutrients externally from plant roots. For example, *Xiphinema* (California dagger nematode) transmits Grapevine fanleaf virus, causing significant economic losses in grapes worldwide (Villate et al. 2008). Endoparasitic nematodes are further divided into migratory and sedentary. Migratory endoparasitic nematodes move inside the roots and feed up on cytoplasm and the kill the host cell, whereas sedentary nematodes develop a feeding site within the host tissue and then become active (Williamson and Hussey 1996). Examples of migratory endoparasitic nematodes include *Pratylenchus* spp., *Radopholus* spp., and *Hirschmanniella*.

9.4 The Impact of Plant-Parasitic Nematodes on Crop Plants

Plant-parasitic nematodes cause huge economic losses to crop production. Approximately 4100 species of PPNs have been identified that infect plants (Decraemer and Hunt 2006). Out of these, 15% are grouped as most economically damaging species that act directly on plant roots and prevent water and nutrient absorption by plants resulting in tremendous losses through declination in agronomic parameters, quality, and yield. The estimated loss of $80–118 billion dollars per year in crops is caused by nematodes (Nicol et al. 2011; Sasser 1987). Order Tylenchida is considered to be the most important, which are composed of pathogens of plants, invertebrates, and fungi (Sasser 1987). The major phytoparasitic nematode genera reported to cause crop losses were *Heterodera*, *Hoplolaimus*, *Meloidogyne*, *Pratylenchus*, *Rotylenchulus*, and *Xiphinema* (Koenning et al. 1999). A recent survey on most threatening nematodes on crop plants gives the list as below: (1) root-knot nematodes (*Meloidogyne* spp.), (2) root lesion nematodes (*Pratylenchus* spp.), (3) cyst nematodes (*Heterodera* and *Globodera* spp.), (4) pine wilt nematode (*Bursaphelenchus xylophilus*), (5) burrowing nematode (*Radopholus similis*), (6) *Ditylenchus dipsaci*, (7) the reniform nematode (*Rotylenchulus reniformis*), (8) *Xiphinema index* (the only virus vector nematode), (9) *Nacobbus aberrans*, and (10) *Aphelenchoides besseyi* (Jones et al. 2013). In wheat (*Triticum aestivum*), significant losses are estimated by parasitic nematodes like cereal cyst nematodes (*Heterodera* spp.) of the *Heterodera avenae* group that also infect barley (*Hordeum vulgare*) and oat (*Avena sativa*). In few cases, the losses in the wheat fields by *H. avenae* can vary from 30 to 100% (Bonfil et al. 2004; Nicol et al. 2004). In rice, the nematode *Meloidogyne graminicola* may cause a yield reduction of up to 80% (Soriano et al. 2000). In maize, the most devastating genera include *Meloidogyne* spp., *Pratylenchus* spp., root lesion nematodes, and *Heterodera* spp. (Nicol et al. 2011). The needle nematode *Longidorus breviannulatus* leads to an economic yield loss of up to 60% (Norton and Hoffmann 1975). In case of potato, *Globodera rostochiensis*, *Globodera pallid*, *Meloidogyne* spp., and *Ditylenchus* destroy the crop vigorously (Turner and Evans 1998; Santo et al. 1980). In sweet potato, 10.2% yield loss per year has been recorded because of the attack by plant-parasitic nematodes. Root-knot nematodes (RKNs) are significant pests of sweet potato (Decraemer and Hunt 2006). Hence, it is of serious concern to identify various management strategies against PPNs for important crop plants.

9.5 Nematode Management Methods

The management of plant-parasitic nematodes has been a great confront for the farmers and the researchers. It is a big issue to keep up the sustainable agriculture and maintain food security worldwide. Nematode management mainly focuses on reduction of nematode load to below the threshold level than on its eradication. A good number of ecofriendly nematode management approaches is available currently (Ansari and Mahmood 2017a, 2019a, b; Ansari et al. 2019). However, nematode management is of utmost importance since once they get established could not be removed from the soil and also, they act as predisposing factor to many invading pathogens and environmental stresses. The main strategies used to control plant nematodes include cultural, chemical, biological, and transgenic methods.

9.5.1 Cultural Methods

Cultural methods adopted for nematode management include various activities. (a) Organic manuring: Large populations of free-living nematodes can control many plant-parasitic nematodes in the soil, so enough organic matter is applied to increase free-living nematode populations; (b) Flooding (MacGuidwin 1993); (c) Solarization (Katan 1981), which is very effective for control of many nematodes and other soil-borne pathogens; (d) Fallowing (Brodie and Murphy 1975); (e) Cover crops (Kirkegaard and Sarwar 1998), such as sudan grass, Marigold, and *Brassica* spp. (green manure crops)—the glucosinolate or isothiocyanate content in many *Brassica* species is known to control many parasitic nematodes and also boost free-living nematode populations in the soil; (f) Crop rotation (Westphal 2011; Dababat et al. 2015).

9.5.2 Chemical Control

Till recent years chemical control measures have been the main focus in almost every crop to control the pests, and thus chemicals are being sold and used widely all over the world. The main advantages of using chemicals include their relatively low cost, ease of application, effectiveness in control of the pest, ease of their availability, and their stability. Chemical pesticides are usually fast-acting, which helps to prevent the crop damage at earliest. In chemical method, organophosphates, fumigants, and carbamates are widely used for control of PPNs. Recently, many nematicides were banned as they incurred huge costs and caused health and environment hazard (Sorribas et al. 2005). For example, organophosphate and carbamate compounds, that is, fosthiazate, oxamyl, and ethoprophos, are at risk of withdrawal by EU Directive 91/414/EEC (Clayton et al. 2008). The drawbacks of using chemical pesticides are quite huge affecting environment and health of living organisms. The major drawback is that they act on nontarget pests also when applied, due to nonselective nature of many pesticides. They can act on even harmless and the useful or biofriendly species. Another significant disadvantage is the development of resistance for chemical pesticides by the pest, which lead to pest resurgence and the development of heritable resistance (Van Emden 2007). Main reason for resistance development in the pest is the mutation of the organism for that particular pesticide that makes it immune, and hence there is a need to change the pesticide frequently. Another drawback is the residual effect or accumulation of the pesticide in the biological food chain. Since predators or humans/animals are at the top of the food chain, they will have higher chance of developing toxicity due to the build-up of pesticide residue in their system. The most significant threat is the remains or residues of pesticides left over on the harvested crops or on fruit or vegetables, which could be consumed. Further, these pesticide residues may get soaked down in the soil or groundwater and thus contaminate water, affecting the organisms consuming it. Due to unceasing problems and drawbacks associated with chemicals to control pests, it is crucial to search for more ecofriendly method that will serve as an alternative to chemicals (biopesticides). Biocontrol method is considered to be the best alternative for the chemical control method. Biopesticides are quite effective, biodegradable, and leave no residual traces in the environment. Due to the adverse effects of chemicals, biopesticide use is being adopted widely, and their efficiency against pests is found to be significant (Salma and Jogen 2011).

9.5.3 Biocontrol of Pxhytonematodes

The expanding hazardous effects of nematicides have led to check for ecofriendly biological control of PPNs, and hence biological nematicides are becoming more popular and adding as major component of biomanagement practices for PPNs (Ansari et al. 2017a, b; Abd-Elgawad and Askary 2018; Ansari and Mahmood 2019a, b). Recent trend for use of organic food and its enormous demand are adding pressure on organic agriculture for production of more food while concerns for environmental welfare is utmost priority. Less hazardous, economical, and ecofriendly pest management methods like biological control is on high selection by farmers. Due to these reasons, the uses of chemical pesticides have decreased at present and also looks the same for future. Bio-products containing fungi and bacteria antagonists rank high among the bio-nematicides used against PPNs (Askary 2015a, b; Eissa and Abd-Elgawad 2015). Their effect on root-knot nematodes, cyst nematode, has been well demonstrated in previous studies (Stirling 1991; Meyer 2003).

9.6 Biopesticides

Biopesticides is a combination of biological pesticides, which includes pest management strategies like predatory, parasitic, or chemical relationships.

9.6.1 *Definition of Biopesticides*

Biopesticides are the substances extracted from the natural materials such as living organisms (natural enemies) or their products (microbial products, phytochemicals) or they may be their by-products (semi-chemical), which are used in crop protection and management (Dybas 1989). Definition by the U.S. Environmental Protection Agency (EPA) states: Biopesticides "include naturally occurring substances that control pests (biochemical pesticides), microorganisms that control pests (microbial pesticides), and pesticidal substances produced by plants containing added genetic material (plant-incorporated protectants) or PIPs" (https://www.epa.gov/pesticides/biopesticides, p. 1). These biopesticides include broad area of microbial pesticides and biochemicals derived from the natural sources and microorganisms (bacteria, fungi, and nematodes). Biopesticides have a significant role in control and management of nematode population under the economic thresholds. They are released in masses to control pest and pathogens. In the integrated pest management programs, biopesticides are an important component and are considered as effective and ecofriendly compared to synthetic materials, and also biopesticides do not persist in high number in the crop environment.

9.6.2 *Biopesticide Formulations*

The formulation of biopesticides consists of the active ingredients similar to synthetic pesticides, which is most easy to apply and to be used by farmers (Slavica and Brankica 2013). The biopesticides consist mostly of living organisms and it should be taken care of their viability while preparing the formulation by mixing with carrier materials and storing them to the standards that are accepted (Boyetchko et al. 1998). Major concerns during preparation of biofertilizers include stabilization of the bioagents during distribution and storage, protecting them from adverse environmental conditions, and improving their activity by increasing contact and interaction with the target pest. There are different types of biopesticide formulations (Mollet and Grubenmann 2001). Based on the physical states of preparation, the biopesticide formulations are classified as dry and liquid formulations, while we can enhance its activity by addition of few stickers, synergists, stabilizers, spreads, surfactants, coloring agents, anti-freezing compounds, extra nutrients, dispersants, and melting agents (Brar et al. 2006; Knowles 2008).

9.6.2.1 Dry Formulation (for Direct Applications)

There are different forms in which these formulations are prepared and used. The following are the types: dustable powders, granules (GR), seed dressing, wettable powders, water dispersible granules (WDG).

Dustable Powders (DP)

Dustable powders consists of active inert ingredient (a.i) of 10% and formulated by sorption of a.i on finely ground solid mineral powder such as talc or clay of particle size 50–100 mm. The inert ingredients selected are ultraviolet (UV) protectants and adhesives (stickers) to improve adsorption (Slavica and Brankica 2013).

Granules (GR)

Granules are mostly applied to control insects living in soils, weeds, and nematodes for uptake by roots. Active inert ingredient (a.i) concentration ranges from 2 to 20% and these a.i are either coated outside or absorbed into the granules. Further, granules can be coated with resins or polymers to improve their effectiveness. Granules have particle size ranging from 100 to 600 microns made with materials such as starch, polymers, kaolin, silica, etc. (Slavica and Brankica 2013; Tadros 2005).

Seed Dressing

This formulation is obtained by mixing active ingredient carrier with adhesive inert material to promote adherence to seed coats. Also, coloring red pigment as a safety maker for treated seed can be used in this treatment (Woods 2003).

Wettable Powders

These are dry formulation finely ground and applied after suspension in water. Wettable powders consist of blending active ingredients with melting and dispersing agents, synergist, or surfactants. Due to the dusty nature it can cause serious health hazards, hence strict safety measures should be ensured while using them. These have longer storage stability, good miscibility, and are easy to apply (Brar et al. 2006; Knowles 2008).

Water Dixspersible Granules (WDG)

These are water suspended and overcome problems associated with wettable powder, are dust free, and have good storage stability, which is an added advantage (Knowles 2008; Slavica and Brankica 2013).

9.6.2.2 Liquid Formulations

Emulsion

Emulsion formulations are to be mixed with water and used, and it could be normal emulsion oil in water (O/W) or an inert emulsion water in oil (W/O). Most importantly, the proper choice of emulsifiers for stabilization to avoid instability is necessary (Brar et al. 2006; Slavica and Brankica 2013).

Suspension Concentrate (SC)

These formulations are prepared by mixing finely ground, solid active ingredient dispersed in liquid phase (water). Agitation is always required before application to keep particles evenly distributed. It is a popular formulation because of safety to user and environment (Knowles 2005; Woods 2003).

Suspo-Emulsion (SE)

It is a mixture of emulsion and suspension concentrate and a highly demanding formulation, because a homogenous emulsion component with a particle suspension is the product that is obtained and the final product stays stable. In addition, it is necessary that it is carried out using storage stability resting (Knowles 2008).

Oil Dispersion (OD)

The product is obtained similar to suspension concentrate. Here, instability problems can be avoided with selection of good inert ingredients (Vernner and Bauer 2007).

Capsule Suspension (CS)

Active ingredients are formulated as microencapsulated stable suspension. Capsules are made to protect from harsh environmental conditions. The materials used here are cellulose, gelation, or starch. It is a highly efficient formulation, typically fungal biopesticide (Winder et al. 2003).

Ultra Low Volume Liquids (ULV)

These are formulations not intended to be diluted in water before use and have concentration of active ingredients. It is easy to transport and can be formulated

Table 9.1 Effect of biopesticides in the management of plant-parasitic nematodes

S. No.	Plant-parasitic nematodes	Source of biopesticides	References
1.	*Meloidogyne incognita*	*Aspergillus niger*, *P. lilacinus*	Bhat and Wani (2012), Baidoo et al. (2017)
2.	*Meloidogyne javanica*	*A. niger*, *P. lilacinus*	Askary (2012), Ganaie and Khan (2010)
3.	*Meloidogyne graminicola*	*Paecilomyces lilacinus*	Narasimhamurthy et al. (2017a, b)
4.	*M. incognita*	*P. lilacinus*	Kumar et al. (2017)
5.	*R. reniformis*	*P. lilacinus*	Ashraf and Khan (2008)
6.	*Tylenchulus semipenetrans*	*P. lilacinus*	Mahanta et al. (2016)
7.	*M. incognita*	*Pochonia chlamydosporia*	Silva et al. (2017)
8.	*M. javanica*	*Trichoderma harzianum*	Feyisa et al. (2016)
9.	*M. incognita*	*T. Harzianum*	Kumar and Chand (2015)
10.	*M. incognita*	*T. Harzianum*	Deori and Borah (2016)
11.	*M. javanica*	*T. Harzianum*	Jamshidnejad et al. (2013)
12.	*Meloidogyne* spp.	*T. harzianum*, *Trichoderma viride*	Jegathambigai et al. (2011)
13.	*M. incognita*	*T. viride*	Muthulakshmi et al. (2010)
14.	*M. graminicola*	*T. viride*	Priya (2015)
15.	*Meloidogyne incognita*	*Pasteuria penetrans*	Kokalis-Burelle (2015)
16.	*Meloidogyne arenaria*	*P. penetrans*	Kokalis-Burelle (2015)
17.	*M. graminicola*	*Pseudomonas fluorescens*	Narasimhamurthy et al. (2017a, b)
18.	*M. incognita*	*Bacillus subtilis*	Gao et al. (2016)

using a suspended biocontrol agent as an active ingredient (Woods 2003) (Table 9.1).

9.7 Application Methodologies

Good selection of application techniques/methods is quite important for effective control of pests. This is possible by application of the technique at an appropriate time and on the basis of frequency of biopesticides. The following are various methods of application of biopesticides:

9.7.1 Seed Treatment

This method is the most effective method. Powder formulations with adhesive nature are used to treat seeds (Matthews et al. 2014; Wood 2003).

9.7.2 *Foliar Application*

Biopesticides when applied onto leaves' surface as sprays is called foliar application.

9.7.3 *Seedling Dipping*

It involves dipping roots of the seedlings in biopesticide suspension for definite time (minutes or hours) before transplanting. For example, *Trichoderma* spp. are applied by seedling dip method.

9.8 Mechanisms of Action of Bioagent/Biopesticides for Pest Control

There are four different mechanisms by which pesticides act, viz., competition, antibiosis, hyperparasitism, and synergism.

9.8.1 *Competition*

The mechanism of control operated by biopesticides lies in their ability to compete aggressively with other organisms or pest for food and space. Due to this competitive nature, they grow rapidly and colonize substrate and exclude pathogens. For example, *Trichoderma*. spp. are competitors of *M. javanica* in tomato (Feyisa et al. 2016; Naserinasab et al. 2011).

9.8.2 *Antibiosis*

Due to interaction with other microbes (microorganisms), bioagents produce specific metabolite, volatile compounds, lytic enzymes, or other toxic substances (Rikita and Utpal 2014). For example, the bacterial bioagents produce antibiotics, bacteriocin, volatile compounds, and metabolites.

9.8.3 *Hyperparasitism*

Hyperparasitism refers to lysis and death of pest by bioagent or by direct parasitism of bioagent (Rikita and Utpal 2014).

9.8.4 Synergism

The ability of some bioagents to combine actions of hydrolytic enzymes and antibiotic secondary metabolites is called synergism. For example, the role of *Trichoderma* sp. as a biocontrol agent and its fitness in the environment are due to synergistic effects of antimicrobial compounds. Example includes pyrones and coumarins (Rikita and Utpal 2014).

9.9 Types of Biopesticides

Biopesticides can be categorized into mainly four groups, which are categorized as under.

9.9.1 Microbial Pesticides

Microbial pesticides include microbes that are bacterium, fungus, virus, and protozoans, which act as biological control for plant, which are specific to pest species or entomopathogenic nematodes as an active ingredient. Rhizosphere microorganism is considered to provide defense against pathogen attack (Weller 1988). Many microorganisms colonize in the rhizoplane and rhizosphere of plants. Plant growth promoting microbes (PGPM) produce plant promoting substance as well as antibiotics, which are having capability to protect the plants from nematode disease (Siddiqui and Mahmood 1996). Most of the research has been carried out among the nematode-antagonistic organisms including nematophagous fungi and bacteria (Askary and Martinelli 2015). Among the nematophagous fungi, *Paecilomyces lilacinus* has antagonistic effects on root-knot nematodes (Sharon et al. 2001). In cotton and some vegetables crops grown by seed treatment, *Streptomyces avermitilis* is used as biopesticide. Abamectin, which is a mixture of macrocyclic lactone metabolites of fungus *Streptomyces avermitilis*, is used to control plant-parasitic nematode. Abamectin is active against root-knot nematodes (*M. incognita* and *M. reniformis*) and lesion nematodes (*Pratylenchus* spp.; Faske and Starr 2007). *Trichoderma* for management of plant-parasitic nematode has been confirmed (Haseeb and Khan 2012). Also, few nematophagous fungi can be used as potential biological control agent, for example, *Pochonia chlamydosporia* for *M. incognita* in vegetable crops. In a root-knot nematode susceptible tomato, the use of *P. chlamydosporia*, along with crop rotational methods, demonstrated a reduction in nematode levels (Atkins et al. 2003). Nematophagous fungal products have great potential for biopesticide development, including chitinases. Purified chitinase LPCHI1 derived from *Lecanicillium psalliotae* has reported to degrade *M. incognita* eggs by acting on transparent protective chitin-containing shells of nematode (Gan et al. 2007). Another parasitic bacterium, *Pasteuria* spp., is found to be effective on 323 nematode species (both plant-parasitic nematodes and free-living), which adds a plus point for its use as biopesticide (Chen and Dickson 1998). For example, in a greenhouse study conducted on cucumber, treatment with three species of *Pasteuria* reduced gall caused by *M. incognita* and also reduced the number and reproduction of nematodes. Bacteria belonging to *Bacillus* spp. have shown greater potential in nematode management. *Bacillus cereus* strain S2 treatment in *M. incognita* resulted in a mortality of 90.96% (Gao et al. 2016). Treatment with *Bacillus firmus* YBf-10 showed nematocidal activity by exhibiting an inhibition of egg hatching and motility against *M. incognita* (Xiong et al. 2015). *Bacillus thuringienis* (Bt) produces ∂ -endotoxin and exotoxin that exhibit nematocidal activities against nematodes occurring in soil and plant, and also against endoparasitic and free-living forms of plant-parasitic nematodes. The major genera include *Meloidogyne*, *Heterodera*, *Globodera*, *Helicotylenchus*, *Tylenchulus*, and *Radopholus*, *Pratylenchus* (Zuckerman et al. 1994). Microbial biopesticide K84 strain of *Agrobacterium radiobacter* is used to control crown gall (Chandler et al. 2011).

9.9.2 Plant-Incorporated Protectants (PIPs)

Plant-incorporated protectants (PIPs) are known as genetically engineered or modified crops. These plants produce biopesticidal materials itself using genetic material that has been incorporated into their genetic content. The transgenic plants expressing two proteins (CpTI and Bt toxin) were developed against parasitic nematodes. The plants with Bt toxin have reduced the nematode fertility and viability (Marroquin et al. 2000), while CpTI protein expressing plants affected the sexual development in nematodes (Urwin et al. 1998). Tomato plants expressing Cry6A protein in roots have shown the resistance to root-knot nematode (*M. incognita*). Genetic engineering of transgenic plants is done by insertion of few resistant genes like serine protease inhibitors (Cai et al. 2003; Vishnudasan et al. 2005), cystatins (Urwin et al. 1997; Atkinson et al. 2004), and the snow drop lectin (Ripoll et al. 2003). Proteinase inhibitors are the most promising ones in development of nematode resistance, and potatoes have been developed expressing cystatins (Urwin et al. 2001).

9.9.3 RNAi Pesticides

It involves expression of a double-stranded ribonucleic acid (dsRNA) targeting parasitism or housekeeping genes in the root-knot nematode in resistance to nematode infection in a host plant (Gheysen and Vanholme 2007). Transgenic tobacco plants having 95% more resistant to root-knot nematode as compared to wild tobacco was developed by expressing dsRNA targeting two root-knot nematode housekeeping gene (Yadav et al. 2006). Similar work has been done in *Arabidopsis thaliana*, confirming resistance to four major species of root-knot nematode by RNA interference (RNAi; Huang et al. 2006), carried out by inducing plant-delivered RNAi in cyst nematodes (Steeves et al. 2006; Sindhu et al. 2009).

9.9.4 Biochemical Pesticides

They are also called as herbal pesticides (Pal and Kumar 2013) as they are naturally occurring substances used for controlling pests through a nontoxic mechanism, and because it is difficult to assess whether a natural pesticide can control the pest by a nontoxic method, Environmental Protection Agency (EPA) has established a committee to determine the standard specific criteria for the pesticides. Biochemicals are the extracts of plants having antimicrobial properties that have been proven effective to decrease plant-parasitic nematode population (Ferris and Zheng 1999). There has been study undertaken to find the allelochemicals in all kind of faunas, which has included annual to perennial grasses, herbs, and woody trees, which have included altogether 46 families of plant kingdom (Ferris and Zheng 1999). Allelochemicals can be defined as plant metabolites or their products that are released into the surrounding environment through exudation from roots, volatilization, leaching from plants or plant residues, and decomposition of residues (Waller 1989; Putnam and Tang 1986; Einhellig 1995; Halbrendt 1996). Many allelochemicals were found to be very effective in controlling plant-parasitic nematodes, which are mainly as follows.

9.9.4.1 Glucosinolates

Glucosinolates, which are primary components of the Brassicaceae family, mainly rapeseed, when extracted have showed nematocidal activity at various stages, which was due to two reasons, that is, concentration and duration of the exposure to glucosinolates (Lazzeri et al. 1993). Allyl isothiocyanate and allylnitrile derived from the glucosinolate play a role in transforming the physical condition of the soil and spread at rapid rate in the soil. Allyl isothiocyanate is found to be highly toxic to the *Caenorhabditis elegans* (Donkin et al. 1995).

9.9.4.2 Benzaldehyde (Benzoic Aldehyde)

Benzaldehyde mainly extracted from the bitter almond (*Prunus dulcis*) and used as fungicide is well known (Flor 1926), but it also reduces the population of the *M. incognita* by ovicidal activity (Kokalis-Burelle et al. 2002). It also has an effect on chemotaxis kinetics of *C. elegans* (Nuttley et al. 2001). Under in vitro conditions, application of benzaldehyde and furfural on *M. javanica* cause immobilization and hatching inhibition (Oka 2001), and these two compounds also reduce galling (Oka 2001). When benzaldehyde is combined with organic amendments, it reduces damages caused by nematodes and Gram-positive rhizosphere bacteria (Siddiqui and Shaukat 2003).

9.9.4.3 Furfural

Furfural (C4H3O-CHO), also known as 2-furaldehyde, they are best known member of the furan family and the source of the other technically important furans. Application of furfurals improved plant growth and reduced the of population of *M. arenaria* and *R. reniformis* in ground nut (Rajendran et al. 2003). Furfurals were unaffected on free-living nematodes while plant-parasitic nematodes showed susceptibility with species of *Paratrichodorus* and *Xiphinema*, which have high susceptibility compared to *Helicotylenchus* and *Tylenchorhynchus* (Spaull 1997). Furfural is highly effective against *M. arenaria*, *M. incognita*, *Heterodera glycines*, and *Pratylenchus* spp. on crops like ladies' finger, soya bean, and squash (Rodriguez-Kabana et al. 1993). *M. incognita* population reduced in fields of cotton with application of furfural (Bauske et al. 1994).

9.9.4.4 Thymol

Thymol is isopropyl-m-cresol, a volatile, phenolic monoterpene produced by several plants, mainly plant thyme (*Thymus vulgaris*; Baerheim Svenden and Scheffer 1985). Combination of benzaldehyde and thymol has controlled the root-knot and cyst nematodes (soya beans), *Meloidogyne* spp., and Dorylaimid nematodes, and also led to gall formation reduction (Soler-Serratosa et al. 1996).

9.9.4.5 Saponins

Alkaloids, saponins, terpenes, flavonoid, and glycosides show antiparasitic activity against gastrointestinal nematodes (Botura et al. 2013). Saponins have been associated with the nematocidal effects against nematode juveniles (Francis et al. 2002). Saponins have ability to interact and form complexes and disrupt the protein of the nematode cuticle (Argentieri et al. 2008).

9.9.4.6 Citral

Citral is an aldehyde of geraniol and extracted from the volatile oils of lemon grass, lemon, orange, limetta. and pimento (Harborne and Baxter 1993). When citral was tested in vitro in tomato against root-knot nematode eggs, egg viability reduction was shown to be 80% (Kokalis-Buerelle et al. 2002). When it was applied on cotton-growing soil that has root-knot nematode juveniles, galling on roots of the cotton reduced, which increased the growth of the cotton plant.

9.10 Commercial Biopesticides

In the U.S., approximately 279 biopesticides were approved by 2009. In France, 77 biopesticides were approved, the reason for lesser number being a more restrictive policy of the European Union. Fourteen bacteria and 12 fungi have been registered with the EPA for the control of plant diseases (Fravel 2005).

9.11 General Advantages of Biopesticides

- Biopesticides usually have minimum toxicity/harm and environmental hazards.
- They are particularly designed for few pests or one specific target pest as compared to chemicals that have broad spectrum of activity.
- Have lower cost of production and are economically feasible than synthetic chemical pesticides.
- Their nature of control is preventive and not curative.

9.12 Disadvantages of Biopesticides

- They are highly specific and hence usually require specific identification of the target pest/pathogen.
- They are slow in action, hence biopesticides are often not suitable if the pest outbreak is massive with greater threat to crops.
- For better efficiency, biopesticides need to be used along with other control strategies.
- Resistance development by the pest is common for biological, chemical, or physical method of control.

9.13 Conclusions and Future Prospects

Nematodes are one of the significant pests in crop production causing heavy economic losses and destroying crop yield every year. Pest management using chemical methods is most widely used. Since chemical methods cause environmental hazards and are costly, the need for ecofriendly and cost-effective methods like biological control has taken an upper hand in recent years. In this regard, the use of biopesticides as supplement has emerged as promising alternative to chemical pesticides and their demand is rising steadily in all parts of the world. Biopesticides used for the control of nematodes are gaining major attention and in future there is a need to be exploit much in this area. There are various mechanisms involved in the management of plant-parasitic nematodes. Biopesticides could serve as a very effective alternative method for nematode control in future as they are safe for the environment. Application of these biopesticides not only reduces the pest/pathogen population but also improves the soil health. However, large-scale commercialization is sometimes a limiting factor in the exploitation of biopesticides. Henceforth, robust studies are recommended to be conducted in order to provide more holistic information.

References

Ali MA, Abbas A, Azeem F, Javed N, Bohlmann H (2015) Plantnematode interactions: from genomics to metabolomics. Int J Agric Biol 17:1071–1082

Ansari RA, Mahmood I (2017b) Determination of disease incidence caused by *Meloidogyne* spp. and or *Fusarium udum* on pigeonpea in Aligarh district: a survey. Trends Biosci 10 (24):5239–5243

Ansari RA, Mahmood I (2019a) Plant health under biotic stress: volume 2: microbial interactions. Springer, Singapore. https://doi.org/10.1007/978-981-13-6040-4

Ansari RA, Mahmood I (2019b) Plant health under biotic stress: volume 1: organic strategies. Springer, Singapore. https://doi.org/10.1007/978-981-13-6043-5

Ansari RA, Khan TA (2012a) Parasitic association of root-knot nematode, *Meloidogyne incognita* on guava. e-J Sci Technol 5:65–67

Ansari RA, Khan TA (2012b) Diversity and community structure of phytonematodes associated with guava in and around Aligarh, Uttar Pradesh, India. Trends Biosci 5(3):202–204

Ansari RA, Mahmood I (2017a) Optimization of organic and bio-organic fertilizers on soil properties and growth of pigeon pea. Sci Hortic 226:1–9

Ansari RA, Mahmood I, Rizvi R, Sumbul A (2017b) Siderophores: augmentation of soil health and crop productivity. In: Kumar V, Kumar M, Sharma S, Prasad R (eds) Probiotics in agroecosystem. Springer, Singapore, pp 291–312

Ansari RA, Rizvi R, Sumbul A, Mahmood I (2017a) PGPR: current vogue in sustainable crop production. In: Kumar V, Kumar M, Sharma S, Prasad R (eds) Probiotics and plant health. Springer, Singapore, pp 455–472

Ansari RA, Sumbul A, Rizvi R, Mahmood I (2019) Organic soil amendments: potential tool for soil and plant health management. In: Ansari RA, Mahmood I (eds) Plant health under biotic stress. Springer, Singapore, pp 1–35

Argentieri MP, D'Addabbo T, Tava A, Agostinelli A, Jurzysta M, Avato P (2008) Evaluation of nematicidal properties of saponins from Medicago spp. Eur J Plant Pathol 120(2):189–197

Ashraf MS, Khan TA (2008) Biomanagement of reniform nematode, *Rotylenchulus reniformis* by fruit wastes and *Paecilomyces lilacinus* on chickpea. World J Agric Sci 4:492–494

Askary TH (2012) Management of root-knot nematode *Meloidogyne javanica* in pigeonpea through seed treatment. Indian J Ecol 39:151–152

Askary TH (2015a) Nematophagous fungi as biocontrol agents of phytonematodes. In: Askary TH, Martinelli PRP (eds) Biocontrol agents of phytonematodes. CAB International, Wallingford, pp 81–125

Askary TH (2015b) Limitations, research needs and future prospects in the biological control of phytonematodes. In: Askary TH, Martinelli PRP (eds) Biocontrol agents of phytonematodes. CAB International, Wallingford, pp 446–454

Askary TH, Martinelli PRP (2015) Biocontrol agents of phytonematodes. CAB International, Wallingford, p 470

Atkins S, Hidalgo-Diaz L, Kalisz H, Mauchline T, Hirsch P, Kerry B (2003) Development of a new management strategy for the control of root-knot nematodes (Meloidogyne spp) in organic vegetable production. Pest Manag Sci 59:183–189

Atkinson HJ, Grimwood S, Johnston K, Green J (2004) Prototype demonstration of transgenic resistance to the nematode Radopholus similis conferred on banana by a cystatin. Transgenic Res 13(2):135–142

Baidoo R, Mengistu T, McSorley R, Stamps RH, Brito J, Crow WT (2017) Management of root-knot nematode (*Meloidogyne incognita*) on *Pittosporum tobira* under greenhouse, field, and on-farm conditions in Florida. J Nematol 49:133–139

Baerheim-Svenden A, Scheffer JJC (1985) Essential oils and aromatic plants. Martinus Nijhoff/Dr. W. Junk Publishers, Dordrecht

Bastiaans L, Paolini R, Baumann DT (2008) Focus on ecological weed management: what is hindering adoption? Weed Res 48(6):481–491

Bauske EM, Rodríguez-Kábana R, Estaun V, Kloepper JW, Robertson DG, Weaver CF, King PS (1994). Management of Meloidogyne incognita on cotton by use of botanical aromatic compounds

Berkeley M (1855) Vibrio forming cysts on the roots of cucumbers. Gardeners' Chronicle 7:220

Bhat MY, Wani AH (2012) Bio-activity of fungal culture filtrates against root-knot nematode egg hatch and juvenile motility and their effects on growth of mung bean (*Vigna radiata* L. Wilczek) infected with the root-knot nematode, *Meloidogyne incognita*. Arch Phytopathol Plant Prot 45:1059–1069

Bonfil D, Dolgin B, Mufradi I, Asido S (2004) Bioassay to forecast cereal cyst nematode damage to wheat in fields. Precis Agric 5:329–344. https://doi.org/10.1023/B:PRAG.0000040804.97462.02

Botura MB, dos Santos JDG, da Silva GD, de Lima HG, de Oliveira JVA, de Almeida MAO et al (2013) In vitro ovicidal and larvicidal activity of Agave sisalana Perr.(sisal) on gastrointestinal nematodes of goats. Vet Parasitol 192(1–3):211–217

Boyetchko S, Pedersen E, Punja Z, Reddy M (1998) Formulation of biopesticides. In: Hall FR, Menn JJ (eds) Biopesticides: use and delivery methods in biotechnology. Humana Press, Totowa, NJ, pp 487–508

Brar SK, Verma M, Tyagi RD, Valero JR (2006) Recent advances in downstream processing and formulations of Bacillus thuringiensis based biopesticides. Process Biochem 41:323–342

Brodie BB, Murphy WS (1975) Population dynamics of plant nematodes as affected by combinations of fallow and cropping sequence. J Nematol 7:91–92

Cai D, Thurau T, Tian Y, Lange T, Yeh KW, Jung C (2003) Sporamin-mediated resistance to beet cyst nematodes (Heterodera schachtii Schm.) is dependent on trypsin inhibitory activity in sugar beet (Beta vulgaris L.) hairy roots. Plant Mol Biol 51(6):839–849

Chen Z, Dickson D (1998) Review of Pasteuria penetrans: biology, ecology, and biological control potential. J Nematol 30:313–340

Clayton R, Storey M, Parker B, Ballingall M, Davies K (2008) Impact of reduced pesticide availability on control of potato cyst nematodes and weeds in potato crops. Potato Council Ltd., Oxford

Dababat A, Imren M, Erginbas-Orakci G, Ashrafi S, Yavuzaslanoglu E, Toktay H et al (2015) The importance and management strategies of cereal cyst nematodes, Heterodera spp., in Turkey. Euphytica 202:173–188. https://doi.org/10.1007/s10681-014-1269-z

Donkin SG, Eiteman MA, Williams PL (1995) Toxicity of glucosinolates and their enzymatic decomposition products to Caenorhabditis elegans. J Nematol 27(3):258

Chandler D, Bailey AS, Tatchell GM, Davidson G, Greaves J, Grant WP (2011) The development, regulation and use of biopesticides for integrated pest management. Philos Trans R Soc B Biol Sci 366(1573):1987–1998

Davis EL, Hussey RS, Mitchum MG, Baum TJ (2008) Parasitism proteins in nematode plant interactions. Curr Opin Plant Biol 11:360–366. https://doi.org/10.1016/j.pbi.2008.04.003

Decraemer W, Hunt D (2006) Structure and classification. In: Perry R, Moens M (eds) Plant nematology. CAB International, Oxfordshire, pp 3–32. https://doi.org/10.1079/9781845930561.0000

Deori R, Borah A (2016) Efficacy of *Glomus fasciculatum*, *Trichoderma harzianum* for the management of *Meloidogyne incognita* and *Rhizoctonia solani* disease complex in green gram. Indian J Nematol 46:61–64

Dybas RA (1989) Abamectin use in crop protection. In: Campbell WC (ed) Ivermectin and abamectin. Springer, New York, NY, pp 287–310

Einhellig FA (1995) Allelopathy: organisms, processes, and applications. In: Allelopathy: current status and future goals. ACS Publications, pp 1–24

Eissa MFM, Abd-Elgawad MMM (2015) Nematophagous bacteria as biocontrol agents of phytonematodes. In: Askary TH, Martinelli PRP (eds) Biocontrol agents of phytonematodes. CAB International, Wallingford, pp 217–243

Faske TR, Starr JL (2007) Cotton root protection from plant parasitic nematodes by abamectin treated seed. J Nematol 39(1):27–30

Ferris H, Zheng L (1999) Plant sources of Chinese herbal remedies: effects on Pratylenchus vulnus and Meloidogyne javanica. J Nematol 31(3):241

Feyisa B, Lencho A, Selvaraj T, Getaneh G (2016) Evaluation of some botanicals and *Trichoderma harzianum* against root-knot nematode *Meloidogyne incognita* (Kofoid and white) Chitwood in tomato. J Entomol Nematol 8:11–18

Flor HH (1926) Fungicidal activity of furfural. Iowa State Coll J Sci 1:199–227

Francis G, Kerem Z, Makkar HP, Becker K (2002) The biological action of saponins in animal systems: a review. Br J Nutr 88(6):587–605

Fravel DR (2005) Commercialization and implementation of biocontrol. Annu Rev Phytopathol 43:337–359

Gan Z, Yang J, Tao N, Liang L, Mi Q, Li J, Zhang K (2007) Cloning of the gene Lecanicillium psalliotae chitinase Lpchi1 and identification of its potential role in the biocontrol of root-knot nematode Meloidogyne incognita. Appl Microbiol Biotechnol 76:1309–1317. https://doi.org/10.1007/s00253-007-1111-9

Ganaie MA, Khan TA (2010) Biological potential of *Paecilomyces lilacinus* on pathogenesis of *Meloidogyne javanica* infecting tomato plant. Eur J Appl Sci 2:80–84

Gao H, Qi G, Yin R, Zhang H, Li C, Zhao X (2016) Bacillus cereus strain S2 shows high nematicidal activity against Meloidogyne incognita by producing sphingosine. Sci Rep 6:28,756. https://doi.org/10.1038/srep28756

Gheysen G, Vanholme B (2007) RNAi from plants to nematodes. Trends Biotechnol 25(3):89–92

Halbrendt JM (1996) Allelopathy in the management of plant-parasitic nematodes. J Nematol 28 (1):8

Harborne JB, Baxter H (eds) (1993) Phytochemical dictionary- a handbook of bioactive compounds from plants. Taylor and Francis, UK

Haseeb A, Khan MR (2012) Prospects of Trichoderma in nematode management. In: Nagesh M, Rajkumar BSB, Krishnakumar NK (eds) Status and prospects for enhancing the uptake of antagonistic organisms for nematode management in India. NBAII (ICAR), Bangalore, pp 42, 187p–56

Huang G, Dong R, Allen R, Davis EL, Baum TJ, Hussey RS (2006) A root-knot nematode secretory peptide functions as a ligand for a plant transcription factor. Mol Plant-Microbe Interact 19 (5):463–470

Jamshidnejad V, Sahebani N, Etebarian H (2013) Potential biocontrol activity of *Arthrobotrys oligospora* and *Trichoderma harzianum* BI against *Meloidogyne javanica* on tomato in the greenhouse and laboratory studies. Arch Phytopathol Plant Prot 46:1632–1640

Jegathambigai V, Wilson-Wijeratnam RS, Wijesundera RLC (2011) Effect of *Trichoderma viride* strain NRRL 6418 and *Trichoderma harzianum* (Hypocrea lixii TWC1) on *Livistona rotundifolia* root-knot nematode *Meloidogyne incognita*. J Entomol 8:229–239

Jones JT, Haegeman A, Danchin EG, Gaur HS, Helder J, Jones MG et al (2013) Top 10 plant-parasitic nematodes in molecular plant pathology. Mol Plant Pathol 14(9):946–961

Katan J (1981) Solar heating (solarization) of soil for control of soilborne pests. Annu Rev Phytopathol 19:211–236. https://doi.org/10.1146/annurev.py.19.090181.001235

Kiontke K, Fitch D (2013) Nematodes. Curr Biol 23:862–864. https://doi.org/10.1016/j.cub.2013.08.009

Kirkegaard JA, Sarwar M (1998) Biofumigation potential of brassicas. I. Variation in glucosinolate profiles of diverse field-grown brassicas. Plant and Soil 201:71–89. https://doi.org/10.1023/A:1004364713152

Knowles A (2005) New developments in crop protection product formulation. (p. 153–156). Agrow Reports UK: Tand F Informa UK Ltd. Formulations of agrochemicals. Environmentalist 28:35–44

Knowles A (2008) Recent development of safer formulations of agrochemicals. Environmentalist 28:3544

Koenning S, Overstreet C, Noling J, Donald P, Becker J, Fortnum B (1999) Survey of crop losses in response to phytoparasitic nematodes in the United States for 1994. J Nematol 31(4S):587–618

Kokalis-Burelle N, Martinez-Ochoa N, Rodríguez-Kábana R, Kloepper JW (2002) Development of multi-component transplant mixes for suppression of Meloidogyne incognita on tomato (Lycopersicon esculentum). J Nematol 34(4):362

Kokalis-Burelle N, Rodríguez-Kábana R (2006) Allelochemicals as biopesticides for management of plant-parasitic nematodes. In: Allelochemicals: biological control of plant pathogens and diseases. Springer, Dordrecht, pp 15–29

Kokalis-Burelle N (2015) *Pasteuria penetrans* for control of *Meloidogyne incognita* on tomato and cucumber, and *M. arenaria* on snapdragon. J Nematol 47:207–213

Kumar D, Bhatt J, Sharma RL (2017) Efficacy of different bio control agents against *Meloidogyne incognita* and *Fusarium oxysporum* on black gram (*Vigna mungo* L). Int J Curr Microbiol Appl Sci 6:2287–2291

Kumar P, Chand R (2015) Bioefficacy of *trichoderma harzianum* against root-knot nematode *Meloidogyne incognita* on brinjal. Ann Plant Prot Sci 23:361–364

Lazzeri L, Tacconi R, Palmieri S (1993) In vitro activity of some glucosinolates and their reaction products toward a population of the nematode Heterodera schachtii. J Agric Food Chem 41 (5):825–829

MacGuidwin AE (1993) Management of nematodes. In: Rowe RC (ed) Potato health management. APS Press, St. Paul, MN

Mahanta B, Borbora AC, Konwar B (2016) Effect of *Paecilomyces lilacinus* on *Tylenchulus semipenetrans* population in Khasi mandarin. Indian J Nematol 46:55–81

Abd-Elgawad MMM, Askary TH (2018) Fungal and bacterial nematicides in integrated nematode management strategies. Egypt J Biol Pest Control 28:74. https://doi.org/10.1186/s41938-018-0080-x

Maier T, Hewezi T, Peng J, Baum TJ (2013) Isolation of whole esophageal gland cells from plant parasitic nematodes for transcriptome analyses and effector identification. Mol Plant-Microbe Interact 26:31–35. https://doi.org/10.1094/MPMI-05-12-0121-FI

Marroquin LD, Elyassnia D, Griffitts JS, Feitelson JS, Aroian RV (2000) Bacillus thuringiensis (Bt) toxin susceptibility and isolation of resistance mutants in the nematode Caenorhabditis elegans. Genetics 155(4):1693–1699

Matthews GA, Bateman R, Lacey P, Miller PHC (2014) Pesticide application methods, 4th edn. Wiley, Chichester

Meyer S (2003) United States Department of Agriculture – Agricultural Research Service research programs on microbes for management of plant-parasitic nematodes. Pest Manag Sci 59:665–670

Mollet H, Grubenmann A (2001) Formulation technology, 5th edn. Wiley-VCH, Berlin, pp 247–259

Muthulakshmi M, Devrajan K, Jonathan EI (2010) Biocontrol of root-knot nematode, *Meloidogyne incognita* (Kofoid and white) Chitwood in mulberry (*Morus alba* L.). J Biopest 3:479–482

Narasimhamurthy HB, Ravindra H, Sehgal M (2017a) Management of rice root-knot nematode, *Meloidogyne graminicola*. Int J Pure Appl Biosci 5:268–276

Narasimhamurthy HB, Ravindra H, Sehgal M, Ekabote SD, Ganapathi (2017b) Management of rice root-knot nematode, *Meloidogyne graminicola*. J Entomol Zool Stud 5:1433–1439

Naserinasab F, Sahebani N, Etebarian HR (2011) Biological control of *Meloidogyne javanica* by *Trichoderma harzianum* BI and salicylic acid on tomato. African J Food Sci 5:276–280

Needham T (1742) A letter from Mr. Turbevil Needham, to the president; concerning certain chalky tubulous concretions, called Malm: with some microscopical observations on the farina of the red lily, and of worms discovered in smutty corn. Philos Trans R Soc Lond 42:462–471. https://doi.org/10.1098/rstl.1742.0101

Nicol J, Elekçioğlu H, Bolat N, Rivoal R (2004) The global importance of the cereal cyst nematode (Heterodera spp.) on wheat and international approaches to its control. Commun Agric Appl Biol Sci 72:677–686

Nicol J, Turner D, Coyne L, den Nijs L, Hockland S, Maafi Z (2011) Current nematode threats to world agriculture. In: Jones J, Gheysen G, Fenoll C (eds) Genomics and molecular genetics of plant-nematode interactions. Springer Science Business Media, Berlin, pp 21–43. https://doi.org/10.1007/978-94-007-0434-3_2

Noel GR (1992) History, distribution, and economics. In: Riggs R, Wrather JA (eds) Biology and management of the soybean cyst nematode. American Phytopathological Society, pp 1–13

Norton D, Hoffmann J (1975) Longidorus breviannulatus n. sp. (Nematoda: Longidoridae) associated with stunted corn in Iowa. J Nematol 7:168–171

Nuttley WM, Harbinder S, van der Kooy D (2001) Regulation of distinct attractive and aversive mechanisms mediating benzaldehyde chemotaxis in Caenorhabditis elegans. Learn Mem 8 (3):170–181

Oka Y (2001) Nematicidal activity of essential oil components against the root-knot nematode Meloidogyne javanica. Nematology 3(2):159–164

Oerke EC, Dehne HW, Schoenbeck F, Weber A (1994) Crop production and crop protection: estimated losses in major and cash crops. Elsevier, Amsterdam, the Netherlands

Pal GK, Kumar B, Shahi SK (2013) Antifungal activity of some common weed extracts against phytopathogenic fungi Alternaria spp. Int J Universal Pharm Life Sci 3(2):6–14

Poinar GO (1983) The natural history of nematodes. Prentice Hall, Englewood Cliffs, NJ

Priya MS (2015) Biomanagement of rice root knot nematode, *Meloidogyne graminicola* Golden and Brichfield in aerobic rice. Int J Manag Soc Sci 3:591–598

Putnam AR, Tang CS (eds) (1986) The science of Allelopathy. Wiley, New York

Quentin M, Abad P, Favery B (2013) Plant parasitic nematode effectors target host defense and nuclear functions to establish feeding cells. Front Plant Sci 4:1–7. https://doi.org/10.3389/fpls.2013.00053

Rajendran G, Ramakrishnan S, Subramanian S (2003) Cropguard-a botanical nematicide for the management of Meloidogyne arenaria and Rotylenchulus reniformis in groundnut. In: Proceedings of national symposium on biodiversity and management of nematodes in cropping systems for sustainable agriculture, Jaipur, India, pp 122–125

Rikita B, Utpal D (2014) An overview of fungal and bacterial biopesticides to control plant pathogens/diseases. Afr J Microbiol Res 8:1749–1762

Rodriguez-Kabana R, Kloepper JW, Weaver CF, Robertson DG (1993). Control of plant parasitic nematodes with furfural-a naturally occurring fumigant

Ripoll C, Favery B, Lecomte P, Van Damme E, Peumans W, Abad P, Jouanin L (2003) Evaluation of the ability of lectin from snowdrop (Galanthus nivalis) to protect plants against root-knot nematodes. Plant Sci 164(4):517–523

Salma M, Jogen CK (2011) A review on the use of biopesticides in insect pest management. Int J Sci Adv Technol 1:169178

Santo G, O'Bannon J, Finley A, Golden A (1980) Occurrence and host range of a new root-knot nematode (Meloidogyne chitwoodi) in the Pacific northwest. Plant Dis 64:951952. https://doi.org/10.1094/PD-64-951

Sasser JN (1987) A world perspective on nematology: the role of the society. In: Veech J, Dickson D (eds) Vistas on nematology. Society of Nematologists, Inc., Hyattsville, MD, pp 7–14

Sharon E, Bar-Eyal M, Chet I, Herrera-Estrella A, Kleifeld O, Spiegel Y (2001) Biological control of the root-knot nematode Meloidogyne javanica by Trichoderma harzianum. Biol Control 91 (7):687–693

Siddiqui ZA, Mahmood I (1996) Biological control of plant parasitic nematodes by fungi: a review. Bioresour Technol 58:229–239

Siddiqui IA, Shaukat SS (2003) Factors influencing the effectiveness of non-pathogenic Fusarium solani strain Fs5 in the suppression of root-knot nematode in tomato. Phytopathol Mediterr 42 (1):17–26

Silva JO, Santana MV, Freire LL, Ferreira BS, Rocha MR (2017) Biocontrol agents in the management of *Meloidogyne incognita* in tomato. Cienc Rural 47. https://doi.org/10.1590/0103-8478cr20161053

Sindhu AS, Maier TR, Mitchum MG, Hussey RS, Davis EL, Baum TJ (2009) Effective and specific in planta RNAi in cyst nematodes: expression interference of four parasitism genes reduces parasitic success. J Exp Bot 60(1):315–324

Slavica G, Brankica T (2013) Biopesticide formulations, possibility of application and future trends. Pesticidi i fitomedicina 28(2):97–102

Soler-Serratosa A, Kokalis-Burelle N, Rodríguez-Kábana, R, Weaver, CF, King P S (1996). Allelochemicals for control of plant-parasitic nematodes. 1. In vivo nematicidal efficacy of thymol and thymol/benzaldehyde combinations. Nematropica (USA)

Soriano I, Prot J, Matias D (2000) Expression of tolerance for Meloidogyne graminicola in rice cultivars as affected by soil type and flooding. J Nematol 32:309–317

Sorribas FJ, Ornat C, Verdejo-Lucas S, Galeano M, Valero J (2005) Effectiveness and profitability of the Mi-resistant tomatoes to control root-knot nematodes. Eur J Plant Pathol 111(1):29–38

Spaull VW (1997) On the use of furfural to control nematodes in sugarcane. Proc S Afr Sugar Techn Ass 71:96

Steeves RM, Todd TC, Essig JS, Trick HN (2006) Transgenic soybeans expressing siRNAs specific to a major sperm protein gene suppress Heterodera glycines reproduction. Funct Plant Biol 33:991–999

Stirling G (1991) Biological control of plant-parasitic nematodes. CAB International, Wallingford, 282 p

Tadros F (2005) Applied surfactants, principles and applications. Wiley, Weinheim

Turner S, Evans K (1998) The origins, global distribution and biology of potato cyst nematodes (Globodera rostochiensis (Woll.) and G. pallida Stone). In: Marks R, Brodie B (eds) Potato cyst nematodes: biology, distribution, and control. University Press, Cambridge, pp 7–26

Urwin PE, Lilley CJ, McPherson MJ, Atkinson HJ (1997) Resistance to both cyst and root-knot nematodes conferred by transgenic Arabidopsis expressing a modified plant cystatin. Plant J 12 (2):455–461

Urwin PE, McPherson MJ, Atkinson HJ (1998) Enhanced transgenic plant resistance to nematodes by dual proteinase inhibitor constructs. Planta 204(4):472–479

Urwin PE, Troth KM, Zubko EI, Atkinson HJ (2001) Effective transgenic resistance to Globodera pallida in potato field trials. Mol Breed 8(1):95–101

Van Emden HF (2007) 17 Host-plant resistance. In: Van Emden HF, Harrington R (eds) Aphids as crop pests. CABI, Wallingford, p 447

Vernner R, Bauer P (2007) Q-TEO, a formulation concept that overcomes the incompatibility between water and oil. Pfalzenschutz-Nachrichten Bayer 60:7–26
Villate L, Fievet V, Hanse B, Delemarre F, Plantard O, Esmenjaud D, van Helden M (2008) Spatial distribution of the dagger nematode *Xiphinema index* and its associated grapevine fanleaf virus in French vineyard. Phytopathology 98:942–948. https://doi.org/10.1094/PHYTO-98-8-0942
Vishnudasan D, Tripathi MN, Rao U, Khurana P (2005) Assessment of nematode resistance in wheat transgenic plants expressing potato proteinase inhibitor (PIN2) gene. Transgenic Res 14 (5):665–675
Waller GR (1989) Allelochemical action of some natural products, in phytochemical ecology: allelochemicals, mycotoxins, and insect pheromones and allomones. In: Chou CH, Waller GR (eds) Academia Sinica Monograph Ser, vol 9. Academia Sinica, Taipei, ROC, pp 129–154
Webster JM (1995) The host-parasite relationships of plant-parasitic nematodes. In: Dawes B (ed) Advances in parasitology. Academic Press Inc, Cambridge, MA
Weller DM (1988) Biological control of soilborne plant pathogens in the rhizosphere with bacteria. Annu Rev Phytopathol 26(1):379–407
Westphal A (2011) Sustainable approaches to the management of plant-parasitic nematodes and disease complexes. J Nematol 43:122–125
Williamson V, Hussey R (1996) Nematode pathogenesis and resistance in plants. Plant Cell 8:1735–1745. https://doi.org/10.1105/tpc.8.10.1735
Winder RS, Wheeler JJ, Conder N, Otvos SS, Nevill R, Duan L (2003) Microencapsulation: a strategy for formulation of inoculum. Biocontrol Sci Tech 13:155–169
Woods TS (2003) Pesticide formulations, in AGR 185 in encyclopedia of agrochemicals. Van Nostrand Reinhold, New York, pp 1–11
Xiong J, Zhou Q, Luo H, Xia L, Li L, Sun M, Yu Z (2015) Systemic nematicidal activity and biocontrol efficacy of Bacillus firmus against the root-knot nematode Meloidogyne incognita. World J Microbiol Biotechnol 31:661–667. https://doi.org/10.1007/s11274-015-1820-7
Yadav BC, Veluthambi K, Subramaniam K (2006) Host-generated double stranded RNA induces RNAi in plant-parasitic nematodes and protects the host from infection. Mol Biochem Parasitol 148:219–222
Zuckerman BM, Matheny M, Acosta N (1994) Control of plant-parasitic nematodes by a nematicidal strain of Aspergillus niger. J Chem Ecol 20(1):33–43

Chapter 10
Efficacy of Microbial Biocontrol Agents in Integration with Other Managing Methods against Phytoparasitic Nematodes

Mohammad Reza Moosavi

Abstract Biological control can be a safe alternative to detrimental chemical nematicides if its persistence and performance increase to a satisfactory level. But at present, no biocontrol agent (BCA) can provide adequate nematode control when applied alone. One approach to improve their controlling importance is to use them in integration with one or more compatible practices that enhance BCAs' population, diversity, durability and efficacy. This goal may be achieved by combined use of BCAs with measures aiming at manipulating the soil environment in favour of BCAs, reducing nematode population and enhancing BCAs' activity. Here a brief outline of some measures for controlling phytonematode is illustrated with extra attention to those that can be applied combinedly with biological control. Their advantages and disadvantages as well as their effects on altering biocontrol activity are demonstrated along with selected examples of each tactic. The reviewed strategies in combination with biocontrol are using host plant resistance (tolerance, resistance and induced resistance); agronomic practices (rotation, trap crops, antagonistic crops, cover crops, fallow, flooding, organic amendments and tillage); decrease in phytonematode populations (soil solarisation, biofumigation and chemical nematicides); and application more than one BCA. Finally, the future paths of integrated nematode management are designed.

Keywords Antagonists · Farming practices · IPM · Nematophagous bacteria · Nematophagous fungi

---

M. R. Moosavi (✉)
Department of Plant Pathology, Marvdasht Branch, Islamic Azad University, Marvdasht, Iran
e-mail: rmmoosavi@miau.ac.ir

R. A. Ansari et al. (eds.), *Management of Phytonematodes: Recent Advances and Future Challenges*, https://doi.org/10.1007/978-981-15-4087-5_10

10.1 Introduction

Food production currently suffers an unprecedented pressure due to increase in the global population and inability to feed the world. It seems that the pressure will be exceedingly increased over the coming years when the population projected to reach 9.3 billion in 2050 (FAO, IFAD and WFP 2013). So, the most important challenge for agricultural industry is to provide more food in spite of diminishing fertile land, water supply and energy resources (Godfray et al. 2010; Gomiero et al. 2011; Ansari and Mahmood 2017a; Ansari and Mahmood 2019a, b). More food can be imagined in three main ways: cultivate more land; intensify the frequency of cropping; and increase the harvested yield in a given area. Several studies have evaluated the contribution of these three ways to the increase in universal food production. In a study, the contribution of more harvested yield, cultivation more land and intensified agriculture to enhancement of world food supply during 1960–1999 was estimated as 78%, 15% and 5%, respectively (Bruinsma 2003). One of the most important components of yield improvement is reducing the damage of the pests to agricultural plants. The estimated amount of losses to global agricultural production implies that plant diseases are a problem of great significance. They cause a serious threat to agricultural productivity and sustainability, and may endanger food security (Strange and Scott 2005). Phytonematodes are a group of plant pathogens whose importance is progressively understood in the last few decades. It is estimated that they now impose about 12% loss to global food production (Nicol et al. 2011). However, it should not be forgotten that indigenous phytonematodes in traditional cropping are not normally an agricultural problem. They usually turn into noxious pests in consequence of alteration in the cropping system, agricultural practices, climate, or being introduced to new regions (Sikora et al. 2005). There is a serious need to control the phytonematodes; however, the task is not easy to achieve. Many approaches can be employed to either increase the plant resistance/tolerance to phytonematodes or reduce nematode populations below its economic injury level. The main controlling method is now based on chemical nematicides which should be apparently substituted with other safer methods (Moosavi and Askary 2015). Biological control offers a safe opportunity for nematode control and many organisms have been identified with hostile activity against phytonematodes (Moosavi and Zare 2015). However, it has been recently shown that phytonematodes can defend themselves by producing antimicrobial peptides whose secretion is regulated by the nematode immunity system (Liang et al. 2019). One strategy to improve the general levels of phytonematode biocontrol is to manipulate the environment in favour of resident microbiota so enhanced their diversity and populations. In spite of its great potential, biological control takes little or no part in present phytonematode management programmes (Stirling 2014). This may be partly due to our insufficient understanding about the traits involved in efficacy of biocontrol agents (BCAs) which in turn leads to inconsistency of their effectiveness. It seems that, at least for now, no BCA can provide sufficient nematode control when implemented by itself. Therefore, a successful nematode-biocontrol programme needs more approaches than the application of a BCA (Viaene et al. 2013). Using BCAs in an integrated programme can improve their performance against phytonematodes. The main principle of integrated pest management (IPM) is to decrease the pest population under damage threshold by using an integration of measures. These measures are chosen according to our information about pest and plant biology, environmental factors and ecological principles (Stirling 1999). Decrease in phytonematode damage in an IPM programme can be achieved by practices targeting either reducing nematode population or enhancing BCAs' activity (Hildalgo-Diaz and Kerry 2008). This chapter tries to illustrate the measures which can be used in integration with BCAs to enhance nematode control.

10.2 Using Host Plant Tolerance/Resistance

Resistant cultivars present an efficient, environmentally safe, persistent and inexpensive tactic for phytonematode control (Koenning et al. 2001; Castagnone-Sereno 2002; Starr et al. 2013) but with one disadvantage that is effective only when the soil is infested with one (important) species (Hildalgo-Diaz and Kerry 2008). This restriction is because of specificity in resistance which acts against one species of phytonematodes or even more specifically against one race of a species without any effect on other existing parasitic species (Roberts 2002). Therefore, a number of researchers believe that resistance could not be a worldwide solution to phytonematode management. Tolerant cultivars are able to endure severe pathogen infection and produce an acceptable yield (Agrios 2005). The tolerant host plants do not restrict the development of phytonematodes, so their population densities increase during growing season but the yield does not remarkably reduce (Dalmasso et al. 1992). The extent of yield loss corresponds to phytonematode initial population densities (Sikora et al. 2005). Integration of host plant tolerance or resistance with other controlling measures such as biological control could positively increase the efficacy of nematode management.

10.2.1 Host Plant Tolerance

Planting tolerant hosts in combination with BCAs makes the controlling programme of phytonematodes more efficient. The tolerant host could produce acceptable yield while robust BCAs reduce nematode populations. This will decrease the residual nematode population density in soil and consequently the initial nematode population for the next successive crop. On the other hand, decline in females' fecundity on tolerant host might delay the occurrence or selection of virulent nematode species or pathotypes (Hildalgo-Diaz and Kerry 2008; Ansari and Khan 2012a, b). Obligate parasites of phytonematodes have better efficacy on tolerant cultivars than on resistant ones. Contrary to resistant plants, the population density of nematode which supports the establishment and multiplication of obligate parasites is higher on tolerant plants. For example, densities in soil and parasitising abilities of *Nematophthora gynophila* on cereal and sugar beet cyst nematode increased when the tolerant host of each nematode was cultivated successively (Kerry 1987). Soil suppressiveness to cereal cyst nematodes in monocultures in Western Europe may occur because of concomitant presence of fungal BCAs and partial resistance (Cook and Starr 2006). Suitable integration of tolerant cultivars of sugar beet with facultative parasites of nematode, short crop rotation and nematicide increased the efficacy of the BCAs as well. Concomitant use of mentioned measures could successfully control the first generation of sugar beet nematode (*Heterodera schachtii*) with *Pochonia chlamydosporia* and the second generation with *Cylindrocarpon destructans* (Crump 1989, 1991).

10.2.2 Host Plant Resistance

Resistance can be generated either by selecting plants with natural resistance gene (s) or by inserting new resistance gene into proper crops (Davies and Elling 2015). Several approaches are usually used for engineering resistance including expression of the inserted natural resistance genes in transgenic plants; aiming at and disruption of the initial nematode–plant interaction; aiming directly at the nematode (*Bt* & Cry proteins, plantibodies, lectins, protease inhibitors and RNA interference); and interfere with feeding site formation (Cottage and Urwin 2013). A number of major resistant (*R*) genes with the ability of conferring resistance to sedentary endoparasitic nematodes have been either cloned or mapped from agronomic or wild plants (Davies and Elling 2015). Resistance to phytonematodes may be provided by dominant, recessive, co-dominant *R* genes or by quantitative trait loci (Molinari 2011). Most of the *R* genes that have been already identified are translated to resistant proteins with a similar structure. These proteins consist of a central nucleotide-binding site (NBS) and a C-terminal leucine-rich repeat (LRR) region (Kaloshian et al. 2011). Resistant high-yielding crops offer an idle basis upon which further controlling tactics can be built (Sikora et al. 2005). Since resistance is considered as a major method for nematode management, a few researches have examined the combined effect of host resistance and BCAs. Despite scarce references, it seems that application of BCAs on resistant plants resulted in better nematode management. However, it must be taken into consideration that continuous planting of resistant cultivars often resulted in development and establishment of nematode races which could feed and colonise those resistant plants. Application of *Hirsutella rhossiliensis* on a resistant cultivar of potato to *Pratylenchus penetrans* produced synergistic effect and better nematode control was achieved compared with when each treatment was used alone (Timper and Brodie 1994).

10.2.3 Induction of Host Plant Resistance

Host plants could defend against pathogen infection through a number of different constitutive, induced and systemic defence mechanisms (Walters 2011). Induction of plant immunity system is a fascinating method for controlling phytonematodes (Leadbeater and Staub 2014; Alesadi et al. 2017). The symbiotic fungi-mediated resistance, systemic acquired resistance, induced systemic resistance, and β-aminobutyric acid-induced resistance are the suggested methods for controlling the pathogenic nematodes (Bakker et al. 2006). Plant hormones or their derivatives are considered as defence inducer molecules which could trigger the inducible defence mechanisms. Salicylic acid, jasmonic acid and ethylene are the most famous defence inducers (Conrath 2011; Takur and Sohal 2013), However, it has been illustrated that other hormones like auxins, abscisic acid, cytokinins, gibberellins and brassinosteroids have regulatory role on plant defence (Denancé et al. 2013; Moosavi 2017). Plant defences are induced by nematode attack but can also be started by exogenous application of defence inducers. Application of these molecules may decrease phytonematode invasion, development and reproduction. BCAs may control the pathogens by direct or indirect mechanisms. It has been frequently reported that BCAs can indirectly antagonise the pathogens by improving plant photosynthetic and respiratory activities (Shoresh et al. 2010) or by inducing the host plant resistance (Walters and Bennett 2014). It has been reported that endophytic fungi (non-pathogenic *Fusarium oxysporum*) could stimulate defence mechanisms in tolerant cultivars of bananas against *Radopholus similis* (Paparu et al. 2007). Similarly, *P. chlamydosporia* induced zucchini defence responses to *Meloidogyne javanica* (Lalezar et al. 2016). It has been also reported that induced resistance to nematode by BCAs is inheritable. For example, *Trichoderma atroviride* could stimulate tomato resistance against *M. javanica* which could be inherited to tomato progeny (de Medeiros et al. 2017). The combined application of BCA and defence inducer molecules will satisfactorily control the nematode while defence inducer molecule prevents initial nematode damage and the BCA provides long-term protection (Moosavi and Ghani 2019). Soil application of *P. chlamydosporia* in combination with foliar sprays of benzothiadiazole (BTH) or *cis*-jasmone reduces *Meloidogyne chitwoodi* reproduction on potato. The number of eggs per egg mass was lesser and the proportion of parasitised eggs was greater in plants treated with both the defence activators and the fungus than in the plants treated only with one of the treatments (Vieira Dos Santos et al. 2014). Conversely, integrated use of *P. chlamydosporia* with spraying tomato plants with methyl jasmonate, acibenzolar-S-methyl or ethephon reduced the ability of the fungus to lessen the gall number of *M. javanica* or activate local induced resistance (de Medeiros et al. 2015). Applying salicylic acid as a defence inducer and *Arthrobotrys oligospora* as a BCA against *M. javanica* considerably decreased diameter of nematode galls, number of galls per plant, number of egg masses per plant and number of eggs per egg mass, but their simultaneous application reduced these indices to a greater extent (Mostafanezhad et al. 2014).

10.3 Agronomic Practices

Agronomic practices have been among the most important concepts in managing phytonematodes for a long time. The involved mechanisms for nematode suppression are starvation, entrapping, antagonism, induction of soil antagonistic activity and/or biofumigation (Sikora et al. 2005). Biological control and agronomic measures can synergistically contribute in management of phytonematodes. Here a concised outline of some cultural-based tactics for controlling phytonematode is presented with emphasis on those that can be implemented integratedly with biological control. Their advantages and disadvantages along with selected examples of each tactic are included.

10.3.1 Rotation

The oldest and one of the most operational measures to manage phytonematodes is crop rotation. Succession of the crops can be temporal or spatial. Rotation of susceptible plants with non-host or poor-host provides adequate time after each susceptible host to decrease nematode population under the level which allows the next crop to grow and yield acceptably (Trivedi and Barker 1986). A crop rotation system can be successful only when resistant or tolerant plants are available for cultivation in the target agricultural systems. The host range of parasitic nematodes must also be considered for devising a practical rotational programme. Choosing a proper host plant is so difficult when the host range of the target nematode is an extensive one. Differences in host resistance to various populations of a species are another problem. However, the presence of multiple or polyphagous species of phytonematodes in a given soil restricts the ability of selecting appropriate plants for rotation (Viaene et al. 2013). When multiple species exist in soil, a non-host plant for one species might possibly be a proper host for the non-target species. Despite its deceptive simplicity, devising a specific crop rotation system for controlling phytonematodes needs much knowledge. The nematode population density is to same degree under the influence of both individual crops and their sequence in time and space (Trivedi and Barker 1986). As well, the BCA population is also affected by different plant species. Therefore, more reduction in nematode population will occur if poor or non-host plant for nematode can support the growth of BCAs (Timper 2014). Various plants secrete different root diffusates which are effective on soil and rhizosphere microbial construction. However, our knowledge about the effect of root diffusates on the performance of BCAs is so limited (Viaene et al. 2013). The needed time span between susceptible hosts or the needed number of rotational plants may be varied in relation to many factors. To recommend a rotational scheme we first need information about species, race and the host range of the local nematodes in a particular region. The host status of different crop cultivars and weeds must be determined. An understanding of nematode (species and/or races) population dynamics on different crops, relationship of nematode populations to crop loss and the environmental effect on the population of nematodes and BCAs is also vital in developing efficient rotation schemes. The attenuation of nematode population densities during rotation differs significantly with the time, place, pathogens, weed hosts, and the nature and length of the rotation (Kratochvil et al. 2004; Viaene et al. 2013). A two- to four-year rotation programme of cultivating non-host plants usually resulted in acceptable management of economically important phytonematodes (Trivedi and Barker 1986). However, rotation is not useful where a large spectrum of nematode species exists and no rotation scheme can adequately control a wide range of nematode species on different plants (Barker 1991). On the other hand, nematode population decline in consequence of cultivating non-host plants for a long time may be detrimental to BCAs especially for obligate ones (Timper 2011; Stirling 2014). Combining rotation with biological control could reciprocally improve efficacy of both tactics when non- or poor-host rotating plants reduce nematode population densities to levels that prevent or lessen initial infection of successive susceptible host. Then application of BCAs in an augmentation or inundation strategy manages the phytonematode during the cultivation of susceptible host. Otherwise the rotation plants must be carefully selected to support the survival of the existent BCAs and therefore to maintain phytonematode suppression. There are contradictory results about the survival and maintenance of BCAs during rotation programme that emphasise on our inadequate knowledge about interactional, ecological and environmental factors which impact on BCAs' potency. It has been demonstrated that parasitism ability of *H. rhossiliensis* on motile phytonematodes depended on nematode population density (Jaffee et al. 1992). Planting susceptible cultivars of soybean to *Heterodera glycines* resulted in more parasitised juveniles (Chen and Reese 1999). But when maize- or soybean-resistant cultivars were cultivated in rotation with susceptible cultivar, the proportion of infected juveniles decreased (Chen and Liu 2007). Planting switchgrass (*Panicum virgatum*) in rotation with peanut for controlling *Meloidogyne arenaria* caused shifting in rhizosphere bacterial diversity. The shifts in bacterial community structure were connected to fluctuations in phytonematode populations (Kokalis-Burelle et al. 2002). In another experiment, *P. chlamydosporia* was applied to a field with high populations of *M. incognita,* and then bean and cabbage (poor-hosts for the nematode) were cultivated prior to tomato (susceptible host). The population densities of *M. incognita* declined considerably during rotation period. But when tomato was again planted in the field, the nematode population density increased only in control treatment and remained low in plots where *P. chlamydosporia* was used several months earlier. This showed that the fungus population maintained at sufficient levels during rotation that could adequately control the nematode on tomato (Atkins et al. 2003). Maize, bean and cabbage are good choice for rotation programme where *M. incognita* is the main pest. These crops are poor hosts for the nematode while *P. chlamydosporia* could grow well in their rhizosphere (Puertas and Hidalgo-Díaz 2007). Contrasting to *P. chlamydosporia*, the efficiency of *Purpureocillium lilacinum* is less affected by the host plant. So, *P. lilacinum* persistence in soil is not related to particular rotation schemes. Among 12 plant species, the decline of fungus population was seen only when *Phaseolus vulgaris* was planted (Rumbos and Kiewnick 2006). The nematode antagonistic bacteria or fungi with endophytic life style depend on their exclusive hosts on which they can endophytically establish and develop. Therefore, certain crop rotation system could support their survival (Hildalgo-Diaz and Kerry 2008). The host plants can evidently affect the abundance of BCAs through root exudates and other plant characteristics (Timper 2014). Higher populations of BCAs could support host plants against soilborne pathogens (Weller et al. 2012). Therefore, continuous cultivation of a plant species may sometimes resulted in a selective enhancement in population densities of particular BCAs and the formation of suppressive soils (Stirling 1988), but it is not surely evidenced that crop rotation always annihilates suppressiveness in a soil (Timper 2014).

It is too hard to correlate the general suppression of phytonematodes in soil to any specific BCA. However, choosing the appropriate plants for rotation has undeniable effect on preserving nematode suppression by BCAs. Suppressiveness to *Heterodera schachtii* was endangered when wheat was cultivated in a soil but not when an *H. schachtii*-resistant sugar beet was planted. It was assumed that the BCAs could colonise and reproduce in the rhizospheric soil of resistant sugar beet rather than wheat (Westphal and Becker 2001).

The performance of *P. chlamydosporia* in a double-cropping system of lettuce and tomato was examined in soil infested with *M. javanica* for two successive growing seasons. The fungus was detected from *M. javanica* eggs up to 9 months after application to soil and survived at low densities in the rhizospheric soil for the entire growing season (Verdejo-Lucas et al. 2003). The efficacy of *P. chlamydosporia* in reducing *M. javanica* population densities or enhancing yield was different when it was applied in two different cropping systems. Cropping system I included eggplant–okra–tomato–okra–eggplant–tomato–okra while cropping system II composed of eggplant–broad bean–tomato–broad bean–cabbage–tomato–cabbage. Both the nematode population density and yield in cropping system II decreased respectively by 12 and 65% than in cropping system I (Amer-Zareen et al. 2004). The proportion of involvement in nematode decline was different among various cropping systems. During a two-year crop rotation in the absence of potato plants, between 76 and 80% of the *Globodera rostochiensis* decline occurred in relation to spontaneous hatch while only 10% was due to bacterial BCAs (Devine et al. 1999). But when *Purpureocillium* sp. (=*Paecilimyces* sp.) was applied against *G. rostochiensis* in combination with two different legominous plant, the proportion of *Purpureocillium* involvement in *G. rostochiensis* decline was more than rotation. When the fungus was applied in combination with two different legominous plants, 89% of nematode population reduced but the decline percent for rotation alone was 31% (López-Lima et al. 2013). There was meaningful difference among various plants (oilseed rape, sugarbeet and wheat in the potato rotation) in supporting the survival or abundance of *P. lilacinum*, *Monographella cucumerina* and *P. chlamydosporia* var. *chlamydosporia* and *P. chlamydosporia* var. *catenulata* (Manzanilla-López et al. 2011).

10.3.2 Trap Crops

Usually, sedentary endoparasitic nematodes are the target of trap cropping. The concept of this tactic involves planting a susceptible host with fast and extensive root growth in a nematode-infested soil during a short period of time (Sikora et al. 2005). The motile juveniles of sedentary nematodes invade the roots and soon start their development to sedentary growth stage. Thereafter, the non-motile juveniles are annihilated via destruction of trap crop with herbicide, ploughed in, or physical removal before the juveniles could complete their life cycle and reproduce (Viaene et al. 2013). A crop trap is considered as an idle one if it stimulates the nematode eggs to hatch, be very attractive to juveniles, being heavily infected but not supporting nematode reproduction (Trivedi and Barker 1986). The trap crop should be cultivated quite densely in infested soil so that root system and/or its diffusates can contact as many juveniles as possible. The success of this tactic depends on appropriate cultivation methods, accurate timing and complete destruction of the crop when it is a susceptible one (Viaene et al. 2013). Though proper implementation of trap cropping can be an extremely helpful measure for phytonematode management, it has not been welcomed by the farmers due to the time and cost involved. I found only one literature in which trap cropping was used in combination with BCAs. Dandurand and Knudsen (2016) applied *Solanum sisymbriifolium* as a trap crop against *Globodera pallida* alone or together with *Trichoderma harzianum* or *Plectosphaerella cucumerina*. Soil of three different ‘cropping systems’ including potato (*Solanum tuberosum*), *S. sisymbriifolium*, or soil only (fallow) was amended with *P. cucumerina*, *T. harzianum* or left unamended. The trap crop *S. sisymbriifolium* significantly decreased nematode reproduction rate by 99% with or without BCAs assistance. Since the outcome of *S. sisymbriifolium* on reproduction rate of the nematode was large, addition of the BCAs made no further reduction.

10.3.3 Antagonistic Crops

These plants usually could produce detrimental compounds which have antagonistic effects on phytonematodes. The antagonistic substances may be secreted when the roots are growing or be released after biodegradation in the soil (Hildalgo-Diaz and Kerry 2008). Non-host status of antagonistic crops for some phytonematode species is another mechanism for control (Viaene et al. 2013). Phytonematodes do not usually invade the roots of antagonistic plant, but if penetration occurs, little progress is seen in their life cycle (Trivedi and Barker 1986). The most famous antagonistic plants are marigold, neem, sunn hemp, castorbean, partridge pea, asparagus, rape seed, velvet bean, some grass species (family Poaceae) and sesame (Grubišić et al. 2018). The application of this approach may practically be achieved by pre-plant cover crops, intercropping or green manures (Viaene et al. 2013). The antagonistic crops that could make money have greater potential to be accepted by the growers.

Acceptability and application of *Tagetes* spp. have been greatly increased when food industry began to use them as a food colourant (Hooks et al. 2010). Cultivation of antagonistic plants may offer a substitute for chemical nematicides if integrated with other measures such as BCAs. However, few researches have studied the combined effect of antagonistic plants and biological control on phytonematodes. Forty-nine species of endophytic bacteria were recovered from *Tagetis erecta* and *Tagetis patula*, of those *Microbacterium esteraromaticum* (recovered from *T. patula*) and *Kocuria varians* (recovered from *T. erecta*) significantly reduced the population density of *Pratylenchus penetrans* in the potato rhizosphere without decreasing the tuber fresh weight (Sturz and Kimpinski 2004). There are also reports on the adverse effect of marigolds on useful microorganisms. It was suggested that antifungal molecules were present in fresh marigold tissue (Baker 1981) and Owino (1992) demonstrated that extracts of *T. patula* could prevent *Fusarium solani* and *F. oxysporum* to prarasitise *M. javanica* and *M. incognita* eggs on water agar. Therefore, it is feasible to use antagonistic plants in combination with augmented releases of BCAs to enhance phytonematode suppression but this needs more investigation.

10.3.4 Cover Crops

Cover crops refer to plants that are not cultivated for their commercial value but for management of phytonematodes, suppression of weed growth and soil conservation throughout the off season (winter or dry period). Noteworthy, little or no nematode control might be obtained if off season overlaps with nematode low activity or over-seasoning period. The cover crops may finally be incorporated to soil as a green manure or utilised for livestock fodder. Cover crops decrease nematode population densities simply by their non-host status or by their antagonistic, suppressive or damaging properties (Sikora et al. 2005; Viaene et al. 2013). Incorporating the cover crops into the soil usually leads to considerable enhancement in BCAs activity and increase in the host plant growth. Cover crops have the potency to regulate soil microbial structure and ecosystem services (DuPont et al. 2009).

Applying *P. chlamydosporia* simultaneously with planting cover crop had interesting results. The population densities of *P. chlamydosporia* increased in soil where black oat or oil radish was cultivated, but decreased in fallow soil or in soil where tomato was planted. Despite increase in population, the fungus fails to efficiently control *M. javanica* on tomato plant that is being cultivated after either black oats or oil radish. Contradictorily, galling and egg production of the nematode decreased when tomato plants were cultivated after fallow or tomato (Dallemole-Giaretta et al. 2011). In another approach used in warm climate, organic mulch is prepared from intact remainder of the cover plants which provide several advantages than when residues are incorporated into soil. Cover mulch can enhance BCAs activity by supplying a carbon resource, and by preventing excessive changes in soil temperature and moisture. The population densities of nematophagous bacteria were increased when sunn hemp is used as organic mulch (Wang et al. 2008). Leaving sugarcane residues on soil resulted in greater suppressiveness to *M. javanica* and *Pratylenchus zeae* than adding the residue into soil (Stirling et al. 2011).

10.3.5 Fallow and Flooding

Fallow is a simple tactic in which no crop is planted in a varying period of time, but no measure is employed against weed growth. 'Clean fallow' or 'black fallow' refers to a period of time when no plants (including weeds) are permitted to grow on farm by using herbicide or frequent tillage (Viaene et al. 2013). The phytonematode controlling strategy in fallow is to reduce populations by starvation, while desiccation and exposure to heat emanated from sun may assist (Trivedi and Barker 1986). Fallow has many problems which make it an unpractical method in many countries. The adverse effect on soil conservation via erosion; no contribution to farm income, being less-effective or ineffective in dry weathers, and no or low controlling impact on cyst or some lesion nematodes (like *Pratylenchus brachyurus*) progressively diminish the palatability of this measure. The growth of BCAs may be limited under fallow periods since they need the root diffusates or their phytonematode host to grow. During fallow period, the BCAs usually remain in resting phase without vegetation (Kerry 2000). The abundance of *P. lilacinum*, *Monographella cucumerina* and *P. chlamydosporia* var. *chlamydosporia* (Pc280, potato cyst nematode biotype) and *P. chlamydosporia* var. *catenulata* (Pc392, root-knot nematode biotype) was significantly lower in fallow than their abundance in the presence of different plants (oilseed rape, sugarbeet and wheat) in the potato rotation (Manzanilla-López et al. 2011). The population of *G. rostochiensis* declined by 84% when *Purpureocillium* sp. applied to soil left in fallow while the decline percent in soil left in fallow (without BCA) was as low as 7% (López-Lima et al. 2013). When it is possible, flooding for an extended time may kill phytonematodes because of high moisture level, soil anaerobic conditions and the production of toxic substances (Trivedi and Barker 1986). This measure can be used in non-sloping fields where water is not a limiting factor. In the paddy rice fields where the soil was flooded for 3 months or more, the root-knot nematodes were undetectable on the succeeding tomato crop (Sikora et al. 2005).

When both sterilised and non-sterilised soils were water-saturated, the populations of nematodes significantly reduced in non-sterile soil compared with those reduced in sterile one. This illustrated the potential role of microbial activity in decrease of nematode numbers in flooded soil (Hollis and Rodriguez-Kabana 1966). Though flooding may control phytonematodes, it is not a feasible tactic in many parts of the world due to water deficiency. Another disadvantage is to exclude the inundated farm from cultivation so produce no revenue (Viaene et al. 2013).

10.3.6 Organic Amendments

The growers have been practising this tactic for centuries without noticing its influences on phytonematodes (Renčo 2013). Studying the impact of organic matter amendment on biocontrol and soil microbial populations is a progressively growing research area (Ciancio et al. 2016; Ansari et al. 2019; Ansari and Mahmood 2019a, b). Organic amendments usually cause an enhancement in diversity and populations of soil inhabitant microorganisms and lead to more phytonematode suppression. However, neutral or negative impacts of organic amendment on BCAs have also been reported. It seems that the change in biocontrol activity of antagonists is depended on the kind and amount of organic matter as well as on the types of BCAs (Timper 2014). For example, trapping activity and populations were mutually related for *Dactylellina haptotyla* but not for *Arthrobotrys oligospora* when soil was amended with organic matter (Jaffee et al. 1998; Jaffee 2004). It has been frequently reported that addition of organic matter to soil alone could suppress phytonematode populations (Fatemy and Moosavi 2019) but increase saprophytic ones (Viaene et al. 2013). The added organic matters usually originate from animal dung, chitinous materials, industrial wastes, composts, processing residues and green manure (Sikora et al. 2005; Timper 2014). They may be applied as composted or fresh material or as the exudates from the roots of growing plants (Hildalgo-Diaz and Kerry 2008). Management typically ascribes to any one or more mechanisms as liberation of nematicidal molecules (like glucosinolates) from organic matters; release of allelochemicals (like antibiotics and hydrolytic enzymes) from microorganisms whose populations were enhanced by amendments; augmentation of the BCAs potency; and better plant growth and improved nematode-tolerance (by improving soil structure and water-holding capacity, and by increasing the activity of plant growth-promoting organisms) (Widmer et al. 2002; Thoden et al. 2011). It is rational to expect that organic amendment increases the abundance and performance of facultative parasites of nematode but not of obligate ones. However, the conducted experiments have not supported the hypothesis. Many indirect effects may involve in this rebuttal. For example, incorporations of organic matter into soil may increase the population level of bacteriophagous nematodes which are accessible prey for obligate parasites such as *P. penetrans* (Gomes et al. 2002) or *Drechmeria coniospora* (Van den Boogert et al. 1994). Application of organic matters can enhance the activity of soil resident antagonists. It seems that their application in combination with promising BCAs has potency to manage nematode on a large-field basis. Amending soil with castor oil cakes increased the parasitisation of *Tylenchulus semipenetrans* females by *T. harzianum* (Reddy et al. 1996). The infection rate of *M. javanica* eggs by *Trichoderma longibrachiatum* (Sajadi et al. 2016) and *T. harzianum* (Amir-Ahmadi et al. 2017) was improved by increase in the soil organic matter content. The efficiency of *P. lilacinum* and *Cladosporium oxysporum* increased when applied combinedly with oil cakes. The best eggplant growth and *M. javanica* management was seen in treatments which received *P. lilacinum* and groundnut cake simultaneously (Ashraf and Khan 2010).

Populations of *P. chlamydosporia* and the parasitised percent of *M. incognita* eggs increased when neem (*Azadirachta indica*) leaves were incorporated into soil (Reddy et al. 1999). Similarly, adding dry neem leaves to field soil increased the antagonistic activity of *P. chlamydosporia*, *P. lilacinum* and *T. harzianum* against female and egg masses compared with the antagonistic activity when fungi were applied alone (Khan et al. 2012). The effect of *T. patula* residual on increasing activity of BCAs was reported. However, the enhancement was not adequate to control *Rotylenchulus reniformis* on pineapple (Ko and Schmitt 1996). To examine the rise or fall of biocontrol level of a BCA after specific amendment, we need to design an experiment with a factorial structure. Two winter cover crops (rye and crimson clover) were cultivated for 1 month, then killed and their above ground part was left on soil or removed. Afterward *P. lilacinum* was applied to soil. Suppression of *M. incognita* on cotton in the presence of the fungus and residues was 60% for rye and 49% for crimson clover compared with 35% for soil left in follow. Greater decrease in the nematode reproduction rate was seen when the above-ground residues were left on the soil surface than when it was eliminated. In the treatments where the above-ground residues were removed, nematode suppression was lower than in the fallow soil (Timper and Parajuli 2012). To separate the effect of nematicidal metabolites from the effect of enhancing BCAs activity, it is better to assess the biocontrol level of phytonematodes several months after adding organic matters into soil. Five months after incorporating sugarcane residue into soil, biological suppression of *P. zeae* was observed (Stirling et al. 2005). Likewise, after one and 2 years of combined adding of poultry manure and sawdust, biological suppression of *M. javanica* was demonstrated (Stirling et al. 2012).

Amending soil with specific substances such as chitin will enhance the populations of chitinase-producing microorganisms which in turn may increase the degradation of chitinous layer of nematode eggshell, especially those aggregated in gelatinous matrixes or cysts. Amending soil with chitin stimulated the activity of antagonistic fungi in soil and resulted in reduction in population of *M. arenaria* (Godoy et al. 1983). As well, incorporating 1% (w/w) chitin into soil could efficiently suppress *M. incognita* on cotton and enhance the population level of chitinolytic bacteria (Hallmann et al. 1999). The same result was obtained when chitin was applied in *M. arenaria*-infested soil in combination with *P. lilacinum* (Culbreath et al. 1986; Rodriguez-Kabana et al. 1987). Though the abundance of antagonistic BCAs is increased after chitin amendments, the main impact of chitin on nematodes attributed to the liberation of ammonia, at least near to application time (Viaene et al. 2013). As said before, enhancing response may not be always seen by BCAs to organic amendments. The impact of organic amendments on the performance of BCAs may be nematode species-, organic matter type- and organic matter amount-dependent. Compared to bare soil, the vermiform stages of *R. reniformis* were more parasitised by amending soil with sunn hemp (*Crotalaria juncea*) and pineapple (*Ananas comosus*) but not with rapeseed (*Brassica napus*) and marigold (*Tagetes erecta*). More eggs of *R. reniformis* were parasitised only when soil was amended with sunn hemp (Wang et al. 2001). Amending soil with composted leaves of *T. minuta* induced parasitic activity of *P. lilacinum* on

M. javanica eggs but the percent of infected eggs was too low to establish a considerable nematode suppression (Oduor-Owino 2003). Soil amendments may sometimes adversely affect the population densities of feeble saprophytic BCAs which surrender the competition to other strong saprotroph microorganisms whose population is much increased. However, more secreted fungicidal molecules by enhanced populations of soil microbiota should not be neglected (Timper 2014). Controlling ability of *H. rhossiliensis* did not increase subsequent to applying huge amounts of chicken manure, wheat straw or composted cow manure (Jaffee et al. 1994). Organic soil amendments may increase the nutrient level of soil and make the BCAs less aggressive by decreasing their tendency to switch from saprophytic to parasitic phase. In the presence of glucose and quickly absorbable nitrogen sources, the production of serine proteases by *P. chlamydosporia* and *P. lilacinum* was suppressed (Viaene et al. 2013). Supplemented soil with either grape (*Vitis vinifera*) or alfalfa (*Medicago sativa*) leaves resulted in increasing populations of *Arthrobotrys oligospora* and *Dactylellina candidum*, but trap formation was only enhanced with *D. candidum* (Jaffee 2004). It seems that their trophic status involves in this difference. Network-producing nematophagous fungi (such as *A. oligospora*) usually are good saprophytes and by increase in soil nutrient level, their reliance on phytonematodes will decrease. But, reliance of weak saprophytes (such as *D. candidum*) on phytonematodes is more (Moosavi and Zare 2012). Predatory behaviour of nematode-trapper fungi is considered as a tactic of surviving competition when microbial activity increased consequent to organic amendments and is not correlated to nematode population levels (Stirling 1988). Population densities of nematode-trapping fungi were increased 1 month after *T. erecta* residues were incorporated into soil but the effect was not durable and disappeared soon (Wang et al. 2002). Incorporating plant parts which may produce allelopathic molecules into soil can cause adverse effect on BCAs. Amending soil by organic matter with low C/N ratio may also cause negative effect on BCAs due to enhancing ammonia (NH_3) concentrations during decomposition (Rodriguez-Kabana et al. 1987; Oka 2010). Though soil amendment could reduce plant host infection by phytonematode, this task might be achieved only by incorporating a huge quantity of amending matters into soil. The impact of soil amendment on BCAs populations is complicated and cannot be interpreted easily. Critical investigations are needed to reveal the suppression mechanisms of organic amendments; to determine the amounts of organic matter which can practically and efficiently be applied to soil; and to better comprehend the interaction of organic matter with soil biota, host plant and phytonematodes.

10.3.7 Tillage Effect

Routine tillage practice may adversely affect the soil community by mechanical damage (ripping and inverting the soil), being buried deep or by brining to soil surface. Few studies have been found to investigate the effect of combined application of tillage and BCAs. But investigating the effect of tillage on BCAs has been resulted in inconsistent results. The parasitism of *H. glycines* eggs was more in the soil that was disc ploughed compared with those that were moldboard ploughed or no-tilled (Bernard et al. 1996). No significant differences were detected in parasitism of *H. glycines* juveniles by *H. rhossiliensis* and *H. minnesotensis* in soybean fields with conventional tillage and no-tillage (Chen and Liu 2007), but the parasitised percent of *H. glycines* juveniles by *H. rhossiliensis* was lower in simulated tillage (sieving the soil) than in minimal tillage (Bao et al. 2011). Rotary tilling soil caused 28% decrease in percentage of *P. penetrans*-infected juveniles of *M. incognita* than in no-tilled (Talavera et al. 2002b). However, no significant differences were observed in parasitising *H. glycines* on soybean by *Pasteuria nishizawae* when the plots were subjected to conventional and no tillage (Noel et al. 2010). When conservation tillage was adopted in wheat fields of Australia, the cyst populations of *Heterodera avenae* on roots and its damage to yield were decreased (Roget and Rovira 1987). The population densities of nematodes were a little more in no-tilled regime than in conservation tillage (Fortnum and Karlen 1985). More study is required to understand the effect of tillage on BCAs efficacy.

10.4 Decreasing Phytonematode Populations

Any measure which involves in reducing phytonematode populations may help BCAs to complete their task better. Some of these tactics may be used in combination with BCAs and some may be employed prior to applying BCAs. Soil disinfestations by steaming, soil solarisation, biofumigation or broad-spectrum biocides may reduce the populations and activity of soil microbial competitors and applying BCAs after these treatments resulted in easier establishment and survival of the introduced BCAs. On the other hand, biocontrol of lesser populations of nematode is too easier (Viaene et al. 2013). No BCA can suppress phytonematode populations as rapidly as nematicides (Cumagun and Moosavi 2015), so several post-planting nematicides can be used in combination with BCAs where the chemical nematicide impedes initial nematode damage and the BCAs provide long-term protection (Moosavi and Zare 2012; Sokhandani et al. 2016).

10.4.1 Soil Solarisation

Soil solarisation can successfully disinfest soil of weeds, soilborne pests and pathogens where the climate is hot, water is available and the soil depth is shallow. Exposing moistened soil below plastic mulches to solar radiation for at least 4–6 weeks will kill phytonematodes (Hildalgo-Diaz and Kerry 2008); however, the cost of polyethylene sheeting and the needed length of time may be deterrent. The greenhouse effect under polyethylene sheeting causes an increase in soil temperature up to 35–50 °C in 30 cm upper depth. The efficacy of this tactic to disinfest soil of nematodes depends on soil type, solar intensity, soil moisture content and preceding tillage (Viaene et al. 2013). Successive cultivation of susceptible plants (tomato-melon) to root-knot nematodes became possible by solarising soil for 30 days in summer (Sano 2002). Biological control may integratedly be implemented with soil solarisation if the BCAs possess heat resistance propagules. Spores of *Pasteuria* are moderately resistant to heat and drought (Sikora 1992) so would survive solarisation. The use of soil solarisation and *P. penetrans* had synergistic effect on controlling *M. javanica* on grapevine (Walker and Wachtel 1988) and cucumber (Tzortzakakis and Goewn 1994). However, application of *P. lilacinum* after a 15-day soil solarisation could not adequately reduce root-knot nematode populations (Anastasiadis et al. 2008).

10.4.2 Biofumigation

Producing volatile molecules during hydrolysing of glucosinolate-containing plant residues or organic matters by soil microbiota is called biofumigation. The majority of glucosinolate-containing plants are clustered within the Brassicaceae, Capparaceae and Caricaceae families. These biocidal/nematicidal molecules have high toxicity against soilborne pests and pathogens (Kruger et al. 2013). However, biofumigation usage is now restricted mainly because of the bulky amount of organic materials to be incorporated to soil (Hildalgo-Diaz and Kerry 2008). For better control of *M. incognita*, it has been recommended to replace cover crop (such as small grains) by crops with biofumigation potential (Westphal 2011).

Despite appearing as a fair promising tactic for the control of soilborne diseases, it is improbable that biofumigation could sufficiently suppress the phytonematode by itself. But its application in combination with soil solarisation or resistant hosts could improve its efficacy (Ploeg 2008). The activity of nematode antagonist may increase after biofumigation (Sikora et al. 2005), but little literature has been found to study the combined effect of biofumigation and BCAs.

10.4.3 Chemical Nematicides

The main method to reduce the damage of phytonematodes is now based on chemical nematicides. These compounds could reduce the numbers of nematodes, stop or decrease nematode reproduction, or paralyse them (nematostatic) for a while (Haydock et al. 2013). Present chemical nematicides in the market can be divided to fumigants and non-fumigants and their classification usually is based on their mode of action (Ebone et al. 2019). Chemical nematicides severely endanger human health and environment (Moosavi and Zare 2016). It has been reported that phytonematode populations at the end of the season in the nematicide-treated plots are usually more than in the untreated plots (Sipes and Schmitt 1998) due to more spread root system which could support more nematode when the effect of nematicicde diminishes. However, growth in phytonematode populations may be for lower activity of BCAs in nematicide treated soil (Timper 2014). Notwithstanding theses disadvantages, nematicides cannot be eliminated, at least for now, from nematode management programmes. However, it seems that the best approach is to integrate the nematicides with other management methods instead of applying the nematicides as the only control measure (Hillocks 2012). The integrated application of synthetic chemicals and BCAs has drawn much interest for its probable additive or synergistic effects (Spadaro and Gullino 2005). Theoretically, non-fumigant nematicides could increase BCAs' potency if they are applied combinedly. It is highly recommended that nematicides should be applied when more than one phytonematode species exist in soil. The nematicide will suppress the species that might not be susceptible to applied BCA or even the populations that do not encounter with applied BCA. A few investigations have tested the effect of combined application of nematicides and BCAs on phytonematode management, but it has been reported that fungal BCAs from different groups (trapping fungi, *P. lilacinum* and *P. chlamydosporia*) were little affected when they were exposed to standard doses of several different pesticides applied to soil (Kerry 1987). Thus, it is feasible to apply these BCAs with nematicides to extend and enhance nematode management (Sokhandani et al. 2016). The biological suppression of nematodes decreased immediately after nematicide application, but in the next spring biological suppression increased wherever nematicide applied (Timper et al. 2012). Applying *P. lilacinum* and *Trichoderma viride* in combination with mustard cake and furadan (nematicide) resulted in least *M. incognita* reproduction rate as compared to untreated treatments (Goswami et al. 2006). Applying granule formulations containing fosthiazate and *Monacrosporium ellipsosporum* resulted in better *M. incognita* control and increased the establishment of fungus BCA in soil (Taba et al. 2006). Contradictorily, the populations of nematode-trapping fungi in the fields were greatly enhanced subsequent to sunn hemp incorporation wherever no fumigant nematicide (1, 3-D) had been applied (Wang et al. 2003). This may be due to broader spectrum of organisms (including BCAs) that can be affected by fumigant nematicides. The growth and activity of *P. penetrans* are not affected by many chemical pesticides, except for chloropicrin that is directly toxic to this bacterium (Chen and Dickson 1998). Gall formation was reduced by 50 and 63% when *P. penetrans* was applied in combination with carbofuran on tomato to control *M. javanica* (Brown and Nordmeyer 1985) and *M. incognita* (Somasekhar and Gill 1991), respectively. As well, integrated application of carbofuran and *P. penetrans* reduced J2 penetration of *Heterodera cajani* into pigeonpea roots. A total number of females, eggs per cysts and final population were reduced compared with control treatment (Gogoi and Gill 2001). Aldicarb and ethoprop could not change the percentage of infected *M. arenaria* juveniles by *P. penetrans* (Timper 1999; Timper et al. 2001), but the controlling effect of *P. penetrans* against *M. javanica* on tomato and cucumber crops was additively increased when it was applied in combination with oxamyl (Tzortzakakis and Goewn 1994). *Meloidogyne graminicola* population was decreased by 79% on rice when *Pseudomonas fluorescens* was combinedly used with carbofuran. The decreased amount of *M. graminicola* population on rice was 69% in simultaneous use of *T. viridae* and carbofuran (Narasimhamurthy et al. 2017). The in vitro response of three different fungal BCAs was examined to frequently used pesticides in potato fields. No prevention was seen in *P. lilacinum* and *Plectosphaerella cucumerina* growth in response to pencycuron or oxamyl, but their growth was slowed down in response to fenpiclonil and tolclofos-methyl. All mentioned chemicals partially inhibited *P. chlamydosporia* growth (Jacobs et al. 2003). Combined application of *T. harzianum* and carbofuran resulted in more prominent suppression of *M. incognita* on french bean (Gogoi and Mahanta 2013), brinjal (Devi et al. 2016), *Mentha arvensis* (Haseeb et al. 2007) and pea (Brahma and Borah 2016).

Integrated application of *P. chlamydosporia* with carbofuran (Gopinatha et al. 2002), with neem cake and carbofuran (Dhawan and Singh 2009) and with dazomat (Nagesh and Jankiram 2004) increased suppression of *M. incognita*. But its integration with fosthiazate did not provide any additional decrease in potato cyst nematodes reproduction rate (Tobin et al. 2008). Optimum levels of *Trichoderma longibrachiatum* concentration and cadusafos dose were determined in their combined application for the control of *M. javania* on zucchini plants. The data were analysed using a custom response surface regression model and the optimum levels of the cadusafos and *Trichoderma* concentration that caused the best plant growth and lowest nematode reproduction were determined as 1.7 mg a.i./kg soil and 10^8 conidia/ml suspension, respectively (Sokhandani et al. 2016).

10.5 Application of Multiple BCA

Little studies directly investigate or compare the effect of combined application of several BCAs against phytonematodes in contrast with application one BCA. It is globally accepted that inconsistency in performance of BCAs is a main barrier to the broad usage of BCAs (Moosavi and Zare 2015); however, integration of two or more robust BCAs may help in overcoming this disadvantage. Arguably, integrated use of BCAs may increase the achieved control especially if the combined BCAs are compatible or have different mode of actions (Moosavi and Zare 2016). Fluctuation in populations of BCAs is hardly happened when several BCAs exist in or introduce into soil. Combined application of BCAs provides several benefits such as more and better colonisation of the rhizosphere; parasitising more than one stage of the life cycle of a nematode as well as parasitising more than one target species; being active in more extensive time span during growing season; possessing various mode of action; and performing more consistent under more broaden range of environmental and soil condition (Crump 1998; Siddiqui and Mahmood 1996; Meyer and Roberts 2002; Ansari and Mahmood 2017b; Ansari et al. 2017a, b). Applying BCAs with diverse mode of action such as *Fusarium oxysporum* (endophyte), *P. lilacinum* (egg parasite) and *Bacillus firmus* (antagonistic bacteria) increased biomanagement of

R. similis on banana (Mendoza and Sikora 2009). The population density of added BCAs is a matter of importance. For example in the conducted experiments, the combined effect of two BCAs with a specific application rate (x) has been usually compared with the effect of each BCA used at the same rate alone (x) and not with one BCA used at twice the rate ($2x$). On the other hand, analysing the data which have presented in conducted investigations usually demonstrated additive effect more than synergistic one (Hildalgo-Diaz and Kerry 2008). Further experiments are required to evidently elucidate the phenomena after combined application of two or more BCAs especially on the ecological basis. BCAs have to survive and act in biologically complex environments where they are subjected to numerous inter- and intera-specific interactions (Knudsen and Dandurand 2014). Consequent to those interactions, the combined application of BCAs may result in competence, independence, additive or synergistic effect. Though analysing the published paper on combined application of BCAs indicated that antagonistic interactions are more probable (Xu et al. 2011), it has been suggested that introducing BCAs to soil as consortium might imitate natural soil condition and could stimulate host plant defence responses and enhance its growth (Sarma et al. 2015). More understanding on host plants' responses to phytonematodes in the presence of several BCAs and the probable interactions is required for sound judgement.

The suppression effect of *A. oligospora* in combination with each of 11 different strains of bacteria on J2 populations of *Meloidogyne mayaguensis* was assessed in tomato rhizosphere. Greater control achieved when *A. oligospora* was integratedly applied with three different unidentified bacteria (Duponnois et al. 1998). When *Embellisia chlamydospora, P. chlamydosporia* and a sterile fungus were applied alone and in different binary combinations, the lowest number of females and cysts of *H. schachtii* was observed when either *Embellisia* or *Pochonia* was applied in combination with the sterile fungus. No nematode suppression was observed by individual application of the mentioned fungi or by the *Embellisia–Pochonia* combination (Hojat Jalali et al. 1998). Applying *T. harzianum* simultaneously with *Monacrosporium lysipagum* (nematode trapper fungus) resulted in better control of *M. javanica* and *H. avenae* (Khan et al. 2006). Together application of *Acremonium strictum* and *T. harzianum* enhanced biocontrol activity against *M. incognita* on tomato significantly (Goswami et al. 2008). When 90 combinations of different *Trichoderma* (18 strains from five species) and nematode-trapping fungi (six strains from four species) were examined against *Caenorhabditis elegans*, integration of *T. harzianum* and *Monacrosporium cionopagum* had the best compatibility (Szabó et al. 2012). Five different fungi (*Pochonia bulbillosa* (Pb), *Pochonia chlamydosporia* var. *catenulata* (Pccat), *Pochonia chlamydosporia* var. *chlamydosporia* (Pcc), *Lecanicillium aphanocladii* (La) and *T. harzianum* (Th)) whose in vitro pathogenicity on the *M. javanica* eggs was similar used alone and in binary or trinary combination for the control of *M. javanica* on eggplant. Combined application of these BCAs had no significant increasing effect on plant growth compared with growth in the pots treated by one BCA. But integrated application of BCAs increased the egg infection rate significantly. None of the treatment could manage *M. javanica* as the same level as cadusafos nematicide (96%); however, the integrated application of Pccat, La with Pcc (85%) or Th (83%) controlled *M. javanica* acceptably (Moosavi et al. 2015). The bacterium *P. penetrans* was compatibly combined with *P. chlamydosporia* (de Leij et al. 1992), *P. lilacinum* (Gautam et al. 1995) and *Glomus* sp. (Talavera et al. 2002a), and their integrated usage could control *M. incognita* on tomato plants better. Sometimes BCAs may be assisted by soil microbiota to fulfil their tasks. The endospores adherence of *P. penetrans* to *M. graminicola* increased with the help of rhizospheral resident bacteria (Duponnois et al. 1997). The population density of *M. javanica* on a cropping system consisted of eggplant–okra–tomato–okra–eggplant–tomato–okra was further reduced by 24% when *P. chlamydosporia* was applied in combination with *P. penetrans* rather than their solo application (Amer-Zareen et al. 2004). There are several reports that combination usage of BCAs makes no advantages over their individual application. Combined application of *H. rhossiliensis* and *P. chlamydosporia* against *Meloidogyne hapla* on lettuce did not enhance controlling activity compared with their individual usage (Viaene and Abawi 2000). Combined application of *P. penetrans*, *P. lilacinum*, *B subtilis* and *Talaromyces flavus* caused the same control level of root-knot nematode as their individual application did (Zaki and Maqbool 1991). Mixed usage of *M. lysipagum* with *T. harzianum* made no significant additive control of *R. similis* on banana compared with applying *M. lysipagum* alone (Khan et al. 2006). Combining BCAs sometimes results in lower management potential. The combination of *Bacillus thuringiensis*, *Paecilomyces marquandii* and *Streptomyces costaricanus* was not as effective as individual treatments for decreasing *R. similis* and *Helicotylenchus multicinctus* populations on banana (Esnard et al. 1998). Similarly, integration of *S. costaricanus* and *B. thuringiensis* resulted in lower biocontrol level of *M. hapla* on lettuce compared with biocontrol level achieved with their individual usage (Chen et al. 2000). While solitary application of *Trichoderma virens* and *Burkholderia cepacia* could successfully control *M. incognita* on bell pepper, their combined application was not as successful (Meyer et al. 2001). Integrated application of *Calothrix parietina* with either *P. lilacinum* or *Pichia guilliermondii* reduced their potency against *M. incognita* maybe due to antagonistic activity of *C. parietina* (Hashem and Abo-Elyousr 2011).

10.6 Conclusions and Future Prospects

Though phytonematode control is now based on chemical nematicides, it seems that in the near future it should inevitably be changed to an integrated multi-based management system. It must be also taken into consideration that inundative release of BCAs could not manage the phytonematodes consistently and must be applied in combination with other approaches. But careful consideration is required in choosing the effective measures that should be integrated. The selected ones must have several characteristics such as being compatible, provide synergistic effect, being economically affordable and impose the least environmental hazard. Maybe multidisciplinary studies could be of help to develop effective combinable measures.

From growers' viewpoint, chemical nematicides have several priorities to biocontrol in characteristics such as cost, transportation, delivery technology, performance and simplicity of storage and application. Therefore, consolidated use of BCAs with other tactics that provide synergistic or additive effect could be of help in accepting biocontrol as a potent substitute. If biological control wants to become an important part of phytonematode management, more improvement is also needed in their producing procedures, formulation, efficacy, consistency and application methods. The improvement must primarily occur in reducing production cost and enhancing potency. Moreover, production of BCA derivatives such as bionematicides is another measure of choice which has both advantages of effectiveness and being eco-friendly. The preparation of natural nematicides whose bioactive ingredients derive from BCAs should be oriented to products with low amount but strongly effective natural molecules which could effectively control the phytonematodes. Another probable approach is host plants transformation with effective genes against phytonematodes supplied by different sources include BCAs. As many people prefer not to consume genetically modified crops, the exploitation of this approach is hesitated. However, transforming BCAs to enhance their detrimental effect against phytonematodes is another option. The efficiency, consistency, virulence, ecological adaptability and easier mass production and formulation of BCAs can be improved by genetic engineering. In fact, numerous nematode-antagonists exist in arable lands and their persistence or activity mainly depends on our chosen cultural practices. However, the possible lack of suitable native BCAs can be compensated by introducing potent antagonists. Integration of biocontrol with compatible measures surely increases nematode suppression to an acceptable level. The farmers should be informed that the increase in populations of beneficial BCAs and adequate nematode suppression may need more than one growing season. It seems that the proper combination of plant-BCA (indigenous or introduced) provides a liable foundation that other locally available measures can be constructed upon it.

References

Agrios GN (ed) (2005) Plant pathology, 5th edn. Academic Press, New York

Alesadi GA, Moosavi MR, Basirnia T (2017) Effect of nano-K, potassium sulphate and salicylic acid on tomato growth and control of root-knot nematode (*Meloidogyne javanica*). J Plant Prot 40(3):71–82; in Persian with English abstract

Amer-Zareen ZMJ, Abid M, Gowen SR, Kerry BR (2004) Management of root knot nematode (*Meloidogyne javanica*) by biocontrol agents in two crop rotations. Int J Biol Biotechnol 1 (1):67–73

Amir-Ahmadi N, Moosavi MR, Moaf-Poorian GR (2017) Investigating the effect of soil texture and its organic content on the efficacy of *Trichoderma harzianum* in controlling *Meloidogyne javanica* and stimulating the growth of kidney bean. Biocontrol Sci Tech 27(1):115–127

Anastasiadis IA, Giannakou IO, Prophetou-Athanasiadou DA, Gowen SR (2008) The combined effect of the application of a biocontrol agent *Paecilomyces lilacinus*, with various practices for the control of root-knot nematodes. Crop Prot 27:352–361
Ansari RA, Khan TA (2012a) Parasitic association of root-knot nematode, *Meloidogyne incognita* on guava. e-J Sci Technol 5:65–67
Ansari RA, Khan TA (2012b) Diversity and community structure of phytonematodes associated with guava in and around Aligarh, Uttar Pradesh, India. Trends Biosci 5(3):202–204
Ansari RA, Mahmood I (2017a) Optimization of organic and bio-organic fertilizers on soil properties and growth of pigeon pea. Sci Hortic 226:1–9
Ansari RA, Mahmood I (2017b) Determination of disease incidence caused by *Meloidogyne* spp. and or *Fusarium udum* on pigeonpea in Aligarh district: a survey. Trends Biosci 10 (24):5239–5243
Ansari RA, Mahmood I (2019a) Plant health under biotic stress: volume 2: microbial interactions. Springer, Singapore. https://doi.org/10.1007/978-981-13-6040-4
Ansari RA, Mahmood I (2019b) Plant health under biotic stress: volume 1: organic strategies. Springer, Singapore. https://doi.org/10.1007/978-981-13-6043-5
Ansari RA, Rizvi R, Sumbul A, Mahmood I (2017a) PGPR: current vogue in sustainable crop production. In: Kumar V, Kumar M, Sharma S, Prasad R (eds) Probiotics and plant health. Springer, Singapore, pp 455–472
Ansari RA, Mahmood I, Rizvi R, Sumbul A (2017b) Siderophores: augmentation of soil health and crop productivity. In: Kumar V, Kumar M, Sharma S, Prasad R (eds) Probiotics in agroecosystem. Springer, Singapore, pp 291–312
Ansari RA, Sumbul A, Rizvi R, Mahmood I (2019) Organic soil amendments: potential tool for soil and plant health management. In: Ansari RA, Mahmood I (eds) Plant health under biotic stress. Springer, Singapore, pp 1–35
Ashraf MS, Khan TA (2010) Integrated approach for the management of *Meloidogyne javanica* on eggplant using oil cakes and biocontrol agents. Arch Phytopathol Plant Protect 43(6):609–614
Atkins SD, Hidalgo-Diaz L, Kalisz H, Mauchline TH, Hirsch PR, Kerry BR (2003) Development of a new management strategy for the control of root-knot nematodes (*Meloidogyne* spp.) in organic vegetable production. Pest Manag Sci 59:183–189
Baker KF (1981) Biological control. In: Mace E, Bell AA, Beckman CH (eds) Fungal wilt diseases of plants. Academic Press, London, pp 523–561
Bakker E, Dees R, Bakker J, Goverse A (2006) Mechanisms involved in plant resistance to nematodes. In: Tuzun S, Bent E (eds) Multigenic and induced systemic resistance in plants. Springer Science + Business Media, New York, pp 314–334
Bao Y, Neher DA, Chen SY (2011) Effect of soil disturbance and biocides on nematode communities and extracellular enzyme activity in soybean cyst nematode suppressive soil. Nematology 13:687–699
Barker KR (1991) Rotation and cropping systems for nematode control: the North Carolina experience-introduction. J Nematol 23(3):342–343
Bernard EC, Self LH, Tyler DD (1996) Fungal parasitism of soybean cyst nematode, Heterodera glycines (Nemata: Heteroderidae), in differing cropping-tillage regimes. Appl Soil Ecol 5:57–70
Brahma U, Borah A (2016) Management of *Meloidogyne incognita* on pea with bioagents and organic amendment. Indian J Nematol 46:58–61
Brown SM, Nordmeyer D (1985) Synergistic reduction in root galling by *Meloidogyne javanica* with *Pasteuria penetrans* and nematicide. Rev Nematol 8:285–286
Bruinsma J (ed) (2003) World agriculture: towards 2015/2030: an FAO perspective. Earthscan Publications, London
Castagnone-Sereno P (2002) Genetic variability in parthenogenesis root-knot nematodes, *Meloidogynes* pp., and their ability to overcome plant resistance genes. Nematology 4:605–608
Chen ZX, Dickson DM (1998) Review of *Pasteuria penetrans*: biology, ecology, and biological control potential. J Nematol 30:313–340

Chen S, Liu S (2007) Effects of tillage and crop sequence on parasitism of *Heterodera glycines* juveniles by *Hirsutella* spp. and on juvenile population density. Nematropica 37:93–106

Chen SY, Reese CD (1999) Parasitism of the nematode *Heterodera glycines* by the fungus *Hirsutella rhossiliensis* as influenced by crop sequence. J Nematol 31:437–444

Chen J, Abawi GS, Zuckerman BM (2000) Efficacy of Bacillus thuringiensis, *Paecilomyces marquandii,* and *Streptomyces costaricanus* with and without organic amendments against *Meloidogyne hapla* infecting lettuce. J Nematol 32:70–77

Ciancio A, Pieterse CMJ, Mercado-Blanco J (2016) Editorial: harnessing useful rhizosphere microorganisms for pathogen and pest biocontrol. Front Microbiol 7:1620

Conrath U (2011) Molecular aspects of defense priming. Trends Plant Sci 16:524–531

Cook R, Starr JL (2006) Resistant cultivars. In: Perry RN, Moens M (eds) Plant nematology. CABI Publishing, Wallingford, UK, pp 370–389

Cottage A, Urwin P (2013) Genetic engineering for resistance. In: Perry RN, Moens M (eds) Plant nematology, 2nd edn. CABI Publishing, Wallingford, UK, pp 437–458

Crump DH (1989) Interaction of cyst nematodes with their natural antagonists. Asp Appl Biol 22:135–140

Crump DH (1991) Biological control of the beet cyst nematode. Br Sugar Beet Rev 59:54–55

Crump DH (1998) Biological control of potato and beet cyst nematodes. Asp Appl Biol 53:383–386

Culbreath AK, Rodriguez-Kabana R, Morgan-Jones G (1986) Chitin and *Paecilomyces lilacinus* for control of *Meloidogyne arenaria*. Nematropica 16:153–166

Cumagun CJR, Moosavi MR (2015) Significance of biocontrol agents of phytonematodes. In: Askary TH, Martinelli PRP (eds) Biocontrol agents of phytonematodes. CABI Publishing, Wallingford, UK, pp 50–78

Dallemole-Giaretta R, de Freitas LG, Lopes EA, Ferraz S, de Podesta GS, Agnes EL (2011) Cover crops and *Pochonia chlamydosporia* for the control of *Meloidogyne javanica*. Nematology 13:919–926

Dalmasso A, Castagnone-Sereno P, Abad P (1992) Seminar: tolerance and resistance of plants to nematodes-knowledge, needs and prospects. Nematologica 38:466–472

Dandurand L-M, Knudsen GR (2016) Effect of the trap crop *Solanum sisymbriifolium* and two biocontrol fungi on reproduction of the potato cyst nematode, *Globodera pallida*. Ann Appl Biol 169(2):180–189

Davies LJ, Elling AA (2015) Resistance genes against plant-parasitic nematodes: a durable control strategy? Nematology 17:249–263

Denancé N, Sánchez-Vallet A, Goffner D, Molina A (2013) Disease resistance or growth: the role of plant hormones in balancing immune responses and fitness costs. Front Plant Sci 4:155

Devi TS, Mahanta B, Borah A (2016) Comparative efficacy of *Glomus fasciculatum*, *Trichoderma harzianum*, carbofuran and carbendazim in management of *Meloidogyne incognita* and *Rhizoctonia solani* disease complex on brinjal. Indian J Nematol 46:161–164

Devine KJ, Dunne C, O'Gara F, Jones PW (1999) The influence of in-egg mortality and spontaneous hatching on the decline of *Globodera rostochiensis* during crop rotation in the absence of the host potato crop in the field. Nematology 1:637–645

Dhawan SC, Singh S (2009) Compatibility of *Pochonia chlamydosporia* with nematicide and neem cake against root-knot nematode, *Meloidogyne incognita* infesting okra. Indian J Nematol 39:85–89

Duponnois R, Netscher C, Mateille T (1997) Effect of the rhizosphere microflora on *Pasteuria penetrans* parasitizing *Meloidogyne graminicola*. Nematol Mediterr 25:99–103

Duponnois R, Ba AM, Mateille T (1998) Effects of some rhizosphere bacteria for the biocontrol of nematodes of the genus *Meloidogyne* with *Arthrobotrys oligospora*. Fundam Appl Nematol 21:157–163

DuPont ST, Ferris H, Van Horn M (2009) Effects of cover crop quality and quantity on nematode-based soil food webs and nutrient cycling. Appl Soil Ecol 41:157–167

Ebone LA, Kovaleski M, Deuner CC (2019) Nematicides: history, mode, and mechanism action. Plant Sci Today 6(2):91–97

Esnard J, Marban-Mendoza N, Zuckerman BM (1998) Effects of three microbial broth cultures and an organic amendment on growth and populations of free-living and plant-parasitic nematodes on banana. Eur J Plant Pathol 104:457–463

FAO (Agriculture Organization of the United Nations), IFAD (International Fund for Agricultural Development) & WFP (World Food Programme) (2013) The state of food insecurity in the world 2013, the multiple dimensions of food security. FAO, Rome, Italy

Fatemy S, Moosavi MR (2019) Nematotoxic potential of daikon, chinaberry and purslane herbal green manures against *Globodera rostochiensis in vitro* and microplot. J Crop Prot 8(1):69–80

Fortnum BA, Karlen DL (1985) Effects of tillage systems and irrigation on population densities of plant nematodes in field corn. J Nematol 17:25–28

Gautam A, Siddiqui ZA, Mahmood I (1995) Integrated management of *Meloidogyne incognita* on tomato. Nematol Mediterr 23:245–247

Godfray HCJ, Beddington JR, Crute IR, Haddad L, Lawrence D, Muir JF, Pretty J, Robinson S, Thomas SM, Toulmin C (2010) Food security: the challenge of feeding 9 billion people. Science 327:812–818

Godoy G, Rodriguez-Kabana R, Shelby RA, Morgan-Jones G (1983) Chitin amendments for control of *Meloidogyne arenaria* infested soil. II. Effects on microbial population. Nematropica 13:63–74

Gogoi BB, Gill JS (2001) Compatibility of *Pasteuria penetrans* with carbofuran and organic amendments, its effect on *Heterodera cajani*. Ann Plant Prot Sci 9(2):254–257

Gogoi D, Mahanta B (2013) Comparative efficacy of *Glomus fasciculatum*, *Trichoderma harzianum*, carbofuran and carbendazim in management of *Meloidogyne incognita* and *Rhizoctonia solani* disease complex on French bean. Ann Plant Prot Sci 21:172–175

Gomes CB, De Freitas LG, Ferraz S, Oliveira RDDL, Da Silva RV (2002) Influence of cattle manure content in the substrate on the multiplication of *Pasteuria penetrans* in tomato. Nematol Brasil 26:59–65

Gomiero T, Pimentel D, Paoletti MG (2011) Is there a need for a more sustainable agriculture? Crit Rev Plant Sci 30:6–23

Gopinatha KV, Gowda DN, Nagesh M (2002) Management of root-knot nematode *Meloidogyne incognita* on tomato using bioagent *Verticillium chlamydosporium*, neem cake, marigold and carbofuran. Indian J Nematol 32:179–181

Goswami BK, Pandey RK, Rathour KS, Bhattacharya C, Singh L (2006) Integrated application of some compatible biocontrol agents along with mustard oil seed cake and furadan on *Meloidogyne incognita* infecting tomato plants. J Zhejiang Univ Sci B 7(11):873–875

Goswami J, Pandey RK, Tewari JP, Goswami BK (2008) Management of root knot nematode on tomato through application of fungal antagonists, *Acremonium strictum* and *Trichoderma harzianum*. J Environ Sci Health B 43(3):237–240

Grubišić D, Uroić G, Ivošević A, Grdiša M (2018) Nematode control by the use of antagonistic plants. Agric Conspec Sci 83(4):269–275

Hallmann J, Rodriguez-Kabana R, Kloepper JW (1999) Chitin-mediated changes in bacterial communities of the soil, rhizosphere and within roots of cotton in relation to nematode control. Soil Biol Biochem 31:551–560

Haseeb A, Kumar V, Abuzar S, Sharma A (2007) Integrated management of *Meloidogyne incognita-Sclerotinia sclerotiorum* disease complex of *Mentha arvensis* cv. Gomti by using *Trichoderma* species, neem seed powder, carbofuran and topsin-M. In: 7th national symposium on plant protection options implementation and feasibility. 20–22 Dec, p. 102

Hashem M, Abo-Elyousr KA (2011) Management of the root-knot nematode *Meloidogyne incognita* on tomato with combinations of different biocontrol organisms. Crop Prot 30:285–292

Haydock PPJ, Woods SR, Grove IG, Hare MC (2013) Chemical control of nematodes. In: Perry RN, Moens M (eds) Plant nematology, 2nd edn. CABI Publishing, Wallingford, UK, pp 459–479

Hildalgo-Diaz L, Kerry BR (2008) Integration of biological control with other methods of nematode management. In: Ciancio A, Mukerji KG (eds) Integrated management and biocontrol of vegetableand grain crops nematodes. Springer, Dordrecht, the Netherlands, pp 29–49

Hillocks RJ (2012) Farming with fewer pesticides: EU pesticide review and resulting challenges for UK agriculture. Crop Prot 31:85–93

Hojat Jalali AA, Segers R, Coosemans J (1998) Biocontrol of *Heterodera schachtii* using combinations of the sterile fungus, StFCH1-l, *Embellisia chlamydospora* and *Verticillium chlamydosporium*. Nematologica 44:345–355

Hollis JP, Rodriguez-Kabana RA (1966) Rapid kill of nematodes in flooded soil. Phytopathology 56:1015–1019

Hooks CRR, Wang K-H, Ploeg A, McSorley R (2010) Using marigold (*Tagetes* spp.) as a cover crop to protect crops from plant-parasitic nematodes. Appl Soil Ecol 46:307–320

Jacobs H, Gray SN, Crump DH (2003) Interactions between nematophagous fungi and consequences for their potential as biological agents for the control of potato cyst nematodes. Mycol Res 107(1):47–56

Jaffee BA (2004) Do organic amendments enhance the nematode-trapping fungi *Dactylellina haptotyla* and *Arthrobotrys oligospora*? J Nematol 36:267–275

Jaffee B, Phillips R, Muldoon A, Mangel M (1992) Density dependent host-pathogen dynamics in soil microcosms. Ecology 73:495–506

Jaffee BA, Ferris H, Stapleton JJ, Norton MVK, Muldoon AE (1994) Parasitism of nematodes by the fungus *Hirsutella rhossiliensis* as affected by certain organic amendments. J Nematol 26:152–161

Jaffee BA, Ferris H, Scow KM (1998) Nematode-trapping fungi in organic and conventional cropping systems. Phytopathology 88:344–350

Kaloshian I, Desmond OJ, Atamian HS (2011) Disease resistance-genes and defense responses during incompatible interactions. In: Jones J, Gheysen G, Fenoll C (eds) Genomics and molecular genetics of plant-nematode interactions. Springer, Dordrecht, the Netherlands, pp 309–325

Kerry BR (1987) Biological control. In: Brown RH, Kerry BR (eds) Principles and practice of nematode control in crops. Academic Press, Sydney, Australia, pp 233–263

Kerry BR (2000) Rhizosphere interactions and the exploitation of microbial agents for the biological control of plant-parasitic nematodes. Annu Rev Phytopathol 38:423–441

Khan A, Williams KL, Nevalainen HKM (2006) Control of plant-parasitic nematodes by *Paecilomyces lilacinus* and *Monacrosporium lysipagum* in pot trials. BioControl 51:643–658

Khan MR, Mohiddin FA, Ejaz MN, Khan MM (2012) Management of root-knot disease in eggplant through the application of biocontrol fungi and dry neem leaves. Turk J Biol 36:161–169

Knudsen GR, Dandurand LMC (2014) Ecological complexity and the success of fungal biological control agents. Adv Agric 2014:542703, 11 pages

Ko MP, Schmitt DP (1996) Changes in plant-parasitic nematode populations in pineapple fields following inter-cycle cover crops. J Nematol 28:546–556

Koenning SR, Barker KR, Bowman DT (2001) Resistance as tactic for management of *Meloidogyne incognita* on cotton in North Carolina. J Nematol 33:126–131

Kokalis-Burelle N, Mahaffee WF, Rodriguez-Kabana J, Kloepper W, Bowen KL (2002) Effects of switchgrass (*Panicum virgatum*) rotations with peanut (*Arachis hypogaea* L.) on nematode populations and soil microflora. J Nematol 34:98–105

Kratochvil RJ, Sardanelli S, Everts K, Gallagher E (2004) Evaluation of crop rotation and other cultural practices for management of root-knot and lesion nematodes. Agron J 96:1419–1428

Kruger DHM, Fourie JC, Malan AP (2013) Cover crops with biofumigation properties for the suppression of plant-parasitic nematodes: a review. South Afr J Enol Viticulture 34(2):287–295

Lalezar M, Moosavi MR, Hesami A (2016) Changes in zucchini defense responses against *Meloidogyne javanica* (Rhabditida: Meloidogynidae) induced by *Pochonia chlamydosporia*. Munis Entomol Zool 11(1):151–159

Leadbeater A, Staub T (2014) Exploitation of induced resistance: a commercial perspective. In: Walters DR, Newton AC, Lyon GD (eds) Induced resistance for plant defense: a sustainable approach to crop protection. Blackwell, Oxford, pp 300–315

de Leij FAAM, Davies KG, Kerry BR (1992) The use of *Verticillium chlamydosporium* Goddard and *Pasteuria penetrans* (Thorne) Sayre & Starr alone and in combination to control *Meloidogyne incognita* on tomato plants. Fundam Appl Nematol 15:235–242

Liang L-M, Zou C-G, Xu J, Zhang KQ (2019) Signal pathways involved in microbe–nematode interactions provide new insights into the biocontrol of plant-parasitic nematodes. Philos Trans Roy Soc B 374:20180317

López-Lima D, Sánchez-Nava P, Carrión G, Núñez-Sánchez A (2013) 89% reduction of a potato cyst nematode population using biological control and rotation. Agron Sustain Dev 33 (2):425–431

Manzanilla-López RH, Esteves I, Powers SJ, Kerry BR (2011) Effects of crop plants on abundance of *Pochonia chlamydosporia* and other fungal parasites of root-knot and potato cyst nematodes. Ann Appl Biol 159:118–129

de Medeiros HA, Resende RS, Ferreira FC, Freitas LG, Rodrigues FÁ (2015) Induction of resistance in tomato against *Meloidogyne javanica* by *Pochonia chlamydosporia*. Nematoda 2:e10015. https://doi.org/10.4322/nematoda.10015

de Medeiros HA, de Araújo Filho JV, de Freitas LG, Castillo P, Rubio MB, Hermosa R, Monte E (2017) Tomato progeny inherit resistance to the nematode *Meloidogyne javanica* linked to plant growth induced by the biocontrol fungus *Trichoderma atroviride*. Sci Rep 7:40,216. https://doi.org/10.1038/srep40216

Mendoza AR, Sikora RA (2009) Biological control of Radopholus similis in banana by combined application of the mutualistic endophyte *Fusarium oxysporum* strain 162, the egg pathogen *Paecilomyces lilacinus* strain 251 and the antagonistic bacteria *Bacillus firmus*. BioControl 54:263–272

Meyer SLF, Roberts DP (2002) Combinations of biocontrol agents for management of plant-parasitic nematodes and soilborne plant-pathogenic fungi. J Nematol 34(1):1–8

Meyer SLF, Roberts DP, Chitwood DJ, Carta LK, Lumsden RD, Mao W (2001) Application of *Burkholderia cepacia* and *Trichoderma virens,* alone and in combinations, against *Meloidogyne incognita* on bell pepper. Nematropica 31:75–86

Molinari S (2011) Natural genetic and induced plant resistance, as a control strategy to plant-parasitic nematodes alternative to pesticides. Plant Cell Rep 30:311–323

Moosavi MR (2017) The effect of gibberellin and abscisic acid on plant defense responses and on disease amount caused by *Meloidogyne javanica* on tomato plants. J Gen Plant Pathol 83 (3):173–184

Moosavi MR, Askary TH (2015) Nematophagous fungi- commercialization. In: Askary TH, Martinelli PRP (eds) Biocontrol agents of phytonematodes. CABI Publishing, Wallingford, UK, pp 187–202

Moosavi MR, Ghani M (2019) The optimal concentrations of *Purpureocillium lilacinum* and jasmonic acid in controlling *Meloidogyne javanica* on tomato. Arch Phytopathology Plant Protect 52(6–7):582–600

Moosavi MR, Zare R (2012) Fungi as biological control agents of plant-parasitic nematodes. In: Merillon JM, Ramawat KG (eds) Plant defence: biological control, progress in biological control 12. Dordrecht, Netherlands, Springer Science + Business Media, pp 67–107

Moosavi MR, Zare R (2015) Factors affecting commercial success of biocontrol agents of phytonematodes. In: Askary TH, Martinelli PRP (eds) Biocontrol agents of phytonematodes. CABI Publishing, Wallingford, UK, pp 423–445

Moosavi MR, Zare R (2016) Present status and the future prospects of microbial biopesticides in Iran. In: Singh HB, Sarma BK, Keswani C (eds) Agriculturally important microorganisms: commercialization and regulatory requirements in Asia. Springer, Singapore, pp 293–305

Moosavi MR, Shakeri S, Mohammadi S (2015) The ability of separate and combined application of five nematopathogenic fungi against *Meloidogyne javanica*. Iran J Plant Prot Sci 46 (1):179–190; in Persian with English abstract

Mostafanezhad H, Sahebani N, Nourinejhad Zarghani S (2014) Control of root-knot nematode (*Meloidogyne javanica*) with combination of *Arthrobotrys oligospora* and salicylic acid and study of some plant defense responses. Biocontrol Sci Tech 24(2):203–215

Nagesh M, Jankiram T (2004) Root-knot nematode problem in polyhouse roses and its management using dazomat, neem cake and *Pochonia chlamydosporia* (*Verticillium chlamydosporium*). J Ornamental Hortic New Series 7(2):147–152

Narasimhamurthy HB, Ravindra H, Sehgal M (2017) Management of rice root-knot nematode, *Meloidogyne graminicola*. Int J Pure Appl Biosci 5:268–276

Nicol JM, Turner SJ, Coyne DL, den Nijs L, Hockland S, Tahna Maafi Z (2011) Current nematode threats to world agriculture. In: Jones J, Gheysen G, Fenoll C (eds) Genomics and molecular genetics of plant-nematode interactions. Springer Science + Business Media, Dordrecht, the Netherlands, pp 21–43

Noel GR, Atibalentja N, Bauer SJ (2010) Suppression of *Heterodera glycines* in a soybean field artificially infested with *Pasteuria nishizawae*. Nematropica 40:41–52

Oduor-Owino P (2003) Integrated management of root-knot nematodes using agrochemicals, organic matter and the antagonistic fungus, *Paecilomyces lilacinus* in natural field soil. Nematol Mediterr 31:121–123

Oka Y (2010) Mechanisms of nematode suppression by organic soil amendments—a review. Appl Soil Ecol 44:101–115

Owino OP (1992) Effect of marigold leaf extract and captafol on fungal parasitism of root knot nematode eggs—Kenyan isolates. Nematol Mediterr 20:211–213

Paparu P, Dubois T, Coyne D, Viljoen A (2007) Defense-related gene expression in susceptible and tolerant bananas (*Musa* spp.) following inoculation with non-pathogenic *Fusarium oxysporum* endophytes and challenge with *Radopholus similis*. Physiol Mol Plant Pathol 71:149–157

Ploeg A (2008) Biofumigation to manage plant-parasitic nematodes. In: Ciancio A, Mukerji KG (eds) Integrated management and biocontrol of vegetable and grain crops nematodes. Springer, Dordrecht, the Netherlands, pp 239–248

Puertas A, Hidalgo-Díaz L (2007) Influence of the host plant and its interaction with *Meloidogyne incognita* on the effectiveness of *Pochonia chlamydosporia* var. *catenulata*. Revista de Protección Vegetal 22(2):104–109; in Spanish with English abstract

Reddy PP, Rao MS, Nagesh M (1996) Management of the citrus nematode, *Tylenchulus semipenetrans*, by integration of *Trichoderma harzianum* with oil cakes. Nematol Mediterr 24:265–267

Reddy PP, Rao MS, Nagesh M (1999) Eco-friendly management of *Meloidogyne incognita* on tomato by integration of *Verticillium chlamydosporium* with neem and calotropis leaves. J Plant Dis Prot 106(5):530–533

Renčo M (2013) Organic amendments of soil as useful tools of plant parasitic nematodes control. Helminthologia 50(1):3–14

Roberts PA (2002) Concepts and consequences of resistance. In: Starr JL, Cook R, Bridge J (eds) Plant resistance to parasitic nematodes. CABI Publishing, Wallingford, pp 23–41

Rodriguez-Kabana R, Morgan-Jones G, Chet I (1987) Biological control of nematodes: soil amendments and microbial antagonists. Plant Soil 100(1–3):237–247

Roget DK, Rovira AD (1987) A review on the effect of tillage on cereal cyst nematode. Wheat Research Council of Australia, Workshop Report Series, No. 1, pp 31–35

Rumbos CI, Kiewnick S (2006) Effect of plant species on persistence of *Paecilomyces lilacinus* strain 251 in soil and on root colonization by the fungus. Plant Soil 283:25–31

Sajadi Z, Moosavi MR, Moaf-Poorian GR (2016) The effect of soil texture and organic matter on ability of *Trichoderma longibrachiatum* in controlling *Meloidogyne javanica* and growth promoting of kidney bean. Iran J Plant Prot Sci 46(2):227–240

Sano Z (2002) Nematode management strategies in east Asian countries. Nematology 4:129–130

Sarma BK, Yadav SK, Singh S, Singh HB (2015) Microbial consortium-mediated plant defense against phytopathogens: readdressing for enhancing efficacy. Soil Biol Biochem 87:25–33

Shoresh M, Harman GE, Mastouri F (2010) Induced systemic resistance and plant responses to fungal biocontrol agents. Annu Rev Phytopathol 48:21–43

Siddiqui ZA, Mahmood I (1996) Biological control of plant-parasitic nematodes by fungi: a review. Bioresour Technol 58:229–239

Sikora RA (1992) Management of the antagonistic potential in agricultural ecosystems for the biological control of plant parasitic nematodes. Annu Rev Phytopathol 30:245–270

Sikora RA, Bridge J, Starr JL (2005) Management practices: an overview of integrated nematode management technologies. In: Luc M, Sikora RA, Bridge J (eds) Plant parasitic nematodes in subtropical and tropical agriculture, 2nd edn. Wallingford, CABI Publishing, pp 793–825

Sipes BS, Schmitt DP (1998) Nematode-pesticide interactions. In: Barker KR, Pederson GA, Windham GL (eds) Plant and nematode interactions. American Society of Agronomy, Madison, WI, pp 173–185

Sokhandani Z, Moosavi MR, Basirnia T (2016) Optimum levels of *Trichoderma longibrachiatum* concentration and cadusafos dose in controlling *Meloidogyne javanica* on zucchini plants. J Nematol 48(1):54–63

Somasekhar N, Gill JS (1991) Efficacy of *Pasteuria penetrans* alone and in combination with carbofuran controlling *Meloidogyne incognita*. Indian J Nematol 21:61–65

Spadaro D, Gullino ML (2005) Improving the efficacy of biocontrol agents against soilborne pathogens. Crop Prot 24:601–613

Starr JL, McDonald AH, Claudius-Cole AO (2013) Nematode resistance in crops. In: Perry RN, Moens M (eds) Plant nematology, 2nd edn. CABI Publishing, Wallingford, pp 411–436

Stirling GR (1988) Biological control of plant-parasitic nematodes. In: Poinar GO Jr, Jansson H-B (eds) Diseases of nematodes, volume II. CRC Press, Boca Raton, Fl, pp 93–139

Stirling GR (1999) Increasing the adoption of sustainable, integrated management strategies for soilborne diseases of high-value annual crops. Australas Plant Pathol 28:72–79

Stirling GR (2014) The soil environment and the soil–root interface. In: Stirling GR (ed) Biological control of plant-parasitic nematodes, soil ecosystem management in sustainable agriculture, 2nd edn. CABI Publishing, Wallingford, UK, pp 15–47

Stirling GR, Wilson EJ, Stirling AM, Pankurst CE, Moody PW, Bell MJ, Halpin N (2005) Amendments of sugarcane trash induce suppressiveness to plant-parasitic nematodes in a sugarcane soil. Australas Plant Pathol 34:203–211

Stirling GR, Halpin NV, Bell MJ (2011) A surface mulch of crop residues enhances suppressiveness to plant-parasitic nematodes in sugarcane soils. Nematropica 41:109–121

Stirling GR, Smith MK, Smith JP, Stirling AM, Hamill SD (2012) Organic inputs, tillage and rotation practices influence soil health and suppressiveness to soilborne pests and pathogens of ginger. Australas Plant Pathol 41:99–112

Strange RN, Scott PR (2005) Plant disease: a threat to global food security. Annu Rev Phytopathol 43:83–116

Sturz AV, Kimpinski J (2004) Endoroot bacteria derived from marigold (*Tagetes* spp.) can decrease soil population densities of root-lesion nematodes in the potato root zone. Plant Soil 262 (1–2):241–249

Szabó M, Csepregi K, Gálber M, Virányi F, Fekete C (2012) Control plant-parasitic nematodes with *Trichoderma* species and nematode-trapping fungi: the role of *chi18-5* and *chi18-12* genes in nematode egg-parasitism. Biol Control 63:121–128

Taba S, Moromizato K, Takaesu Z, Ooshiru A, Nasu K (2006) Control of the southern root-knot nematode, *Meloidogyne incognita* using granule formulations containing nematode-trapping fungus, *Monacrosporium ellipsosporum* and a nematicide. Jap J Appl Entomol Zool 50:115–122

Takur M, Sohal BS (2013) Role of elicitors in inducing resistance in plants against pathogen infection: a review. ISRN Biochem 2013:762412. https://doi.org/10.1155/2013/762412

Talavera M, Itou K, Mizukubo T (2002a) Combined application of *Glomus* sp. and *Pasteuria penetrans* for reducing *Meloidogyne incognita* (Tylenchida: Meloidogynidae) populations and improving tomato growth. Appl Entomol Zool 37:61–67

Talavera M, Mizukubo T, Itou K, Aiba S (2002b) Effect of spore inoculum and agricultural practices on the vertical distribution of the biocontrol plant-growth-promoting bacterium *Pasteuria penetrans* and growth of *Meloidogyne incognita*-infected tomato. Biol Fertil Soils 35:435–440

Thoden TC, Korthals G, Termorshuizen A (2011) Organic amendments and their influences on plant-parasitic and free-living nematodes: a promising method for nematode management? Nematology 13:133–153

Timper P (1999) Effect of crop rotation and nematicide use on abundance of *Pasteuria penetrans*. J Nematol 31:575; abstract

Timper P (2011) Utilization of biological control for managing plant-parasitic nematodes. In: Davies K, Spiegel Y (eds) Biological control of plant-parasitic nematodes: building coherence between microbial ecology and molecular mechanisms. Springer, London, pp 259–289

Timper P (2014) Conserving and enhancing biological control of nematodes. J Nematol 46 (2):75–89

Timper P, Brodie BB (1994) Effect of host-plant resistance and a nematode pathogenic fungus on *Pratylenchus penetrans*. Phytopathology 84:1090; abstract

Timper P, Parajuli G (2012) Suppression of *Meloidogyne incognita* by *Paecilomyces lilacinus* is enhanced by planting cover crops. J Nematol 44:494–495; abstract

Timper P, Minton NA, Johnson AW, Brenneman TB, Culbreath AK, Burton GW, Baker SH, Gascho GJ (2001) Influence of cropping systems on stem rot (*Sclerotium rolfsii*), *Meloidogyne arenaria* and the nematode antagonist *Pasteuria penetrans* in peanut. Plant Dis 85:767–772

Timper P, Davis R, Jagdale G, Herbert J (2012) Resiliency of a nematode community and suppressive service to tillage and nematicide application. Appl Soil Ecol 59:48–59

Tobin JD, Haydock PPJ, Hare MC, Woods SR, Crump DH (2008) Effect of the fungus *Pochonia chlamydosporia* and fosthiazate on the multiplication rate of potato cyst nematodes (*Globodera pallid* and *G. rostochiensis*) in potato crops grown under UK field conditions. Biol Control 46:194–201

Trivedi PC, Barker KR (1986) Management of nematodes by cultural practices. Nematropica 16:213–236

Tzortzakakis EA, Goewn SR (1994) Evaluation of *Pasteuria penetrans* alone and in combination with oxamyl, plant resistance and solarization for control of *Meloidogyne* spp. on vegetables grown in greenhouses in Crete. Crop Prot 13:455–462

Van den Boogert PHJF, Velvis H, Ettema CH, Bouwman LA (1994) The role of organic matter in the population dynamics of the endoparasitic nematophagous fungus *Drechmeria coniospora* in microcosms. Nematologica 40:249–257

Verdejo-Lucas S, Sorribas FJ, Ornat C, Galeano M (2003) Evaluating *Pochonia chlamydosporia* in a double-cropping system of lettuce and tomato in plastic houses infested with *Meloidogyne javanica*. Plant Pathol 52:521–528

Viaene NM, Abawi GS (2000) *Hirsutella rhossiliensis* and *Verticillium chlamydosporium* as biocontrol agents of the root-knot nematode Meloidogyne hapla on lettuce. J Nematol 32 (1):85–100

Viaene N, Coyne DL, Davies KG (2013) Biological and cultural management. In: Perry RN, Moens M (eds) Plant nematology, 2nd edn. CABI Publishing, Wallingford, UK, pp 383–410

Vieira Dos Santos MC, Curtis RHC, Abrantes I (2014) The combined use of *Pochonia chlamydosporia* and plant defence activators—a potential sustainable control strategy for *Meloidogyne chitwoodi*. Phytopathol Mediterr 53(1):66–74

Walker GE, Wachtel MF (1988) The influence of soil solarization and non-fumigant nematicides on infection of *Meloidogyne javanica* by *Pasteuria penetrans*. Nematologica 34:477–483

Walters DR (2011) Sounding the alarm: signaling and communication in plant defense. In: Walters DR (ed) Plant defense: warding off attack by pathogens, herbivores, and parasitic plants. Blackwell Publishing, Oxford, pp 77–124

Walters DR, Bennett AE (2014) Microbial induction of resistance to pathogens. In: Walters DR, Newton AC, Lyon GD (eds) Induced resistance for plant defense: a sustainable approach to crop protection. Blackwell Publishing, Oxford, pp 149–170

Wang KH, Sipes BS, Schmitt DP (2001) Suppression of *Rotylenchulus reniformis* by *Crotalaria juncea*, *Brassica napus*, and *Tagetes erecta*. Nematropica 31:235–249

Wang K-H, Sipes BS, Schmitt DP (2002) Management of *Rotylenchulus reniformis* in pineapple, *Ananas comosus*, by intercycle cover crops. J Nematol 34:106–114

Wang KH, Sipes BS, Schmitt DP (2003) Enhancement of *Rotylenchulus reniformis* suppressiveness by *Crotalaria juncea* amendment in pineapple soils. Agric Ecosyst Environ 94:197–203

Wang K-H, Mcsorley R, Gallaher RN, Kokalis-Burelle N (2008) Cover crops and organic mulches for nematode, weed and plant health management. Nematology 10(2):231–242

Weller DM, Mavrodi DV, van Pelt JA, Pieterse CMJ, van Loon LC, Bakker PAHM (2012) Induced systemic resistance in *Arabidopsis thaliana* against *Pseudomonas syringae* pv. *tomato* by 2,4-diacetylphloroglucinol-producing *Pseudomonas fluorescens*. Phytopathology 102:403–412

Westphal A (2011) Sustainable approaches to the management of plant-parasitic nematodes and disease complexes. J Nematol 43(2):122–125

Westphal A, Becker JO (2001) Soil suppressiveness to *Heterodera schachtii* under different cropping sequences. Nematology 3:551–558

Widmer TL, Mitkowski NA, Abawi GS (2002) Soil organic matter and management of plant-parasitic nematodes. J Nematol 34:289–295

Xu X-M, Jeffries P, Pautasso M, Jeger MJ (2011) Combined use of biocontrol agents to manage plant diseases in theory and practice. Phytopathology 101:1024–1031

Zaki MJ, Maqbool MA (1991) Combined efficacy of *Pasteuria penetrans* and other biocontrol agents on the control of root-knot nematode on okra. Pak J Nematol 9:49–52

Chapter 11
Role of *Trichoderma* spp. in the Management of Plant-Parasitic Nematodes Infesting Important Crops

Dina S. S. Ibrahim, Marwa M. Elderiny, Rizwan Ali Ansari, Rose Rizvi, Aisha Sumbul, and Irshad Mahmood

Abstract Phytonematodes can be considered as one of the main obstacles of quantitative and qualitative production of many important crops throughout the world. The strident importance of biological approaches for the management of phytonematodes might be because of friendly nature with reference to agroecosystem sustainability. *Trichoderma*, a mycoparasite, is an abundant bioagent that kills various plant pathogens and enhances crop productivity. Various mechanisms are implicated behind the effective management of plant disease such as induction of defense responses in plants, antibiosis, competition, direct parasitism, enzymatic hydrolysis, etc. These biocontrol agents offer synergistic effects when introduced through integrated approaches for the management of nematodal diseases. Identification of research priorities for the utilization of bio-nematicides especially *Trichoderma* spp. in sustainable agriculture, as well as understanding of their mechanisms of action and interaction with other agricultural inputs, is still needed. Application of *Trichoderma* spp. in controlling various phytopathogenic

D. S. S. Ibrahim (✉) · M. M. Elderiny
Department of Nematode Diseases and Central Lab of Biotechnology, Plant Pathology Institute, Agriculture Research Center (ARC), Cairo, Egypt

R. A. Ansari
Section of Plant Pathology and Nematology, Department of Botany, Aligarh Muslim University, Aligarh, India

Present Address: Department of Plant Protection, Faculty of Agricultural Sciences, Aligarh Muslim University, Aligarh, India

R. Rizvi · I. Mahmood
Section of Plant Pathology and Nematology, Department of Botany, Aligarh Muslim University, Aligarh, India

A. Sumbul
Section of Plant Pathology and Nematology, Department of Botany, Aligarh Muslim University, Aligarh, India

Present Address: Department of Biosciences, Jamia Millia Islamia, New Delhi, India

R. A. Ansari et al. (eds.), *Management of Phytonematodes: Recent Advances and Future Challenges*, https://doi.org/10.1007/978-981-15-4087-5_11

nematodes leading to improved plant health and yield is the current brainstorming session among the researchers.

Keywords Plant-parasitic nematodes · *Trichoderma* · Fungal nematicides · Antibiosis and enzymatic hydrolysis

11.1 Introduction

Nowadays, the excessive introduction of inorganic fertilizers and pesticides of chemical origin has created serious environmental issues (Hermosa et al. 2012; Meena and Meena 2015; Ansari and Khan 2012a, b; Ansari and Mahmood 2017a). Therefore, scientists have focused on devising some alternatives which can be used in the management of phytonematodes. In addition, complex biological as well as ecological processes occur in the rhizosphere surrounding the plant roots (Bais et al. 2006). Plants continuously release some chemicals which directly or indirectly become the source of food for various microbial agents (Nannipieri et al. 2007; Ansari et al. 2019). Antagonistic activities of various microbial agents including *Trichoderma* spp. improve the plant growth and yield attributes (Weller 1988; Weller et al. 2002; Whipps 2001; Whipps 1997; Berg et al. 2005; Ansari et al. 2017a, b). Therefore, putative strains of biological control agents are the need of the hour which could be used in the management of plant nematodes infesting various agricultural crops (Dong et al. 2004). Certain groups of fungi may assist the plants to rescue them from pathogen attack by using various mechanisms and improve the productivity of the plants. Moreover, *Trichoderma* spp. can easily be encountered in the diverse array of soil. These biocontrol agents are helpful in the inhibition of pathogen activity through various mechanistic approaches (Saldajeno et al. 2014; Viterbo and Horwitz 2010). *Trichoderma* species also have positive effects on different plant growth parameters, which help in increasing the proliferation of secondary roots, leaf area, shoot length, dry weight, and crop yield (Hermosa et al. 2013; Mukherjee et al. 2013). *Trichoderma* spp. are also known as plant growth-promoting fungi (PGPF) as they can secrete phosphate-solubilizing enzymes, siderophores, and a wide range of plant growth-promoting organic molecules, i.e., phytohormones (Doni et al. 2013). An attempt was given to shed light on the biocontrol activity and mechanism of *Trichoderma* species against plant-parasitic nematodes.

11.2 Biocontrol Potential of *Trichoderma* spp. Against Plant-Parasitic Nematodes

Management of plant nematodes through biological control agents is very important as it is considered to be cheap, economical, and eco-friendly (Ansari and Khan 2012a, b; Ansari and Mahmood 2017b, 2019a, b). *Trichoderma* species not only have different strategies for nematode antagonism but also improve the plant growth effectively as they are able to colonize the rhizosphere as well as plant roots. Additionally, they are very effective in producing the plant growth-promoting organic molecules and phytohormones and solubilizing raw phosphate leading to enhanced plant growth and yield (Carvalho et al. 2011; Ansari and Mahmood 2019b). A notable count of microbial agents have been screened out and found to be effective in the management of plant nematodes including root-knot nematodes *Meloidogyne* spp. (Arevalo et al. 2009; Aminuzzaman et al. 2013; Aminuzzaman et al. 2018). Various beneficial roles including plant growth-promoting activity have been found in *Trichoderma* spp.; therefore, they have received major attention as excellent candidates for successful exploitation (Chet 1987). Several members of this genus are well known to be effectively used in controlling *Meloidogyne* spp. (Samuels et al. 2012). *Trichoderma* spp. have been reported to reduce the reproduction of *Meloidogyne* spp. in various economically important crops including tomato (Sharon et al. 2007). For example, in a study *T. harzianum* BI and its filtrates were used as a biocontrol agent for the management of *M. javanica* through the inhibition of egg hatching, direct parasitism, and production of secondary metabolites which are lethal to J2 of *M. javanica* infesting tomato (Naserinasab et al. 2011). Also, in greenhouse test, cucumber plants were given conidial suspensions of *Trichoderma* sp. before and after introduction of *M. incognita*. Surprisingly, a 50% reduction in nematode multiplication was obtained (Mascarin et al. 2012). Culture filtrates of *Trichoderma* spp. like *T. viride* (S-1 and S-3), *T. harzianum*, and *T. koningii* registered more than 50% death of juveniles of *M. javanica*. Maximum mortality was recorded in the plants treated with *T. viride* S-3 (90%) followed by *T. harzianum* (88%) (Qureshi et al. 2012). Fifteen isolates of *T. harzianum* caused an average of 66.6% J2 mortality of *M. javanica* with a sight to genetic relatedness among the indigenous isolates *of T. harzianum* (Khattak et al. 2018). Besides, *T. asperellum* M2RT4 successfully reduced galls, egg mass, and eggs in pineapple roots in Kenya (Kirigaa et al. 2018). Several *Trichoderma* species, viz., *T. asperelloides*, *T. hamatum*, *T. harzianum*, and *T. viride*, showed high qualitative and quantitative chitinase activity and recorded remarkable antagonistic activity against *M. incognita* infecting tomato. For example, the *Trichoderma* species *T. asperelloides* successfully reduced the root galling to up to 92.81% and final nematode population to up to 97.81% (Sayed et al. 2019).

11.3 Mechanisms of Action of *Trichoderma* spp.

Trichoderma species can act against different plant pathogens through several mechanisms such as direct parasitism, antibiosis, food competition, induction of disease resistance, and enzymatic hydrolysis (Elad and Freeman 2002; Harman et al. 2004; Howell 2003). Furthermore, *Trichoderma* spp. may also help out the plant to improve its productivity in case of absence of pest and pathogens (Sharon et al. 2001; Yedidia et al. 1999).

11.3.1 Direct Parasitism

Trichoderma spp. can easily be isolated from a diverse range of soils which contain highly branched conidiophores producing conidia that may attach to different parts of the nematode's body at various stages of life cycle. *Trichoderma* spp. also develop coil and appressorium-like structure which help in the penetration of the nematode cuticles or eggshells. *Trichoderma harzianum* and *T. asperelloides* parasitize eggs and juveniles of *Meloidogyne* spp. which might be due to peptaibiotics and glycolytic and chitinolytic enzymes (Spiegel et al. 2005; Sharon et al. 2009). A high degree of parasitism (83.8%) and immobility (95.2%) of J_2 of *M. javanica* (Golzari et al. 2011) was demonstrated after 48 h of incubation with *Trichoderma harzianum*. In addition, conidia of *T. harzianum* adhere to and immobilize 64% of eggs and second-stage juveniles of *Meloidogyne incognita* race 4 (Mascarin et al. 2012).

11.3.2 Induction of Defense Responses in Plants

The vital role of biocontrol agents in the induction of resistance in the host plant can be considered another important mechanism in controlling phytopathogens (Leonetti et al. 2014). The plant response in case of nematode attack can be detected by an increased peroxidase activity. Ground nut treated with *T. harzianum* showed enhanced levels of defense enzymes like peroxidase and polyphenol oxidase and some important defense compounds like ortho-dihydric phenol and total phenol (Sreedevi et al. 2011). Chitinases, which are important defense enzymes in plants against phytopathogens, have markedly increased activities in both roots and leaves of the oil palm treated with *Trichoderma* sp. (Naher et al. 2012; Pusztahelyi 2018). The potential of *T. harzianum* for control of *M. incognita* induced systemic resistance in tomato which showed an obvious increase in the accumulation of total chlorophyll and enzymes, viz., chitinase, phenylalanine ammonia-lyase (PAL), and peroxidase, which are known to confer systemic resistance (Singh et al. 2017) besides causing infection on the eggs and juveniles of root-knot nematode in in vitro conditions and in plant under greenhouse conditions.

11.3.3 Competition

The high and rapid growth capacity of *Trichoderma* spp. has the chance to compete successfully for place and nutrients and it is an important feature to act as a biocontrol agent (Contreras-Cornejo et al. 2016). For example, Sivan and Chet (1989a, b) discussed the inhibition activity of *Trichoderma* when applied as conidial suspension against the germination of soil-borne pathogens which resulted to competition for space and nutrients in the rhizosphere. Besides, *Trichoderma* spp. are introduced through various modes of application such as soil drenching or seed treatments. Such application allows them to grow rapidly along with the plant root system. However, there are inconsistent results in the identification of mechanism (competition) involved in different biological strains of *Trichoderma* spp. (Howell 2003).

11.3.4 Enzymatic Hydrolysis (Enzyme Producers)

Nematode egg shell chemical structure indicated the presence of proteases and chitinases and their role in causing the infection of nematode eggs by nematophagous fungi. *Trichoderma* species successfully secrete several hydrolytic enzymes, i.e., chitinases, cellulases, xylanases, glucanases, and proteases, that cause degradation of nematode cell wall. These enzymes are generally extracellular, of low molecular weight, and highly stable which are produced in various forms or isozymes differing in size, regulation, and ability (Cheng et al. 2017). The chitinase production of *Trichoderma* can be manually improved through mutation with γ-radiation which can help in the management of plant diseases—*T. harzianum* mutants were used for chitinolytic enzyme production which was manifested in T. h M15 that contains a large amount of endochitinase (24.5, 26, and 42 kDa) and β-1,4-N-acetylglucosaminidase (68 kDa) (Baharvand et al. 2015). *Trichoderma* species have the mycolytic enzyme having high potentiality to combat with destructive plant pathogens (Hyder et al. 2017). Moreover, antagonistic activity depends on the coiling ability of *Trichoderma* spp. around the host hyphae by the formation of appressorium and also by the production of significant hydrolytic enzymes, viz., chitinases (Anand and Reddy 2009) and β-1,3-glucanase and protease (Ahmed 2008; Gajera et al. 2012). Since chitin is the basic component of nematode egg shell, chitinases which are secreted as secondary metabolites by *Trichoderma* spp. are considered the most effective component against pathogenic nematodes (Haggag and Amin, 2001; Morton et al. 2004; Jin et al. 2005; Abo-Elyousr et al. 2010). *T. harzianum* enzymatically penetrated cysts and eggs resulting in the death of *G. rostochiensis* (Saifullah and Khan 2014). *T. harzianum* with the aid of several lytic enzymes such as chitinases, glucanases, and proteases showed successful parasitism of *Meloidogyne* and *Globodera* eggs as the chitin layer is dissolved through microbial enzymes. The hyphae of *T. harzianum* penetrate the cuticle of nematode eggs and juveniles, proliferate within the organism, and secrete toxic metabolites (Askary and Martinelli 2015). *Trichoderma* spp. are considered as potential biocontrol agents for the management of a wide array of pathogens including fungi, nematodes, etc. by producing the lytic enzymes (Hussain et al. 2017). For example, *T. viride* produces lipase enzymes in a specific medium containing olive oil at 30–31 °C for 4 days (Kashmiri et al. 2006; Mehta et al. 2017). Many *Trichoderma* spp. such as *T. longibrachiatum*, *T. harzianum*, *T. koningii*, and *T. viride* are well-considered for their cellulolytic potentiality.

Such biocontrol members have been isolated from various kinds of soil and found to be very effective in disease management (Strakowska et al. 2014).

11.3.5 Antibiosis

Various effective antibiotics such as trichodermin, dermadin, trichoviridin, and sesquiterpene heptalic acid are produced by *T. viride* which have a positive effect in the suppression of plant-parasitic nematodes (Abd-Elgawad and Askary 2020).

11.3.6 Production of Toxic Compounds

Trichoderma spp. are the potential biocontrol fungi which are used in the management of plant-parasitic nematodes (Sharon et al. 2011; Javeed et al. 2016; Al-Hazmi et al. 2017; Abdelrafaa et al. 2018; Migunova et al. 2018). The role of *Trichoderma* spp. (*T. harzianum*, *T. hamatum*, and *T. koningii*) and their culture filtrates in control of both reniform and root-knot nematodes can be assumed to be due to the direct effect of toxic metabolites (Bokhary 2009). *Trichoderma* culture filtrate was greatly significant on root-knot nematode (*Meloidogyne javanica*) eggs than on juveniles. Production of nematicidal compounds is the key for root-knot nematode management (Sharon et al. 2011). *Trichoderma* isolates, i.e., *Trichoderma asperellum* M2RT4, *T. atroviride* F5S21, and *Trichoderma* sp. MK4, have been reported to cause successful endophytical colonization of pineapple root (Kirigaa et al. 2018). *T. asperellum* M2RT4 and *Trichoderma* sp. MK4 showed significant reduction in production of nematode egg and egg mass as well as suppression in root galling rate. On the other hand, the tested isolates improved the plant root mass growth when compared to the untreated control. *Trichoderma viride* inhibited the egg-hatching rate (Hallman et al. 2009) and trade formulations have also been proven to be efficacious in tropical greenhouse conditions (Akhtar 2000). There are certain species of *Trichoderma* which are used against soil-borne plant pathogens. These members also have the ability to check the potentiality of some root-knot nematodes (Goswami and Mittal 2004; Meyer et al. 2001; Sharon et al. 2001; Al Kader 2008; Migunova et al. 2018; Nagachandrabose 2018; Herrera-Parra et al. 2018). *T. harzianum* has shown significant results in the reduction of *M. javanica* egg hatching (Al-Hazmia et al. 2019). The two antagonistic fungi like *Trichoderma harzianum* and *T. viride* caused remarkable reductions in nematode-related parameters such as galls, egg masses, and fecundity and reproduction factors of *M. incognita* (Mukhtar 2018).

11.4 *Trichoderma* Species as Plant Growth Enhancer

Trichoderma species exhibit other characteristics that may play a role in disease resistance or tolerance. *Trichoderma* spp. can impart disease resistance, improve plant growth, and enable the plants to fight against various harmful plant pathogens. They are considered as plant growth promoters as they are able to enhance the nutritional efficiency of the plants and make them competent against various stresses (Kashyap et al. 2017).

11.4.1 Root Colonizers

Trichoderma spp. are among the most researched biocontrol fungal agents which have the ability to multiply themselves rapidly and colonize the rhizosphere and various plant parts (Carvalho et al. 2011). *Trichoderma* () spp. colonization has been linked to the successful inhibition of root-knot multiplication (Harman 2000; Howell et al. 2000; Harman et al. 2004; Sharon et al. 2001; Siddiqui and Shaukat 2003; Yedidia et al. 1999). Application of *Trichoderma* spp. before seedling transplantation enhances the root colonization (Dababat and Sikora 2006; Van Damme et al. 2005). Microscopic analysis revealed that *T. harzianum* ThzID1-M3 colonized the cyst nematode *Globodera pallida* and remarkably multiplied in the surrounding of potato roots (Contina et al. 2017).

11.4.2 Phosphate-Solubilizing Producers

Phosphate-solubilizing microorganisms (PSMs) are organisms that produce some enzymes which are helpful in the conversion of inaccessible phosphate to soluble forms making them available to the plants for its absorption. PSMs have a remarkable role in the enhancement of plant growth yield attributes of various important crops. Inoculation of seeds/crops/soil with PSMs is an important strategy to intensify the world food production leaving no toxic residue to the environment (Walpola and Yoon 2012). *Trichoderma* spp. have some features that make them able to touch long distances in the rhizospheric soil; henceforth, they are effective in solubilization of inorganic phosphate as they can secrete more compounds like gluconic, citric, lactic, 2-ketogluconic, oxalic, tartaric, and acetic acid; such fungi can also impart in plant growth promotion (Carvalho et al. 2011). *Trichoderma* spp. also actively participate in the solubilization of phosphate which is very helpful in the promotion of tomato plant health (França et al. 2017). In addition, *T. harzianum* (T-22) has also been found to solubilize insoluble rock phosphate; however, no organic acids were detected during the culture filtrate analysis which was correlated with chelation and reduction processes (Altomare et al. 1999). Another important member of *Trichoderma* spp., i.e., *T. viride*, was reported as one of the microorganisms which are also having phosphate solubilization activity (Shrivastava et al. 2018). *Trichoderma* spp. hold the ability to produce indole-3-acetic acid (IAA) and solubilize phosphate (Migunova et al. 2018) (Table 11.1).

Table 11.1 Effect of *Trichoderma* spp. in the management of phytoparasitic nematode

S. no.	*Trichoderma* spp.	Commercial crops	Phytoparasitic nematodes infesting the crops	References
1.	*Trichoderma* spp.	Sugarcane	*Meloidogyne incognita*	Freitas et al. (2012)
2.	*T. viride*, *T. Harzianum*	Sugarbeet	*M. javanica*, *M. incognita*	Maareg and Badr (2000), Maareg et al. (2003)
3.	*T. harzianum*, *T. viride*	Red kidney bean	*M. incognita*	Sharf et al. (2014), Sharf and Hisamuddin (2016)
4.	*Trichoderma* spp., *T. harzianum*, *T. hamatum*, *T. lignorum*, *T. Harzianum* *T. harzianum*, *T. viride*, *T. koningii*, *T. reesei*, *T. hamatum*	Soybean	*Meloidogyne* spp., *Heterodera glycines*, *Pratylenchus brachyurus* *Rotylenchulus reniformis* *M. javanica* or *R. reniformis*	Elhady et al. (2018), El-Sherif and Ismail (2009), Ahmed (2010), and Izuogu and Abiri (2015)
	T. viride, *T. harzianum*	Sunflower	*M. incognita*	Haggag and Amin (2001)
	T. Harzianum		*M. javanica*	Amin and Mostafa (2000)
	T. harzianum, *T. glaucum*, and *T. hamatum*		*M. incognita*	Dawar et al. (2008) Ahmed (2010)
	T. Harzianum		*M. incognita*	EL-Sherif and Ismail (2010)
	T. viride		*R. reniformis*	Hesamedin and Mohammad (2014)
5.	*T. viride*	Potato	*Globodera rostochiensis* and *G. pallida*	Umamaheswari et al. (2012), Bairwa et al. (2017), and Lima et al. (2018)
6.	*T. Harzianum*	Tobacco	*M. incognita*	Khan and Haque (2011)
	Trichoderma spp.		*Globodera tabacum* and *M. incognita*	Goswami et al. (2008), Prasad et al. (2014)
	T. viride		*M. incognita*	Motha et al. (2010)
7.	*T. harzianum*	Chickpea	*M. incognita*	Rizvi et al. (2018)
8.	*T. harzianum*, *T. Koningii*	Maize	*M. arenaria*	Windham (1989)
9.	*T. Harzianum*	Rice	*M. graminicola*	Narasimhamurthy et al. (2017a, b)

11.5 Management of Phytoparasitic Nematodes Infesting some Important Crops

Growing some commercial crops using traditional crop cultivation is an important sect for the commercialization of agriculture. Undoubtedly, these commercial crops are a good income source for the marginal-level farmers. On the other hand, there are strident chances of pathogen and pest attack on these crops including phytonematodes.

11.5.1 Sugarcane

Sugarcane is one of the important commercial crops having highest bioconversion efficiency and able to fix solar energy efficiently that can yield up to 55 tons of dry matter on an annual basis. Sugarcane has also been observed to be affected by high populations of plant-parasitic nematodes, i.e., *Mesocriconema*, *Paratrichodorus*, *Pratylenchus*, and *Tylenchorhynchus* (Bond et al. 2000). Thus, Freitas et al. (2012) evaluated the potentiality of *Trichoderma* spp. against *M. incognita* which induced the resistance in the sugarcane crop. All the filtrates of *Trichoderma* spp. were effective in promoting juvenile mortality. In assessments with nematode eggs, 16 among 22 strains were significant in relation to control for parasitism of eggs with emphasis on strains 8M, 11M, 13M, 15M, and 17M as the most promising. The strains 4M, 14M, A18, and 4077T showed potential in enzymatic action and mortality of juveniles after hatching.

11.5.2 Sugarbeet

Beet (*Beta vulgaris* L.) is also a source of sugar and another important commercial crop which is also used in plant breeding program and responsible for 44% of sugar production globally. Beet is a temperate crop and also suffers from pests and pathogens including plant-parasitic nematodes. Among the plant-parasitic nematodes, the sugarbeet cyst nematode *Heterodera schachtii* is a major pest affecting sugarbeet production wherever sugarbeet is cultivated (Khan et al. 2016). Cyst, stubby-root, root-lesion, pin, spiral, stunt, dagger, ring, and lance nematodes have also been encountered in sugarbeet roots (Yan and Baidoo 2018). Maareg and Badr (2000) investigated the potentiality of two fungi, *A. niger* and *T. viride*, against *M. javanica*-infected sugarbeet. The combination of *A. niger* and *T. viride* greatly reduced the galls, females, and egg masses on sugarbeet roots than did each organism alone. Maareg et al. (2003) determined the inhibitory effects of *A. niger*, *A. teerreus*, *Fusarium solani*, *R. solani*, *Sclerotium rolfsii*, *T. harzianum*, and *T. viride* filtrates in comparison with the nematicides fenamiphos and oxamyl on

M. incognita under laboratory and greenhouse conditions. All the tested materials were toxic to varying extents against *M. incognita* infecting sugarbeet.

11.5.3 Red Kidney Bean

Phaseolus vulgaris or red kidney bean is currently being cultivated throughout the world because of its high protein content. Seeds of the bean keep 22% of dietary protein amounting to 22% of the total seed weight. Like other crops, plant-parasitic nematode *M. incognita* is considered to be one of the important pathogens of *Phaseolus vulgaris* affecting the productivity markedly. The nematodes also hamper the nodulation and affect directly the metabolic process of the plants related to nitrogen content (Karanja 1988; Siddiqui and Mahmood 1994). Sharf et al. (2014) reported that combinatorial application of potassium plus *T. harzianum* improved the plant health by enabling the plant to absorb the nutrients and water in judicious manner. Sharf and Hisamuddin (2016) reported that plant growth and yield parameters were significantly increased when the plants were treated with different doses of *T. harzianum* and *T. viride*. Biocontrol agents also reduced the nematode-related parameters such as population and eggs per root system over nematode-inoculated control.

11.5.4 Oilseed Crops

11.5.4.1 Soybean (*Glycine max* (L.) Merr)

Soybean is another important leguminous crop being grown in tropical, sub-tropical, and temperate regions suffering from nematode attack (Oyedunmade 2003). Various phytonematode species associated with soybean including *Meloidogyne* spp., *Heterodera glycines*, *Pratylenchus brachyurus*, and *Rotylenchulus reniformis* are very common in occurrence (Elhady et al. 2018). Meanwhile, El-Sherif and Ismail (2009) reported that applications of *T. harzianum* filtrate exhibited enhanced plant growth and suppressed nematode development. Ahmed (2010) reported that various strains of *Trichoderma* spp. at two different concentrations showed significant suppression in nematode population. The highest reduction in total nematode population was recorded when the plants were treated with *T. lignorum*. However, *T. harzianum* 100% ranked next and showed RF = 0.42. On the other hand, root galling was significantly reduced in all treatments receiving the fungal antagonist *Trichoderma* species and oxamyl. The highest reduction in root galling was recorded with *T. hamatum* (100%), *T. lignorum* (50%), *T. harzianum* (100% and 50%), *T. glaucum* (100%), and oxamyl with percent of reduction equal to 92.11 and RGI = 1.00. However, no egg masses were observed in soybean roots following the application of *Trichoderma* species and oxamyl. Izuogu and Abiri (2015) reported that application of *Trichoderma harzianum* T22 significantly enhanced the plant growth and yield. The same antagonist also minimized the development of parasitic activity of nematodes such as such as *Pratylenchus*, *Helicotylenchus*, *Radopholus*, *Rotylenchulus*, and *Xiphinema*.

11.5.4.2 Sunflowers (*Helianthus* Sp.)

Sunflower is the most important oil crop in the world that produces high-quality oil for human consumption and technical purposes as well. Sunflower is a highly cost-effective crop because of full application of machinery in production practices and has high market value (Crnobarac et al. 2006). Various phytonematodes have been found to attack the sunflower including root-knot, lesion, spiral, and soybean cyst nematodes (Bernard and Keyserling 1985). Treatments of *Trichoderma* species, viz., *Trichoderma harzianum*, *T. viride*, *T. koningii*, *T. reesei*, or *T. hamatum*, decreased the *Fusarium* CFU counts in soil infested with either *Meloidogyne javanica* or *Rotylenchulus reniformis* and also markedly enhanced the plant growth and yield-related parameters (Haggag and Amin, 2001). Potentialities of *Glomus mosseae*, *T. viride*, *T. harzianum*, and *Aspergillus oligospora* singly or concomitantly as biological control agents against root-knot nematode *M. incognita* on sunflower were evaluated by Amin and Mostafa (2000). They reported that *T. harzianum* and *T. viride* significantly reduced the *M. incognita* mortality percentage and suppressed the number of galls by 53.13%. Ahmed (2010) reported that fungal filtrates of *Trichoderma* in addition to oxamyl significantly reduced the nematode population infesting sunflower. The highest reduction in nematode population (soil and root) was observed with plants receiving *T. harzianum* (100%) fungal filtrates. However, introduction of fungal filtrates of *T. glaucum* as well as *T. hamatum* effectively controlled the nematode population infesting sunflower. Hesamedin and Mohammad (2014) evaluated the effectiveness of *T. viride* against reniform nematode *Rotylenchulus reniformis* on sunflower under greenhouse condition. It was found that application (seed and soil) of *T. viride* markedly controlled the *R. reniformis* population when treated with *T. viride* on sunflower.

11.5.5 *Potato (*Solanum tuberosum*)*

Potato is an important crop being cultivated in a diverse range of agroecosystems in temperate, sub-tropical, and tropical regions. Currently, potato is cultivated at a large scale in about 130 countries (Askew 2001). Several phytonematode species such as potato cyst nematodes *Globodera rostochiensis* and *G. pallida*, root-knot nematodes *Meloidogyne* spp., root lesion nematodes *Pratylenchus* spp., the potato rot nematode *Ditylenchus destructor*, and the false root-knot nematode *Nacobbus aberrans* are major constraints which have significantly limited the productivity of crop (Lima et al. 2018). Potato cyst nematodes (PCN), viz., *Globodera rostochiensis* and

G. pallida, are one of the most destructive pests of quarantine significance hindering the potato production. *Trichoderma viride* recorded significantly higher plant growth parameters and tubers and lowered the PCN population in soil and roots (Umamaheswari et al. 2012). Bairwa et al. (2017) reported that neem cake (at 5 t/ ha) in combination with *T. viride* (at 5 kg/ha) produced significantly maximum yield of potato and reduced the PCN multiplication ratio.

*11.5.6 Tobacco (*Nicotiana tabacum*)*

Tobacco is well known to be attacked by a wide range of plant-parasitic nematode genera and species; among them, *Meloidogyne* spp., *Pratylenchus* spp., *Globodera tabacum*, *Ditylenchus dipsaci*, and *Aphelenchoides ritzemabosi* are common in occurrence (Shepherd and Barker 1990). *Globodera tabacum*, also called as the tobacco cyst nematode, chiefly infests the tobacco plant (LaMondia and Taylor 1987). *Trichoderma harzianum* was found to be effective in controlling root-knot nematodes as well as reducing galling and egg mass production in tobacco (Khan and Haque 2011), and this reduction effect was suggested to be a result of chitinase activity of *Trichoderma* spp. which might have caused premature hatching of nematode eggs and could be used in control of nematodes (Prasad et al. 2014). Goswami et al. (2008) demonstrated that gall and egg masses were reduced due to high rhizospheric competition of bioagents as they can easily colonize roots and may reduce feeding sites for nematodes.

*11.5.7 Chickpea (*Cicer arietinum *L.)*

India produces 75% of chickpea and chickpea ranks third among all pulses in terms of productivity (Khan et al. 2014). Mycoparasite *Trichoderma harzianum* has been found to reduce the multiplication of plant-parasitic nematodes (*Meloidogyne incog nita*) which might be due to production of some nematostatic compounds leading to improved plant growth and yield (Rizvi et al. 2018).

*11.5.8 Maize (*Zea mays*)*

Maize plays an important role in human life due to its use in multiple purposes like fodder and biofuel production. Prior application of *T. harzianum* or *T. koningii* did not allow the plant nematode (*M. arenaria*) to infest the maize crop significantly and improved the plant yield, when the plants were grown in soil infested with (Windham 1989).

*11.5.9 Rice (*Oryza sativa *L.)*

Rice is one of the most important staple crops of Asian countries including India which provides major calories and influences the livelihoods of billions throughout the world. The grain crop is highly susceptible to *M. incognita*, *M. graminicola*, *M. triticoryzae*, *M. javanica*, *M. oryzae*, *and M. arenaria* which cause great economic losses (Gaur and Pankaj 2010). Narasimhamurthy et al. (2017a, b) reported that application of *P. fluorescens* and *T. harzianum* significantly reduced the population of *M. graminicola* and enhanced the plant height, root length, and maximum grain yield.

11.6 Combinatorial Effect of *Trichoderma* Species with Other Strategies in the Nematode Management

Application of *Trichoderma viride* plus organic additives exhibited surprising results in terms of plant health amelioration of different crops especially those growing under biotic stresses (Chang et al. 1986; Ansari and Mahmood 2019a). Soil application of *P. fluorescens* and *T. viride* singly or jointly controlled the nematode population and enhanced the mulberry leaf yield and nutritional standards (Muthulakshmi et al. 2010). Narasimhamurthy et al. (2017a, b) reported that application of *T. viride* (at 20 g/m^2) plus carbofuran effectively managed the rice root-knot nematode *Meloidogyne graminicola* leading to enhanced plant growth and yield parameters (height, root length, grain yield).

11.7 Conclusions and Future Prospects

Phytonematode infestations to various commercial crops are very common nowadays due to a large number of factors including climate change. Different important commercial crops have significantly reduced the yield remunerations due to poor yield growing under nematode-influenced fields. A wide number of phytoparasitic nematodes have been encountered that are associated with/infest commercial crops. In order to manage these pests effectively and in an eco-friendly manner, biocontrol is the appropriate alternative which could be considered in time ahead. *Trichoderma* species are the most widespread fungi in a wide array of agroecosystems and can easily be isolated from decaying wood, soil, and other forms of organic additives. *Trichoderma* spp. are important antagonists which are currently used in the management of plant pathogens including phytonematodes. However, their performance significantly varies due to an uncountable number of reasons. Application of *Trichoderma* exhibited poor population of phytonematodes, galling, and female fecundity leading to ameliorated plant health. Some species of *Trichoderma* are known as parasites of nematodes that also secrete metabolites which can paralyze or kill infective stages of the nematodes. Some compounds of metabolites and their mechanisms of action against plant-parasitic nematodes are identified; however, others need further investigation. Combination of *Trichoderma* species with other strategies would serve as a promising tool for nematode management in the upcoming era. Accurate identification of potential candidate of *Trichoderma* spp. is the need of the hour as the appropriate strain will exert better results. Accurate identification of *Trichoderma* spp. can be done by using some advanced molecular techniques. Current understanding of *Trichoderma* genome may also reflect a new road map of research to the researchers.

References

Abdelrafaa OAM, Ahmed EHM, Haroun NE, Dawabah AAM, Al-Yahya FA, Lafi HA (2018) Effect of the introduction time of *Trichoderma harzianum* into soil on its biocontrol potential against *Meloidogyne javanica* on tomato plants under greenhouse conditions. Int J Curr Microbiol App Sci 7(5):1836–1843

Abd-Elgawad MM, Askary TH (2020) Factors affecting success of biological agents used in controlling the plant-parasitic nematodes. Egyptian Journal of Biological Pest Control 30 (1):1–11

Abo-Elyousr KA, Zakaullah K, Award ME, Abedel-Moneim MF (2010) Evaluation of plant extracts and *Pseudomonas* spp. for control of root-knot nematode, *Meloidogyne incognita* on tomato. Nematropica 40:289–299

Ahmed M (2008) Management of Fusarium wilt of tomato by soil amendment with *Trichoderma koningii* and a white sterile fungus. Indian Journal of Research 5:35–38

Ahmed, SMM (2010). Studies on the impact of certain soil fungi in controlling root-knot nematodes. PhD. Thesis, Mansoura Univ. Faculty Agri., p 100

Akhtar M (2000) Approaches to biological control of nematode pests by natural products and enemies. J Crop Prod 3:367–395

Al Kader MAA (2008) In vitro studies on nematode interaction with their antagonistic fungi in the rhizosphere of various plants. Ph.D. Thesis 58. Albert-Ludwigs-Universität, Germany

Al-Hazmi AS, Dawabah AAM, Al-Nadhari SN, Al-Yahya FA (2017) Comparative efficacy of different approaches to managing *Meloidogyne incognita* on green bean. Saudi J Biol Sci 24:149–154

Al-Hazmia AS, Al-Yahyaa FA, AbdelRafaab OA, Lafia AH (2019) Effects of humic acid, *Trichoderma harzianum* and *Paecilomyces lilacinus* on *Meloidogyne javanica*. Int J Agric Environ Biores 4(1):61–74. www.ijaeb.org, ISSN: 2456-8643

Altomare C, Norvell WA, Borjkman T, Harman GE (1999) Solubilization of phosphates and micronutrients by the plant growth promoting and biocontrol fungus *Trichoderma harzianum* Rifai 1295–22. Appl Environ Microbiol 65:2926–2933

Amin AW, Mostafa FAM (2000) Management of *Meloidogyne incognita* infecting sunflower by integration of *Glomus mosseae* with *Trichoderma viride*, *T. harzianum* and *Arthrobotrys oligospora*. Egypt J Agronematol 4(1&2):21–30

Aminuzzaman FM, Xie HY, Duan WJ, Sun BD, Liu XZ (2013) Isolation of nematophagous fungi from eggs and females of *Meloidogyne* spp. and evaluation of their biological control potential. Biocontrol Sci Tech 23:170–182

Aminuzzaman FM, Jahan SN, Shammi J, Mitu AI, Liu XZ (2018) Isolation and screening of fungi associated with eggs and females of root-knot nematodes and their biocontrol potential against

Meloidogyne incognita in Bangladesh. Arch Phytopathol Plant Protect 51:288. https://doi.org/10.1080/03235408.2018.1472359

Anand S, Reddy J (2009) Bio-control potential of *Trichoderma* sp. against plant pathogens. Int J Agric Sci 1(2):30–39

Ansari RA, Khan TA (2012a) Parasitic association of root-knot nematode, *Meloidogyne incognita* on guava. e-J Sci Technol 5:65–67

Ansari RA, Khan TA (2012b) Diversity and community structure of phytonematodes associated with guava in and around Aligarh, Uttar Pradesh, India. Trends Biosci 5(3):202–204

Ansari RA, Mahmood I (2017a) Optimization of organic and bio-organic fertilizers on soil properties and growth of pigeon pea. Sci Hortic 226:1–9

Ansari RA, Mahmood I (2017b) Determination of disease incidence caused by *Meloidogyne* spp. and or *Fusarium udum* on pigeonpea in Aligarh district: a survey. Trends Biosci 10 (24):5239–5243

Ansari RA, Mahmood I (2019a) Plant health under biotic stress: volume 2: microbial interactions. Springer, Singapore. https://doi.org/10.1007/978-981-13-6040-4

Ansari RA, Mahmood I (2019b) Plant health under biotic stress: volume 1: organic strategies. Springer, Singapore. https://doi.org/10.1007/978-981-13-6043-5

Ansari RA, Rizvi R, Sumbul A, Mahmood I (2017a) PGPR: current vogue in sustainable crop production. In: Kumar V, Kumar M, Sharma S, Prasad R (eds) Probiotics and plant health. Springer, Singapore, pp 455–472

Ansari RA, Mahmood I, Rizvi R, Sumbul A (2017b) Siderophores: augmentation of soil health and crop productivity. In: Kumar V, Kumar M, Sharma S, Prasad R (eds) Probiotics in agroecosystem. Springer, Singapore, pp 291–312

Ansari RA, Sumbul A, Rizvi R, Mahmood I (2019) Organic soil amendments: potential tool for soil and plant health management. In: Ansari RA, Mahmood I (eds) Plant health under biotic stress. Springer, Singapore, pp 1–35

Arevalo J, Hidalgo-Diaz L, Martins I, Souza JF, Castro JMC, Carneiro RMDG, Tigano MS (2009) Cultural and morphological characterization of *Pochonia chlamydosporia* and *Lecanicillium psalliotae* isolated from *Meloidogyne mayaguensis* eggs in Brazil. Trop Plant Pathol 34:158–163

Askary TH, Martinelli PRP (2015) Biocontrol agents of phytonematodes. CAB International, Wallingford, p 470

Askew MF (2001) The economic importance of the potato. In: Loebenstein G, Berger PH, Brunt AA, Lawson RH (eds) Virus and virus-like diseases of potatoes and production of seed-potatoes. Springer, Dordrecht

Baharvand A, Shahbazi S, Afsharmanesh H, Ebrahimi MA, Askari H (2015) The enhancement of chitinase enzyme production in *Trichoderma harzianum* by induced mutation by γ-irradiation. The 4th National Congress on nuclear technology application in agricultural & natural resource sciences (19–20 May, 2015, Nuclear Agriculture Research School

Bairwa A, Venkatasalam EP, Umamaheswari R, Singh BP, Colleg J (2017) Effect of cultural practices on potato cyst nematode population dynamics and potato tuber yield. Indian J Hortic 74(1):91–96

Bais HP, Weir TL, Perry LG, Gilroy S, Vivanco JM (2006) The role of root exudates in rhizosphere interactions with plants and other organisms. Annu Rev Plant Biol 57:233–266

Berg G, Zachow C, Lottmann J, Gotz M, Costa R, Smalla K (2005) Impact of plant species and site on rhizosphere associated fungi antagonistic to *Verticillium dahliae* Kleb. Appl Environ Microbiol 71:4203–4213

Bernard EC, Keyserling ML (1985) Reproduction of root knot, lesion, spiral, and soybean cyst nematodes on sunflower. Plant Dis 69:103–105

Bokhary FA (2009) Efficacy of some *Trichoderma* species in the control of *Rotylenchulus reniformis* and *Meloidogyne javanica*. Arch Phytopathol Plant Protect 42(4):361–369

Bond JP, McGawley EC, Hoy JW (2000) Distribution of plant-parasitic nematodes on sugarcane in Louisiana and efficacy of Nematicides. Suppl J Nematol 32(4S):493–501

Carvalho DDC, Mello SCM, Lobo Júnior M, Silva MC (2011) Control of *Fusarium oxysporum* f. sp phaseoli in vitro and on seeds and growth promotion of common bean in early stages by *Trichoderma harzianum*. Trop Plant Pathol 36(1):028–034

Chang YC, Baker R, Kleifeld O, Chet I (1986) Increased growth of plants in the presence of the biological control agent. *Trichoderma harzianum*. Plant Dis 70:145–148

Cheng P, Liu B, Su Y, Hu Y, Hong Y, Yi X, Chen L, Su S, Chu SC, Chen N, Xiong X (2017) Genomics insights into different cellobiose hydrolysis activities in two *Trichoderma hamatum* strains. Microb Cell Factories 16:63

Chet L (1987) *Trichoderma*-application, mode of action, and potential as a biocontrol agent of soilborne plant pathogenic fungi. In: Chet I (ed) Innovative approaches to plant disease control. John Wiley & Sons, New York, pp 147–160

Contina JB, Dandurand LM, Knudsen GR (2017) Use of GFP-tagged *Trichoderma harzianum* as a tool to study the biological control of the potato cyst nematode *Globodera pallida*. Appl Soil Ecol 115:31–37

Contreras-Cornejo HA, Macías-Rodríguez L, del Val E, Larsen J (2016) Ecological functions of *Trichoderma* spp. and their secondary metabolites in the rhizosphere: interactions with plants. FEMS Microbiol Ecol 92(4):1–17. https://doi.org/10.1093/femsec/fiw036

Crnobarac J, Skoric D, Dusanic N, Miklic V (2006) Economic importance, biological properties, cultivars and production practices of sunflower. Biljni lekar 34:285–298

Dababat AA, Sikora RA (2006) Use of *Trichoderma harzianum* and *Trichoderma viride* for the biological control of *Meloidogyne incognita* on tomato. Jordan J Agric Sci 3:953–961

Dawar S, Sattar A, Zaki MJ (2008) Seed dressing with biocontrol agents and Nematicides for the control of root knot nematode on sunflower and okra. Pak J Bot 40(6):2683–2691

Dong JY, Zhao ZX, Cai L, Liu SQ, Zhang HR, Duan M, Zhang KQ (2004) Nematicidal effect of freshwater fungal culture against the pine-wood nematode, *Bursaphelenchus xylophilus*. Fungal Divers 15:125–135

Doni F, Al-Shorgani NKN, Tibin EMM, Abuelhassan NN, Anizan I, CheRadziah CMZ, Wan Mohtar WY (2013) Microbial involvement in growth of paddy. Curr Res J Biol Sci 5 (6):285–290

Elad Y, Freeman S (2002) Biological control of fungal plant pathogens. In: Esser K, Benett JW (eds) The mycota A comprehensive treatise on fungi as experimental systems for basic and applied research. Springer, Germany, pp 92–109

Elhady A, Heuer H, Hallmann J (2018) Plant parasitic nematodes on soybean in expanding production areas of temperate regions. J Plant Dis Protect 125:567. https://doi.org/10.1007/s41348-018-0188-y

El-Sherif AG, Ismail AFA (2009) Integrated management of *Meloidogyne incognita* infecting soybean by certain organic amendments, *Bacillus thuringiensis*, *Trichoderma harzianum* and oxamyl with reference to NPK and total chlorophyll status. Plant Pathol J 8(4):159–164

El-Sherif AG, Ismail AFA (2010) Integrated management of *Meloidogyne incognita* infecting sunflower by certain organic amendments, *Bacillus thuringiensis* and oxamyl with reference to NPK and total chlorophyll status. Egypt J Agronematol 9(1):40–51

França DVC, Kupper KC, Magri MMR, Gomes TM, Rossi F (2017) Trichoderma spp. isolates with potential of phosphate solubilization and growth promotion in cherry tomato. Pesq Agropec Trop 47(4):360–368. e-ISSN: 1983-4063. www.agro.ufg.br/pat

Freitas MA, Pedrosa EMR, Mariano RLR, Maranhão SRVL (2012) Screening *Trichoderma* spp. as potential agents for biocontrol of *Meloidogyne incognita* in sugarcane. Nematropica 42:115–122

Gajera HP, Bambharolia RP, Patel SV, Khatrani TJ, Goalkiya BA (2012) Antagonism of *Trichoderma* spp. against *Macrophomina phaseolina*: evaluation of coiling and cell wall degrading enzymatic activities. J Plant Pathol Microb 3:7

Gaur HS, Pankaj (2010) Root-knot nematode infestation in rice. In: M.R. Khan, M.S. Jairajpuri (Eds.), Nematode infestations, Part I: Food Crop, NASI, pp. 72–90

Golzari H, Panjehkeh N, Ahmadzadeh M, Salari M, Sedaghatikhoravi E (2011) Elucidating the parasitic capabilities of *Trichoderma* against *Meloidogyne javanica* on tomato. Insight Plant Dis 1:12–19

Goswami BK, Mittal A (2004) Management of root-knot nematode infecting tomato by *Trichoderma viride* and *Paecilomyces lilacinus*. Indian Phytopathol 57:235–236

Goswami J, Pandey RK, Tewari JP, Goswami BK (2008) Management of Root-Knot Nematode, *Meloidogyne incognita*, in tomato with two *Trichoderma* species. J environ Sci Hlth 43:237–240

Haggag WM, Amin AW (2001) Efficiency of Trichoderma species in control of Fusarium-rot, root knot and reniform nematodes disease complex on sunflower. Pak J Biol Sci 4:314–318

Hallman J, Davies KG, Sikora R (2009) Biological control using microbial pathogens, endophytes and antagonists. In: Perry RN, Moens M, Starr JL (eds) Root-knot nematodes. CAB International, Wallingford, pp 380–411

Harman GE (2000) Myths and dogmas of biocontrol: changes in perceptions derived from research on Trichoderma harzianum T-22. Plant Dis 84:377–391

Harman GE, Howell CR, Viterbo A, Chet I, Lorito M (2004) *Trichoderma* species opportunistic, avirulent plant symbionts. Nat Rev Microbiol 2:43–56

Hermosa R, Viterbo A, Chet I, Monte E (2012) Plant beneficial effects of *Trichoderma* and of its genes. Microbiology 158:17–25

Hermosa R, Rubio MB, Cardoza RE, Nicolas C, Monte E, Gutierrez S (2013) The contribution of *Trichoderma* to balancing the costs of plant growth and defense. Int Microbiol 16:69–80

Herrera-Parra E, Ramos-Zapata J, Cristobal-Alejo J, Tun-Suarez J, Reyes-Ramirez A (2018) Species of *Trichoderma* antagonistic to the root knot nematode (*Meloidogyne incognita*) in habanero pepper. FYTON 87:7–13. ISSN: 0031 9457

Hesamedin R, Mohammad AE (2014) Biological control of reniform nematode on sunflower through some bioagents. Int J Biol Pharm Res 5(10):786–789

Howell CR (2003) Mechanisms employed by *Trichoderma* species in the biological control of plant diseases: the history and evolution of current concepts. Plant Dis 87:4–10

Howell CR, Hanson LE, Stipanovic RD, Puckhaber LS (2000) Induction of terpenoid synthesis in cotton roots and control of *Rhizoctonia solani* by seed treatment with *Trichoderma virens*. Phytopathology 90:248–252

Hussain M, Zouhar M, Ryšnek P (2017) Effects of nematophagous fungi on viability of eggs and juveniles of *Meloidogyne incognita*. J Anim Plant Sci 27(1):252–258

Hyder S, Inam-ul-Haq M, Bibi S, Malik AH, Ghuffar S, Iqbal S (2017) Novel potential of *Trichoderma* Spp. as biocontrol agent. J Entomol Zool Stud 5(4):214–222

Izuogu NB, Abiri TO (2015) Efficacy of *Trichoderma harzianum* T22 as a biocontrol agent against root-knot nematode (*Meloidogyne incognita*) on some soybean varieties. Croat J Food Sci Technol 7(2):47–51

Javeed MT, Al-Hazmi AS, Molan YY (2016) Antagonistic effects of some indigenous isolates of *Trichoderma* spp. against *Meloidogyne javanica*. Pak J Nematol 34:183–191

Jin DE, Won SR, Dong PJ, Woong KR, Krishnan Y, Kim HB (2005) Effect of chitin compost and broth on biological control of *Meloidogyne Incognita* on tomato (*Lycopersicon esculentum* mill.). Nematology 7:125–132

Karanja NK (1988). Selecting rhizobium phaseoli strains for use with beans (*Phaseolus vulgaris* L.) in Kenya. Ph.D. Thesis, University of Reading, UK, p 221

Kashmiri MA, Adnan A, Butt BW (2006) Production, purification and partial characterization of lipase from *Trichoderma viride*. Afr J Biotech 5(10):878–882

Kashyap PL, Rai P, Srivastava AK, Kumar S (2017) Trichoderma for climate resilient agriculture. World J Microbiol Biotechnol 33(8):155

Khan MR, Haque Z (2011) Soil application of *Pseudomonas fluorescens* and *Trichoderma harzianum* reduces root-knot nematode, *Meloidogyne incognita*, on tobacco. Phytopathol Mediterr 50:257–266

Khan MR, Ashraf S, Rasool F, Salati KM, Mohiddin FA, Haque Z (2014) Field performance of *Trichoderma* species against wilt disease complex of chickpea caused by *Fusarium oxysporum* f. sp. ciceri and *Rhizoctonia solani*. Turk J Agric For 38:447–454

Khan M, Arabiat S, Chanda AK, Yan GP (2016) Sugar beet cyst nematode. North Dakota extension bulletin PP1788. North Dakota State Univ, Fargo, ND. 2 p

Khattak B, Saifullah, Hussain S, Ahmad M, Ali A, Junaid M, Khan IA, Khan TA, Hussain M (2018) Genetic relatedness among the indigenous isolates of *Trichoderma harzianum*, using RAPD and their nematocidal capabilities against *Meloidogyne javanica*. Sarhad J Agric 34 (2):486–493

Kirigaa AW, Haukeland S, Kariukia GM, Coynec DL, Beekd NV (2018) Effect of *Trichoderma* spp. and *Purpureocillium lilacinum* on *Meloidogyne javanica* in commercial pineapple production in Kenya. Biol Control 119(2018):27–32

LaMondia JA, Taylor GS (1987) Influence of the tobacco cyst nematode *Globodera tabacum* on Fusarium wilt of Connecticut broadleaf tobacco. Plant Dis 71:1129–1132

Leonetti P, Costanz A, Zonno MC, Molinari S, Altomare C (2014) How fungi interact with nematode to activate the plant defense response to tomato plants. Comm Appl Biol Sci, Ghent University 79(3):357–363

Lima SO, Mattos VS, Silva ES, Carvalho AS, Teixeira RA, Silva JC, Correa R (2018) Nematodes affecting potato and sustainable practices for their management. Potato: From Incas to All Over the World. p. 107

Maareg MF, Badr ST (2000) The effects of certain biocontrol organisms, oxamyl and their combinations of *Meloidogyne javanica* infecting sugarbeet. Egypt J Agronematol 4 (1/2):95–104

Maareg MF, Hassanein MA, Hussein MY (2003) Inhibitory effect of certain soil fungal filtrates on *Meloidogyne incognita* nematode on sugar beet. Ann Agric Sci, Moshtohor, Egypt 41 (2):987–998

Mascarin GM, Bonfim Junior MF, Filho JA (2012) *Trichoderma harzianum* reduces population of *Meloidogyne incognita* in cucumber plants under greenhouse conditions. J Entomol Nematol 4 (6):54–57

Meena AK, Meena AK (2015) Characterization and antagonistic effect of isolated *Trichoderma* sp. against pathogens under clusterbean (*Cyamopsis tetragonoloba* L.). Indian J Agric Res 50 (3):249–253

Mehta A, Bodh U, Gupta R (2017) Fungal lipase: a review. J Biotech Res 8:58–77

Meyer SLF, Roberts DP, Chitwood DJ, Carta LK, Lumsden RD (2001) Application of *Burkholderia cepacia* and *Trichoderma virens*, alone and in combinations, against *Meloidogyne incognita* on bell pepper. Nematropica 31:75–86

Migunova V, Sasanelli N, Kurakov A (2018) Effect of microscopic fungi on larval mortality of the root-knot nematodes *Meloidogyne incognita* and *Meloidogyne javanica*. Biol Integr Contr Plant Path 133:27–31

Morton CO, Hisch PR, Kerry BR (2004) Infection of plant parasitic nematodes by nematoghagous fungi—a review of the application of molecular biology to understand infection processes and to improve biological control. Nematology 6:161–170

Motha KF, Abeysekara R, Kottearachchi NS (2010) Effect of biological agents and botanicals in controlling root-knot nematodes, *Meloidogyne* spp., in *Nicotiana tabacum*. Trop Agric Res Exten 13(1). https://doi.org/10.4038/tare.v13i1.3129

Mukherjee PK, Horwitz BA, Herrera-Estrella A, Schmoll M, Kenerley CM (2013) *Trichoderma* research in the genome era. Annu Rev Phytopathol 51:105–129

Mukhtar T (2018) Management of root-knot nematode, *Meloidogyne incognita*, in tomato with two *Trichoderma* species. Pak J Zool 50(4):1589–1592

Muthulakshmi M, Devrajan K, Jonathan EI (2010) Biocontrol of root knot nematode, *Meloidogyne incognita* (Kofoid and white) Chitwood in mulberry (*Morus alba* L.). J. Biopesticides 3 (2):479–482

Nagachandrabose S (2018) Liquid bioformulations for the management of root-knot nematode, *Meloidogyne hapla* that infects carrot. Crop Prot 114:155–161

Naher L, Tan SG, Yusuf K, Ho CL, Siddiquee S (2012) Activities of chitinase enzymes in the oil palm (*Elaeis guineensis* Jacq.) in interactions with pathogenic and non-pathogenic fungi. POJ 5 (4):333–336

Nannipieri P, Ascher J, Ceccherini MT, Landi L, Pietramellara G, Renella G, Valori F (2007) Microbial diversity and microbial activity in the rhizosphere. Suelo (Argentina) 25:89–97

Narasimhamurthy HB, Ravindra H, Mukesh S (2017a) Management of rice root-knot nematode, *Meloidogyne graminicola*. Int J Pure Appl Biosci 5(1):268–276

Narasimhamurthy HB, Ravindra H, Sehgal M, Ekabote SD, Ganapathi (2017b) Bio-management of rice root-knot nematode (*Meloidogyne graminicola*). J Entomol Zool Stud 5(4):1433–1439

Naserinasab F, Sahebani N, Etebarian HR (2011) Biological control of *Meloidogyne javanica* by *Trichoderma harzianum* BI and salicylic acid on tomato. Afr J Food Sci 5(3):276–280

Oyedunmade EEA (2003) The efficacy of phorate (Thimet) used in nematode control in some biochemical component of seeds of treated soybeans (*Glycine max*). Biosci Res Commun 15 (13):191–193

Prasad GRG, Ravichandra NG, Narasimhamurthy TN, Kumar CHP, Yadahalli P (2014) Management of *Meloidogyne incognita* infecting carrot by using bioagents. J Biopest 7(2):144–150

Pusztahelyi T (2018) Chitin and chitin-related compounds in plant–fungal interactions. Mycology 9 (3):189–201

Qureshi SA, Ruqqia, Sultana V, ARA J, Ehteshamul-Haque S (2012) Nematicidal potential of culture filtrates of soil fungi associated with rhizosphere and rhizoplane of cultivated and wild plants. Pak J Bot 44(3):1041–1046

Rizvi R, Mahmood I, Ansari S (2018) Interaction between plant symbionts, bio-organic waste and antagonistic fungi in the management of *Meloidogyne incognita* infecting chickpea. J Saudi Soc Agric Sci 17(4):424–434

Saifullah, Khan NU (2014) Low temperature scanning electron microscopic studies on the interaction of *Globodera rostochiensis* Woll. and *Trichoderma harzianum* Rifai. Pak J Bot 46 (1):357–361

Saldajeno MGB, Naznin HA, Elsharkawy MM, Shimizu M, Hyakumachi M (2014) Enhanced resistance of plants to disease using *Trichoderma* spp. Biotechnol Biol Trichoderma:477–487. https://doi.org/10.1016/B978-0-444-59576-8.00035-7

Samuels GJ, Ismaiel A, Mulaw TB, Szakacs G, Druzhinina IS, Kubicek CP, Jaklitsch WM (2012) The *Longibrachiatum* clade of *Trichoderma*: a revision with new species. Fungal Divers 55:77–108

Sayed MA, Abdel-rahman TMA, Ragab AA, Abdellatif AAM (2019) Biocontrol of root-knot nematode *Meloidogyne incognita* by Chitinolytic *Trichoderma* spp. Egypt J Agronematol 18 (1):30–47

Sharf R, Hisamuddin (2016) Potential for biological control of nematode by *Trichoderma* spp. on growth yield of *Phaseolus vulgaris*. World J Pharm. Res. 5(6):1044–1064

Sharf R, Hisamuddin A, Akhtar A (2014) Management of root-knot disease in *Phaseolus vulgaris* using potassium fertilizer and biocontrol agents. J Plant Pathol Microb 5:242. https://doi.org/10.4172/2157-7471.1000242

Sharon E, Bar-Eyal M, Chet I, Herra-Estrella A, Kleifeld O, Spiegel Y (2001) Biological control of the root-knot nematode *Meloidogyne javanica* by *Trichoderma harzianum*. Phytopathology 91:687–693

Sharon E, Chet I, Viterbo A, Bar-Eyal M, Nagan H, Samuels GJ, Spiegel Y (2007) Parasitism of *Trichoderma* on *Meloidogyne javanica* and role of the gelatinous matrix. Eur J Plant Pathol 118:247–258

Sharon E, Chet I, Spiegel Y (2009) Improved attachment and parasitism of *Trichoderma* on *Meloidogyne javanica in vitro*. Eur J Plant Pathol 123(3):291–299

Sharon E, Chet I, Spiegel Y (2011) *Trichoderma* as a biological control agent. In: Davies K, Spiegel Y (eds) Biological control of plant-parasitic nematodes: building coherence between microbial ecology and molecular mechanisms, Progress in biological control. Springer, Netherlands. https://doi.org/10.1007/978-1-4020-9648-8_8

Shepherd JA, Barker KR (1990) Nematode parasites of tobacco. In: Luc M, Sikora RA, Bridge J (eds) Plant-parasitic nematodes in subtropical and tropical agriculture. CAB International, Wallingford, pp 493–517

Shrivastava M, Srivastava PC, D'Souza SF (2018) Phosphate-solubilizing microbes: diversity and phosphates solubilization mechanism. In: Meena VS (ed) Role of rhizospheric microbes in soil. Springer Nature, Singapore, pp 137–165

Siddiqui ZA, Mahmood I (1994) Interaction of *Meloidogyne javanica*, *Rotylenchulus reniformis*, *Fusarium oxysporum* f. sp. ciceri and *Bradyrhizobium japonicum* on the wilt disease complex of chickpea. Nematol Mediterr 22:135–140

Siddiqui ZA, Shaukat S (2003) Endophytic bacteria: prospects and opportunities for the biological control of plant-parasitic nematodes. Nematol Mediterr 31:111–120

Singh UB, Singh S, Malviya D, Chaurasia R, Imrani M, Rai A, Sharma A (2017) Harnessing biocontrol potential of *Trichoderma harzianum* for control of *Meloidogyne incognita* in tomato. Indian Phytopath 70(3):331–335

Spiegel Y, Sharon E, Chet I (2005) Mechanisms and improved biocontrol of the root-knot nematodes by *Trichoderma* spp. Acta Hortic 698:225–228

Sreedevi B, Charitha M, Saigopal DVR (2011) Induction of defense enzymes in Trichoderma harzianum treated groundnut plants against *Macrophomina phaseolina*. J Biol Control 25 (1):33–39

Strakowska J, Błaszczyk L, Chełkowski J (2014) The significance of cellulolytic enzymes produced by *Trichoderma* in opportunistic lifestyle of this fungus. J Basic Microbiol 54:1–12

Sivan A, Chet I (1989a) The possible role of competition between *Trichoderma harzianum* and *Fusarium oxysporum* on rhizosphere colonization. Phytopathology 79(2). https://doi.org/10.1094/Phyto-79-198

Sivan A, Chet I (1989b) Degradation of fungal cell walls by lytic enzymes of Trichoderma harzianum. Microbiology 135(3):675–682

Umamaheswari R, Somasekhar N, Manorama K, Joseph TA (2012) Ecofriendly management of potato cyst nematodes in the Nilgiris of Tamil Nadu. Potato J 39(2):185–190

Van Damme V, Hoedekie A, Viaene N (2005) Long-term efficacy of *Pochonia chlamydosporia* for management of *Meloidogyne javanica* in glasshouse crops. Nematology 7:727–736

Viterbo A, Horwitz BA (2010) Mycoparasitism. In: Borkovich KA, Ebbole DJ (eds) Cellular and molecular biology of filamentous fungi, vol 42. American Society for Microbiology, Washington, pp 676–693

Walpola BC, Yoon MH (2012) Prospectus of phosphate solubilizing microorganisms and phosphorus availability in agricultural soils: a review. Afr J Microbiol Res 6(37):6600–6605

Weller DM (1988) Biological control of soilborne plant pathogens in the rhizosphere with bacteria. Annu Rev Phytopathol 26:379–407

Weller DM, Raaijimakers JM, Gardener BBM, Thomashow LS (2002) Microbial population responsible for specific soil suppressiveness to plant pathogens. Annu Rev Phytopathol 40:309–348

Whipps JM (1997) Interactions between fungi and plant pathogens in soil and rhizosphere. In: Gange AC, Brown VK (eds) Multitrophic interactions in terrestrial systems. Blackwell Science, Oxford, pp 47–63

Whipps JM (2001) Microbial interactions and biocontrol in the rhizosphere. J Expt Bot 52:487–511

Windham GL (1989) Effects of *Trichoderma* spp. on maize growth and Meloidogyne arenaria reproduction. Plant Dis 73(6). https://doi.org/10.1094/PD-73-0493

Yan G, Baidoo R (2018) Plant parasitic nematodes of North Dakota and South Dakota, In plant parasitic nematodes in sustainable agriculture of North America (pp. 181–207). Springer, Cham

Yedidia I, Benhamou N, Chet I (1999) Induction of defense response in cucumber plants (*Cucumis sativus* L.) by the biocontrol agent *Trichoderma harzianum*. Appl Environ Microbiol 65:1061–1070

Chapter 12
Role of Organic Additives in the Sustainable Management of Phytoparasitic Nematodes

Thangjam Sunita Devi, Debanand Das, Rizwan Ali Ansari, Rose Rizvi, Aisha Sumbul, and Irshad Mahmood

Abstract Phytoparasitic nematodes are recognized as one of the important limiting factors for successful cultivation of crops. Compared to other pathogens, plant parasitic nematodes are sometimes difficult to control because they live in the soil and possess a great ability to attack almost every part of the plant. They can be effectively managed by application of chemical nematicides, but at the cost of environmental pollution, toxicity, as well as residual hazards and effect on food chain. Plant protectionists have been evaluating many nematode management strategies like soil solarization, use of fungal and bacterial biological control agents, trap as well as antagonistic crops, and GAP for management of plant parasitic nematodes in crops. Application of organic amendments therefore is a successful alternative strategy for managing plant parasitic nematodes in soil. A good number of organic amendments such as vermicompost, nematicidal plants, farm yard manure, protein-rich wastes, and animal and green manures have been effective in the management of

T. S. Devi (✉)
Department of Nematology, Assam Agricultural University, Jorhat, Assam, India

Department of Plant Pathology, Pandit Deen Dayal Upadhyay Institute of Agricultural Sciences, Bishnupur District, Utlou, Manipur, India

D. Das
Department of Nematology, Assam Agricultural University, Jorhat, Assam, India

R. A. Ansari
Section of Plant Pathology and Nematology, Department of Botany, Aligarh Muslim University, Aligarh, India

Present Address: Department of Plant Protection, Faculty of Agricultural Sciences, Aligarh Muslim University, Aligarh, India

R. Rizvi · I. Mahmood
Section of Plant Pathology and Nematology, Department of Botany, Aligarh Muslim University, Aligarh, India

A. Sumbul
Section of Plant Pathology and Nematology, Department of Botany, Aligarh Muslim University, Aligarh, India

Present Address: Department of Biosciences, Jamia Millia Islamia, New Delhi, India

R. A. Ansari et al. (eds.), *Management of Phytonematodes: Recent Advances and Future Challenges*, https://doi.org/10.1007/978-981-15-4087-5_12

plant parasitic nematodes infesting a wide range of agricultural crops. The use of organic amendments can reduce disease caused by nematodes directly by affecting soil properties and indirectly by improving plant growth, changing root physiology, and enhancing populations of antagonistic microorganisms and productivity. Several nematotoxic chemicals have been reported to be released during decomposition of the organic amendments in soil which are detrimental to nematodes. Also, various factors such as soil type, agronomic practices, rate of application of organic additives, etc. are greatly implicated in bringing down the population below economic threshold. Application of organic amendment improves the soil physicochemical properties, alters the C/N ratio, adds different micronutrients, and thereby improves the plant growth, which ultimately help the plant to withstand nematode attack. We have discussed different types of organic amendments that are used during the management process. We also give emphasis on how they can reduce the nematode population. Due to the chemical released during the process, the nematode population was found to be less harmful to the crop. It has been discussed how this organic matter is important to soil health and the environment of the soil rhizosphere through changes which make the nematodes to thrive less in the soil and how it increases the soil's properties and the yield of the crops. Many reports indicated that the performance of organic amendments as bionematicides is not satisfactory. In the present chapter, an emphasis has been given to discuss the current understanding pertaining to application of organic additives in the management of phytoparasitic nematodes. Recent progress in plant nematology in terms of their management through organic approaches has been put forth.

Keywords Plant parasitic nematodes · Organic amendments · Plant growth · Soil physicochemical properties · C/N ratio

12.1 Introduction

Phytopathogenic nematodes are considered as major constraints for successful cultivation of crops throughout the world (Ansari and Khan 2012a, 2012b; Ansari and Mahmood 2017b). Nematodes, commonly known as thread worms, round worms, or eel worms, mostly dwell in soil with a thin film of water. Most of them attack the root system and derive their food from the plant either as ecto- or endoparasite. Few of them are aerial feeders, attacking the leaves, buds, bulbs, or inflorescence causing substantial damage to the crops. Being ubiquitous, they occupy all possible types of habitats and climatic conditions. However, crops grown in tropical and subtropical countries are highly prone to the attack by these tiny creatures. Plant parasitic nematodes are responsible for 12.3% average annual loss of the world's major crops. Yield loss mainly depends upon the crop, nematode species, population of nematode, and climatic condition. One estimate reveals that yield loss to different crops due to plant parasitic nematodes ranges from 5% to 15%, and in some particular crops nematodes become major limiting factors in production of such crops (Stirling 2013). Phytopathogenic nematodes cause estimated annual crop losses of $8 billion in the USA and $78 billion worldwide (Barker et al. 1998). Phytopathogenic nematodes cause a significant damage on 24 vegetable crops in the USA of about 11%. Van Berkum and Seshadri (1970) were the first to have calculated monetary losses to crops caused by nematode parasites in India. They estimated annual losses of $10 million on wheat due to ear cockle disease caused by ear cockle nematode, *Anguina tritici*, $3 million on coffee due to *Pratylenchus coffeae*, and $8 million due to "Molya disease" caused by *Heterodera avenae* in Rajasthan province alone. Jain et al. (2007) estimated that about 16.67, 18.20, 21.35, 14.10, 27.21, and 10.54% yield loss occurs in brinjal, cucurbits, jute, okra, tomato, and rice, respectively, due to plant parasitic nematodes in India. Plant parasitic nematodes can be effectively managed by application of various chemical pesticides. But, their indiscriminate use may give rise to many serious problems like soil and groundwater pollution, food contamination, adverse effect on non-target organisms, development of pesticidal resistance, etc. These demerits of chemical pesticides lead the nematologist to search for alternative methods for management of these noxious pests of crops. Present-day agriculture is focused on going back to nature by adopting organic modes of cultivation practices, and it will be not wrong to say that use of organic amendments as well as botanical biopesticides is the most economic and ecologically approachable method. Besides being inexpensive and easy to apply in comparison to the chemical pesticides, this approach will be environmentally friendly and suitable for human health as well as soil health (Neher 2001; Adegbite and Adesiyan 2005). Moreover, use of organic amendments may be helpful in managing the green wastes that are generated due to human settlements and agro-industrial setups (Ansari et al. 2019; Ansari and Mahmood 2019a, 2019b). The green wastes contain many essential plant nutrients such as nitrogen, phosphorus, and calcium and other elements such as zinc, copper, and magnesium (Tester 1990), which in turn are returned back to the soil while being used as organic amendments. While reviewing the role of organic amendments on the management of nematodes, Muller and Gooch (1982) acknowledged that though the concept of applying organic amendments was developed in developed countries, more work on this line was done in the developing countries like India. The positive effect of organic amendment in combating plant parasitic nematodes in different crops at different agro-ecological situations has been documented with some inconsistent results (Szczech et al. 1993; Mcsorley and Gallaher 1995; Akhtar and Malik 2000; Hassan et al. 2001). Efficacy of organic amendments against nematodes may depend on the composting material, method of composting, time of application, season, dose, and nematode species present (Rivera and Aballay 2008; Renco et al. 2007, 2010). However, contradictory reports are also available such as Gergon et al. (2001) which did not observe any suppressive effect while using organic amendments on the population of rice root-knot nematode *M. graminicola*.

12.2 Types of Organic Amendments

Organic amendments are of two types: plant origin and animal origin. Different types of organic amendments that help in the management of phytonematode population in soil will be discussed below.

12.2.1 Vermicompost

Vermicompost is a product prepared by composting which is an accelerated biodegradation process of organic materials using earthworms and various organisms through non-thermophilic decomposition. It contains higher concentration of nutrients and is basically used as organic fertilizer. Vermicompost-treated plants show accelerated growth which might be due to the presence of growth hormones, micronutrients such as carotenoids, flavones, and phenolic compounds in vermicompost (Kumar et al. 2011). It was seen that vermicompost not only exhibited more potentiality as nematicide but probably also acted as better growth promoter. Vermicompost can also exert some beneficial effects in the management of plant parasitic nematodes. It was seen that the introduction of vermicompost at 1 kg m^2 considerably impaired the reproduction of *Meloidogyne incognita* in tobacco plants (Swathi et al. 1998). Morra et al. (1998) reported that application of solid vermicompost effectively suppressed the population as well as attack of *Meloidogyne incognita*. Arancon et al. (2002, 2003) conducted a series of field experiments on suppression of plant parasitic nematodes by application of solid vermicompost in tomato, pepper, strawberries, and grapes. They revealed that application of solid vermicompost ranging from 2 to 8 kg per hectare can significantly reduce the population of plant parasitic nematodes. They also noticed an increased population of fungivorous and bacteriovorous nematodes in solid vermicompost-treated plots. However, Mishra et al. (2017) observed that suppression of root-knot nematode (*Meloidogyne* spp.) by vermicompost tea has been inconsistent. Application of vermicompost tea reduced penetration and hatching of nematode but not reproduction over a period of time. Effectiveness of vermicompost for suppression of plant parasitic nematodes such as root-knot nematode (*Meloidogyne* spp.) in particular has been significantly observed (Ribeiro et al. 1998; D'Addabbo et al. 2011; Mahalik and Sahu 2018). In some reports, vermicompost does not show a positive result in reducing the population of plant parasitic nematodes. For example, Szczech et al. (1993) observed that vermicompost did not reduce *Heterodera schachtii*. Vermicompost exhibited no suppressive effects on the number of *M. hapla* galls infesting cabbage and tomato roots (Kimpinski et al. 2003). Various mechanisms might be implicated in the suppression of phytoparasitic nematodes such as decomposition of the compost into the soil and ammonia production, stimulation of soil microbial biomass, and release of biocidal substances having nematicidal activity (Oka and Yermiyahu 2002).

12.2.2 Organic Amendments with Botanicals

Plant parts have been effectively used as organic amendments for bringing down the population of phytopathogenic nematodes below economic threshold. For the management of plant parasitic nematodes, a good number of plant parts exhibited nematicidal properties. Some of the early records in this direction included used of chopped pineapple leaves against root-knot nematode. Plant parts of *Melilotus alba* var. *annua* and Sorghum vulgare are effective in reducing the population of *Meloidogyne* spp. (Patel and Desai 1964). Fifty percent of gall number was found to be reduced in tomato formed by *Meloidogyne javanica* when chopped leaves of *karanj* are applied in soil (Singh 1965). Leaves of *Cassia fistula*, *Crotalaria juncia*, and *Sesbania aculeata* at 5–10% (w/w) were effective in reducing infestation of root-knot nematode in tomato and okra. Applications of powdered form of plant parts of *Crotalaria*, *Tagetes*, *Parthenium hysterophorus*, *Cannabis sativa*, *Calotropis procera*, *Anagallis arvensis*, *Catharanthus roseus*, *Euphorbia hirta*, *Canna indica*, *Ricinus communis*, *Clerodendrum inerme*, and *Lantana camara* were effective against *Meloidogyne hapla*, *M. incognita*, *Helicotylenchus dihystera*, *Tylenchorhynchus annulatus*, *Pratylenchus zeae*, and *Rotylenchulus reniformis* (Thakur 2014; Alsayed et al. 1992). Parts of different plants like *Datura* leaves (Kaliram and Gupta 1982), castor leaves (Dutt and Bhatti 1986), *Argemone mexicana*, *Eucalyptus globolus*, *Datura metel*, *Phyllanthus niruri* (Goswami and Vijayalakshmi 1986), *Polygonum hydropiper*, *Ageratum* sp., *Mikania* sp., rice straw, and water hyacinth (Das et al. 1999) were found effective against different species of root-knot nematode, *Meloidogyne* spp. Siddiqui et al. (1987, 1992) reported that incorporation of chopped shoots of latex-bearing plants (*Ficus elastica*) significantly suppressed the population build-up of *Rotylenchulus reniformis Tylenchorhynchus brassicae*, and *Meloidogyne incognita* as well as significantly increased the plant growth parameters of tomato and brinjal. Incorporation of plant parts in soil results in the release of toxic volatile substances which may be lethal to phytoparasitic nematodes. Incorporation of Brassicaceae plant parts into soil for suppression of different plant parasitic nematodes as well as increasing soil fertility status and physicochemical properties had been attempted in many parts of the world with significant success (Mojtahedi et al. 1993; Walker and Morey 1999; Ploeg and Stapleton 2001; Stirling and Stirling 2003; Zasada and Ferris 2004; Rahman and Somers 2005). It had been observed that plant members belonging to the family Brassicaceae contains secondary metabolites called glucosinolates, which might be responsible for enhancing the defense mechanisms of plants against nematodes. The breakdown products of glucosinolates were nitriles, thiocyanates, and isothiocyanates (Cole 1976; Fenwick et al. 1983). Among these breakdown products, isothiocyanates (ITCs) had the biocidal properties (Kirkegaard and Sarwar 1998; Harvey et al. 2002; Bello et al. 2004). However, in contradiction it was also reported that isothiocyanates released due to conversion of glucosinolates from the plant parts were very low and might not be predicted as a nematode suppressive component (Potter et al. 1998; Bending and Lincoln 1999; McLeod and Steele 1999;

Harvey et al. 2002; Morra and Kirkegaard 2002). But, Kirkegaard and Matthiessen (2004) confirmed that isothiocyanate was responsible for killing of nematodes in plant part-amended soil. Neem (*Azadirachta indica*) is another promising source of biopesticides of botanical origin. Application of neem plant parts is considered as one of the effective, inexpensive, and environmentally safe methods of nematode management. Both fresh and dried leaves were used as soil amendment. Singh and Sitaramaiah (1966) reported that incorporation of fresh neem leaves at 5–10% (w/w) of soil could effectively control *Meloidogyne javanica* on tomato. Similar reports had been published by Goswami and Vijayalakshmi (1987), Abolusoro and Oyedunmade (2005), and Singh and Patel (2015). Efficacy of neem plant parts against nematodes had also been reported by Mehta and Sundararaj (1995), Engunjobi and Larinde (1975), and Hussain et al. (2011). Yadav et al. (2018) observed that application of neem leaf at 30 g/kg of soil can effectively suppress the galling and final nematode population of *Meloidogyne incognita* on tomato. Therefore, the use of neem products stimulated and changed the physiology of plant cells and tissue to repel the nematode parasites. Various parts of the neem tree contain over 40 bitter principles which belong to diterpenoid, triterpenoid, limonoid, and flavonoid groups of natural products (Thakur et al. 1981). The most important so far known compounds are the azadirachtins. The other limonoids which have been found in traces are meliantriol, salannin, nimbin, and nimbidin. In the USA, neem pesticides are permitted on certain greenhouse and ornamental crops. Nematicidal properties of neem plants and their products as soil amendments were thoroughly reviewed by Akhtar (2000). Marigold (*Tagetes erecta*) was among one of the important plant additives which are still used rigorously for the management of phytopathogenic nematodes due to the presence of α-terthienyl in its roots (Uhlenbroek and Bijloo 1958). It showed nematicidal activity in 6-day tests in vitro against three nematodes, viz., *Ditylenchus dipsaci* (applied at 5 IJ.g/mL), *Anguina tritici* (applied at 0.5 IJ.g/mL), and *Globodera rostochiensis* (applied at 0.1–0.2 Ilg/mL). In 1959, Uhlenbroek and Bijloo reported the importance of 5-(3-buten-1-ynyl)-2,2′-bithienyl against the plant nematodes. The hydrogenated derivative (5-butyl-2,2′-bithienyl) was found to be highly nematotoxic against nematode species. In addition, Gommers and Voorin'tholt (1976) reported the application of 110 different Asteraceae against the populations of the root lesion nematode, *Pratylenchus penetrans*. In greenhouse, more than 40 species significantly suppressed the plant parasitic nematode population levels. Gommers (1981) and Gommers and Bakker (1988) thoroughly reviewed the nematicidal activity of *Tagetes* spp.

12.2.3 Organic Amendments with Products of Botanicals

Application of organic additives and their derived products has been extensively used in the management of plant parasitic nematodes on various important crops like tomato, eggplant, okra, etc. (Muller and Gooch 1982; Akhtar and Alam 1993).

Moreover, the fertilization effects of neem by-products are well known for a long time as they contain plant nutrients such as N (5.5–7.1%), P (1.1%), and K (1.5%). The suppressive efficacy of neem cake on nematode populations was often associated with the observation of better growth of plants. In field trials oil cakes and leaves of neem are generally applied at 110 kg N/ha. In many studies effective rates of organic additives were found to be 4–10 t/ha. Rodriguez-Kábana and Morgan-Jones (1987) reported that 150 kg N/ha is required to control the plant parasitic nematodes significantly. In oil cake-amended soil, the nematode population, galling, and egg mass production were reduced, while the growth of tomato was improved in a greenhouse experiment (Goswami and Vijayalakshmi 1987). Sarma et al. (1971), Gowda (1972), Bhattacharrya and Goswami (1987), Sarma (1989), Gaur and Mishra (1990), Tiyagi and Alam (1995), Ramakhrishna et al. (1997), and El-Sherif et al. (2008) critically evaluated the efficacy of oil cakes against nematodes and reported that amended soil with oil cakes effectively controlled the nematodes. The nematode control effects of neem products occur after incorporation into soil and during their decomposition, presumably due to the release of nematotoxic compounds (Stirling 1991). There are numerous reports on the usefulness of neem by-products in providing effective nematode control, but they are of limited availability in bulk quantities and this poses problems in satisfactory nematode control in field level. Nematode populations may have been affected by leached chemicals, either from a coating on the seeds or having being absorbed by the roots during bare root dip treatment into rhizosphere, which repelled or killed the juveniles that attacked the host's root. The presence of azadirachtins, phenols, fatty acids, and tannins at certain concentrations in neem products and in some amendments at high level has effects on nematode mortality. The decomposition products of neem include organic chemicals that curb nematode populations. It is known that many of these compounds such as the phenols, fatty acids, and water-soluble volatiles released during decomposition have nematicidal properties. It has been recorded that application of neem cake at 1000 kg/ha gave maximum yield of cucumber and increased the plant growth parameter (Devi 2016). It also has antifungal and antifeedant properties which immobilize the growth of juveniles and reduce the population resulting to better yield of the crop. It has ammonia and fatty acids which are released during degradation which help in the growth of the plant and protect it from phytonematodes Combined applications of oil cakes with biocontrol agents such as *Trichoderma viride*, *T. harzianum*, *Glomus fasciculatum*, *G. mosseae*, and *Paecilomyces lilacinus* were always effective in managing nematodes in soil (Mahanta and Phukan 2004; Borah and Phukan 2004; Borah et al. 2007).

12.2.4 Poultry Manure

It is manure which is used as fertilizer that is applied as bedding material in field and kitchen garden to improve the fertility of the soil and the crop. It rejuvenates the soil texture and the environment around it. It is one of the cheapest and very effective forms of fertilizer which gives good returns to the farmers. The low content of C/N ratio in poultry manure results in high suppression of nematode population. Its high content on soil increased the bacteriovorous nematode which reduced the juveniles. Due to high content of carbon in poultry manure, it enhanced the thriving communities of the decomposers and predators which make up the soil digestive system. It also supplies food to the trapping fungi. It was found that the high content of nitrogen in manure suppressed the infection of nematode, phosphorus increased the plant vigor and avoidance of diseases, and potassium encouraged root growth and increased plant resistance to diseases (Devi 2016). Application of poultry refuse in the management of root-knot nematodes infesting okra (Bari et al. 1999), brinjal (Bari et al. 2004; Ahamad et al. 1987), potato (Hossain et al. 1989), and jute (Mishra et al. 1987) has been performed.

12.2.5 Ashes

They are solid end products of fire which are easily available that are used by farmers to achieve good porosity in soil and are also a cheap source of potassium. Ashes like rice husk ash, rice hull ash, and banana pseudostem ash are important wastes of agriculture that are produced in large quantities. They are easily available to our farmers since most of the people feed on rice. Their application on plants resulted to improved growth and yield. Fewer people know the value of ashes on crops as they also work as insecticides. Sen and Dasgupta (1981) have tried using rice hull ash for the first time for managing the root-knot nematode infesting tomato. They have reported that rice hull ash significantly reduced the incidence of root-knot nematode (46–100%) in three sites of their experimentation and increased the yield of tomato. Addition of rice husk ash at 30 g/plant on cucumber increased the plant resistance which prevented the nematode attack (Devi 2016). It also mitigated the severity which reduced the nematode population. It increased the nutrients which enhanced the root development and overall increased the plant growth. Ashes contain more potassium which maintained the plant water balance and increased the plant resistance and virulence against the nematode. In acidic soil, ashes act as liming agents that help in the replenishment of some nutrients. These ashes serve as correctives of soil acidity which react faster than the conventional limestone. They also block the movement of juveniles in the soil giving them fewer opportunities to enter the roots resulting to better yield. Silica present in the rice husk ash can prevent the crop from the invasion of nematode in roots. The use of ashes in managing the nematode pest of crops will be an economically and environmentally safe option.

12.2.6 Farm Yard Manure

Farm yard manure is mostly derived from the feces of animals or farm slurry. It adds nutrients to the soil which can be used by soil organisms. The animal feces is mixed with straw and used as bedding material which is a good form of organic fertilizer. It is easily available for the farmers giving a good cost-benefit ratio. Cattle manure which is rich in nitrogen and carbon has been mainly used in fields and kitchen gardens. Low-nutrient manures have longer residual effects in the soil health amelioration besides controlling the nematode populations. Application of FYM (farm yard manure) improved soil health and reduced the disease severity of complex disease caused by *Meloidogyne incognita* and root wilt fungus, *Fusarium udum*, infesting pigeon pea under field conditions (Goswami et al. 2007). On an average well-developed farm yard manure contains 0.5% nitrogen, 0.2% P2O5, and 0.5% K2O. In balsam, application of farm yard manure was found to significantly reduce root-knot nematode population (Khan 2003). In polyhouse, the addition of well-decomposed farm yard manure at 20 t/ha was able to enrich the soil and to control the nematode population. Application of manure was found to suppress the nematode population and also increase the growth and yield of tomato plants (Maareg et al. 2000). Yield was found to be increased by the application of manure in tomato plants (Korayen 2003). Plant growth was increased in Chinese cabbage and less nematode population was observed (Wang et al. 2010). Cultivation of important vegetable crops such as potato, radish, cucumber, carrot, tomato, sweet potato, onion, etc. with FYM showed good results. The entire amount of nutrients present in farm yard manure is not available immediately. About 30% of nitrogen, 60–70% of phosphorus, and 70% of potassium are available to the first crop. The current methods for the preparation of farm yard manure are defective. Urine is generally wasted although it contains 1% N and 1.35% K.

12.3 Benefits of Organic Amendments

The use of organic matter and its significant effect in bringing down the population of root-knot nematode during decomposition were reported by Linford et al. Introduction of organic matter has been found to enhance the soil nutrient and physicochemical and biological properties of the soil (Ansari et al. 2017a, 2017b). Application of organic additives in the soil may improve the environment and provide adequate assistance in the nutrient recycling and promoting of plant health. Organic amendments also help in maintaining the water retention capacity and affecting the nematode population badly through direct or indirect approaches. The impact of organic additives in nematode management is unpredictable and inconsistent. High organic amendment content supports thriving communities of the decomposers and predators that make up the soil digestive system. Decomposition of organic matter released nematicides and increased fungal activity and persistence. It released nematicidal compounds like ammonia and fatty acids during degradation. It enhanced the antagonistic organisms as well as produced changes in the soil environment that are unsuitable for nematode behavior. Sometimes, introduction of organic additives indirectly affects the physical properties of soil. Reduction in nematode population and increase in crop yield have been reported due to the application of organic amendments (Kimpinski et al. 2003). It provides food for the detritivores and saprovores and energy for soil biochemical processes and acts as reservoir of nutrients. Chemically it regulates soil pH by providing acids, bases, and buffers. It modifies the thermal properties and compaction. It forms and stabilizes the soil aggregation. Its most important beneficial feature is pest suppression and it also helps in nitrogen mineralization along with supporting the microbial community. On the other hand, plants hold some cyanogenic glycosides or pyrrolizidine alkaloids having nematicidal activity against root-knot nematode (Widmer and Abawi 2002; Thoden et al. 2009). Consequently, a drastic increase of Rhabditidae, Panagrolaimidae, or Diplogastridae, all bacteria-feeding nematode families favored by nutrient-enriched conditions, has been reported by several authors (Bulluck III et al. 2002; Nahar et al. 2006; Yeates et al. 2006; Hu and Qi 2010). Besides higher numbers of active life stage, manured plots also showed higher number of dauer juveniles, a dormant stage of bacteria-feeding nematodes. Likewise, those manured plots show high numbers of omnivorous nematodes (Diepeningen et al. 2006; Hu and Cao 2008; Leroy et al. 2009; Hu and Qi 2010), while the numbers of carnivorous nematodes were undisturbed.

12.4 Organics and Soil Health

Organic matter plays a crucial role in the enhancement of various nutrients into soil and in making soil healthier (Ansari and Mahmood 2017a; Ansari and Mahmood 2019a, 2019b). Introduction of organic additives into soil has often significantly improved the soil physicochemical properties (Thangarajan et al. 2013; Khaliq and Abbasi 2015), such as soil aggregate stability, water holding capacity, and soil porosity (Celik et al. 2004). Application of organic additives like farm yard manure, cow manure, rice husk, and reeds enhanced soil aggregate stability and impaired soil bulk density (Karami et al. 2012; Ansari et al. 2019). Zhao et al. (2009) found farm yard manure and straw application to be very helpful in decreasing the soil bulk density due to increased soil organic carbon and porosity when compared with untreated control. Various organic inputs like compost, manure, and organic and natural mineral fertilizers show an additional role and improve soil health (Delate et al. 2015; Hooks et al. 2015; Tavantriz et al. 2012). The C/N ratio can significantly impact the soil health. Mixture of organic additives can be very effective in building up the nutrient reservoir which improves the soil quality thereby enhancing the crop productivity (Jackson and Bowles 2013). Integrated organic weed management practices can lessen the need for tillage which degrades the soil and cultivation in organic crop rotation. In addition, organic matter containing large content of cellulose-derived carbon may be helpful in the multiplication of some beneficial fungal spores which may later on improve the soil health. There might be 50 different species of nematodes in a handful of soil and millions can occupy in 1 m^2. Nematodes respond quickly to the environmental perturbation and also influence the microbial activities of soil agroecosystem. Soil nematodes are also sometimes found to be ecological indicators as their community reacts rapidly to the environmental disturbances.

12.5 Conclusion and Future Prospects

The conclusions which can be drawn from this chapter are derivatives from the logic that soil is an essential part of the ecosystem's functions and organic amendments play a vital role in uplifting the nutrients present in the soil and depriving the nematode population resulting in crop yield promotion. Intensification of agricultural systems such as use of non-judicious pesticides, chemical fertilizers, power tillage, and inorganic sources of nutrients leaves a poor impact on the agroecosystem. Application of organic additives in a judicious manner can improve the storage of various important nutrients needed for ecological balance. Application of organic amendments is an important way to improve soil property that reduces the pesticide demand. Application of organic additives has also mitigated the leaching of chemicals into groundwater and helps in the detoxification through various microbial inocula. Various phytochemicals are nematostatic in nature released after the decomposition of organic additives. Rate of application, content, and chemical constituent are the important factors; therefore, recommendation for a long-term application should be repeated under diverse agroclimatic conditions. Also, better soil physicochemical properties such as N, P, K, C, C/N ratio, and pH may improve the plant health and curtail the nematode multiplication. The suppressive soil is also considered to accelerate the managing capacity of organic additives of plant parasitic nematodes. People are going out of the inorganic fertilizer fantasy since it has various residual effects that harm the crops. Farmers are turning into organic cultivation nowadays which has low cost as compared to chemical fertilizer that is hazardous to health. Nowadays people are giving more emphasis in organic farming which is less harmful and less hazardous to health. Non-application of chemicals is highly being recommended due to the rise of various side effects on ecosystems and human health. A focus is being given to organic farming since it helps in the multiplication of soil microbial agents which play an indirect role in phytonematode management. Due to the high price of pesticides (nematicides) and synthetic fertilizers, farmers are currently being convinced to go for organic cultivation.

References

Abolusoro SA, Oyedunmade EEA (2005) Field assessment of the nematicidal properties of neem (Azadirachta indica) against the root-knot nematode, *Meloidogyne incognita* on infected tomato (Lycopersicon esculentum). Agrosearch 7:28–33

Adegbite AA, Adesiyan SO (2005) Root extracts of plants to control root-knot nematode on edible soybean. World J Agric Sci 1:18–21

Ahamad MU, Biswas AK, Hossain J (1987) Effect of organic soil amendments on the development of root knot disease at the early growth stage of brinjal. Bangladesh J Plant Pathol 3:31–36

Akhtar M (2000) Nematicidal potential of the neem tree Azadirachta indica (A. Juss). Integr Pest Manage Rev 5:57–66

Akhtar M, Alam MM (1993) Utilization of waste materials in nematode control: a review. Bioresour Technol 45:1–7

Akhtar M, Malik A (2000) Roles of organic amendments and soil organisms in the biological control of plant parasitic nematodes: a review. Bioresour Technol 74:35–47

Alsayed A, Ahmed SS, Abdel-Hameed SH (1992) Influence of decomposting dry leaf powders on *Rotylenchulus reniformis* and growth of response on soybean. Ann Agric Sci Moshtohor 30:615–620

Ansari RA, Khan TA (2012a) Parasitic association of root-knot nematode, *Meloidogyne incognita* on guava. e-J Sci Technol (e-JST) 5:65–67

Ansari RA, Khan TA (2012b) Diversity and community structure of phytonematodes associated with guava in and around Aligarh, Uttar Pradesh, India. Trends Biosci 5(3):202–204

Ansari RA, Mahmood I (2017a) Optimization of organic and bio-organic fertilizers on soil properties and growth of pigeon pea. Sci Hortic 226:1–9

Ansari RA, Mahmood I (2017b) Determination of disease incidence caused by *Meloidogyne* spp. and or *Fusarium udum* on pigeonpea in Aligarh district: a survey. Trends Biosci 10 (24):5239–5243

Ansari RA, Mahmood I (2019a) Plant health under biotic stress: volume 2: microbial interactions. Springer, Singapore. https://doi.org/10.1007/978-981-13-6040-4

Ansari RA, Mahmood I (2019b) Plant health under biotic stress: volume 1: organic strategies. Springer, Singapore. https://doi.org/10.1007/978-981-13-6043-5

Ansari RA, Rizvi R, Sumbul A, Mahmood I (2017a) PGPR: current vogue in sustainable crop production. In: Kumar V, Kumar M, Sharma S, Prasad R (eds) Probiotics and plant health. Springer, Singapore, pp 455–472

Ansari RA, Mahmood I, Rizvi R, Sumbul A (2017b) Siderophores: augmentation of soil health and crop productivity. In: Kumar V, Kumar M, Sharma S, Prasad R (eds) Probiotics in agroecosystem. Springer, Singapore, pp 291–312

Ansari RA, Sumbul A, Rizvi R, Mahmood I (2019) Organic soil amendments: potential tool for soil and plant health management. In: Ansari RA, Mahmood I (eds) Plant health under biotic stress. Springer, Singapore, pp 1–35

Arancon NQ, Edwards CA, Lee S (2002) Management of plant parasitic nematode population by use of vermicomposts. Proc Brighton Crop Protect Conf Pests Dis 2:705–710

Arancon NQ, Galvis PA, Edwards CA, Yardim EN (2003) Trophic diversity of nematode communities in soils treated with vermicompost. Pedobiologia 47:736–740

Bari MA, Nahar MS, Alam MF, Hossain IH (1999) Efficacy of preplant soil treatment with four organic amendments and two nematicides to control root knot of Okra. Bangladesh J Plant Pathol 15:27–30

Bari MA, Faruk MI, Rahman ML, Ali MR (2004) Effect of organic soil amendments and nematicide on root-knot nematode of brinjal. Bangladesh J Plant Pathol 20:27–30

Barker KR, Pederson GA, Windham GL (1998) Plant and nematode interactions. Agronomy monograph 36. American Society of Agronomy, Madison, WI

Bello A, Lopez-Perez JA, Garcia-Alvarez A, Sanz R, Lacasa A (2004) Biofumigation and nematode control in the Mediterranean region. In: Cook RC, Hunt DJ (eds) Nematology monographs and perspectives; Proceedings of the Fourth International Congress of Nematology; 8–13 June, 2002; Tenerife, Spain. Brill, Leiden and Boston, pp 133–149

Bending GD, Lincoln SD (1999) Characterization of volatile sulphur-containing compounds produced during decomposition of Brassica juncea tissues in soil. Soil Biol Biochem 31:695–703

Bhattacharrya D, Goswami BK (1987) A study of the comparative efficacy of neem amd ground nut oil cakes against root-knot nematode, *Meloidogyne incognita* as influenced by micro organisms on sterilized and unsterilized soil. Indian J Nematol 17(1):81–83

Borah A, Phukan PN (2004) Comparative efficacy of *Glomus fasciculatum* with neem cake and carbofuran for the management of *Meloidogyne incognita* on brinjal. Indian J Nematol 34:129–132

Borah A, Neog PP, Sinha AK (2007) Efficacy of Glomus fasciculatum in comparison with neem cake and carbofuran for management of *Meloidogyne incognita* on blackgram and greengram. Ann Plant Protect Sci 15(2):445–448

Bulluck LR III, Barker KR, Ristaino JB (2002) Influences of organic and synthetic soil fertility amendments on nematode trophic groups and community dynamics under tomatoes. Appl Soil Ecol 21:233–250

Celik I, Ortas I, Kilic S (2004) Effect of compost, mycorrhiza, manure and fertilizer on some physical properties of Chromoxerert soil. Soil Tillage Res 78:59–67

Cole RA (1976) Isothiocyanates, nitriles and thiocyanates as products of autolysis of glucosinolates in Cruciferae. Phytochemistry 15:759–762

D'Addabbo T, Papajova I, Sasanelli N, Radicci V, Renco M (2011) Suppression of root-knot nematodes in potting mixes amended with different composted biowastes. Helminthologia 48:278

Das D, Saikia MK, Sarmah DK (1999) Comparative efficacy of botanicals and antagonistic fungi for the management of rice root-knot nematode, Meloidogyne graminicola. Int J Trop Agric 17 (1–4):287–290

Delate KC, Cambardella, Chase C (2015) Effects of cover crops, soil amendments and reduced tillage on carbon sequestration and soil health in a long-term vegetables system. Final report for ORG project. 2010-03956. CRIS abstract

Devi, Th.S (2016) Development of an organic module for the management of root-knot nematode (*Meloidogyne incognita*) in cucumber. Ph.D. Thesis, Assam Agricultural University, Jorhat, Assam

Diepeningen AD, De Vos OJ, Korthals GW, Van Bruggen AHC (2006) Effects of organic versus conventional management on chemical and biological parameters in agricultural soils. Appl Soil Ecol 31:120–135

Dutt R, Bhatti DS (1986) Determination of effective doses and time of application of nematicides and castor leaves for controlling Meloidogyne javanica in tomato. Indian J Nematology 16:8–11

El-Sherif AG, Refaei AR, El-Nagar ME, Salem MMH (2008) Impact of certain oil-seed cakes or powder in comparison with oxamyl or urea on *Meloidogyne incognita* infecting eggplant. Arch Phytopathol Plant Prot 43:88–94

Engunjobi OA, Larinde MA (1975) Nematodes and maize growth in Nigeria II. Effects of some egyptian amendments on populations of and on the growth and production of maize (Zea mays) in Nigeria. Nematol Medit 3:65–73

Fenwick GR, Heaney RK, Mullin WJ (1983) Glucosinolates and their breakdown products in food and food plants. In: Furia TE (ed) Critical reviews in food science and nutrition. CRC Press, Boca Raton, pp 123–201

Gaur HS, Mishra SD (1990) Integrated control of nematode in lentil with aldicarb, neem cake and seed treatment with thimet and its residual effect on the subsequent mung crop. Indian J Entomol 51(3):283–287

Gergon EB, Miller SA, Davide RG, Opina OS, Obien SR (2001) Evaluation of cultural practices (surface burning, deep ploughing, organic amendments) for management of rice root-knot

nematode in rice—onion cropping system and their effect on onion (Allium cepa L.) yield. Int J Pest Manage 47:265–272
Gommers FJ (1981) Biochemical interactions between nematodes and plants and their relevance to control. Helminthol Abstr 50B:9–24
Gommers FJ, Bakker J (1988) Physiological diseases induced by plant responses or products. In: Poinar GO, Jansson H-B (eds) Diseases of nematodes, Volume I. CRC Press, Boca Raton, FL, pp 3–22
Gommers FJ, Voorin'tholt DJM (1976) Chemotaxonomy of compositae related to their host suitability for *Pratylenchus penetrans*. Netherlands J Plant Pathol 82:1–8
Goswami BK, Vijayalakshmi V (1986) Nematicidal properties of some indigenous plant materials against root-knot nematode, *Meloidogyne incognita* on tomato. Indian J Nematol 16(1):65–68
Goswami BK, Vijayalakshmi K (1987) Studies on the effect of some plant and non-edible oil cake extracts on larval hatching of *Meloidogyne incognita*. J Res Assam Agric Univ 8:264
Goswami BK, Pandey RK, Goswami J, Tiwari DD (2007) Management of disease complex caused by root-knot nematode and root wilt fungus on pigeon pea through soil organically enriched with vesicular arbuscular mycorrhiza, Karanj (Pongamia pinnata) oilseed cake and farm yard manure. J Environ Sci Health 42(8):899–904
Gowda DN (1972) Studies on comparative efficacy of oil cakes on control of root- knot nematode *Meloidogyne incognita* on tomato. Mysore J Agric Sci 6(4):524–525
Harvey SG, Hannahan H, Sams CE (2002) Indian mustard and allyl isothiocyanate inhibit *Sclerotium rolfsii*. J Am Soc Hortic Sci 127:27–31
Hassan SME, Rahman MSQ, Amin MR, Hoque ATMR, Islam SMS (2001) Effect of some organic substances on the root knot disease of brinjal. J Biol Sci 1:791–792
Hooks CR, Wang KH, Brust G, Mathew S (2015) Using winter cover crops t0 enhance the organic vegetable industry in the Mid-Atlantic Region. Final report for OREI project 2010-01954. CRIS abstract
Hossain S, Mian IH, Tsuno K (1989) Efficacy of three nematicides and two oilcakes for control of root knot nematode (*Meloidogyne incognita*) in potato seedlings. J Fac Agric Kyushu Univ 34:115–121
Hu C, Cao ZP (2008) Nematode community structure under compost and chemical fertilizer management practices in the North China plain. Exp Agric 44:485–496
Hu C, Qi Y (2010) Abundance and diversity of soil nematodes as influenced by different of organic manure. Helminthologia 47:58–66
Hussain MA, Mukhtar T, Kayani MZ (2011) Efficacy evaluation of Azadirachta indica, Calotropis procera, Datura stramonium and Tagetes erecta against root knot nematodes *Meloidogyne incognita*. Pakistan J Bot 43:197–204
Jackson L, Bowles (2013) Researcher and farmer innovation to increase nitrogen cycling on organic farms (Webinar). http://articles.Extension.org/pages/67391/researcher-and-farmer-innovation-to-icrease-nitrogen-cycling-on-organic-farms-webinar
Jain RK, Mathur KN, Singh RV (2007) Estimation of losses due to plant parasitic nematodes on different crops in India. Indian J Nematol 37:219–220
Kaliram DC, Gupta DC (1982) Efficacy of plant leaves, nematicides and fertilizer alone and in combination against *Meloidogyne javanica* infecting chickpea (*Cicer arietinum* L.). Indian J Nematol. 12:221–225
Karami A, Homaee M, Afzalinia S, Ruhipour H, Basirat S (2012) Organic resource management: impacts on soil aggregate stability and other soil physic-chemical properties. Agric Ecosyst Environ 148:22–28
Khaliq A, Abbasi MK (2015) Improvement in the physical and chemical characteristics of degraded soils supplemented with organic-inorganic amendments in the Himalayan region of Kashmir, Pakistan. Catena 126:209–219
Khan MR (2003) Life cycle, pathogenicity and management of root-knot nematode, *Meloidogyne incognita* Race 2 infecting Balsam (*Impatiens balsamina* L.). Indian J Nematol 33(1):61–88

Kimpinski J, Gallant CE, Henry R, Macleod JA, Sanderson JB, Sturz AV (2003) Effect of compost and manure soil amendments on nematode and on yields of potato and barley: a 7-years study. J Nematol 35:289–293

Kirkegaard JA, Matthiessen J (2004) Developing and refining the biofumigation concept. Agroindustria 3:233–239

Kirkegaard JA, Sarwar M (1998) Biofumigation potential of brassicas. 1. Variations in glucosinoalate profiles of diverse field grown brassicas. Plant and Soil 201:71–89

Korayen AM (2003) Effect of some organic waste of *Meloidogyne incognita* development and tomato tolerance to the nematode. Egypt J Phytonematol 31:119–127

Kumar RK, Nattuthurai KN, Gurusamy R (2011) Influence of vermicompost in root- knot nematode management as a function of soil fortification. Elixir Agric 38:4210–4213

Leroy BLMM, De Shutter N, Ferris H, Moens M, Reheul D (2009) Short term nematode population dynamics as influenced by the quality of exogenous organic matter. Nematology 11:23–38

Maareg MF, Salem FM, Ebieda EM (2000) Effect of certain organic and inorganic amendments on *Meloidogyne javanica* in sandy soil. Egypt J Agronematol 4:82–94

Mahalik JK, Sahu A (2018) Eco friendly management of root-knot nematode (*Meloidogyne incognita*) infecting okra in Odisha, India. Int J Curr Microbiol Appl Sci 6:1999–2004

Mahanta B, Phukan PN (2004) Comparative efficacy of Glomus fasciculatum, neem cake and carbofuran for the management of *Meloidogyne incognita*. Ann Plant Prot Sci 12(3):377–379

McLeod RW, Steele CC (1999) Effects of brassica-leaf green manures and crops on the activity and reproduction of *Meloidogyne javanica*. Nematology 1:613–624

Mcsorley R, Gallaher RN (1995) Effect of yard waste compost on plant-parasitic nematode densities in vegetable crops. J Nematol 27:545–549

Mehta UK, Sundararaj P (1995) Efficacy of some new soil amendments for the control of the lesion nematode in sugarcane. Nematol Medit 23:321–323

Mishra CB, Singh B, Laha SK (1987) Integrated approach for root knot management in Jute. Indian J Nematol 17:285–287

Mishra S, Wang KH, Sipes BS, Tian M (2017) Suppression of root-knot nematode by vermicompost tea prepared from different curing ages of vermicompost. Plant Dis 101:734–737

Mojtahedi H, Santo GS, Wilson JH, Hang AN (1993) Managing *Meloidogyne chitwoodi* on potato with rapeseed as green manure. Plant Dis 77:42–46

Morra MJ, Kirkegaard JA (2002) Isothiocyanate release from soil incorporated Brassica tissues. Soil Biol Biochem 34:1683–1690

Morra L, Palumbo AD, Bilotto M, Ovieno P, Picascia S (1998) Soil solarization: organic fertilization and grafting contribute to build an integrated production system in a tomato-zucchini sequence. Colture-Protte 27:63–70

Muller R, Gooch PS (1982) Organic amendments in nematode control. An examination of literature. Nematropica 12:319–326

Nahar MS, Grewal PS, Miller SA, Stinner D, Stinner BR, Lleinhenz MD, Wszelaki A, Doohan D (2006) Differential effects of raw and composted manure on nematode community and its indicative value for soil microbial, physical and chemical properties. Appl Soil Ecol 34:140–151

Neher DA (2001) Role of nematodes in soil health and their use as bioindicators. J Nematol 33:161–168

Oka Y, Yermiyahu U (2002) Suppressive effect of compost against the root-knot nematode *Meloidogyne javanica* on tomato. Nematology 4:891–898

Patel RM, Desai MV (1964) Possible biological control of root knot nematode. Plant Dis Reptr 48:167–178

Ploeg AT, Stapleton JJ (2001) Glasshouse studies on the effect of time, temperature and amendment of soil with broccoli plant residues on the infestation of melon plants by *Meloidogyne incognita* and M. javanica. Nematology 3:855–861

Potter MJ, Davies K, Rathjen AJ (1998) Supressive impact of glucosinolates in *Brassica* vegetative tissues on root lesion nematode, Pratylenchus neglectus. J Chem Ecol 24:67–80

Rahman L, Somers T (2005) Suppression of root knot nematode (*Meloidogyne javanica*) after incorporation of Indian mustard cv. Nemfix as green manure and seed meal in vineyards. Aust Plant Pathol 34:77–83

Ramakhrishna S, Gunasekharan CR, Vadivelu S (1997) Efficacy of organics in control of *Meloidogyne incognita* on okra. Indian J Nematol 27(1):74–78

Renco M, D'Addabbo T, Sasanelli N, Papajova I (2007) The effect of five compost of different origin on the survival and reproduction of *Globodera rostochiensis*. Nematology 9:537–543

Renco M, Sasanelli N, D'Addabbo T, Papajova I (2010) Soil nematode community changes associated with compost amendments. Nematology 12:681–692

Ribeiro CF, Mizobutsi EH, Silva DG, Pereira JCR, Zambolim L (1998) Control of *Meloidogyne javanica* on lettuce with organic amendments. Fitopatologia Brasileira 23:42–44

Rivera L, Aballay E (2008) Nematicide effect of various organic soil amendments on *Meloidogyne ethiopica* Whitehead, 1968, on potted vine plants. Chil J Agric Res 68(3):290–296

Rodriguez-Kábana R, Morgan-Jones G (1987) Biological control of nematodes: soil amendments and microbial antagonists. Plant and Soil 100:237–247

Sarma B (1989) Pathogenicity and control of *Meloidogyne incognita* (Kofoid and *White*, 1919). Chitwood, 1949 on cowpea (Vigna sinensis). M.Sc. Thesis, Assam Agricultural University, Jorhat

Sarma AN, Sarma VK, Singh RS (1971) Oil cakes and saw dust control root-knot nematode. Indian Fmg 20(12):17–20

Sen K, Dasgupta MK (1981) Effect of rice hull ash on root-knot nematode in tomato. Trop Pest Manage 27(4):518

Siddiqui MA, Haseeb A, Alam MM (1987) Evaluation of nematicidal properties of some latex bearing plants. Indian J Nematol 17:99–102

Siddiqui MA, Haseeb A, Alam MM (1992) Control of plant parasitic nematodes by soil amendments with latex bearing plants. Indian J Nematol 22:25–28

Singh RS (1965) Control of root-knot of tomato with organic soil amendments. FAO Plant Prot Bull 13:35–37

Singh T, Patel BA (2015) Management of root-knot nematode (*Meloidogyne incognita*) in bottle gourd using different botanicals in pots. J Parasit Dis 39(3):441–445

Singh RS, Sitaramaiah K (1966) Incidence of root-knot of okra and tomatoes in oilcake soil. Plant Dis Reptr 50:668–672

Stirling GR (1991) Biological control of plant parasitic nematodes. CAB International, Wallingford, UK

Stirling GR (2013) Ecosystem service and the concept of "Integrated Soil Biology management". In: Graham R (ed) Biological control of plant parasitic nematodes second edition: soil ecosystem management in sustainable agriculture. Stirling Biological Crop Protection Pty Limited, Brisbane, Australia, p 501

Stirling GR, Stirling AM (2003) The potential of *Brassica* green manure crops for controlling root-knot nematode (*Meloidogyne javanica*) on horticultural crops in a subtropical environment. Aust J Exp Agric 43:623–630

Swathi P, Rao KT, Rao PA (1998) Studies on control of root-knot nematode *Meloidogyne incognita* in tobacco miniseries. Tobacco Research 1:26–30

Szczech M, Rondomanski W, Brzeski MW, Smolinska U, Kotowski JF (1993) Suppressive effect of a commercial earthworm compost on some root infecting pathogens of cabbage and tomato. Biol Agric Hortic 10:47–52

Tavantriz SMRP, Larkin AV, Alyokhin MSE, Jemison JM (2012) A systems approach to optimize organic production; Enhancing soil functionality and plant health to suppress plant diseases and pests. Final report for ORG project 2007-01405 CRIS abstract

Tester CF (1990) Organic amendment effects on physical and chemical properties of a sandy soil. Soil Sci Soc Am J 65:827–831

Thakur SK (2014) Effect of green manuring and plant dry powder on soil properties and nematode infecting maize. Agric Sci Digest 34(1):56–59

Thakur RS, Singh SB, And Goswami A (1981) Article E.I. Azadirachta indica A. Juss—a review. CROMAP 3:135–140
Thangarajan R, Bolan NS, Tian G, Naidu R, Kunhikrishnan A (2013) Role of organic amendment application on greenhouse gas emission from soil. Sci Total Environ 465:72–96
Thoden T, Hallmann J, Boppre M (2009) Effects of plants containing pyrrolizidine alkaloids on the root-knot nematode Meloidogyne hapla. Eur J Plant Pathol 123:27–36
Tiyagi SA, Alam MM (1995) Efficacy of oil seed cakes against plant parasitic nematodes and oil-inhibiting fungi on mung bean and chickpea. Bioresour Technol 51:233–239
Uhlenbroek JH, Bijloo JD (1958) Investigations on nematicides. I. Isolation and structure of a nematicidal principle occurring in Tagetes roots. Recueil des Travaux Chimiques des Pays-Bas 77:1004–1009
Van Berkum JA, Seshadri AR (1970) Some important nematode problems in India. In: Proc 10th Int Nematol Symp, Pescara, Italy, pp 136–137
Walker GE, Morey BG (1999) Effect of brassica and weed manures on abundance of Tylenchulus semipenetrans and fungi in citrus orchard soil. Anim Prod Sci 39(1):65–72
Wang D, Shi O, Wang X, Wei M, Hu J, Liu J, Yang F (2010) Influence of cow manure, vermicompost on the growth metabolite contents and antioxidant activities of Chinese cabbage. Biol Fertil Soils 46:689–694
Widmer TL, Abawi GS (2002) Relationship between levels of cyanide in sudangrass hybrids incorporated into soil and suppression of Meloidogyne hapla. J Nematol 34:16–22
Yadav S, Patil J, Kumar A (2018) Bio-nematicidal effect of Azadirachta indica, against *Meloidogyne incognita* in tomato. Int J Chem Stud 6(3):2757–2761
Yeates GW, Speir TW, Taylor MD, Clucas L, Schail AP (2006) Nematode responses to biosolids incorporation I five soil types. Biol Fertil Soils 42:550–555
Zasada IA, Ferris H (2004) Nematode suppression with Brassicaceous amendments. Soil Biol Biochem 36:1017–1024
Zhao Y, Wang P, Li J, Chen Y, Ying X, Liu S (2009) The effect of two organic manures on soil properties and crop yields on a temperate calcareous soil under a wheat-maize cropping system. Eur J Agron 31:36–42

Chapter 13
Plant Parasitic Nematodes Management Through Natural Products: Current Progress and Challenges

Olubunmi Atolani and Oluwatoyin Adenike Fabiyi

Abstract The devastating effect of huge crop losses to field and storage pest brings about a downturn in the availability of food, thus subjecting the citizens to poverty and hunger. The plant-parasitic nematodes (PPN) are important agricultural pests, which are hidden enemies of the farmers. Their existence in soil and the gravity of the potential loss which accrues to them are not known to the lay farmers. Significant quantitative and qualitative economic losses are attributed to their infestation worldwide on a yearly basis. The damage posed by plant-parasitic nematodes is steadily on the increase and is at times not reported, but oftentimes associated with factors such as nutrient deficiency and extreme weather conditions. Nematodes are a big challenge to food production in an overpopulated world. They are listed as an important crop pest in the world, largely because of their wide geographical distribution and their ability to parasitize virtually all crops. Export trade restrictions are often imposed due to the presence of quarantine nematode pests. Different integrated pest management (IPM) strategies have been employed in the control of plant-parasitic nematodes with various degrees of success. The use of synthetic nematicides, though an established internationally recognized effective approach, is largely negated by the toxicity and adverse effects on humans and the environment. Thus, the need for concerted worldwide intensive research for safer alternatives has become imperative. However, there are several challenges in the control of plant-parasitic nematodes, majorly the adoption of new approach to control by the farmers, large-scale production hiccups, government policies, and potential toxicity of the new approach. This work reviews the various strategies used in combating plant-parasitic nematodes. Based on the resultant toxicities reported for chemical control and the effectiveness of biological control (bio-control) which is considered safer, more discussions were made on bio-control. Biological control, application of botanicals, and soil amendment methods rank high among others owing to their

O. Atolani
Department of Chemistry, University of Ilorin, Ilorin, Nigeria

O. A. Fabiyi (✉)
Plant Nematology Unit, Department of Crop Protection, Faculty of Agriculture, University of Ilorin, Ilorin, Nigeria

R. A. Ansari et al. (eds.), *Management of Phytonematodes: Recent Advances and Future Challenges*, https://doi.org/10.1007/978-981-15-4087-5_13

environmental safety. Over 2400 plant species worldwide are known to be pesticidal. Some tested plants and phytochemical isolates with reported in vitro or in vivo nematicidal properties were listed. The merits and demerits of the bio-control method were also discussed. It was suggested that although the bio-control method is affordable and effective, it should be performed with caution as many emerging reports indicated that many plant materials contain phytochemicals that could induce carcinogenicity, endocrine disruption, and various degrees of toxicities with end-points such as hepatotoxicity, cytotoxicity, ecotoxicity, carcinogenicity, mutagenicity, and immunotoxicity which finally result into morbidity and mortality.

Keywords Chemical control · Biological control · Losses · Yield · Phytonematodes

13.1 Introduction

Nematodes which inhabit the entire ecosystem are known to cause various plant and animal diseases leading to enormous losses annually. In the plant kingdom, a whopping US$80 billion or more in loss is associated with the effect of plant-parasitic nematodes globally (Coyne et al. 2018). The root-knot nematodes of the genus *Meloidogyne* are the most important as they have a wide range of hosts and also induce several crop damages to a wide range of both annual and perennial crops as well as horticultural and field crops (Ansari and Khan 2012a, b; Ansari and Mahmood 2017b). The genus *Meloidogyne* has over 90 species of which *M. incognita*, *M. javanica*, *M. arenaria*, and M. *hapla* are of paramount economic importance (Maurice et al. 2009). *M. incognita* alone is reported to cause about 30% crop yield losses under field condition (Jain et al. 1994; Renčo et al. 2014). In the tropics and semi-warm temperate regions, crop losses of up to 20% are sometimes recorded despite the use of nematicides, while total crop loss could be obtained when no nematicides are used against *Meloidogyne* infestation (Stirlin and Stanton 1997). However, *Meloidogyne chitwoodi* and *M. fallax* are also becoming a major threat to field vegetables in some regions due to their potential to provoke massive damage (Wesemael et al. 2006; Wesemael and Moens 2008). They were resultantly placed in the A2 list of EPPO (European and Mediterranean Plant Protection Organization) as quarantine organisms (Holgador and Magnusson 2012). Therefore, emphasis on the genus *Meloidogyne* cannot be overemphasized. *Meloidogyne* spp. infest many important food and economic crops like vegetables such as, tomatoes, bean, beetroot, carrot, celery, cucumber, eggplant, lettuce, okra, onion, potato, sweet potato, and pumpkin; field crops such as peanut, clover, tobacco, cowpea, kenaf, soybean, lucerne, lupin, pigeon pea, tea, and sugarcane; ornamental crops including *Ozothamnus*, *Gladiolus*, *Chrysanthemum*, *Dahlia*, *Gerbera*, *Protea*, and rose; and horticultural crops such as passion fruit, kiwi fruit, almond, nectarine, peach, plum, banana, ginger, pineapple, strawberry, *Aloe vera*, grape, and pawpaw (Stirlin and Stanton 1997). The control of plant-parasitic nematodes (PPN) is an excruciating task. Various techniques have been adopted. These techniques are generally classified as (1) cultural practices, (2) chemical control, (3) biological control, (4) host resistance methods, and (5) mechanical methods (Bakker 1993; Kratochvil et al. 2004; Viaene et al. 2013a, b; Dutta et al. 2019). This chapter reviews the various strategies adopted in combating plant-parasitic nematodes while examining the merits and challenges of the options.

13.2 Cultural Practices

This involves crop rotation, use of tolerant varieties, cover crop plantation, resistant species, fallowing, spread prevention, and other techniques. The crop which is highly susceptible to a particular nematode is replaced from the field with a less or non-susceptible crop for a number of years. While this has been successfully applied, it prevents monocropping. Also when the soil is multi-parasite (nematode) infested, crop rotation success is minimized. Some cultural practices also adopt the cultivation of plants secreting compounds that are toxic to the nematodes in the soil. However, this cultural method is less successful as there is no single plant that may inhibit or minimize the growth of varieties of nematodes (Coyne et al. 2018). Mechanical methods which involve the repeated tilling of fallow soil yield some success, though their implementation on a large scale is quite cumbersome and not economically viable (Kratochvil et al. 2004; Dutta et al. 2019).

13.3 Chemical Control

This method adopts the application of chemical substances known as fumigants (gas) and non-fumigants (liquid or solid) to soil to inhibit or kill the nematodes within a certain depth of the soil (Lambert and Bekal 2002). Since a large amount of chemical substances is involved in this method, the associated risks are also huge especially in the treatment of root-knot nematodes such as *Meloidogyne incognita* which survives at depth below 90 cm in the soil in harsh conditions (Stapleton and DeVay 1983; Starr and Jeger 1985). Depending on the soil type and other environmental factors, some persistent nematode species survive for many months and years without food supply (Steiner and Albin 1946; McSorley 2003) thereby warranting the application of a large amount of chemical nematicide with extended retention in the soil. While many broad-spectrum nematicidal agents such as dibromochloropropane (DBCP), methyl bromide, methyl isothiocyanate, and dibenzofuran are banned from use as nematicides for food products, many other synthetic nematicidal agents such as carbofuran, aldicarb, and oxamyl are still in use in many developing countries. Many of the synthetics are reported to possess medium or average to fatal, toxic, and harmful effects to the entire ecosystem especially when used at increased concentration (Kottegoda 1985; USAEPA 2019). Water bodies, soil, and food are easily contaminated with the application of some of the chemical nematicides. When the chemicals are passed on to humans via the consumption of nematicide-contaminated foods (plant), they could induce toxicity with endpoints such as hepatotoxicity, cytotoxicity, carcinogenicity, mutagenicity, and immunotoxicity, among others (USEPA 2009; WHO 2009; Bouagga et al. 2019). Beside humans, many other non-target organisms in the soil, land, and air which are important for ecosystem balancing are destroyed, denatured, and eliminated with the application of high doses of the chemicals (Loveridge et al. 2019).

13.3.1 Toxicities Challenge of Chemical Control: Carbofuran as a Case Study

Carbofuran, an anticholinesterase carbamate, is one of the most used pesticides for nematode control. Chemically, it is known as 2,3-dihydro-2,2-dimethyl-1-benzofuran-7-yl methylcarbamate. It is widely used in many developing countries due to its potency but has been banned in many countries as a result of associated toxicities to human and non-target organisms (Goad et al. 2004; USEPA 2009; Seth et al. 2019). It is in most cases labeled as moderately toxic to highly toxic. In fact, the World Health Organization labeled carbofuran as a highly dangerous product (WHO 2009). Some reports indicated that carbofuran persists in soil, is leached to water bodies, and also gets absorbed by plants (Sharom et al. 1980). In human, it is metabolized through oxidation reaction to 3-hydroxycarbofuran which is further oxidized to 3-ketocarbofuran (USEPA 1990). Carbofuran is also known to undergo various other metabolic reactions including hydrolysis under alkaline conditions to produce toxic metabolites such as 3-ketocarbofuran, N-hydroxymethylcarbofuran, carbofuran phenol, 3-ketocarbofuran-7-phenol, and 3-hydrocarbofuran-7-phenol (Fig. 13.1) (Eisler 1985; Goad et al. 2004; USEPA 2009; Seth et al. 2019; Fabiyi et al. 2012b).

Carbofuran, a broad-spectrum carbamate nematicide, is further reported to induce acute intoxication in cattle (Choez and Evaristo 2018). The accidental inclusion in food, uptake by plants, discharge in water bodies, and overdose have led to many reported cases of toxicities of the chemical (Novtoný et al. 2011; Zeinsteger et al. 2014; Ferré et al. 2015; Caloni et al. 2016). The rapid absorption by roots of plants and transport to site of high transpiration is a subject of major concern with respect to food poisoning and air pollution. As demonstrated in various in vivo studies, acute carbamate poisoning results into morbidity and mortality induced by respiratory failure caused by pulmonary edema (Ferslew et al. 1992). Carbofuran is also known to exhibit acute toxicity to both aquatic and terrestrial organisms (Dobšíková 2003). It degrades in 1 to 8 weeks in alkaline environment via chemical hydrolysis but it remains stable in acidic condition (Gupta 1994). Apparently, as a result of the high water solubility, it has high potential to migrate and contaminate water bodies

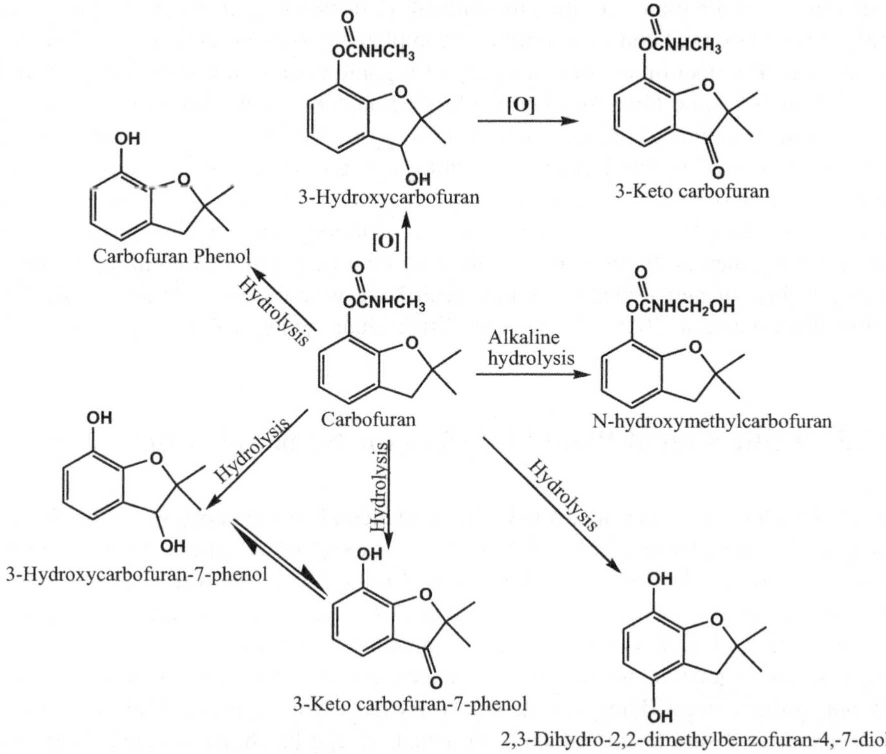

Fig. 13.1 Some metabolic products of carbofuran

through leaching (Sharom et al. 1980). The various reports of the toxicities of the synthetic pesticides obviously underscore the need for safer and greener alternatives.

13.4 Biological Control of Phytonematodes

This control measure involves the adoption of methods such as soil amendments, use of natural nematicides, use of predatory fungi, and application of green manure. Sometimes, the bio-control method also make use of natural techniques that improve the health of the plant which invariably increases the resistance of the plant to nematode infestation (Viaene et al. 2013a; Ansari et al. 2017a, b; Coyne et al. 2018; Ansari and Mahmood 2019b). Nematode resistance genes are crossbred into cultivated plant species to improve their resistance to nematodes. Natural substances such as seed oils and extracts are prepared and applied to control the spread and multiplication of nematodes at various growth stages of the targeted plant. Sometimes, some predatory fungi are also used to trap nematodes especially the root-knot nematodes (Stirlin 1991). The method is generally known to be benign, having little side effects on the plant and the environment (Viaene et al. 2013a, b; Coyne et al. 2018). The biological control is used as a complement to other methods for effective result. The introduction of some quantity of organic matter into the soil impairs the populations of some types of plant-feeding nematodes as the decomposing matter releases some chemical products which destroy or retard the growth of nematodes. The method is considered easy, affordable, and pollution-free. For instance, the decomposition of rye (*Secale cereale*) produces butyric acid, while the decomposition of rapeseed (*Brassica napus*) produces isothiocyanates which possess anti-nematicidal potential particularly against root-knot nematodes (*Meloidogyne* spp.), a major class of nematodes associated with huge global economic losses (Bridge 1996; Talwana et al. 2016; Coyne et al. 2018; Sikora et al. 2018).

13.5 Utilization of Plant Materials for Nematode Control

Several plants have been identified with nematicidal or nematostatic properties in their seeds, fruits, leaves, barks, and roots or root exudates (Fuglie 1998; Stoll 2000; Atolani et al. 2014; Ansari and Mahmood 2017b, 2019a, b; Ansari et al. 2019). Extracts from *Parinari polyandra* were found to be useful in ameliorating the multiplication of *Meloidogyne incognita* infesting *Celosia argentea*; there was an increase in vegetative growth of the leafy vegetable with the highest dose of *P. polyandra* extract (Fabiyi et al. 2014). Extracts from cypress shrub, *Lawsonia inermis*, caused significant reduction in nematode egg hatch and juvenile mortality of *M. incognita*, while *Daniellia oliveri* leaf extracts with ozoic and daniellic acids as main constituents showed viable nematicidal action and reduction in root galling of okra plants under *M. incognita* infection (Fabiyi and Atolani 2011; Fabiyi 2014). Similarly, *Enantia chloranta* extract metal complex significantly reduced *M.incognita* population on groundnut with notable increase in yield. (Fabiyi et al. 2019). Application of chopped green leaves of *Chromolaena odorata* is known to be toxic to *Meloidogyne incognita*; at 15 t/ha the population of *Meloidogyne incognita* was reduced, while an increase in yield of about 135% was recorded in okra plants. Increase in yield of *Musa* species was equally recorded with the use of *C. odorata* as soil amendment (Subramaniyan 1985; Ajith and Sheela 1996; Kashaja et al. 1999). Karanj leaves at 20,000, 40,000, and 80,000 ppm compared with carbofuran at 2.0 kg a.i/ha provided a significant reduction in the population of *Aphelenchoides composticola* by 80% (Rao and Pandey 1991). The extracts of *Ipomea fistulosa* effectively controlled *Meloidogyne incognita* in eggplant with improved plant growth (Alam et al. 1995). *Helicotylenchus dihystera* population was reduced by 52–62% with a significant increase in tomato yield with the use of *Azadirachta indica*, *Datura fastuosa*, and *Calotropis procera* leaves as soil amendments; the nematode population decrease was directly proportional to the quantity of leaves used. *Anacardium occidentale* and *Gmelina arborea* exhibited dose-dependent nematicidal activity on *Meloidogyne incognita* infecting cowpea cv. Ife Brown and okra. There was a general increase in growth and vigor without toxicity to the cowpea and okra plants (Firoza and Maqbool 1996; Onifade and Fawole 1996;

Fadeyi et al. 2016). The general practice of incorporating leaves of botanicals into the soil has proved effective in the control of plant-parasitic nematodes. *Brassica campestris*, *Catharanthus roseus*, *Pedilanthus tithymaloides*, *Ricinus communis*, *Azadirachta indica*, and *Calotropis procera* at 80 g/kg soil were compared favorably with carbofuran application at 2 kg a.i/ha, while *Aloe arborescens and Cymbopogon citratus* were compared with fenamiphos; they effectively reduced nematode population by 64.8%, 55.4%, and 87.9%, respectively; moderate nematicidal activity was exhibited by thermally degraded products of *C. citratus* at 90 mL (Sweelam 1989; Rao and Reddy 1991; Fabiyi et al. 2018). Several plant families have provided nematicidal metabolites and phytochemicals. A great variety of potent compounds with elucidated structures have suppressive effect on the proliferation of plant-parasitic nematodes in vitro and in vivo (Gommers and Voorin'tholt 1976; Chitwood 2002; Atolani et al. 2014). α-Terthienyl and related compounds were isolated from *Tagetes* spp. and have been established to be nematicidal (Uhlenbroek and Bijloo 1958). Chromatographic fractions isolated from *Alstonia boonei* and *Bridelia ferruginea* exhibited toxicity to *Meloidogyne incognita* eggs and juveniles under laboratory conditions; the relatively non-polar compounds from *A. boonei* and *B. ferruginea* include n-hexadecanoic acid, tetradecane, and pentadecanone,6,10,14-trimethyl, while eicosanoic acid methyl ester, 1-nonadecanal, 1-cyclohexene-1-butanol,2,6,6-trimethyl, and octadecane1-(etheneloxy) fall within the relatively medium polar compounds. Friedlan-3-one, dichloroacetic acid, heptadecyl ester, and tetratetracontane are relatively polar compounds, and this explains partly the basis for the comparatively higher toxicity of the fractions to *Meloidogyne incognita* juveniles and eggs (Fabiyi et al. 2012a, b). Indoles and substituted methoxy-indoles are the mainframe of most of the compounds identified in *A. boonei* (Cook 1990). Nonanoic acid, a C_9 fatty acid, exhibited substantial toxicity against root-knot nematode and soybean cyst nematode. Methyl ester derivative of nonanoic acid reduced significantly tomato root galling of *M. javanica*; the physical properties of methyl esters of fatty acids as compared to ethyl esters promote increased permeation into nematodes with a resultant increase in toxicity. The methyl esters were substantially more effective, producing reduced galling of plants at one-eighth the concentration at which the ethyl ester was active (Dijan et al. 1994). A variety of fatty acid esters have been used to control nematodes in vitro and in vivo (Pinkerton and Kitner 2006). A mixture of sodium lauryl sulfate and citric acid immobilized some of the nematodes; this mixture also reduced nematode reproduction significantly when applied at planting. Some fatty acids and fatty acid derivatives in the group of short carbon chains (including C_8 to about C_{14}) that can be in the epoxide, cyclopropane, methylated, or hydroxylated forms have also been confirmed to be toxic to some nematodes (Pinkerton and Kitner 2006). Twelve fatty acids (C_3 to C_{18}) and some of their derivatives, seven methyl esters, and four primary alcohols were investigated in vitro, with a few additional greenhouse tests. Toxicity increased with carbon number (C_3 to C_{11}); decanoic acid killed all second-stage juveniles of the plant-parasitic nematode *Meloidogyne hapla*. In 24 h, at a concentration of 50 ppm for methyl esters and primary alcohols, toxicity increased with chain length up to C_{10}

(Vrain 1980). The inhibitory effects of fatty acids of intermediate chain length that have been observed in biological activities may involve a direct interaction between fatty acids and lipophilic regions of the target nematode plasma membranes; *p*-methoxy isomers of propionic acid derivatives are known to be more active than *o*- and *m*-methoxy isomers on nematodes (Hayashi et al. 1983; Jalal and Reed 1986; Davis et al. 1997). Compounds like acetylenes, terpenoids, aldehydes, sesquiterpenoids, and phenoxypropanonic acid are known to have nematicidal activity against some nematodes (Mori et al. 1982). Lactones, ketones, phenolic aldehydes, and carboxylic acids have been indicated to be nematicidal. Leaf extracts from *Anthocephalus chinensis* and *Eichornia crassipes* and bulb extracts of *Allium sativum* were reported to be effective against *M. incognita* on banana in the field. The active nematicidal ingredients were identified to be a phenolic aldehyde from *A. chinensis*, a carboxylic acid from *E. crassipes*, and a ketone from *A. sativum* (Waele and Romulo 1998). The levels of naturally occurring phenols in roots have been correlated with resistance in certain plant cultivars to nematodes (Narayana and Reddy 1980). A number of phenolic compounds and aromatic acids have been studied for their nematicidal activity and their effect on juvenile and egg hatch of *Meloidogyne incognita.* Phenolic compounds like 2-OH naphthoic acid, 3,4-dihydroxyphenyl acetic acid, chlorogenic acid, caffeic acid, and aromatic acids like trans-cinnamic acid are effective in suppressing egg hatch of *M. incognita* (Mahajan et al. 1986). Plant metabolites such as oleanolic acid, β-sitosterol, lupeol, quercetin, 3, 3-dimethyl hexanone, and octadecanoic acid, 2-hydroxy-1,3-propanediyl ester from *Acanthus ilicifolius* hold promise as natural biodegradable crop protectants against *Pratylenchus* spp. infecting maize (Fabiyi 2015). Plant essential oils, most especially terpenes, triterpenes, and monoterpenes, have been associated with control of plant-parasitic nematodes; sesquiterpenoid lactones such as alantolactone and sesquiterpenic acids which consist of costic and isocostic acids are known to have exhibited nematicidal activities (Mahajan et al. 1986; Bourrel et al. 1993; Oka et al. 2000; Oka 2001). Essential oil from *Hyptis suaveolens* with sabinene, beta-caryophyllene, terpinolene, and 1,8-cineole as major constituents expressed a promising bioactivity on eggs and second-stage juveniles of *Heterodera sacchari* (Fabiyi et al. 2015). Some plant materials possessing potent phytochemical compounds with nematicidal activities are shown in Table 13.1. Phytochemicals such as alkaloids, terpenoids, flavonoids, and tannins play a vital role in the bio-control of many nematodes (Fig. 13.2). Generally, plants with high alkaloid content seem to possess more nematicidal properties (Renčo et al. 2014). Methyl benzoate obtained from the aerial parts of *Buddleja crispa* was reported to exhibit more nematicidal activity than *Azadirachta indica* at the same concentration (Sultana et al. 2010).

Table 13.1 Some confirmed natural plant metabolites with nematicidal activities against *Meloidogyne* species

S. no.	Natural sources	Natural nematicidal constituents	Nematode type	References
1	*Acorus calamus* rhizome	Synergy of compounds	*Meloidogyne incognita*	Wiratno et al. (2009)
2	*Alstonia boonei* leaves	Fatty acids and triterpenes	*Meloidogyne incognita*	Fabiyi et al. (2012b), Fabiyi et al. (2012a, b), Fabiyi et al. (2016)
3	*Armoracia rusticana* roots	Allyl isothiocyanate	*Meloidogyne incognita*	Aissani et al. (2013)
4	*Artemisia herba-alba*	Synergy of compounds	*Meloidogyne incognita*	Al-Banna et al. (2003)
5	*Artemisia vulgaris* rhizome	Synergy of compounds	*Meloidogyne megadora*	Costa et al. (2003)
6	*Azadirachta indica* leaves	Synergy of compounds	*Meloidogyne incognita*	Bawa et al. (2014)
7	*Azadirachta indica* seeds	Synergy of compounds	*Meloidogyne javanica*	Moosavi (2012)
8	*Berberis brevissima* roots	Berberine	*Meloidogyne javanica*	Ali et al. (2019)
9	*Bridelia ferruginea* leaves	Fatty acids and triterpenes	*Meloidogyne incognita*	Fabiyi et al. (2012b), Fabiyi et al. (2012a, b), Fabiyi et al. (2016)
10	*Buddleja crispa* aerial parts	Methyl benzoate	*Meloidogyne incognita*	Sultana et al. (2010)
11	*Calea urticifolia* leaves and roots	Synergy of compounds	*Meloidogyne incognita*	Alejo et al. (2006)
12	*Capsicum annuum* fruits	Synergy of compounds	*Meloidogyne incognita*	Bawa et al. (2014)
13	*Castanea sativa*	Tannins	*Meloidogyne javanica*	Maistrello et al. (2010)
14	*Chenopodium ambrosioides* leaves	Essential oil	*Meloidogyne incognita*	Bai et al. (2011)
15	*Chromolaena odorata* roots	Synergy of compounds	*Meloidogyne incognita*	Adegbite and Adesiyan (2005)
16	*Chrysanthemum cinerariaefolium* flower	Synergy of compounds	*Meloidogyne incognita*	Perez et al. (2003)
17	*Cymbopogon citratus* root	Synergy of compounds	*Meloidogyne incognita*	Adegbite and Adesiyan (2005)
18	*Cymbopogon citratus* leaves	Synergy of compounds	*Meloidogyne incognita*	Fabiyi et al. (2018)
19	*Cymbopogon nardus* leaves and stem	Synergy of compounds	*Meloidogyne incognita*	Varma and Dubey (1999)
20	*Derris elliptica* roots	Synergy of compounds	*Meloidogyne incognita*	Feng (2001)

(continued)

Table 13.1 (continued)

S. no.	Natural sources	Natural nematicidal constituents	Nematode type	References
21	*Eugenia winzerlingii* leaves	Synergy of compounds	*Meloidogyne incognita*	Alejo et al. (2006)
22	*Euphorbia macroclada* leaves	Synergy of compounds	*Meloidogyne incognita*	Al-Banna et al. (2003)
23	*Helianthus annuus* leaves	Synergy of compounds	*Meloidogyne incognita*	Fabiyi and Atolani (2013)
24	*Kigelia africana* leaves	Atolaside	*Meloidogyne incognita*	Atolani et al. (2014a)
25	*Kigelia africana* leaves	Isovitexin	*Meloidogyne incognita*	Atolani et al. (2014b)
26	*Lavandula luisieri* aerial parts	Rosmarinic acid Necrodane	*Meloidogyne javanica*	Julio et al. (2016)
27	*Lawsonia inermis* leaf	Synergy of compounds	*Meloidogyne incognita*	Fabiyi and Atolani (2011)
28	*Meliaazedarach*	Synergy of compounds	*Meloidogyne incognita*	Cavoski et al. (2012)
29	*Mentha pulegium* aerial parts	Salicylic acid, pulegone	*Meloidogyne incognita*	Caboni et al. (2013)
30	*Mentha spicata* aerial parts	Salicylic acid, carvone	*Meloidogyne incognita*	Caboni et al. (2013)
31	*Myrothecium verrucaria* (fungi)	Verrucarin A	*Meloidogyne incognita*	Nguyen et al. (2018)
32	*Myrothecium verrucaria* (fungi)	Roridin A	*Meloidogyne incognita*	Nguyen et al. (2018)
33	*Parkia biglobosa* seeds	Synergy of compounds	*Meloidogyne incognita*	Bawa et al. (2014)
34	*Pelargonium graveolens* leaves	Essential oil	*Meloidogyne incognita*	Leela et al. (1992)
35	*Petroselinum crispum* aerial parts	Xanthotoxol	*Meloidogyne incognita*	Caboni et al. (2014)
36	*Piper betle* leaves	Synergy of compounds	*Meloidogyne incognita*	Wiratno et al. (2009)
37	*Pulsatilla koreana* root	Hederacolchiside E, hederacoside B	*Meloidogyne incognita*	Li et al. (2013)
38	*Ricinus communis* roots	Synergy of compounds	*Meloidogyne incognita*	Adegbite and Adesiyan (2005)
39	*Ricinus communis* seeds	Synergy of compounds	*Meloidogyne incognita*	Cavoski et al. (2012)
40	*Syzygium aromaticum* bud	Synergy of compounds	*Meloidogyne incognita*	Wiratno et al. (2009)

(continued)

Table 13.1 (continued)

S. no.	Natural sources	Natural nematicidal constituents	Nematode type	References
41	*Tephrosia cinerea* stems	Synergy of compounds	*Meloidogyne incognita*	Alejo et al. (2006)
42	*Terminalia nigrovenulosa* bark	3,4-Dihydroxybenzoic acid	*Meloidogyne incognita*	Nguyen et al. (2013)
43	*Triumfetta gradidens* aerial parts	Waltherione E Waltherione A	*Meloidogyne incognita*	Jang et al. (2015)
44	*Xenorhabdus budapestensis* SN84	Rhabdopeptide J	*Meloidogyne incognita*	Bi et al. (2018)
45	*Zanthoxylum alatum*	Synergy of compounds	*Meloidogyne incognita*	Mukhtar et al. (2013)
46	*Zingiber officinale* rhizomes	Synergy of compounds	*Meloidogyne incognita*	Bawa et al. (2014)

13.6 Impact of Some Organic (Natural) Nematicides on the Environment

Obviously, most phytochemicals are generally accepted as being safe for human consumption, animal health, and the environment. This concept has led to a wide range of applications of phytochemicals for various purposes such as pesticides, nutraceuticals, and cosmeceuticals, among others. The phytochemicals in current use are not subjected to rigorous toxicity scrutiny and they are not adequately controlled by law enforcement agents (Bode and Dong 2014). However, there are growing concerns about the safety of the wide use of some bio-nematicides particularly of plant origin. Many phyto-nematicides are reportedly toxic or produce metabolites or degraded products that are toxic to other economically important soil microfauna and microflora. Some of the toxic phyto-constituents in bio-nematicides are leached to lakes and rivers where they contaminate the water bodies and kill aquatic organisms. Some endocrine disruptors, especially phytoestrogens such as genistein and daidzein (both isoflavones), have been identified in some plants (Zin et al. 2013, 2014; Delclos et al. 2001; Ripamonti et al. 2018). Kojic acid obtained in some plant extracts is known to interfere with the thyroid funtion in humams (Higa et al. 2002; Ota et al. 2009). Methylazoxymethanol from *Cycas* seed, safrole from sassafras plant, and ptaquiloside from bracken fern are some examples of phytochemicals reportedly indicated to be carcinogenic (Bode and Dong 2014). Some of the natural products such as capsaicin from *Capsicum* spp. have generated a lot of debate with respect to the claim that they are carcinogenic (Perry et al. 2007). While many natural organic compounds have been proven to have potency as pesticides, some of the compounds are persistent in the environment (not easily degraded) and are potentially carcinogenic, neurotoxic, genotoxic, ecotoxic, mutagenic, and nephrotoxic agents (Bode and Dong 2014; Klaschka 2016). Interestingly, natural crude extracts of many plants such as *Tanacetum vulgare*, *Nicotiana tabacum*, *Eucalyptus* species, *Ricinus communis*, and *Melilotus officinalis* contain toxic

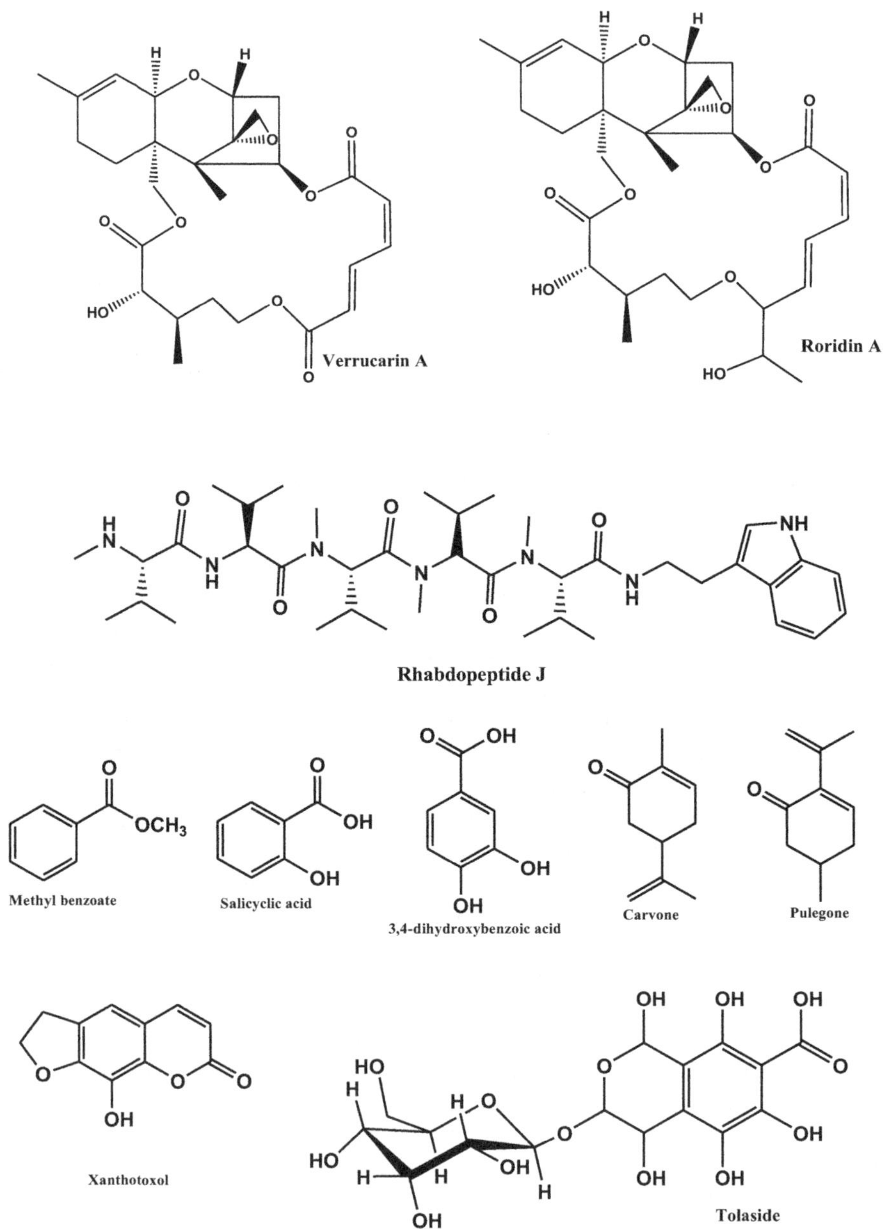

Fig. 13.2 Some potential bio-nematicides from natural sources

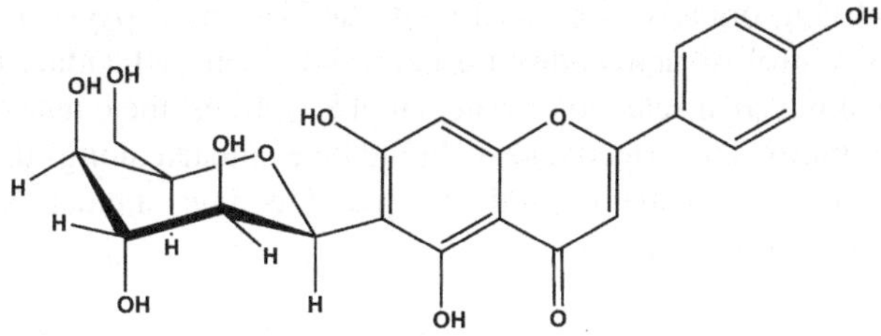

Isovitexin

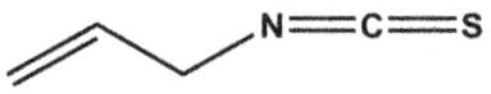

Allylisothiocyanate

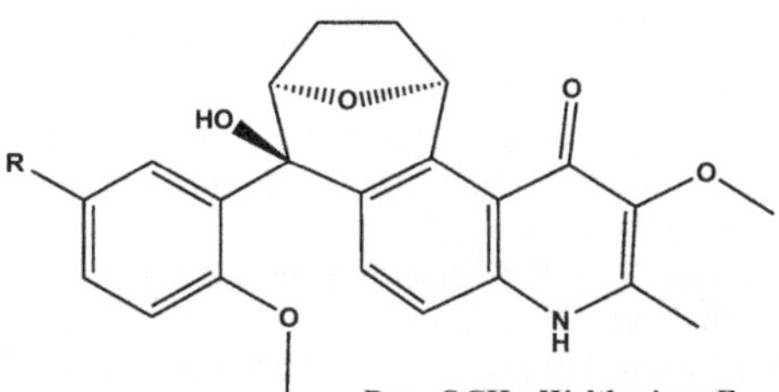

R = - OCH_3, Waltherione E
R = - H, Waltherione A

Fig. 13.2 (continued)

constituents that could affect farm animals, other organisms, and humans (Badawy et al. 2010; Phillips et al. 2010). Likewise, pure natural isolates (compounds) such as digitoxin from *Digitalis purpurea*, ephedrine from *Ephedra* spp., nicotine from *Nicotiana tabacum*, colchicine from *Colchicum* spp., atropine from *Atropa belladonna*, physostigmine from *Physostigma venenosum*, strychnine from *Strychnos nux-vomica*, aconitine from *Aconitum* spp., hyoscyamine from *Hyoscyamus niger* and other Solanaceae species, and papaverine from various Papaveraceae species have been reported to be toxic (Klaschka 2016). These reports clearly reveal that plant materials may not be absolutely safe in the environment if not carefully selected or screened before use.

13.7 Conclusions, Caution, and Future Prospects

Obviously, natural products have numerous compounds with potential as nematicidal agents. While crude extracts or raw plant materials have exhibited nematicidal potential, purified isolates and compounds have also demonstrated such capabilities. However, with the growing and increasing interest in the application of phyto-based nematicides for a greener environment, caution should be exercised to prohibit the indiscriminate proliferation of phytotoxic endocrine disruptors, xenobiotics, and other dangerous natural chemicals which could inadvertently induce micro-resistance and toxicities in living organisms as well as destabilize the ecosystem. Apparently, the careful bio-control of nematode infestation has many advantages over other options. While making such selection for nematicide options, the consideration of affordability, availability, and effectiveness should be evaluated alongside environmental safety. If a bio-nematicide is well adopted, the huge annual or perennial loss could be grossly minimized.

References

Adegbite AA, Adesiyan SO (2005) Root extracts of plants to control root-knot nematode on edible soybean. World J Agric Sci 1(1):18–21

Aissani N, Tedeschi P, Maietti A, Brandolini V, Garau VL, Caboni P (2013) Nematicidal activity of allylisothiocyanate from horseradish (*Armoracia rusticana*) roots against *Meloidogyne incognita*. J Agric Food Chem 61(20):4723–4727

Al-Banna L, Darwish R, Aburjai T (2003) Effect of plant extracts and essential oils on root-knot nematode. Phytopathol Mediterr 42(2):123–128

Ajith K, Sheela MS (1996) Utilization of green leaves of neem and Eupatorium for the management of soil organisms in okra and cowpea. Indian J Nematol 26(2):139–143

Alam KS, Ahmed M, Hossain I, Haque AK (1995) Efficacy of dholkalmi (*Ipomoea fistulosa*) and Furadan to control root knot of okra. Bangladesh J Plant Path 11(1-2):33–35

Alejo JC, Suárez JMT, Catzin SM, Mendoza MN, Baizabal LM, Polanco PS, Sánchez SRP, Angulo MMG (2006) In vitro sensitivity of *Meloidogyne incognita* to extracts from native Yucatanean plants. Nematropica 36:89–97

Ali S, Naz I, Alamzeb M, Urrashid M (2019) Activity guided isolation of nematicidal constituents from the roots of *Berberis brevissima* Jafri and *Berberis parkeriana* Schneid. Tarim Bilimleri Dergisi 25(1):108–114

Ansari RA, Khan TA (2012a) Parasitic association of root-knot nematode, *Meloidogyne incognita* on guava. e-J Sci Technol (e-JST) 5:65–67

Ansari RA, Khan TA (2012b) Diversity and community structure of phytonematodes associated with guava in and around Aligarh, Uttar Pradesh, India. Trends Biosci 5(3):202–204

Ansari RA, Mahmood I (2017a) Optimization of organic and bio-organic fertilizers on soil properties and growth of pigeon pea. Sci Hortic 226:1–9

Ansari RA, Mahmood I (2017b) Determination of disease incidence caused by *Meloidogyne* spp. and or *Fusarium udum* on pigeonpea in Aligarh district: a survey. Trends Biosci 10 (24):5239–5243

Ansari RA, Mahmood I (2019a) Plant health under biotic stress: volume 2: Microbial interactions. Springer, Singapore. https://doi.org/10.1007/978-981-13-6040-4

Ansari RA, Mahmood I (2019b) Plant health under biotic stress: volume 1: Organic strategies. Springer, Singapore. https://doi.org/10.1007/978-981-13-6043-5

Ansari RA, Rizvi R, Sumbul A, Mahmood I (2017a) PGPR: current vogue in sustainable crop production. In: Kumar V, Kumar M, Sharma S, Prasad R (eds) Probiotics and plant health. Springer, Singapore, pp 455–472

Ansari RA, Mahmood I, Rizvi R, Sumbul A (2017b) Siderophores: Augmentation of soil health and crop productivity. In: Kumar V, Kumar M, Sharma S, Prasad R (eds) Probiotics in agroecosystem. Springer, Singapore, pp 291–312

Ansari RA, Sumbul A, Rizvi R, Mahmood I (2019) Organic Soil Amendments: Potential tool for soil and plant health management. In: Ansari RA, Mahmood I (eds) Plant health under biotic stress. Springer, Singapore, pp 1–35

Atolani O, Fabiyi OA, Olatunji GA (2014) Nematicidal isochromane glycoside from Kigelia pinnata leaves. Acta Agric Slov 104(1):25–31. https://doi.org/10.14720/aas.2014.104.1.3

Atolani O, Fabiyi OA, Olatunji GA (2014a) Nematicidal isochromane glycoside from *Kigelia pinnata* leaves. Acta Agri Slovenica 104(1):25–31

Atolani O, Fabiyi OA, Olatunji GA (2014b) Isovitexin from *Kigelia pinnata*, a potential eco-friendly nematicidal agent. Trop Agric 91(2):67–74

Badawy ME, El-Arami SA, Abdelgaleil SA (2010) Acaricidal and quantitative structure activity relationship of monoterpenes against the two-spotted spider mite, *Tetranychus urticae*. Exp Appl Acarol 52:261–274

Bai CQ, Liu ZL, Liu QZ (2011) Nematicidal constituents from the essential oil of *Chenopodium ambrosioides* aerial parts. J Chem 8:143–148

Bakker J (1993) Current state of nematicides. In: Zadoks JC (ed) Modern crop protection: developments and perspectives. Wageningen Agric. Univ. Press, Wageningen, pp 21–26

Bawa JA, Mohammed I, Liadi S (2014) Nematicidal effect of some plants extracts on root-knot nematodes of tomato. World J Life Sci Med Res 3(3):82

Bi Y, Gao C, Yu Z (2018) Rhabdopeptides from *Xenorhabdus budapestensis SN84* and their nematicidal activities against *Meloidogyne incognita*. J Agric Food Chem 66(15):3833–3839

Bode AM, Dong Z (2014) Toxic phytochemicals and their potential risks for human cancer. Cancer Prev Res (Phila) 8(1):1–8

Bouagga A, Chaabane H, Toumi K, Mougou Hamdane A, Nasraoui B, Joly L (2019) Pesticide residues in Tunisian table grapes and associated risk for consumer's health. Food Addit Contam Part B Surveill 12:135–144. https://doi.org/10.1080/19393210.2019.1571532

Bourrel C, Vilarem G, Perineau F (1993) Chemical analysis, bacteriostatic and fungistatic properties of the essential oil of Elecampane (*Inula helenium* L.). J Essent Oil Res 5:411–417

Bridge J (1996) Nematode management in sustainable and subsistence agriculture. Annu Rev Phytopathol 34:201–225

Caboni P, Saba M, Tocco G, Casu L, Murgia A, Maxia A, Menkissoglu-Spiroudi U, Ntalli N (2013) Nematicidal Activity of Mint Aqueous Extracts against the Root-Knot Nematode Meloidogyne incognita. J Agric Food Chem 61(41):9784–9788

Caboni P, Saba M, Oplos C, Aissani N, Maxia A, Menkissoglu-Spiroudi CL, Ntalli N (2014) Nematicidal activity of furanocoumarins from parsley against *Meloidogyne* spp. Pest Manag Sci 71(8):1099–1105

Caloni F, Cortinovis C, Rivolta M, Davanzo F (2016) Suspected poisoning of domestic animals by pesticides. Sci Total Environ 539:331–336

Cavoski I, Chami ZA, Bouzebboudja F, Sasanelli N, Simeone V, Mondelli D, Miano T, Sarais G, Ntalli NG, Caboni P (2012) *Melia azedarach* controls *Meloidogyne incognita* and triggers plants defence mechanisms on cucumber. Crop Prot 35:85–90

Chitwood DJ (2002) Phytochemical-based strategies for nematode control. Annu Rev Phytopathol 40:221–249

Choez AK, Evaristo RR (2018) Acute intoxication by carbofuran in cattle. J Vet Med Surg 2 (1):20–22

Cook PW (1990) Alkaloids in *A. boonei*. J Org Chem 79:631

Costa PA, Bohlmannn F, Schulz J, Buhmannn U (2003) Pathogenicity study and nematoxic properties of some plant extracts on root knot nematodes of tomato. Blackwell Publishing Ltd, New York, p 22

Coyne DL, Cortada L, Dalzell JJ, Claudius-Cole AO, Haukeland S, Luambano N, Talwana H (2018) Plant-parasitic nematodes and food security in sub-Saharan Africa. Annu Rev Phytopathol 56(1):381–403

Davis EL, Meyers DM, Dullum CJ, Feitelson JS (1997) Nematicidal activity of fatty acid esters on soyabean cyst and root-knot nematodes. J Nematol 29(45):677–684

Delclos KB, Bucci TJ, Lomax LG, Latendresse JR, Warbritton A, Weis CC, Newbold RR (2001) Effects of dietary genistein exposure during development on male and female CD (Sprague-Dawley) rats. Reprod Toxicol 15:647–663

Dijan, N; Hasan, S.M. and Madanlar, H. (1994). Nematocidal properties of carboxylic-acids and derivatives. Pestic Biochem Phys 50:229–239. www.faqs.org/faqs/doc. Accessed 4Apr 2012
Dobšíková R (2003) Acute toxicity of carbofuran to selected species of aquatic and terrestrial organisms. Plant Protect Sci 39:103–108
Dutta TK, Khan MR, Phani V (2019) Plant-parasitic nematode management via biofumigation using brassica and non-brassica plants: current status and future prospects. Curr Plant Biol 17:17–32
Eisler R (1985) Carbofuran hazards to fish, wildlife, and invertebrates: a synoptic review. US Fish Wildl Serv Biol Rep 85:36
Fabiyi OA (2014) Characterization, phytochemical analysis and nematicidal activity of *Daniellia oliveri* leaves against Meloidogyne incognita. Pak J Nematol 32(1):91–100
Fabiyi OA (2015) Toxicity of *Acanthus ilicifolius* (L) fractions against *Pratylenchus* spp. on maize (*Zea mays*). Albanian J Agric Sci 14(2):149–156
Fabiyi OA, Atolani O (2011) *Lawsonia inermis* in the control of *Meloidogyne* spp. on *Corchorus olitorius*. J Environ Agric Food Chem 10(3):2000–2006
Fabiyi OA, Atolani O (2013) Varietal response of sunflower *Helianthus annuus* (L) to root-knot nematode *Meloidogyne* spp and other field pathogens. Agric Biol J North Am 4(1):14–18
Fabiyi OA, Olatunji GA, Atolani O (2012a) Nematicidal activities of chromatographic fraction from *Alstonia boonei* and *Bridelia ferruginea* on *Meloidogyne incognita*. Pak J Nematol 30 (2):189–198
Fabiyi OA, Atolani O, Olatunji GA (2012b) The nematicidal activity of *Alstonia boonei* and *Bridelia ferruginea* leaves on *Meloidogyne incognita* infecting *Corchorus olitorius*. Albanian J Agric Sci 11:105–114
Fabiyi OA, Olatunji GA, Olabiyi TI (2014) Toxicity of the constituents of *Parinari polyandra* on *Meloidogyne incognita* infecting Celosia argentea. Nigerian J Pure Appl Sci 27:2552–2561
Fabiyi OA, Osunlola OS, Olatunji GA (2015) In vitro toxicity of extracts from *Hyptis suaveolens* (L.) Poit. on eggs and second-stage juveniles of *Heterodera sacchari*. Agrosearch 15(1):89–99
Fabiyi OA, Atolani O, Olatunji GA (2016) Identification of some chemical constituents of the leaves of *Alstonia boonei* and *Bridelia ferruginea*. Selçuk Tarım ve Gıda Bilimleri Dergisi 30 (1):21–28
Fabiyi OA, Olatunji GA, Adebayo MO, Atolani O (2018) Effect of thermal degraded products of *Cymbopogon citratus* on the in vitro survival of *Meloidogyne incognita* eggs and juveniles. Ceylon J Sci 47(3):235–239Fabiyi OA, Olatunji GA, Daodu IO (2019) Nematicidal effect of organic extract metal complex on *Meloidogyne incognita* infecting groundnuts (*Arachis hypogea*). Scient Agricul Bohem 50(3):191–196
Fadeyi OE, Fabiyi OA, Olatunji GA, Ahmed O (2016) Comparative in vitro and screenhouse response of *Abelmoschus esculentus* infected with *Meloidogyne incognita* to *Anacardium occidentale* fractions. Univ Maurit Res J 22:359–379
Fabiyi OA, Olatunji GA, Daodu IO (2019) Nematicidal effect of organic extract metal complex on Meloidogyne incognita infecting groundnuts (Arachis hypogea). Sci Agric Bohem 50(3):191–196. https://doi.org/10.2478/sab-2019-0026
Feng Y-HW (2001) Plant extracts of several plant pathogenic nematodes kill line screening activity. Huazhong Agric Univ J 3:235–238
Ferré D, Saldeña E, Albarracín L, Neuilly V, Gorla N (2015) Inhibition of butyrylcholinesterase in two intoxicated dogs and analytical confirmation of carbofuran as the causative agent. Rev Vet 26:43–48
Ferslew K, Hagardorn A, McCormick W (1992) Poisoning from oral ingestion of carbofuran (Furadan 4F), a cholinesterase-inhibiting carbamate insecticide, and its effects on cholinesterase activity in various biological fluids. J Forensic Sci 37:337–344
Firoza K, Maqbool MA (1996) Effect of plant extracts in the control of *Helicotylenchus dihystera*. Pak J Nematol 14(1):61–66

Fuglie LJ (1998) Producing food without pesticides: local solutions to crop pest control in West Africa. CTA, Wageningen, The Netherlands, p 158

Goad RT, Goad JT, Atieh BH, Gupta RC (2004) Carbofuran-induced endocrine disruption in adult male rats. Toxicol Mech Methods 14(4):233–239

Gommers FJ, Voorin'tholt DJM (1976) Chemotaxonomy of compositae related to their host suitability for *Pratylenchus penetrans*. Netherlands J Plant Path 82:1–8

Gupta RC (1994) Carbofuran toxicity. J Toxicol Environ Health 43:383–418

Hayashi M, Wada K, Munakata K (1983) Synthesis and nematicidal activity of phenoxypropionic acid derivatives. Agric Biol Chem 47:2653–2655

Higa Y, Ohkubo A, Kitajima S, Moriyasu M, Kariya K (2002) Effects of kojic acid on thyroidal functions in rats by single-dose administration and in cultured rat thyroid cells (FRTL-5 cells). J Toxicol Sci 27:423–431.

Holgador R, Magnusson C (2012) Quarantine nematodes – European legislation, current status and perspectives. Data sheet of quanratine pests for European and Mediterranean plant protection organization. Bioforsk, Belgrade. https://www.eppo.int/RESOURCES/eppo_publications

Jain RK, Dabur KR, Gupta DS (1994) Assessment of available losses in field due to root knot nematodes *Meloidogyne incognita* in a fewer vegetable crops. Indian J Nematol 24:81–184

Jalal GE, Reed SK (1986) Materials and methods for controlling nematodes. Plant and Soil 85:112–118

Jang JY, Le Dang Q, Choi YH, Choi GJ, Jang KS, Cha B, Luu NH, Kim J-C (2015) Nematicidal activities of 4-quinolone alkaloids isolated from the aerial part of *Triumfetta grandidens* against *Meloidogyne incognita*. J Agric Food Chem 63(1):68–74

Julio LF, Barrero AF, Herrador Del Pino MM, Arteaga JF, Burillo J, Andres MF, Díaz CE, González-Coloma A (2016) Phytotoxic and nematicidal components of *Lavandula luisieri*. J Nat Prod 79(2):261–266

Kashaja IN, Fogain R, Speijer PR (1999) Habitat management for control of banana nematodes. In: Frison EA, Gold CS, Karamura EB, Sikora (eds) Mobilizing IPM for sustainable banana production in Africa, p 108. https://cgspace.cgiar.org/handle/10568/105389

Klaschka U (2016) Are natural compounds used in personal care products toxic for the aquatic environment? Sustain Chem Pharm 4(2016):13–20

Kottegoda MB (1985) Safety in use of pesticides and medical treatment. Chem Ind:623–625

Kratochvil RJ, Sardanelli S, Everts K, E. (2004) Gallagher evaluation of crop rotation and other cultural practices for management of root-knot and lesion nematodes. Agron J 96:1419–1428

Lambert K, Bekal S (2002) Introduction to plant-parasitic nematodes. Plant Health Instr. https://doi.org/10.1094/PHI-I-2002-1218-01

Leela NK, Khan RM, Reddy PP, Nidiry ESJ (1992) Nematicidal activity of essential oil of *Pelargonium graveolens* against the root-knot nematode *Meloidogyne incognita*. Nematol Mediterr 20:57–58

Li W, Sun Y, Yan X, Yang S, Lee S, Byun H, Moon CS, Han BS, Kim Y (2013) Isolation of nematicidal triterpenoid saponins from *Pulsatilla koreana* root and their activities against *Meloidogyne incognita*. Molecules 18(5):5306–5316

Loveridge R, Ryan GE, Sum P, Grey-Read O (2019) Poisoning causing the decline in South-East Asia's largest vulture population. Bird Conserv Int 29(1):41–54

Mahajan R, Singh P, Bajaj KL, Kalsi PS (1986) Nematicidal activity of some sesquiterpenoids against root-knot nematode (*M. incognita*). Nematologia 32:119–123

Maistrello L, Vaccari G, Sasanelli N (2010) Effect of chestnut tannins on the root-knot nematode *Meloidogyne javanica*. Helminthologia 47:48–57

Maurice M, Perry RN, Starr JL (2009) Meloidogyne species: a diverse group of novel and important plant parasites. In: Perry RN, Moens M, Starr JL (eds) Root-knot Nematodes. CABI Publishing, Wallingford, pp 1–17

McSorley R (2003) Adaptations of nematodes to environmental extremes. Florida Entomologist 86 (2):138–142

Moosavi MR (2012) Nematicidal effect of some herbal powders and their aqueous extracts against *Meloidogyne javanica*. Nematropica 42:48–56

Mori M, Hyeon S, Kimura Y, Susuki A (1982) The nematicidal activity of acetylene compounds. Agric Biol Chem 46:309–311
Mukhtar T, Kayani MZ, Hussain MA (2013) Nematicidal activities of Cannabis sativa L. and Zanthoxylum alatum Roxb. against Meloidogyne incognita. Ind Crop Prod 42:447–453. https://doi.org/10.1016/j.indcrop.2012.06.027
Narayana YD, Reddy DR (1980) Penetration, development and histopathology of root-knot nematodes in tomato 'NTDR-1'. Nematol Mediterr 8:43–49
Nguyen D-M-C, Seo D-J, Kim K-Y, Park R-D, Kim D-H, Han Y-S, Kim TH, Jung W-J (2013) Nematicidal activity of 3,4-dihydroxybenzoic acid purified from *Terminalia nigrovenulosa* bark against *Meloidogyne incognita*. Microb Pathog 59–60:52–59
Nguyen LTT, Jang JY, Kim TY, Yu NH, Park AR, Lee S, Bae CH, Yeo JH, Hur JS, Park HW, Kim JC (2018) Nematicidal activity of verrucarin A and roridin A isolated from *Myrothecium verrucaria* against *Meloidogyne incognita*. Pestic Biochem Phys 148:133–143
Novtoný L, Misík L, Honzlová A, Ondracek P, Kuca K et al (2011) Incidental poisoning of animals by carbamates in the Czek Republic. J Appl Biomed 9:157–161
Oka Y (2001) Nematicidal activity of essential oil components against the root-knot nematode *Meloidogyne javanica*. Nematology 3:159–164
Oka Y, Nacar S, Putievsky E, Ravid U, Yaniv Z, Spiegel Y (2000) Nematicidal activity of essential oils and their components against the root-knot nematode. Phytopathology 90:710–715
Onifade AK, Fawole B (1996) Effect of some plant extracts on the pathogenicity of *M. incognita* on cowpea. Global J Pure Appl Sci 2(1):9–15
Ota Y, Imai T, Onose J, Takami S, Cho YM, Hirose M, Nishikawa A (2009) A 55-week chronic toxicity study of dietary administered kojic acid (KA) in male F344 rats. J Toxicol Sci 34:305–313.
Perez MP, Navas-Cortes JA, Pascual-Villalobos MJ, Castillo P (2003) Nematicidal activity of essential oils and organic amendments from Asteraceae against root-knot nematodes. Plant Pathol 52(3):395–401
Perry L, Dickau R, Zarrillo S, Holst I, Pearsall DM, Piperno DR et al (2007) Starch fossils and the domestication and dispersal of chili peppers (Capsicum spp. L.) in the Americas. Science 315:986–988
Phillips AK, Appel AG, Sims SR (2010) Topical toxicity of essential oils to the German cockroach (Dictyoptera: Blattellidae). J Econ Entomol 103(2):448–459
Pinkerton CB, Kitner DR (2006) Effects of biologically derived products on motility and reproduction of the root-lesion nematode, Pratylenchus penetrans on strawberry. Nematropica 36:181–196
Rao MS, Reddy PP (1991) Studies on the comparative efficacy of certain plant leaves and carbofuran in the management of *Meloidogyne incognita* on tomato. Curr Nematol 3(1):5–6
Renčo M, Sasanelli N, Maistrello L (2014) Plants as natural sources of nematicides. Nova Science Publishers, Inc, New York, pp 115–142
Ripamonti E, Allifranchini E, Todeschi S, Bocchietto E (2018) Endocrine disruption by mixtures in topical consumer products. Cosmetics 2018(5):61
Seth B, Yadav A, Tandon A, Shankar J, Kumar Chaturvedi R (2019) Carbofuran hampers oligodendrocytes development leading to impaired myelination in the hippocampus of rat brain. Neurotoxicology 70:161–179
Sharom MS, Miles JRW, Harris CR, McEwen FL (1980) Behaviour of 12 insecticides in soil and aqueous suspensions of soil and sediment. Water Res 14:1095–1100
Sikora RA, Coyne DL, Hallman J, Timper P (2018) Reflections on nematology in subtropical and tropical agriculture, vol 125. CABI, Wallingford, pp 1–19
Stapleton JJ, DeVay JE (1983) Response of phytoparasitic and free-living nematodes to soil solarization and 1,3-dichloropropene in California. Phytopathology 73:1429
Starr JL, Jeger MJ (1985) Dynamics of winter survival of eggs and juveniles of *Meloidogyne incognita* and *M. arenaria*. J Nematol 17(3):252–256
Steiner G, Albin FM (1946) Resuscitation of the nematode *Tylenchus polyhypnus*, n. sp., after almost 39 years dormancy. J Wash Acad Sci 36:97–99

Stirlin GR (1991) Antagonists of nematodes. In: Stirling GR (ed) Biological control of plant parasitic nematodes. CAB International, Wallingford, Oxford, pp 50–98

Stirlin G, Stanton J (1997) Nematode diseases and their control. In: Brown JF, Ogle HJ (eds) Plant pathogens and plant diseases. Rockvale Publications, Armidale, Australia, pp 505–517

Stoll G (2000) Natural Crop protection in the tropics: letting information came to life. Margraf Verlag Publication, Weikersheim, Germany, p 376

Subramaniyan S (1985) Effect of *Eupatorium odoratum* extracts on *Meloidogyne incognita*. Indian J Nematol 15:247

Sultana N, Akhter M, Khan RA, Afza N, Tareen RB, Malik A (2010) Nematicidal natural products from the aerial parts of *Buddleja crispa*. Nat Prod Res 24(9):783–788

Sweelam ME (1989) The potential use of some ornamental plants for nematode control in Egypt. Bull Facul Agric Univ Cairo 40(2):391–393

Talwana H, Sibanda Z, Wanjohi W, Kimenju W, Luambano-Nyoni N et al (2016) Agricultural nematology in East and Southern Africa: problems, management strategies and stakeholder linkages. Pest Manag Sci 72:226–245

Uhlenbroek JH, Bijloo JD (1958) Investigations on nematicides. Isolation and structure of a nematicidal principle occurring in Tagetes roots. Recueil des Travaux Chimiques des Pays-Bas 77:1004–1008

United State Environmental Protection Agency (USEPA) (1990) Drinking water criteria document on carbofuran. Prepared under Program No. 1524 for contract 68-C8-0033 by ICAIR, Life Systems, Inc. Cleveland, Ohio for the U.S. Environmental Protection Agency

United State Environmental Protection Agency (USEPA) (2009) Bans carbofuran pesticide residues on food. Environmental News Service

United States Environmental Protection Agency (USAEPA) (2019) Hazard summary. EPA.gov. https://www3.epa.gov/ttn/atw/hlthef/dibromo-.html. Assessed 10 May 2019

Varma J, Dubey NK (1999) Prospectives of botanical and microbial products as pesticides of tomorrow. Curr Sci 76:172–179. http://www.ias.ac.in/currsci/jan25/articles22.html

Viaene N, Coyne DL, Davies K (2013a) Biological and cultural control. In: Perry RA, Moens M (eds) Plant nematology, 2nd edn. CABI, Wallingford, pp 383–410

Viaene N, Coyne D, Davies K, Perry R, Moens M (2013b) Biological and cultural management. In: Perry RN, Moens M (eds) Plant nematology. CABI, Wallingford, pp 383–410

Vrain C (1980) Fatty acids and their derivatives for nematode control. J Nematol 12(4):240

Waele DD, Romulo GD (1998) The root-knot nematodes of banana, Musa pest factsheet no. 3. CGIRA, France. https://cgspace.cgiar.org/bitstream/handle/10568/105398/697.pdf?sequence=7&isAllowed=y

Wesemael WML, Moens M (2008) Quality damage on carrots (*Daucus carota* L.) caused by the root-knot nematode *Meloidogyne chitwoodi*. Nematology 10:261–270

Wesemael WML, Perry RN, Moens M (2006) The influence of root diffusate and host age on hatching of the root-knot nematodes, *Meloidogyne chitwoodi* and *M. fallax*. Nematology 8:895–902

WHO (2009) The World Health Organization (WHO) recommended classification of pesticides by hazard and guidelines to classification. http://www.who.int/ipcs/publications/pesticides_ hazard/en/

Wiratno W, Taniwiryono D, Berg HV, Riksen JAG, Rietjens IMCM, Djiwanti SR, Kammenga JE, Murk AJ (2009) Nematicidal activity of plant extracts against the root-knot nematode, *Meloidogyne incognita*. Open Nat Prod J 2:77–85

Zeinsteger P, Oliver C, Ledri D, Danielle M, Mestorino N (2014) Presence of pesticide carbofuran (carbamato) in bait samples and gastric content of caninos of the city of Urdinarrain, Entre Ríos.

Zin SRM, Omar SZ, Khan NLA, Musameh NI, Das S, Kassim NM (2013) Effects of the phytoestrogen genistein on the development of the reproductive system of Sprague Dawley rats. Clinics 68:253–262

Zin SRM, Omar SZ, Khan NLA, Musameh NI, Das S, Kassim NM (2014) Effects of genistein on male Sprague Dawley rats reproductive development. Biomed Res 25:391–400

Chapter 14
Utilization of Beneficial Microorganisms in Sustainable Control of Phytonematodes

B. D. Narotham Prasad, B. Subramanyam, R. N. Lakshmipathi, Rizwan Ali Ansari, Rose Rizvi, Aisha Sumbul, Irshad Mahmood, N. Susheelamma, and C. M. Rachmi

Abstract Agriculture throughout the globe is subject to both biotic and abiotic stresses. Among biotic stresses that crop plants have to face, plant-parasitic nematodes along with other parasitic microorganisms and insect pests are an age-old destructive force, resulting in the drastic decrease in yields. With the advancement in scientific research, chemical pesticides came into existence. This resolved the problem posed by pests to a great extent. But over the years repeated use of chemical pesticides has given rise to some ecological issues like soil health disruption, groundwater pollution, etc. which cause a greater imbalanced and unproductive agroecosystem. The hazardous effects of these chemicals are creating the hurdles

B. D. N. Prasad (✉) · R. N. Lakshmipathi
Department of Agricultural Microbiology, University of Agricultural Sciences, GKVK, Bangalore, India

B. Subramanyam
Horticultural Research and Extension Centre, Hogalagere, Karnataka, India

R. A. Ansari
Section of Plant Pathology and Nematology, Department of Botany, Aligarh Muslim University, Aligarh, India

Present Address: Department of Plant Protection, Faculty of Agricultural Sciences, Aligarh Muslim University, Aligarh, India

R. Rizvi · I. Mahmood
Section of Plant Pathology and Nematology, Department of Botany, Aligarh Muslim University, Aligarh, India

A. Sumbul
Section of Plant Pathology and Nematology, Department of Botany, Aligarh Muslim University, Aligarh, India

Present Address: Department of Biosciences, Jamia Millia Islamia, New Delhi, India

N. Susheelamma
Department of Rural Development, Jananabarathi University, Bengaluru, India

C. M. Rachmi
Department of Soil Science and Agricultural Chemistry, University of Agricultural Sciences, GKVK, Bangalore, India

R. A. Ansari et al. (eds.), *Management of Phytonematodes: Recent Advances and Future Challenges*, https://doi.org/10.1007/978-981-15-4087-5_14

in the exportation of various food commodities. Therefore, current demand is to search some new alternatives which can provide better solutions and also protect the agroecosystem from non-judicious chemical pesticides. Biological control is considered one of the good options to manage the plant pathogens and pests infesting various crops. Biopesticides do not pose a great threat to the environment and protect the human health from various unexpected diseases. Therefore, the aim of this chapter is to shed light on the importance of biopesticides in relation to management of plant-parasitic nematodes (PPNs).

Keywords Phytonematodes · Microorganisms · Biocontrol · AM fungi and rhizosphere microbes

14.1 Introduction

PPNs cause major damage to many commercial crops which causes a negative impact on the world's food security. As per estimates, the loss incurred due to PPNs is 8.8% and 14.6% annually in various developed and developing countries, respectively (Nicol et al. 2011), which approximately amounts to US$157 billion (Escudero and Lopez-Llorca 2012). This shows the paramount importance of PPNs in agricultural crops (Ansari and Khan 2012a, b; Ansari and Mahmood 2017b). However, the exact loss estimation of PPNs is too difficult (Schomaker and Been 2006). Their microscopic size, underground existence, and non-specific symptoms make their presence often undetected; therefore, the diagnoses of nematode problems are frequently confused with nutritional deficiencies or other soil factors (Perry and Moens 2011). The current available practice for the management of PPNs is the use of chemical nematicides; however, their application has inducted various problems related to the environment as well as human health (Moosavi and Zare 2012). Also, the unbearable cost of the chemical pesticides causes a great economic burden to the marginal farmers. Besides, effective nematicides of chemical origin are unbearable by marginal-level farmers and also create environment and human health problems (Thomason 1987; Quesada-Moraga et al. 2014). Frequent application of other available nematicides can result in emerging resistant nematode races, which in turn will make their control more difficult (Narayanasamy 2013). These problems along with raising social awareness on health concerns and ecological hazards of chemical pesticides have caused an urge to search for other safer alternative methods. Biological control has attained great attention because it promotes sustainable agriculture and environmental protection. "Biological control" is normally used to describe the effect of soil biota in decreasing the nematode populations to lower average levels than would occur in their absence (Schroth and Hancock 1982; Stirling 2011). Several other microorganisms (microfauna such as fungi, bacteria, viruses, and protozoa; mesofauna like rotifers, nematodes, tardigrades, collembolans, mites, and enchytraeids; and macrofauna, i.e., earthworms, termites, and millipedes) are there which influence the nematode populations (Walia and Vats 2000; Costa et al. 2011; Stirling 2011). However, the performance of these bioagents

(BCAs) maximally depends on the agroecosystem. However, predatory soil nematodes have also been found to have biocontrol activity against PPNs. The target of this chapter is focused upon the significance of different BCAs in the management of PPNs with particular emphasis on nematophagous fungi and bacteria to include them in the integrated nematode management (INM) program. The present chapter will discuss the role of antagonistic microorganisms to PPNs and their impact in the plant health ameliorations. The chapter will shed light on recent findings pertaining to the impact of microorganisms in the management of PPNs.

14.2 Fungi Antagonistic to PPN

According to Cayrol et al. (1992), the fungal species antagonistic to PPNs can be dealt under the following types:

1. Predatory fungi
2. Egg-parasitic fungi
3. Nematophagous fungi with adhesive spores
4. Endophytic fungi

14.2.1 Predatory Fungi

Plant-parasitic nematodes fall prey to trapping mechanisms exhibited by some fungi. The fungi differ by their trapping mechanisms as follows: (1) network traps (*Arthrobotrys oligospora*, *A. superba*), (2) constrictive rings (*A. anchonia*, *A. dactyloides*, *Dactylaria brochopaga*, etc.), or (3) adhesive knobs (*Monacrosporium cionopagum*, *Dactylella lobata*). Some fungi belonging to *Hyphomycetes* such as *Dactylaria candida* present two types of trapping mechanisms: (1) adhesive knobs and (2) constrictive but non-adhesive rings (B'Chir 1983; Sosa et al. 2018). Various groups of fungi are there which can trap the nematodes at various stages such as larval or free adult stage. Production of peptides and extracellular proteases hydrolyzing the nematode's cuticles has also been found to induce trap formation (Huang et al. 2004).

14.2.2 Egg-Parasitic Fungi

These kinds of fungi are blessed with the ability to destroy the eggs of PPNs. Generally such fungi are saprophytes in nature and they have the ability to invade already-dead eggs. Examples include *Paecilomyces*, *Pochonia*, and *Verticillium* genera (Rodriguez-Kabana et al. 1984). *Paecilomyces lilacinus* and *Pochonia chlamydosporia* are considered to be the most prominent and effective egg-parasitic fungi. *P. lilacinus* has been found to be very effective in controlling various species of root-knot nematodes infesting tomato and other vegetable crops (Cayrol et al. 1989; Goswami and Mittal 2004; Goswami et al. 2006; Sosa et al. 2018).

14.2.3 Fungi with Adhesive Spores

The fungi producing adhesive spores to parasitize PPNs belong to many classes such as *Oomycetes* (*Catenaria anguillulae*, *Myzocytium lenticulare*, and *M. anomalum* with biflagellated zoospores able to encyst in the nematode's cuticle); *Zygomycetes* (*Meristracum asterospermum* which has spherical structures called conidia which ultimately produce germinating filaments which finally produce new conidiospores); fungi imperfecti, i.e., *Deuteromycetes* (*Meria coniospora* which contains club-like spores, which fix on the host via their anterior part) and *Basidiomycetes* (*Nematoctonus leiosporus*, with adhesive spores); and *Hyphomycetes* (*Hirsutella* sp.).

14.2.4 Endophytic Fungi

There are reports on endophytic fungi such as arbuscular mycorrhizal fungi (AM fungi) and *Fusarium oxysporum*. But, a special emphasis should be made on AM fungi in control of PPNs and it will be dealt separately under the succeeding sub-heading. In vitro experiments suggested secondary metabolite production as a mechanism for nematode control which is used by endophytic, non-pathogenic *Fusarium oxysporum*; however, the production of those toxic metabolites in root cells has not yet been proved (Sikora et al. 2007). Dababat and Sikora (2007) observed that *F. oxysporum*-colonized plants had root exudates that did not attract nematode or the root exudates included compounds repellent to *Meloidogyne incognita*. Another possible mechanism of non-pathogenic *F. oxysporum* in suppression of PPNs is competition for nutritious substances and colonization sites. But, we cannot exclude the role of induced systemic resistance (ISR) in this aspect (Hallmann and Sikora 2011).

14.2.5 AM Fungi: An Antagonist to PPNs

AM fungi are by nature obligate symbionts of plant roots colonizing more than 80% of the terrestrial plant species. As microsymbionts of terrestrial plants, they must be having a certain purpose. First, it gets shelter and photosynthetic carbon from macrosymbiont/plants/crops. In turn they extend their helping hand to plants by improving overall plant growth (Azcon-Aguilar and Barea 1996; Smith et al. 2010).

Besides, they also help in the alleviation of biotic and abiotic stresses as well (Gianinazzi et al. 2010; Singh et al. 2011; Vos et al. 2012). The importance of AM fungi in terms of management of PPNs has been observed in a wide range of host plants (Pozo and Azcón-Aguilar 2007; Jung et al. 2012). According to Veresoglou and Rillig (2012), AM fungi are capable of suppressing nematode diseases by 44–57% by employing one or a combination of several mechanisms against PPNs. The research findings show that it not only controls necrotrophic pathogens but also biotrophs either directly or indirectly. Further, *Acaulosporaceae* has been found to be the weakest family of AM fungi in suppression of PPNs. A review by Pinochet et al. (1996) and Hol and Cook (2005) confirms the application of AM fungi also in the suppression of PPNs. Greenhouse and field experiments indicate that AM fungi act as plant shield against PPN attack (Mayerhofer et al. 2013; Calvet et al. 2001; Alban et al. 2013). Although there are reports which advocate the biocontrol activity of the AM fungi, however, under the natural/field condition, this is not considered a routine agricultural practice due to significant variations in the results (Salvioli and Bonfante 2013, 2013; Dong and Zhang 2006). Therefore, descriptions on modes of action will help to utilize them as efficient biocontrol agents. Different mechanisms involved are dealt under the following headings:

14.2.5.1 Good Plant Stand

In an agroecosystem good crop stand is governed by nutrient status of the soil. It also depends on how plants handle other abiotic and biotic factors acting upon them constantly. AM fungi help hosts by increasing the uptake of water and nutrients especially phosphate (Parniske 2008; Baum et al. 2015; Smith and Smith 2011). On the other hand, AM fungi receive photosynthetic carbon from their host (Gianinazzi et al. 2010). AM fungi alleviate the abiotic stresses like extreme drought and heavy metal toxicity (Singh et al. 2011), and AM fungi could also help out the plants to strengthen themselves so that they can combat with other pathogens and pests including PPNs. AM fungi-mediated biocontrol has been attributed to take up maximum phosphate molecules. Non-mycorrhizal plants were supplied with phosphate which did not show significant control in pathogen multiplication (Bodker et al. 1998). Fritz et al. (2006) showed that tomato plants inoculated with *Rhizophagus irregularis* did not show any remarkable symptoms caused by *Alternaria solani* compared to non-mycorrhizal plants. Plants grown in a suitable environment with a better nutrient status can alleviate the PPN infestation in cotton fields infested with *Rotylenchulus reniformis* (Pettigrew et al. 2005). However, there is no concrete evidence to state that high nutrient status due to AM fungi is the reason behind enhanced resistance against PPNs. In addition to increased nutrient status, the plants colonized by AM fungi often show improved plant health including increased root growth and branching (Gamalero et al. 2010; Gutjahr and Paszkowski 2013). The root morphology in response to AM fungi colonization depends on plant characteristics. Usually tap roots benefit more than fibrous roots in terms of biomass accumulation and nutrient acquisition (Yang et al. 2014). Increased root branching creates more area for nutrient absorption. It also has implications on pathogenesis (Vos et al. 2014). Positive effects can be found with respect to pathogenesis hindrance due to increased root vigor and higher nutrient uptake capacity. AM fungi-mediated root branching will counter the suppressed root growth caused by PPNs. For instance, Elsen et al. (2003) reported poor root branching due to the extensive infestation of *Radopholus similis* and *Pratylenchus coffeae* on banana which was counterbalanced due to enhanced branching in mycorrhizal plants (*Funneliformis mosseae*).

14.2.5.2 Competition for Nutrition and Space

Every creature on the planet will compete for space and nutrition. AM fungi are no different from this rule. Nutrient competition is considered as the mechanism behind AM fungi-mediated biocontrol (Jung et al. 2012). The carbon supply to the AM fungi from the host plants is greatly varied (4–20%) of the total assimilated carbon (Hammer et al. 2011). Therefore, this is generally assumed that that AM fungi create substantial food competitions for the pathogens (Vos et al. 2014). Different AF fungi exhibit various degrees of carbon requirement and, therefore, show different levels of biocontrol efficacy (Ansari et al. 2019; Lerat et al. 2003). However, this assumption does not suggest the relationship between carbon requirement and biocontrol ability (Jung et al. 2012). For example, *Rhizophagus irregularis* neither exerts substantial biocontrol effectiveness on *Radopholus similis* and *P. coffeae* on banana nor on *M. incognita* infesting tomato despite having greater carbon sink strength compared to *Funneliformis mosseae* (Vos et al. 2012). In addition, space competition is also the great factor behind the suppressed population of nematodes as both microorganisms are sheltered by the same root system (Jung et al. 2012). Non-conducive atmosphere for PPNs in the root system due to AM fungi colonization is maintained which do not allow nematodes to multiply effectively. Formation of arbuscules takes place in the cortex of the root where migratory endoparasites generally feed; the avenue of the nematodes and point of feedings are significantly hampered; due to that there is starvation of food and space leading to death or poor multiplication of the nematodes (Khaosaad et al. 2007; Pozo and Azcón-Aguilar 2007).

14.2.5.3 Induced Systemic Resistance

Microorganisms that are beneficial and pathogenic in nature exhibit similar patterns called microbe-associated molecular patterns (MAMPs) (Zamioudis and Pieterse 2012). Since AM fungi are also obligate biotrophs, there is significant overlap in the transcriptional profile of the plant response to AMF and a biotrophic pathogen (Paszkowski 2006). Host plants recognize MAMPs through recognition receptors. This activates MAMP-triggered immunity response (MTI) in host plants which is regarded as the first line of defense to limit further pathogen invasion (Millet et al. 2010).

14.3 Bacteria Antagonistic to PPNs

Bacteria are the most abundant organisms in soil numerically because 1 g of fertile soil contains approximately 10^5–10^8 bacterial colony-forming units (Metting 1993). On weight basis also bacterial biomass is on the higher side as the total weight of bacteria in temperate grassland is estimated to be 1–2 t/ha (Nannipieri et al. 2003). Many of these bacteria inhabit the rhizosphere and their environmental modification may result in direct or indirect effects on PPNs as well as on the host–nematode inter-relationship (Neipp and Becker 1999; Ansari et al. 2017a, b; Ansari and Mahmood 2017a, 2019a, b). Over 99% of the bacteria present in various environmental samples are non-culturable (Sharma et al. 2005), and they still remain unidentified for their ecological functions (Nannipieri et al. 2003). But other culturable ones have been studied extensively for their possible involvement in interference with nematode behavior, feeding, and reproduction (Hallmann et al. 2009). Some nematophagous bacteria have been encountered in various agroclimatic conditions which have been isolated from soil, plant tissues, and cysts and eggs of PPNs. A number of these bacteria have already shown great ability in controlling PPNs (Weller 1988; Stirling 1991; Maheshwari et al. 2013; Trivedi and Malhotra 2013). Some of those bacterial BCAs may cause diseases in humans and must be tested for any adverse effects before registration. The available pathogenicity assays are time-consuming, costly, and unsuitable for facultative pathogens. Therefore a novel, quick, and inexpensive bioassay was developed on the basis of *Caenorhabditis elegans* (Zachow et al. 2009). Different bacteria exhibiting anti-nematode activities are listed in Table 14.1. Nematode-antagonistic bacteria can be grouped into obligate parasites, opportunistic parasites, and plant growth-promoting rhizobacteria (PGPR) according to their mode of operation against PPNs (Siddiqui and Mahmood 1999; Tian et al. 2007a, b).

14.3.1 Obligate Parasites

Pasteuria spp. are considered to be controlling agents of PPNs. They are obligatory parasites that produce very resistant endospores. The results obtained from the most definitive genetic study have placed *P. penetrans* in the low G + C content *Bacillus* group. It belongs to the *Bacillus–Clostridium* clade (Charles et al. 2005). Four commonly occurring species of *Pasteuria* have been isolated and their effectiveness has been evaluated significantly. For example, *P. penetrans* parasitizes on root-knot nematodes (Sayre and Starr 1985), *P. thornei* on lesion nematodes (Starr and Sayre 1988), *P. nishizawae* on cyst nematodes (Sayre et al. 1991), and *P. usgae* on *Belonolaimus* spp. (Davies et al. 2008; Preston et al. 2003). Opportunistic parasites commonly refer to saprophytic bacteria that can utilize nematodes as one of their multiple nutrient resources. Similar to obligate parasitic bacteria, opportunistic ones are capable of breaking through the body wall of their prey. *Brevibacillus*

Table 14.1 Bacterial species antagonistic to PPNs

Bacteria	Targeted PPNs	Host plant	References
Rhizobacteria	*M. ethiopica*, *Xiphinema index*	Grapevine	Aballay et al. (2013)
Bacillus subtilis	*Aphelenchoides besseyi*, *Ditylenchus destructor*, *Bursaphelenchus xylophilus*, *M. javanica*	Various hosts	Xia et al. (2011)
Nitrogen-fixing bacteria, phosphate-solubilizing bacteria, potassium-solubilizing bacteria	*M. incognita*	Various hosts	El-Hadad et al. (2010)
Paenibacillus polymyxa, *P. Lentimorbus*	*M. incognita*	Various hosts	Son et al. (2009)
Pasteuria penetrans	*Meloidogyne* spp.	Vegetables	Davies (2009), Mateille et al. (2009)
Bacillus thuringiensis	*M. incognita*	Various hosts	Jouzani et al. (2008)
Pseudomonas fluorescens	*M. incognita*, *M. javanica*	Tomato, cotton, cucumber	Hallmann et al. (1998), Siddiqui and Shaukat (2005), Siddiqui et al. (2004)
Enterobacter asburiae	*M. incognita*	Cotton	Hallmann et al. (1998)
Stenotrophomonas maltophilia, *Chromobacterium* sp.	*G. rostochiensis*	Potato	Cronin et al. (1997)
Agrobacterium radiobacter, *Pseudomonas* spp.	*M. incognita*	Cotton, cucumber	Hallmann et al. (1998)

laterosporus is considered to possess an opportunistic lifestyle (Tian et al. 2007a, b). This bacterium is a ubiquitous species, which is usually present in a large spectrum of environments such as soil, water, invertebrate bodies, etc. It has biocontrol activity against insects, nematodes, and mollusks (Berdy 2005; Ruiu 2013). Some isolates of *B. laterosporus* could infect plant-parasitic, animal-parasitic, and free-living nematodes (Huang et al. 2005). Different groups of nematode-antagonistic bacteria differ among themselves in their behavior, reproduction, and biocontrol approaches (Siddiqui and Mahmood 1999). Among the groups of BCAs, bacteria exhibit the most diverse mechanisms related to parasitism; toxin, antibiotic, and enzyme production; competition; and induced resistance. In soil, endospores of *P. penetrans* remain dormant and wait for contact with the cuticle of a second-stage juvenile (J2). Once the contact occurs, the endospores adhere to the cuticle and make an entry into the host through the body wall and initiate infection (Sayre et al. 1988; Bishop 2011; Davies et al. 2011). Sticking of endospores does not necessarily result in infection as it has been observed that endospores become attached to cuticle but fail to parasitize (Davies et al. 1990), heat-killed endospores possess the ability of adhering to nematode cuticle (Dutky and Sayre 1978), and a proportion of spores do not germinate at all (Bishop 2011). To guarantee a successful infection of a J2, at least five endospores should bind to its cuticle (Davies et al. 1988). In the case of root-knot nematodes, endospore penetration usually occurs after the nematode has set up its feeding site (Chen et al. 1997). Excluding *P. penetrans*, endospore germination of other phytoparasitic bacterial species has been found to occur without the plant interaction (Giblin-Davis et al. 2003). Opportunistic bacterial parasites are capable of penetrating the body wall of their prey and adversely affect them by producing secondary metabolites. An extracellular protease was found in certain strains of *B. laterosporus*, which restrains both egg hatching and larval development of nematodes. This compound was found to be stable in heat and have low molecular weight protein (Bone and Singer 1991; Huang et al. 2005; Ruiu 2013).

14.3.2 Plant Growth-Promoting Rhizobacteria (PGPR)

Rhizobacteria refer to those bacteria that are capable of colonizing the rhizosphere aggressively. Since bacteria are capable of using root exudates as nutrients, their population densities in the rhizosphere are up to 100 times more than in bulk soil. Application of rhizobacteria in the management of PPNs has been widely studied (Ansari et al. 2017a, b). Rhizobacteria may suppress PPNs by means of one or more mechanisms including direct antagonism or indirect effects (Haque and Gaffar 1993). Mainly, species of *Bacillus*, *Pseudomonas*, and *Serratia* have been investigated and found that these bacteria have the ability to suppress PPNs. However, rhizobacteria may also indirectly reduce nematode population densities when they decompose the organic matter (Beneduzi et al. 2013). After decomposition various organic products such as volatile fatty acids, hydrogen sulfide, and ammonia can result in deleterious effects on nematodes. The subsequent sub-headings under PGPR will cover different mechanisms through which PGPR control PPNs.

14.3.2.1 Antibiosis

Antibiotic is a compound having low molecular weight which is produced by certain microorganisms. It helps a lot in the management of PPNs by various ways such as competition and parasitism (Compant et al. 2005; Raguchander et al. 2011). Fluorescent pseudomonads employ antibiosis as their major weapon in disease suppression. Rhizobacteria produce toxin, metabolic by-products, and enzymes which hinder the nematode hatching, development, survival, and multiplication (Siddiqui and Mahmood 1999). *P. fluorescens* releases secondary metabolites such as

2–4-diacetylphloroglucinol which reduces cyst nematodes (Siddiqui and Shaukat 2003). Also some rhizobacteria keep a favorable environment for plant growth by producing compounds like hydrogen cyanide, which kills deleterious organisms from the rhizosphere (Tian et al. 2007b).

14.3.2.2 Lytic Enzyme Production

Lytic enzymes are degradative in nature and are produced in plants' rhizosphere. Different enzymes active against PPNs are chitinases, peroxidase, phenylalanine ammonia-lyase (PAL), dehydrogenase, lipases, β-glucanase, proteases, phosphatases, etc. Collagenases and chitinases have both been found to affect nematode cuticles and eggs (Abd-Elgawad and Kabeil 2012; Karthik and Arulselvi 2017). Production of hydrogen sulfide and chitinase from *Corynebacterium paurometabolum* is known to suspend nematode egg hatching (Mena and Pimentel 2002). *Stenotrophomonas maltophilia*, *Bacillus mycoides*, and *Pseudomonas* sp. are characterized by nematicidal activity by production of hydrolytic enzymes and HCN and registered 56–74% reduction in the trichodorid nematode population in potato (Insunza et al. 2002).

14.3.2.3 Induced Systemic Resistance (ISR)

Induced systemic resistance (ISR) is considered to provide non-specific protection against various biotic stresses like fungi, bacteria, nematodes, viruses, and insect pests (Beneduzi et al. 2012). Several important defense enzymes have been found to be associated with systemic resistance such as phenylalanine ammonia-lyase, superoxide dismutase, peroxidase, polyphenol oxidase, lipoxygenase, catalase, chitinase, ascorbate peroxidase, β-1,3-glucanase, and proteinase inhibitors (Pokhare et al. 2012). These enzymes help in the resistance induction by secreting of various phytoalexin and phenolic molecules (Viswanathan et al. 2003). Many earlier researchers have supported the fact that rhizobacteria trigger the plant to induce the resistance and thereby reduce the nematode population (Pieterse et al. 2002). Resistance induction brings about some beneficial changes like strengthening of cell wall, callose deposition, and accumulation of some phenolic compounds. Induced resistance also takes place due to synthesis of PR proteins, phytoalexin, lipopolysaccharides, siderophores, salicylic acid, jasmonic acid, PO, chitinase, and various other secondary metabolites (Siddiqui and Mahmood 1999; Ramamoorthy et al. 2001).

14.3.2.4 Phytohormone Production

Several PGPR strains can produce plant growth-promoting molecules such as cytokines (isopentenyl adenosine, isopentenyl adenine riboside, trans-zeatin ribose, and zeatin), auxins (indoleacetic acid, indolebutyric acid, and phenylacetic acid), gibberellic acid, abscisic acid, ethylene, polyamines, brassinosteroids, jasmonates, salicylic acid, strigolactones, etc. (Gopalakrishnan et al. 2015). Key role of the phytohormone is to enhance the plant growth and yield through cell division stimulation, cell elongation, and tissue expansion (Karthik et al. 2016). Among the phytohormones, indoleacetic acid (IAA) is very important in helping the plant to improve its health (Arora et al. 2013; Shaikh and Saraf 2016).

14.3.2.5 Nitrogen Fixation

Nitrogen plays a crucial role in plant health improvement. Agricultural soils contain nitrogen in limited amount due to continuous nitrogen loss. Rhizobacteria help in the nitrogen fixation and supplementation of various nutrients when there is starvation of nitrogen in the plant. Some PGPR strains have both nitrogen fixation and nematicidal activity while others have nitrogen-fixing ability. The latter type help crop plants by providing nitrogen and thus contributing for the good health of the plants. Symbiotic root-nodule bacteria may reduce nematode damage by promoting plant growth or by stimulating changes in their host plants. They assist in plant growth by providing the required nitrogen via fixing atmospheric nitrogen. Rhizobia also induce physiological, biochemical, and histopathological changes in their host plant. Antibiotic and toxic metabolite production ability has been reported for rhizobia, which may adversely affect PPNs together with other plant pathogens (Trivedi and Malhotra 2013). Application of nitrogen-fixing bacteria markedly enhanced the plant growth and yield of banana and significantly suppressed the *Meloidogyne incognita* and *Radopholus similis* population (Aggangan et al. 2013).

14.3.2.6 Siderophore and Ammonia Production

Organic additive decomposition with the help of ammonifying bacteria produces ammonia which has been proven to be helpful in controlling the nematodes. In addition, another element, i.e., iron, is undoubtedly an indispensable part of living organisms which carries out a role in several processes like respiration, electron transport, photosynthesis, cofactor for many enzymes, etc. Under aerobic conditions, iron is unavailable to living organisms. In an iron-deficient atmosphere, rhizobacteria produce siderophores (iron-chelating compounds) which chelate the insoluble iron and make it possible to be available to living organisms (Dell'mour et al. 2012). Some important rhizobacteria like *Aeromonas*, *Azospirillum*, *Azotobacter*, *Bacillus*, *Burkholderia*, *Enterobacter*, *Pseudomonas*, *Rhizobium*, *Serratia*, and *Streptomyces* sp. produce the siderophores and help in the plant health amelioration (Sujatha and Ammani 2013). Rhizobacterium *Streptomyces* sp. was found to produce the siderophore compounds and manage the nematode population (Ruanpanun et al. 2010).

14.4 Predatory Nematodes

In spite of a 100-year-old interest in using predaceous nematodes as biocontrol agents (Cobb 1917), their mass level of testing has recently been begun to be studied extensively. Among a broad range of soil predators which prey on nematodes, predatory nematodes are the most important (Bilgrami and Brey 2005). They eat all types of nematodes and are considered to be an important part of the soil food web. In addition to their biocontrol potential against PPNs, they stimulate cycling of plant nutrients, which in turn may help plants to better withstand any nematode damage (Yeates and Wardle 1996; Stirling 2011). Predaceous nematodes mainly belong to the orders Mononchida, Rhabditida (infraorders Diplogasteromorpha and Rhabditomorpha and superfamily Aphelenchoidea), Dorylaimida (superfamilies Dorylaimoidea, Nygolaimoidea, and Actinolaimoidea), and Enoplida (families Ironidae, Oncholaimidae, Monohysteridae, and Thalassogeneridae). Each group of predaceous nematodes possesses diverse feeding apparatus and also differs in behavior (prey searching and/or catching), food preferences, and feeding mechanisms (Bilgrami 2008). The buccal cavity of the mononchid predators is highly sclerotized with strong musculature, a big piercing dorsal tooth, and small grasping teeth or denticles. Such feeding apparatus enables them to swallow their prey when it is of smaller size, or cut their larger prey into pieces. The second group, which is usually called stylet-bearing predaceous nematodes, is composed of dorylaimids, nyglolaimids, and aphelenchids (Bilgrami and Brey 2005). The feeding apparatus of these predators is of a piercing and sucking type, which enables them to perforate the body wall of their prey and suck the body contents. Dorylaimid predators possess a hollow stylet called the odontostyle, while the feeding apparatus in nyglolaimids is large, solid, and slender, with a protrusible tooth called the mural tooth. Aphelenchid predators have a fine needle-like stylet with a lumen for ingestion. Dorylaimid predators use their long hollow stylet to disrupt the interior organs of the prey to render them motionless very quickly (Linford and Oliviera 1937), whereas aphelenchid predators use their stylet to insert digestive enzymes into the prey's body and paralyze it (Hechler 1963; Wood 1974). The nyglolaimids' tooth has no lumen; therefore, it is solely used for piercing or slitting their prey and ingesting the prey's body contents (Khan and Kim 2007). The third feeding type that is represented by diplogasterids is the cutting and sucking style. The buccal cavity of the diplogasterid predators is small but well equipped with a strong claw-like movable dorsal tooth (Khan and Kim 2007). They may possess teeth or denticles for cutting the cuticle of the prey or crushing the food particles (Jairajpuri and Bilgrami 1990). Among the different groups of predators, diplogasterids seem the most probable candidate to be sold in the future as a commercial product, as they satisfy many requirements of being a good PPN predator, and they also possess short life cycle, easy culture, prey specificity, chemotaxis sense, and resistance to unfavorable environment (Bilgrami and Brey 2005; Khan and Kim 2007). The dorylaimid predators may possibly be the second best candidate, but their long life cycle is a great deterrent. However, only additional research will confirm their effectiveness at a commercial level (Bilgrami 2008).

14.5 Viruses Antagonistic to PPNs

The pathogenic impact of viruses on nematodes is not very clear; however, development of insect viruses to commercial products (Falcon 1976; Flexner and Belnavis 2000; Sun and Peng 2007; Erlandson and Theilmann 2009; Burand et al. 2009; Kamita et al. 2010; Rodriguez et al. 2012) along with similarity of nematodes to insects, which causes both of them being placed in the same superfamily Ecdysozoa (Telford et al. 2008), strengthened the concept of using nematode viruses as a feasible control measure. Several species of PPNs are natural vectors of plant viruses without being infected (Brown et al. 1995; Hull 2009). In addition, there are several reports on viruses that infect nematodes and replicate within their nematode hosts. Loewenberg et al. (1959) for the first time reported sluggish nematode (*Meloidogyne incognita*) in the presence of virus particles. The body of such virus-infected juveniles appeared to be extremely vacuolated or filled with unusual oil-like globules, and they were incapable of forming galls. The suspension of sluggish nematodes was found to transmit the disease even after it passed through bacterial filters and maintained its virulence following serial passages. The disease was supposed to be caused by a virus; however, the probable involvement of a mycoplasma was not eliminated because the virus particles were not detected in diseased nematodes. Foor (1972) observed cellular abnormalities in both males and females of *Trichosomoides crassicauda* in which virus was supposed to be involved. The nuclei of somatic cells in infected nematodes included some spherical virus-like particles with a diameter of 15 nm. Similarly, some polyhedral virus-like particles with a diameter of 20 nm were detected in cytoplasm of intestinal cells of *Dolichodorus heterocephalus* (Zuckerman et al. 1973). Another report on abnormal behavior of nematodes is related to the phenomenon of swarming (aggregation of nematodes in masses) in *Tylenchorhynchus martini*. The swarmers appeared to be more susceptible to chemicals and other unfavorable conditions. Some symmetrical virus-like inclusion bodies were detected in the internal tissues and on the surface of the cuticle of swarming nematodes rather than healthy ones. Likewise, partial disintegration from epicuticle to median zone was only found in swarming nematodes (Ibrahim and Hollis 1973). The exact causal agent of the swarming disease remained unknown as it was not transmittable from swarming to non-swarming nematodes (Ibrahim et al. 1978).

14.6 Conclusions and Future Prospects

Utilization of biological microbes for the management of PPNs has drawn global attention as a safer strategy than conventional/chemical with regard to pest/pathogen control. Utility of chemical nematicides exhibits stunning and satisfactory results with respect to management of PPNs. However, on the other hand, these chemicals cause potential risk to humans and also cause great biodiversity losses. Therefore, development of new biologically derived approaches is the need of the hour because such management tactics are eco-friendly and easily degradable after application into soil. Invention of new such products will certainly help a lot in the formulation and commercialization. The potentiality of such products can be enhanced after its integration with other judicious practices. An efficient strain of microbial-based pesticides could improve the soil as well as plant health even under harsh agroclimatic conditions. Henceforth, the isolation and identification of competent biological organisms which can be used for the identification and characterization of some pesticides is the current focus of research.

References

Aballay E, Ordenes P, Martensson A, Persson P (2013) Effects of rhizobacterial on parasitism by *Meloidoigyne ethiopica* on grapevine. Eur J Plant Pathol 135:137–145

Abd-Elgawad MM, Kabeil SA (2012) Biological control of *Meloidogyne incognita* by *Trichoderma harzianum* and *Serratia marcescens* and their related enzymatic changes in tomato roots. Afr J Biotechnol 11:16247–16252

Aggangan NS, Tamayao PJS, Aguilar EA, Anarna JA, Dizon TO (2013) Arbuscular Mycorrhizal fungi and nitrogen fixing bacteria as growth promoters and as biological control agents against nematodes in tissue-cultured Banana var. Lakatan Philippine J Sci 142(2):153–165

Alban R, Guerrero R, Toro M (2013) Interactions between a root- knot nematode (Meloidogyne exigua) and arbuscular mycorrhizae in coffee plant development (Coffea arabica). Am J Plant Sci 4:19–23

Ansari RA, Khan TA (2012a) Parasitic association of root-knot nematode, *Meloidogyne incognita* on guava. e-J Sci Technol (e-JST) 5:65–67

Ansari RA, Khan TA (2012b) Diversity and community structure of phytonematodes associated with guava in and around Aligarh, Uttar Pradesh, India. Trends Biosci 5(3):202–204

Ansari RA, Mahmood I (2017a) Optimization of organic and bio-organic fertilizers on soil properties and growth of pigeon pea. Sci Hortic 226:1–9

Ansari RA, Mahmood I (2017b) Determination of disease incidence caused by *Meloidogyne* spp. and/or *Fusarium udum* on pigeonpea in Aligarh district: a survey. Trends Biosci 10 (24):5239–5243

Ansari RA, Mahmood I (2019a) Plant health under biotic stress: volume 2: microbial interactions. Springer, Singapore. https://doi.org/10.1007/978-981-13-6040-4

Ansari RA, Mahmood I (2019b) Plant health under biotic stress: volume 1: organic strategies. Springer, Singapore. https://doi.org/10.1007/978-981-13-6043-5

Ansari RA, Rizvi R, Sumbul A, Mahmood I (2017a) PGPR: current vogue in sustainable crop production. In: Kumar V, Kumar M, Sharma S, Prasad R (eds) Probiotics and plant health. Springer, Singapore, pp 455–472

Ansari RA, Mahmood I, Rizvi R, Sumbul A (2017b) Siderophores: Augmentation of soil health and crop productivity. In: Kumar V, Kumar M, Sharma S, Prasad R (eds) Probiotics in agroecosystem. Springer, Singapore, pp 291–312

Ansari RA, Sumbul A, Rizvi R, Mahmood I (2019) Organic soil amendments: potential tool for soil and plant health management. In: Ansari RA, Mahmood I (eds) Plant health under biotic stress. Springer, Singapore, pp 1–35

Arora NK, Tewari S, Singh R (2013) Multifaceted plant-associated microbes and their mechanisms diminish the concept of direct and indirect PGPRs. In: Arora NK (ed) Plant microbe symbiosis: fundamentals and advances. Springer, Singapore, pp 411–449

Azcon-Aguilar C, Barea JM (1996) Arbuscular mycorrhizas and biocontrol of soil-borne plant pathogens—an overview of the mechanisms involved. Mycorrhiza 6:457–464

Baum C, El-Tohamy W, Gruda N (2015) Increasing the productivity and product quality of vegetable crops using arbuscular mycorrhizal fungi: a review. Sci Hortic 87:131–141

B'Chir MM, Horrigue N, Verlodt H (1983) Miseau point d'une method de lutte integree associant unagent biologique it une substance chimique pour combattreles Meloidogyne sous-abris plastiques en Tunisie. Mededelingen van de Faculteit Landbouwwetenschappen Rijksuniversiteit Gent 48:421–432

Beneduzi A, Ambrosini A, Passaglia LMP (2012) Plant growth-promoting rhizobacteria (PGPR): their potential as antagonists and biocontrol agents. Genet Mol Biol 35:1044–1051

Beneduzi A, Ambrosini A, Passaglia LMP (2013) Plant growth-promoting rhizobacteria (PGPR): their potential as antagonists and biocontrol agents. Genet Mol Biol 35:1044–1051

Berdy J (2005) Bioactive microbial metabolites. J Antibiot 58:1–26

Bilgrami AL (2008) Biological control potentials of predatory nematodes. In: Ciancio A, Mukerji KG (eds) Integrated management and biocontrol of vegetable and grain crops nematodes. Springer, Dordrecht, the Netherlands, pp 3–28

Bilgrami AL, Brey C (2005) Potential of predatory nematodes to control plant-parasitic nematodes. In: Grewal PS, Ehlers R-U, Shapiro-Ilan DI (eds) Nematodes as biocontrol agents. CAB International, Wallingford, UK, pp 447–475

Bishop AH (2011) *Pasteuria penetrans* and its parasitic interaction with plant parasitic nematodes. In: Logan NA, De Vos P (eds) Endospore-forming Soil Bacteria, soil biology 27. Springer-Verlag, Berlin, pp 181–201

Bodker L, Kjoller R, Rosendahl S (1998) Effect of phosphate and the arbuscular mycorrhizal fungus *Glomus intraradices* on disease severity of root rot of peas (*Pisum sativum*) caused by *Aphanomyces euteiches*. Mycorrhiza 8:169–174

Bone LW, Singer S (1991) Control of parasitic nematode ova/larvae with a *Bacillus laterosporus*. US Patent 5,045,314, 3 September 1991

Brown DJ, Robertson WM, Trudgill DL (1995) Transmission of viruses by plant nematodes. Annu Rev Phytopathol 33:223–249

Burand JP, Nakai M, Smith I (2009) Host–virus interactions. In: Stock SP, Vanderberg J, Boemare N, Glazer I (eds) Insect pathogens: molecular approaches and techniques. CAB International, Wallingford, UK, pp 195–222

Calvet C, Pinochet J, Hernández-Dorrego A, Estaún V, Camprubí A (2001) Field micro plot performance of the peach-almond hybrid GF-677 after inoculation with arbuscular mycorrhizal fungi in a replant soil infested with root-knot nematodes. Mycorrhiza 10:295–300

Cayrol JC, Dijan C, Pijarowski L (1989) Study of the nematocidal properties of the culture filtrate of the nematophagous fungus *Paecilomyces lilacinus*. Rev Nematol 12(4):331–336

Cayrol JC, Dijan C, Panchaud-Mattei E (1992) La lutte biologique contre les Nématodes phytoparasites. Courrier de la cellule Environnement de l'INRA 17:31–44

Charles L, Carbone I, Davies KG, Bird D, Burke M, Kerry BR, Opperman CH (2005) Phylogenetic analysis of *Pasteuria penetrans* by use of multiple genetic loci. J Bacteriol 187:5700–5708

Chen ZX, Dickson DW, Mitchell DJ (1997) Suppression mechanisms of *Meloidogyne arenaria* race 1 by *Pasteuria penetrans*. J Nematol 29:1–8

Cobb NA (1917) The *Mononchus*: a genus of free living predatory nematodes. Soil Sci 3:431–486

Compant S, Duffy B, Nowak J, Clement C, Barka EA (2005) Use of plant growth-promoting bacteria for biocontrol of plant diseases: principles, mechanisms of action, and future prospects. Appl Environ Microbiol 71:4951–4959

Costa SR, van der Putten WH, Kerry BR (2011) Microbial ecology and nematode control in natural ecosystems. In: Davies KG, Spiegel Y (eds) Biological control of plant-parasitic nematodes: building coherence between microbial ecology and molecular mechanisms, progress in biological control 11. Springer Science + Business Media, Dordrecht, the Netherlands, pp 39–64

Cronin D, Moënne-Loccoz Y, Dunne C, O'Gara F (1997) Inhibition of egg hatch of the potato cyst nematode *Globodera rostochiensis* by chitinase-producing bacteria. Eur J Plant Pathol 103:433–440

Dababat AA, Sikora RA (2007) Influence of the mutualistic endophyte *Fusarium oxysporum* 162 on *Meloidogyne incognita* attraction and invasion. Nematology 9:771–776

Davies KG (2009) Understanding the interaction between an obligate hyperparasitic bacterium, Pasteuria penetrans and its obligate plant-parasitic nematode host, Meloidogyne spp. Adv Parasitol 68:211–245

Davies KG, Kerry BR, Flynn CA (1988) Observations on the pathogenicity of *Pasteuria penetrans*, a parasite of root-knot nematodes. Ann Appl Biol 112:491–501

Davies KG, Flynn CA, Laird V, Kerry BR (1990) The life-cycle, population dynamics, and host specificity of a parasite of *Heterodera avenae*, similar to *Pasteuria penetrans*. Rev Nematol 13:303–309

Davies KG, Rowe J, Williamson VM (2008) Cuticle variation amongst amphimictic and parthenogenetic populations of nematode (*Meloidogyne* spp.) as exhibited by a bacterial parasite (*Pasteuria penetrans*). Int J Parasitol 38:851–860

Davies KG, Rowe J, Manzanilla-López R, Opperman CH (2011) Re-evaluation of the life-cycle of the nematode-parasitic bacterium *Pasteuria penetrans* in root-knot nematodes, *Meloidogyne* spp. Nematology 13:825–835

Dell'mour M, Schenkeveld W, Oburger E, Fischer L, Kraemer S, Puschenreiter M, Lammerhofer M, Koellensperger G, Hann S (2012) Analysis of iron phyto siderophore complexes in soil related samples: LC-ESI-MS/MS versus CE-MS. Electrophoresis 33:726–733

Dong LQ, Zhang KQ (2006) Microbial control of plant-parasitic nematodes: a five-party interaction. Plant and Soil 288:31–45

Dutky EM, Sayre RM (1978) Some factors affecting infection of nematodes by the bacterial spore parasite *Bacillus penetrans*. J Nematol 10:285

El-Hadad M, Magdi M, Shawky S, Mahgoob A, El-Tayeb T, Abdel Aziz N (2010) In vitro evaluation of some bacterial isolates as biofertilizers and biocontrol agents against the second stage juveniles of *Meloidogyne incognita*. World J Microbiol Biotechnol 26:2249–2256

Elsen A, Beeterens R, Swennen R, DeWaele D (2003) Effects of an arbuscular mycorrhizal fungus and two plant-parasitic nematodes on Musa genotypes differing in root morphology. Biol Fertil Soils 38:367–376

Erlandson MA, Theilmann DA (2009) Molecular approaches to virus characterization and detection. In: Stock SP, Vanderberg J, Boemare N, Glazer I (eds) Insect pathogens: molecular approaches and techniques. CAB International, Wallingford, UK, pp 3–31

Escudero N, Lopez-Llorca LV (2012) Effects on plant growth and root-knot nematode infection of an endophytic GFP transformant of the nematophagous fungus *Pochonia chlamydosporia*. Symbiosis 57:33–42

Falcon LA (1976) Problems associated with the use of arthropod viruses in pest control. Annu Rev Entomol 21:305–324

Flexner JL, Belnavis DL (2000) Microbial insecticides. In: Rechcigl JE, Rechcigl NA (eds) Biological and biotechnological control of insect pests. CRC Press, Boca Raton, FL, pp 34–60

Foor WF (1972) Viruslike particles in a nematode. J Parasitol 58:1065–1070

Fritz M, Jakobsen I, Lyngkjaer MF, Thordal-Christensen H, Pons-Kuhnemann J (2006) Arbuscular mycorrhiza reduces susceptibility of tomato to Alternaria solani. Mycorrhiza 16:413–419

Gamalero E, Pivato B, Bona E, Copetta A, Avidano L, Lingua G (2010) Interactions between a fluorescent pseudomonad, an arbuscular mycorrhizal fungus and a hypovirulent isolate of *Rhizoctonia solani* affect plant growth and root architecture of tomato plants. Plant Biosyst Int J Deal Asp Plant Biol 144:582–591

Gianinazzi S, Gollotte A, Binet M-N, vanTuinen D, Redecker D, Wipf D (2010) Agroecology : the key role of arbuscular mycorrhizas in ecosystem services. Mycorrhiza 20:519–530

Giblin-Davis RM, Williams DS, Bekal S, Dickson DW, Brito JA, Becker JO, Preston JF (2003) 'Candidatus *Pasteuria usgae*' sp. nov., an obligate endoparasite of the phytoparasitic nematode *Belonolaimus longicaudatus*. Int J Syst Evol Microbiol 53:197–200

Gopalakrishnan S, Sathya A, Vijayabharathi R, Varshney RK, Gowda CLL, Krishnamurthy L (2015) Plant growth promoting rhizobia: challenges and opportunities. 3 Biotech 5:355–377

Goswami BK, Mittal A (2004) Management of root-knot nematode infecting tomato by *Trichoderma viride* and *Paecilomyces lilacinus*. Indian Phytopathol 57:235–236

Goswami BK, Pandey RK, Rathour KS, Bhattacharya C, Singh L (2006) Integrated application of some compatible biocontrol agents along with mustard oil seed cake and furadan on Meloidogyne incognita infecting tomato plants. J Zhejiang Univ Sci 7:873–875

Gutjahr C, Paszkowski U (2013) Multiple control level soft root system remodeling in arbuscular mycorrhizal symbiosis. Front Plant Sci 4:204

Hallmann J, Sikora RA (2011) Endophytic fungi. In: Davies KG, Spiegel Y (eds) Biological control of plant-parasitic nematodes: building coherence between microbial ecology and molecular mechanisms, progress in biological control 11. Springer Science + Business Media, Dordrecht, the Netherlands, pp 227–258

Hallmann J, Quadt-Hallmann A, Rodrguez-Kabana R, Kloepper JW (1998) Interactions between *Meloidogyne incognita* and endophytic bacteria in cotton and cucumber. Soil Biol Biochem Soil 30:925–937

Hallmann J, Davies KG, Sikora RA (2009) Biological control using microbial pathogens, endophytes and antagonists. In: Perry RN, Moens M, Starr JL (eds) Root-knot nematodes. CAB International, Wallingford, UK, pp 380–411

Hammer EC, Pallon J, Wallander H, Olsson PA (2011) Tit for tat? A mycorrhizal fungus accumulates phosphorus under low plant carbon availability. FEMS Microbiol Ecol 76:236–244

Haque SE, Gaffar A (1993) Use of rhizobia in the control of root rot diseases of sunflower, okra, soybean and mungbean. J Phytopathol 138:157–163

Hechler HC (1963) Description, developmental biology, and feeding habits of *Seinura tenuicaudata* (De man), J.B. Goodey, 1960 (Nematoda: Aphelenchoididae), a nematode predator. Proc Helminthol Soc Wash 30:183–195

Hol WHG, Cook R (2005) An overview of arbuscular mycorrhizal fungi–nematode interactions. Basic Appl Ecol 6:489–503

Huang XW, Zhao NH, Zhang KQ (2004) Extracellular enzymes serving as virulence factors in nematophagous fungi involved in infection of the host. Res Microbiol 155(10):811–816

Huang XW, Tian BY, Niu QH, Yang JK, Zhang LM, Zhang KQ (2005) An extracellular protease from *Brevibacillus laterosporus* G4 without parasporal crystal can serve as a pathogenic factor in infection of nematodes. Res Microbiol 156:719–727

Hull R (2009) Plant-to-plant movement. In: Hull R (ed) Comparative plant virology, 2nd edn. Academic Press, San Diego, California, pp 223–243

Ibrahim IKA, Hollis JP (1973) Electron microscope studies on the cuticle of swarming and nonswarming *Tylenchorhynchus martini*. J Nematol 5:275–281

Ibrahim IKA, Joshi MM, Hollis JP (1978) Swarming disease of nematodes: host range and evidence for a cytoplasmic polyhedral virus in *Tylenchorhynchus martini*. Proc Helminthol Soc Wash 45:233–238

Insunza V, Alstrom S, Eriksson KB (2002) Root bacteria from nematicidal plants and their biocontrol potential against trichodorid nematodes in potato. Plant and Soil 241:271–278

Jairajpuri MS, Bilgrami AL (1990) Predatory nematodes. In: Jairajpuri MS, Alam MM, Ahmad I (eds) Nematode Bio-Control: Aspects and Prospects. CBS Publishers, New Delhi, pp 95–125

Jouzani GS, Seifinejad A, Saeedizadeh A, Nazarian A, Yousefloo M, Soheilivand S, Mousivand M, Jahangiri R, Yazdani M, Amiri RM, Akbari S (2008) Molecular detection of nematicidal crystalliferous *Bacillus thuringiensis* strains of Iran and evaluation of their toxicity on free-living and plant-parasitic nematodes. Can J Microbiol 54:812–822

Jung SC, Martinez-Medina A, Lopez-Raez JA, Pozo MJ (2012) Mycorrhiza-induced resistance and priming of plant defenses. J Chem Ecol 38:651–664

Khan Z, Kim YH (2007) A review on the role of predatory soil nematodes in the biological control of plant parasitic nematodes. Appl Soil Ecol 35:370–379

Kamita SG, Kang K-D, Hammock BD, Inceoglu AB (2010) Genetically modified baculoviruses for pest insect control. In: Gilbert LI, Gill SS (eds) Insect control biological and synthetic agents. Academic Press, London, pp 331–385

Karthik C, Arulselvi PI (2017) Biotoxic effect of chromium (VI) on plant growth promoting traits of novel *Cellulosimicrobium funkei* strain AR8 isolated from *Phaseolus vulgaris* rhizosphere. Geomicrobiol J 34:434–442

Karthik C, Oves M, Thangabalu R, Sharma R, Santhosh SB, Arulselvi PI (2016) *Cellulosimicrobium funkei*-like enhances the growth of *Phaseolus vulgaris* by modulating oxidative damage under chromium (VI) toxicity. J Adv Res 7:839–850

Khaosaad T, García-Garrido JM, Steinkellner S, Vierheilig H (2007) Take-all disease is systemically reduced in roots of mycorrhizal barley plants. Soil Biol Biochem 39:727–734

Lerat S, Lapointe L, Piché Y, Vierheilig H (2003) Strains colonizing barley roots. Can J Bot 81:886–889

Linford MB, Oliviera JM (1937) The feeding of hollow-spear nematodes on other nematodes. Science 85:295–297

Loewenberg JR, Sullivan T, Schuster MI (1959) A virus disease of *Meloidogyne incognita incognita*, the southern root-knot nematode. Nature (London) 184:1896

Maheshwari DK, Shukla S, Aeron A, Kumar T, Jha CK, Patel D, Saraf M, Wahla V (2013) Rhizobacteria for management of nematode disease in plants. In: Maheshwari DK (ed) Bacteria in agrobiology: disease management. Springer-Verlag, Berlin, pp 379–404

Mateille T, Fould S, Dabire KR, Diop MT, Ndiaye S (2009) Spatial distribution of the nematode biocontrol agent *Pasteuria penetrans* as influenced by the soil habitat. Soil Biol Biochem Soil 41:303–308

Mayerhofer MS, Kernaghan G, Harper KA (2013) The effects of fungal root endophytes on plant growth: a meta-analysis. Mycorrhiza 23:119–128

Mena J, Pimentel E (2002) Mechanism of action of *Corynebacterium paurometabolum* strain C-924 on nematodes. Nematology 4:287

Metting FB (1993) Structure and physiological ecology of soil microbial communities. In: Metting FB (ed) Soil microbial ecology. Marcel Dekker, New York, pp 3–25

Millet YA, Danna CH, Clay NK, Songnuan W, Simon MD, Werck Reichhart D (2010) Innate immune responses activated in Arabidopsis roots by microbe-associated molecular patterns. Plant Cell 22:973–990

Moosavi MR, Zare R (2012) Fungi as biological control agents of plant-parasitic nematodes. In: Merillon JM, Ramawat KG (eds) Plant defence: biological control, progress in biological control 12. Springer Science + Business Media, Dordrecht, the Netherlands, pp 67–107

Nannipieri P, Ascher J, Ceccherini MT, Pertramellara G, Renella G (2003) Microbial diversity and soil functions. Eur J Soil Sci 54:655–670

Narayanasamy P (2013) Biological Management of Diseases of Crops, Progress in biological control 16. Springer Science + Business Media, Dordrecht, the Netherlands

Neipp PW, Becker JO (1999) Evaluation of biocontrol activity of rhizobacteria from Beta vulgaris against Heterodera schachtii. J Nematology 31:54–61

Nicol JW, Turner SJ, Coyne DL, den Nijs L, Hockland S, Tahna Z (2011) Current nematode threats to world agriculture. In: Jones J, Gheysen G, Fenoll C (eds) Genomics and molecular genetics of plant-nematode interactions. Springer Science + Business Media, Dordrecht, the Netherlands, pp 21–43

Parniske M (2008) Arbuscular mycorrhiza: the mother of plant root endosymbioses. Nat Rev Microbiol 6:763–775

Paszkowski U (2006) Mutualism and parasitism: the yin and yang of plant symbioses. CurrOpinPlant Biol 9:364–370

Perry RN, Moens M (2011) Introduction to plant-parasitic nematodes; modes of arasitism. In: Jones J, Gheysen G, Fenoll C (eds) Genomics and molecular genetics of plant-nematode interactions. Springer Science + Business Media, Dordrecht, the Netherlands, pp 3–20

Pettigrew WT, Meredith WR, Young LD (2005) Potassium fertilization effects on cotton lint yield, yield components and reniform nematode populations. Agron J 97:1245–1251

Pieterse CMJ, VanWees SCM, Ton J, VanPelt JA, VanLoon LC (2002) Signalling in rhizobacteria-induced systemic resistance in *Arabidopsis thaliana*. Plant Biol 4:535–544

Pinochet J, Calvet C, Camprubí A, Fernández C (1996) Interactions between migratory endoparasitic nematodes and arbuscular mycorrhizal fungi in perennial crops: a review. Plant Soil 185:183–190

Pokhare S, Shakil NA, Kumar J, Singh K (2012) Foliar application of chemicals elicitors induces biochemical changes in wheat against the cereal cyst nematode Heterodera avenae. Nematol Medit 40:181–187

Pozo MJ, Azcón-Aguilar C (2007) Unraveling mycorrhiza-induced resistance. Curr Opin Plant Biol 10:393–398

Preston JF, Dickson DW, Maruniak JE, Nong G, Brito JA, Schmidt LM, Giblin-Davis RM (2003) *Pasteuria* spp. systematics and phylogeny of these bacterial parasites of phytopathogenic nematodes. J Nematol 35:98–207

Quesada-Moraga E, Herrero N, Zabalgogeazcoa I (2014) Entomopathogenic and nematophagous fungal endophytes. In: Verma VC, Gange AC (eds) Advances in endophytic research. Springer India, New Delhi, pp 85–99

Raguchander T, Saravanakumar D, Balasurramanian P (2011) Molecular approaches to improvement of biocontrol agents of plant diseases. J Biol Control 25:71–84

Ramamoorthy V, Viswanathan R, Raguchander T, Prakasam V, Samiyappan R (2001) Induction by systemic resistance by plant growth promoting rhizobacteria in crop plants against pests and diseases. Crop Prot 20:1–11

Rodriguez-Kabana R, Morgan-Jones G, Godroy G, Gintis BO (1984) Effectiveness of species of *Gliocladium*, *Paecilomyces* and *Verticillium* for control of *Meloidogyne arenaria* in field soil. Nematropica 14:155–170

Rodriguez VA, Belaich MN, Ghiringhelli PD (2012) Baculoviruses: members of integrated pest management strategies. In: Larramendy ML, Soloneski S (eds) Integrated pest management and pest control – current and future tactics

Ruanpanun P, Tangchitsomkid N, Hyde KD, Lumyong S (2010) Actinomycetes and fungi isolated from plant-parasitic nematode infested soils: screening of the effective biocontrol potential, indole-3-acetic acid and siderophore production. World J Microbiol Biotechnol 26:1569–1578

Ruiu L (2013) *Brevibacillus laterosporus*, a pathogen of invertebrates and a broad-spectrum antimicrobial species. Insects 4:476–492

Salvioli A, Bonfante P (2013) Systems biology and "omics" tools: a cooperation for next-generation mycorrhizal studies. Plant Sci 203–204:107–114

Sayre RM, Starr MP (1985) Pasteuria penetrans (Ex Thorne, 1940) nom-rev, comb-n, sp-n, a mycelia and endospore-forming bacterium parasitic in plantparasitic nematodes. Proc Helminthol Soc Wash 52:149–165

Sayre RM, Starr MP, Golden AM, Wergin WP, Endo BY (1988) Comparison of *Pasteuria penetrans* from *Meloidogyne incognita* with a related mycelial and endospore-forming bacterial parasite from *Pratylenchus brachyurus*. Proc Helminthol Soc Wash 55:28–49

Sayre RM, Wergin WP, Schmidt JM, Starr MP (1991) *Pasteuria nishizawae* sp. nov., a mycelia and endospore-forming bacterium parasitic on cyst nematodes of genera *Heterodera* and *Globodera*. Res Microbiol 142:551–564

Schomaker CH, Been TH (2006) Plant growth and population dynamics. In: Perry RN, Moens M (eds) Plant nematology. CAB International, Wallingford, UK, pp 275–301

Schroth MN, Hancock JG (1982) Disease suppressive soil and root colonizing bacteria. Science 216:1376–1381

Shaikh S, Saraf M (2016) Biofortification of *Triticum aestivum* through the inoculation of zinc solubilizing plant growth promoting rhizobacteria in field experiment. Biocatal Agric Biotechnol 9:120–126

Sharma R, Ranjan R, Kapardar RK, Grover A (2005) Unculturable bacterial diversity: an untapped resource. Curr Sci 89:72–77

Siddiqui ZA, Mahmood I (1999) Role of bacteria in the management of plant parasitic nematodes: a review. Bioresour Technol 69:167–179

Siddiqui IA, Shaukat SS (2003) Non-pathogenic Fusarium solani represses the biosynthesis of nematicidal compounds in vitro and reduces the biocontrol of Meloidogyne javanica by Pseudomonas fluorescens in tomato. Lett Appl Microbiol 37:109–114

Siddiqui IA, Shaukat SS (2005) Phenylacetic acid-producing *Rhizoctonia solani* represses the biosynthesis of nematicidal compounds *in vitro* and influences biocontrol of *Meloidogyne incognita* in tomato by *Pseudomonas fluorescens* strain CHA0 and its GM derivatives. J Appl Microbiol 9:43–55

Siddiqui IA, Shaukat SS, Khan A (2004) Differential impact of some *Aspergillus* species on *Meloidogyne javanica* biocontrol by *Pseudomonas fluorescens* strain CHA0. Lett Appl Microbiol 39:78–43

Sikora RA, Schäfer K, Dababat AA (2007) Mode of action associated with microbially induced in planta suppression of plant-parasitic nematodes. Australas Plant Pathol 36:124–134

Singh LP, Gill SS, Tuteja N (2011) Unraveling the role of fungal symbionts in plant abiotic stress tolerance. Plant Signal Behav 6:175–191

Smith FA, Smith SE (2011) What is the significance of the arbuscular mycorrhizal colonization of many economically important crop plants? Plant and Soil 348:63–79

Smith SE, Facelli E, Pope S, Smith FA (2010) Plant performance in stressful environments: interpreting new and established knowledge of the roles of arbuscular rmycorrhizas. Plant and Soil 326:3–20

Son SH, Khan Z, Kim SG, Kim YH (2009) Plant growth-promoting rhizobacteria, *Paenibacillus polymyxa* and *Paenibacillus lentimorbus* suppress disease complex caused by root-knot nematode and Fusarium wilt fungus. J Appl Microbiol 107:524–532

Sosa AL, Rosso LC, Salusso FA, Etcheverry MG, Passone MA (2018) Screening and identification of horticultural soil fungi for their evaluation against the plant parasitic nematode *Nacobbus aberrans*. World J Microbiol Biotechnol 34:63

Starr MP, Sayre RM (1988) *Pasteuria thornei* sp. nov. and *Pasteuria penetrans* sensu strict emend., mycelial and endospore-forming bacteria parasitic, respectively, on plant-parasitic nematodes of the genera *Pratylenchus* and *Meloidogyne*. *Annales de l'Institut Pasteur*. Microbiology 139:11–31

Stirling GR (1991) Biological control of plant parasitic nematodes, Progress, Problems and Prospects. CAB International, Wallingford

Stirling GR (2011) Biological control of plant-parasitic nematodes: an ecological perspective, a review of Progress and opportunities for further research. In: Davies KG, Spiegel Y (eds) Biological control of plant-parasitic nematodes: building coherence between microbial ecology and molecular mechanisms, progress in biological control 11. Springer Science + Business Media, Dordrecht, the Netherlands, pp 1–38

Sujatha N, Ammani K (2013) Siderophore production by the isolates of fluorescent pseudomonads. Int J Curr Res Rev 5:1–7

Sun XL, Peng HY (2007) Recent advances in biological control of pest insects by using viruses in China. Virol Sin 22:158–162

Telford MJ, Bourlat SJ, Economou A, Papillon D, Rota-Stabelli O (2008) The evolution of the Ecdysozoa. Philos Trans R Soc B 363:1529–1537

Thomason IJ (1987) Challenges facing nematology: environmental risks with nematicides and the need for new approaches. In: Veech JA, Dickson DW (eds) Vistas on nematology. Society of Nematologists, Hyattsville, Maryland, pp 469–476

Tian B, Yang J, Zhang KG (2007a) Bacteria used in the biological control of plant parasitic nematodes: populations, mechanisms of action and future prospects. FEMS Microbiol Ecol 61:197–213

Tian BY, Yang JK, Lian LH, Wang CY, Zhang KQ (2007b) Role of neutral protease from *Brevibacillus laterosporus* in pathogenesis of nematode. Appl Microbiol Biotechnol 74:372–380

Trivedi PC, Malhotra A (2013) Bacteria in the management of plant-parasitic nematodes. In: Maheshwari DK (ed) Bacteria in agrobiology: disease management. Springer, Berlin, Heidelberg, pp 349–377

Veresoglou SD, Rillig MC (2012) Suppression of fungal and nematode plant pathogens through Arbuscular mycorrhizal fungi. Biol Lett 8:214–217

Viswanathan R, Rajitha R, Ramesh Sundar A, Ramamoorthy V (2003) Isolation and identification of endophytic bacterial strains from sugarcane stalks and their in vitro antagonism against the red rot pathogen. Sugarcane 5:25–29

Vos CM, Tesfahun AN, Panis B, DeWaele D, Elsen A (2012) Arbuscular mycorrhizal fungi induce systemic resistance in tomato against the sedentary nematode *Meloidogyne incognita* and the migratory nematode *Pratylenchus penetrans*. Appl Soil Ecol 61:1–6

Vos CM, Yang Y, De Coninck B, Cammue BPA (2014) Fungal (-like) biocontrol organisms in tomato disease control. Biol Control 74:65–81

Walia RK, Vats R (2000) Fungal antagonists of phytonematodes. In: Upadhyay RG, Mukerji KG, Chamola BP (eds) Biocontrol potential and its exploitation in sustainable agriculture, vol. 1: crop diseases, weeds, and nematodes. Springer Science, Business Media, New York, pp 155–172

Weller DM (1988) Biological control of soil borne plant pathogens in the rhizosphere with bacteria. Annu Rev Phytopathol 26:379–407

Wood FH (1974) Biology of Seinura demani (Nematoda: Aphelenchoididae). Nematologica 20:347–353

Xia Y, Xie S, Ma X, Wu H, Wang X, Gao X (2011) The *purL* gene of *Bacillus subtilis* is associated with nematicidal activity. FEMS Microbiol Lett 322:99–107

Yang H, Zhang Q, Dai Y, Liu Q, Tang J, Bian X (2014) Effects of arbuscular mycorrhizal fungi on plant growth depend on root system: a meta-analysis. Plant and Soil 389:361–374

Yeates GW, Wardle DA (1996) Nematodes as predators and prey: relationships to biological control and soil processes. Pedobiologia 40:43–50

Zachow C, Pirker H, Westendorf C, Tilcher R, Berg G (2009) The *Caenorhabditis elegans* assay: a tool to evaluate the pathogenic potential of bacterial biocontrol agents. Eur J Plant Pathol 125:367–376

Zamioudis C, Pieterse CMJ (2012) Modulation of host immunity by beneficial microbes. Mol Plant-Microbe Interact 25:139–150

Zuckerman BM, Himmelhoch S, Kisiel M (1973) Virus-like particles in *Dolichodorus heterocephalus*. Nematologica 19:117

Chapter 15
Current Management Strategies for Phytoparasitic Nematodes

Rehab Y. Ghareeb, Elsayed E. Hafez, and Dina S. S. Ibrahim

Abstract Plant pathogenic nematodes cause heavy losses in crop yield worldwide. An estimate shows that parasitic nematodes of the plant cause far more damage each year compared to insect pests. A crop yield decrease in different countries due to these tiny invisible pests is immense. Their occurrence led to an expected 12.3% ($157 billion) world yield reduction. Nematode as a plant-parasitic organism causes losses for different crops and therefore needs immediate attention for their management. The studies for newest friendly environmental alternatives with which to control plant-parasitic nematode populations have subsequently become increasingly important. Nematodes can attack and destroy a wide variety of organisms, including animals, microorganisms, and plants. Different types of phytopathogenic nematodes acquired resistance against the traditional man-made chemical nematicides. The research is still on to find out potential natural enemies of plant pathogenic nematodes. Biocontrol agent application in the management of plant pathogenic nematodes and soil-borne plant pathogens may be an alternative. In addition, plant extract of some herbal and medicinal plants, essential oils, chitinases and proteases enzymes could be also an alternative in the management of plant-pathogenic nematodes.

Keywords Biological control · Plant-pathogenic nematodes · Nematicide

---

R. Y. Ghareeb (✉) · E. E. Hafez
Plant Protection and Biomolecular Diagnosis Department, Arid Lands Cultivation Research Institute (ALCRI), City of Scientific Research and Technological Applications, New Borg El-Arab, Alexandria, Egypt

D. S. S. Ibrahim
Nematology Department, Plant Pathology Research Institute, ARC, Giza, Egypt

R. A. Ansari et al. (eds.), *Management of Phytonematodes: Recent Advances and Future Challenges*, https://doi.org/10.1007/978-981-15-4087-5_15

15.1 Introduction

Plant pathogenic nematode causes enormous damages to several economically important crops (Ansari and Khan 2012a, b; Ansari and Mahmood 2017b). The occurrence of these nematodes results in an overall loss to world agriculture of over $100 billion, and an expected $500 million is actually spent on nematode management (Keren-Zur et al. 2000). Many practices such as modifications of soil environment (Maareg et al. 1999, 2008; Ansari et al. 2019) manipulation of biocontrol agents (Maareg and Badr 2000; Youssef et al. 2008; Ansari et al. 2017a) are found to be more useful in reducing environmental unwanted loads. Nematicides of chemical origin are being used in controlling these pests with remarkable results. However, due to environmental issues, search for alternative means have stridently been intensified for the management of nematode induced diseases. Estimate and commercial evolution of such natural chemicals, including nematicide, are pursued by many research groups and biotech companies. Because of considerable studies in this area in the recent past, a lot of compounds have been isolated from fungi and found against nematode effective. There have been several exciting discoveries; however, no major commercial production based on these natural fungal composition have been so far discovered. Management of plant pathogenic nematodes through biological means using putative strains of microorganisms is a good alternative (Stirling 1991)

15.2 Entomopathogenic Nematodes as Biocontrol

Entomopathogenic bacteria which are hosted by some nematodes (entomopathogenic or insect-parasitic nematode) could serve as biological control agents for insects. The entomopathogenic bacteria have been discovered since the seventeenth century, and they belong to the families Heterorhabditidae and Steinernematidae (Gozel and Gozel 2016). These biological control agents could be used in insect killing especially in the early instars of the economic insects such as larvae and pupae. Nematodes infected with nematopathogenic bacteria can kill the insects by two different ways; the first way will be done through an insect-hosted nematode which is a vector for engage bacteria. The second way will be approached through the parasitism of the entomopathogenic bacteria on the insect itself. It is well known that the majority of the modern biological control strategies are related to entomopathogenic nematodes and they are highly recommended. Moreover, entomopathogenic nematodes are propagated in soil and this has a high impact on the agriculture system because they influence the plant health and the insect populations as well (Lu et al. 2017).

Entomopathogenic nematodes have been active as insect parasites in temperate, subtropical, and tropical regions all over the world (Hominick 2002); meanwhile, their population and distribution was affected by soil texture, temperature, and host availability (Hominick and Briscoe 1990; Stock et al. 1999). Now around 25 *Heterorhabditis* species (Plichta et al. 2009) and more or less 70 valid species of *Steinernema* (Orozco et al. 2013; Nguyen and Buss 2011) were identified all over the world, and the search still continues to discover more and more of them. On the other hand, pytopathogenic nematodes cause 5% loss in plant or crop yields because they affect both the photosynthetic processes and plant root growth (Gheysen and Mitchum 2011; Kyndt et al. 2013).

15.3 Biomanagement of Phytonematodes

Biological management of nematodes and soil-borne plant pathogens by inimical microorganisms is a possible non-chemical means of plant disease control (Stirling 1991; Tian et al. 2007). While a lot of efforts have been done to synthesize effective nematicidal substances that can replace traditional chemical nematicides, only a few have been sophisticated for broad use. Anke and Sterner (1997) reported that some higher fungi have nematicidal metabolites and could be used in nematode control. Marketed by Prophyta (acquired in 2013 by Bayer CropScience Biologics GmbH), BioAct® is a biological nematicide product used in the management of the root-knot nematodes (Brand et al. 2009). Biostat® (EPA approved) contains a bioagent of *P. lilacinum*, selected for the management of plant-pathogenic nematodes (RKN). Mycotal® (Koppert 2014) has an inhibiting action on root-knot nematode *M. incognita* on tomatoes, as reported by Meyer et al. (1990). Mahmoud (2009) stated that Met52® (Novozymes 2014) product has shown nematicide potential as well. There are, however, several positive trials underway. Sharma (1994) revealed that the *Pleurotus sajor-caju* culture broth was able to kill pathogenic nematodes in button mushroom *Agaricus bisporus*, and this type of nematode was reduced by 90% in *P. sajor-caju*-inoculated substrates. Xiang and Feng (2000) confirmed the management of *Meloidogyne arenaria* in *Arachis hypogaea* by *Pleurotus ostreatus*, and the reduction in nematode numbers ranged from 86.96% to 94.03% of peanut root knots (*Arachis hypogaea*). Several varieties of *Trichoderma*, on the other hand, have been widely used as biocontrol agents against soil-borne plant diseases. Whipps (2001) reported that *Trichoderma* has been observed to show biocontrol activity against root-knot nematode (Sharon et al. 2001); in addition, *Trichoderma* can improve plant health by using various mechanisms (Meyer et al. 2004). In studies performed on fungi, it has been shown that fungi have convenient characteristics for biological control of nematode; for instance, fungal defense enzymes such as chitins are able to burst nematode egg shells contributing to parasitism of fungi on nematodes (Gortari and Hours 2008). *Trichoderma* spp. have also been found to produce the defense enzyme chitinase into the culture (Chet et al. 1997; Barker and Koenning 1998), which may help to inhibit the hatching of eggs. Dos Santos et al. (1992) recorded that *T. harzianum* was a successful egg parasite of *M. incognita*. *T. harzianum* was able to grow on the surface of the egg and penetrate the egg shell (Saifullah and Thomas 1996). Saifullah and Thomas (1996) have used

T. harzianum with excellent results against *Globodera rostochiensis*. Different *Trichoderma* species have different ways of penetration (Dumas and Boyonowski 1992). Sharon et al. (2001) documented *T. harzianum* reduction of root-knot nematode *M. javanica* galling on tomato plants. Samuel et al. (2001) evaluated the efficacy of *Trichoderma viride*, *T. harzianum*, *T. hamatum*, *T. koningii*, and *T. pseudokoningii* for the management of *M. javanica*, the root-knot nematode in *A. esculentus* and *V. radiata*. Culture filtrate of *Trichoderma* spp. significantly reduced the hatchability of eggs and exhibited nematicidal activity by killing J_2s of *M. javanica*.

Al-Fattah et al. (2007) evaluated the capacity of *Trichoderma harzianum* and *T. viride* to reduce the incidence and pathogenicity of the root-knot nematode *M. incognita* on tomato. In vitro studies explain that all examined isolates were effective in killing J_{2s} compared with the control. Khattak and Saifullah (2008) evaluated in vitro efficacy of *T. harzianum* isolates against *M. javanica* (Treub) Chitwood and stated that the hatchability of *M. javanica* eggs was suppressed by the action of culture filtrates and this suppression (80.36%) was positively correlated with increase in the concentration of culture filtrates. Sahebani and Hadavi (2008) investigated the biological control of *M. javanica* by *Trichoderma harzianum* BI in greenhouse and in vitro experiments and showed that different concentrations (10^2–10^8 CFU/mL) of *T. harzianum* BI decreased nematode infection compared with control. Moghadam et al. (2009) examined ten isolates of *Trichoderma* related to two species *T. harzianum* and *T. virens* in laboratory and greenhouse on eggs of sugarbeet cyst nematode (*Heterodera schachtii*) for 2 years. Results showed that isolates of *Trichoderma* parasitized 60% of eggs on average. Ashoub et al. (2009) evaluated the ability of some fungi species, isolated from certain newly reclaimed areas in Egypt, as bioagents against the root-knot nematode *M. incognita* infecting some vegetable crops. Abd-Elgawad and Kabeil (2010) studied the efficacies of carbofuran at 1 mg a.i./kg soil, *Serratia marcescens* (1×109 bacterium cells/mL water) at 2 mL of the suspension/kg soil, and three different *Trichoderma harzianum* isolates each separately added at 50 mL/kg soil against the root-knot nematode *Meloidogyne incognita* on two tomato cultivars Super Strain B and Alisa which were assessed in the glasshouse. *Serratia marcescens* and *Trichoderma harzianum* isolates can be used to comply with environmental regulations confronting the use of chemicals. Khan et al. (2011) evaluated the effectiveness of certain nematode antagonistic fungi, *Trichoderma harzianum*, *Paecilomyces lilacinus*, and *Arthrobotrys oligospora*, along with natural organic compound (neem compound mix) to control the root-knot nematode *M. incognita*. Naserinasab et al. (2011) evaluated the capacity of *Trichoderma harzianum* BI in reducing the prevalence and virulence factors of *Meloidogyne javanica* root-knot nematode on tomato. In vitro studies showed that the culture filtrates inhibited hatching of *M. javanica* eggs, and this inhibition was strongly correlated with an increase in culture filtrate concentration. Moreover, fungi produce a high number of products as secondary metabolites in their media; these metabolites have different biological activities, including antimicrobial activity (Bills and Gloer 2016; Karwehl and Stadler 2017). It is well known that about 700 different fungal species have nematicidal effects against different species of nematode (Degenkolb et al. 2015). It was observed that there are many calls to diminish the use of toxic nematicides from the market and the advising of growers to use chemical pesticides which have moderate effect on the phytopathogenic organisms especially nematodes. Nowadays, scientists expect that the endophytes could be a good method for controlling the nematodes and reducing the damage caused by them in plants/crops. The plant-colonizing microbes (endophytes) are often facultative and infect the plants without causing any disease symptoms (Bogner et al. 2017). One of these endophytes that are capable to eradicate nematode infection and propagation is *Fusarium oxysporum* species complex (FOSC); this type induces system resistance (Attitalla et al. 2011). And they reside harmlessly within a plant and have the opportunity to participate in the plant's physiological activities (Richardson 1990).

15.4 Natural Plant Extracts as Nematicides

Because of the hazards of chemical pesticides, researchers identified various nematicidal active natural products such as root exudates, volatile organic compounds (Linford et al. 1938), endophytic microbes (Vetrivelkalai et al. 2010), and plant extracts which are supposed to represent 57 families (Muniasamy et al. 2010; Pavaraj et al. 2010); they contain different active compounds that showed high nematicidal activity such as isothiocyanates, thiophenics, glycosides, alkaloids, phenolics, and fatty acids (Pavaraj et al. 2012). Plants also produce secondary metabolites which in most circumstances serve as biopesticides especially when these plants are attacked by one or more pathogens (Ansari and Mahmood 2017a, 2019a, b; Ansari et al. 2019). Evidently, these compounds are gaining a good recognition due to their importance through inducing plant resistance against pests and diseases (Cavoski et al. 2011, 2012). Consequently, plant extracts are considered as an outstanding option for nematode control (Oka et al. 2006). Accordingly, the extract of *Artemisia vulgaris* rhizome showed activity in inhibiting the hatching of nematode egg, causing high mortality rate in juveniles, and reducing the number of root gall initiated by *Meloidogyne megadora* infesting *Phaseolus vulgaris* (Costa et al. 2003). Taylor and Murant (1966) used tannin extracts of *Acacia mollissima* L. and *Schinopsis lorentzii* L. as nematicides against *Longidorus elongatus* and *Meloidogyne arenaria* in soil, respectively. Also, it was reported that tannins of chestnut (*Castanea sativa* L.) are the most potent nematicide used against the root-knot nematode *M. javanica*, the potato cyst nematodes *Globodera rostochiensis* patotype Ro1 and *G. pallida* patotype Pa2, and the carrot cyst nematode *Heterodera carotae* (Maistrello et al. 2003, 2010; Renčo et al. 2012; Renčo 2013). Many researchers reported that there are a huge number of plant extracts which could be used as nematicides or biocontrol agents for *Meloidogyne* spp. such as neem (*Azadirachta indica* L.) (Salawu 1992), African basil (*Ocimum gratissimum*) (Aralepo 1989), bitter leaf (*Vernonia amygdalina* L.) (Jahn 1989), and moringa (*Moringa oleifera* Lam.) (Ajayi 1990; Youssef et al. 2014). Moreover, triterpinc saponin extracted from *Asparagus adscendens, Albizia chinenses*, and *Acacia concinna* showed a high nematicidal activity for *Meloidogyne incognita* (Meher et al. 1988; Rodriguez 1988). Additionally, thymol mixed with benzaldehyde and alkaloid 1,2-dehydropyrrolizidine exhibited nematicidal activity against cyst and root-knot nematode and *Meloidogyne hapla*, respectively (Soler-Serratosa et al. 1996; Thoden et al. 2009). Moreover, the essential oils and the volatile oils of some medicinal and herbal plants are used as pesticides for a wide range of pests and plant diseases and in nematode control as well (Oka et al. 2000, 2003; Oka 2001; Wang et al. 2004; Cohen et al. 2006).

15.5 Bioenzymes as Nematicides: Chitinases and Proteases

It has been observed that chemical nematicides have a bad impact on the environment and on animal and human health; therefore, safer and eco-friendly compounds that could be used as bionematicides are the need of the hour. They found that rhizospheric bacteria could be used as an effective bioagent for controlling the nematode and fixing nitrogen and also produce many plant growth-promoting organic molecules which ameliorate plant health (Saharan and Nehra 2011). The rhizospheric bacteria can attack nematodes through different mechanisms such as producing either metabolites or enzymes (Becker et al. 1988). Chitinases are a group of enzymes that are produced by these bacteria which can hydrolyze chitin which is the major structural component of nematode cell wall (Yang et al. 2013). Many studies confirm that chitinases induce the early hatching of *Meloidogyne* sp. eggs which resulted in incompletely developed juveniles that were not capable to survive in soil (Mercer et al. 1992; Woo-Jin et al. 2002). Other nematophagous fungi and bacteria have the ability to produce several types of enzymes such as chitinases which control several types of nematodes in soil (Siddiqui and Mahmood 1996; Tian et al. 2007a; Nordbring-Hertz et al. 2011). These enzymes can degrade the chemical and physiological integrity of the cuticles of nematodes, as the first step to kill the nematodes (Huang et al. 2004; Yang et al. 2007), and were highly toxic agents in biocontrol of plant-pathogenic nematodes. Nematicidal activity of chitinases was reported to be produced by various fungi, including *Monacrosporium thaumasium* (Ahmed et al. 2014), which killed nematodes through egg shell snooping and cuticle lysis (Yang et al. 2013). As chitin was the main component of the egg shell and cuticle of nematodes and acted as a target for these nematicidal factors (Radwan et al. 2011), proteases came as the second step by which all the important proteins and enzymes of the nematode cells will be broken and degraded, and finally complete degradation will occur for the targeted nematodes (Lopez-Llorca 1990). It was reported that more than 20 different protease enzymes were isolated from different nematophagous fungi (Yang et al. 2007). Luo et al. stated that protease bmp1 acts as a synergistic factor to increase Cry5B activity (by ~7.9-fold) against *C. elegans* (Luo et al. 2013). Moreover, collagenases also have an important role to play in nematode management through nematode-cuticle infiltration and host-cell breakdown

(Tikhonov et al. 2002; Gan et al. 2007a, b; Mi et al. 2010, Huang et al. 2005; Niu et al. 2006a, b; Tian et al. 2007a, b; Stirling 1991). Collagenases are revealed to be important virulence factors that can degrade the main chemical constituents of the nematode cuticle and egg shell (Yang et al. 2013). Also, collagen amendments to soil reduced galling of tomato roots by *Meloidogyne javanica* when compared with other protein and chitin modifications (Galper et al. 1990).

15.6 Biosynthesized Nanomaterials Against Plant-Pathogenic Nematodes

Nanotechnology is an emerging branch that is being integrated for application in plant disease management including phytopathogenic nematodes using nanomaterials which are directly applied either in soil or foliage parts. This trend is considered to be a good substitute for chemical pesticides. Some researchers assumed that the application of nanoparticles either in soil or in the plant parts could affect the beneficial organisms which are living in symbiosis with the plant (Pérez-de-Luqueet 2017). Synthesis of metal nanoparticles is an interesting area in nanoscience. It was reported that different nanoparticles are used in plant protection and plant disease management such as iron oxide, silver nitrate, copper oxide, and zinc oxide (Rouhi et al. 2003; Tiwari and Jain 2018). Additionally, the usage of nanomaterials in plant disease management is considered as a step forward toward organic and clean agriculture, and on the other side a new application strategy is added to the huge number of applications of nanomaterials (Marchiol 2018). It is well known that there are changes in the chemical and physical properties of materials when transformed into nanoparticles (Fernández-García et al. 2004). Gold nanoparticles (AuNPs) which could be synthesized by different methods either chemically or biologically are used in different applications (Raghavendra et al. 2014). Moreover, silver nanoparticles are much like gold nanoparticles as they possess nematicidal activity, and both are applied in soil or foliage plant parts (Kalimuthu et al. 2008; Parikh et al. 2008). This could be approached through irrigation system either by fertigation or tank-mixture; this kind of application showed high control efficacy on the soil infested with high numbers of nematodes (Thakur et al. 2017; Kalishwaralal et al. 2008). Also, a large number of research revealed that AgNPs not only could be used as nematicide but they also possess bactericidal and fungicidal activity against plant-pathogenic microbes (Wright et al. 1999; Yin et al. 1999; Furno et al. 2004; Kim et al. 2007; Roh et al. 2009). The studies so far conducted postulated that the silver nanoparticle activity may result in their toxicity by inducing oxidative stress which directly affects the cells of aimed nematodes (Lim et al. 2012). Due to their activity as nematicide, silver nanoparticles have become an important nematicide product in the market, and there are many methods used in the synthesis of silver nanoparticles (Hardman 2006). AgNP has also shown evidence of being a potentially effective nematicide (Roh et al. 2009).

Many studies report that the toxicity of sublethal doses of AgNP to nematodes can result in reproduction inhibition [with a range of 0.05–0.5 μg/mL of AgNP for 72 h (Roh et al. 2009; Lim et al. 2012)] and growth inhibition [with a range of 5–50 μg/mL of AgNP for 1–3 days (Meyer et al. 2010)]. This proposed that the AgNP effect may be subtle and chronic at low concentrations applied in the field. Nematicidal activity of AgNP against plant-pathogenic fungi and also against plant-parasitic nematodes is not specific but is associated with disrupting multiple cellular modes of action including membrane permeability, ATP synthesis, and response to oxidative stress in both eukaryotic and prokaryotic cells (Roh et al. 2009; Ahmad et al., 2010; Lim et al. 2012). Moreover, AgNP is a wide-ranging antimicrobial agent that is able to affect plant-pathogenic fungi and bacteria (Park et al. 2006; Jo et al. 2009).

15.7 Concluding Remarks and Future Prospects

It can be concluded that nematodes are very serious pathogens for plants and their control is very difficult and expensive and causes disaster for the environment and human health if nematicides of chemical origins are applied. It is high time to search some alternative and safe materials which can bring the nematode population below the threshold level without affecting either the human or environmental health. Researchers are searching alternatives, and the bioagents in which microbes are used in controlling nematodes in the soil are being popularized. Among these microbes bacteria and fungi are considered to be competent enough to control different species of phytoparasitic nematodes. Other agents such as plant extracts and their active compounds have also been used. It was noted that a huge number of plant extracts and active ingredients have high capability to kill nematodes. Biotechnologically derived enzymes are also capable to control nematodes. Besides, synthesis of nanomaterials either biologically or chemically and their application in the management of phytopathogenic nematodes are very important. Biosynthesis of nanomaterials either by using some plant extracts or microbe filtrates to bring down the population of nematodes in soil below the threshold level and to promote plant resistance against nematode infection is the upcoming roadmap which needs to be considered at the earliest.

References

Abd-Elgawad MMM, Kabeil SSA (2010) Management of the root-knot nematode, *Meloidogyne incognita* on tomato in Egypt. J Am Sci 6:256–262

Ahmad N, Sharma S, Alam MK, Singh VN, Shamsi SF, Mehta BR, Fatma A (2010) Rapid synthesis of silver nanoparticles using dried medicinal plant of basil. Colloids Surf B Biointerfaces 81:81–86

Ahmed M, Laing MD, Nsahlai IV (2014) In vivo effect of selected medicinal plants against gastrointestinal nematodes of sheep. Trop Anim Health Prod 46:411–417

Ajayi VA (1990) Comparison of Nematicidal potential of *Vernonia amygdalina* (bitter leaf) and carbofuran (Furadan) on the growth and yield of root-knot infested soybean *Glycine max*. L. M. Sc. Dissertation. Department of Agric. Biology, University of Ibadan, Nigeria, p 283

Al-Fattah A, Dababat A, Sikora A (2007) Use of *Trichoderma harzianum and Trichoderma viride* for the biological control of *Meloidogyne incognita* on tomato. Jordan J Agric Sci 3:297–309

Anke H, Sterner O (1997) Nematicidal metabolites from higher fungi. Curr Org Chem 1:361–374

Ansari RA, Khan TA (2012a) Parasitic association of root-knot nematode, *Meloidogyne incognita* on guava. e-J Sci Technol 5:65–67

Ansari RA, Khan TA (2012b) Diversity and community structure of phytonematodes associated with guava in and around Aligarh, Uttar Pradesh, India. Trends Biosci 5(3):202–204

Ansari RA, Mahmood I (2017a) Optimization of organic and bio-organic fertilizers on soil properties and growth of pigeon pea. Sci Hortic 226:1–9

Ansari RA, Mahmood I (2017b) Determination of disease incidence caused by *Meloidogyne* spp. and or *Fusarium udum* on pigeonpea in Aligarh district: a survey. Trends Biosci 10 (24):5239–5243

Ansari RA, Mahmood I (2019a) Plant health under biotic stress: volume 2: microbial interactions. Springer, Singapore. https://doi.org/10.1007/978-981-13-6040-4

Ansari RA, Mahmood I (2019b) Plant health under biotic stress: volume 1: organic strategies. Springer, Singapore. https://doi.org/10.1007/978-981-13-6043-5

Ansari RA, Rizvi R, Sumbul A, Mahmood I (2017a) PGPR: current vogue in sustainable crop production. In: Kumar V, Kumar M, Sharma S, Prasad R (eds) Probiotics and plant health. Springer, Singapore, pp 455–472

Ansari RA, Sumbul A, Rizvi R, Mahmood I (2019) Organic soil amendments: potential tool for soil and plant health management. In: Ansari RA, Mahmood I (eds) Plant health under biotic stress. Springer, Singapore, pp 1–35

Aralepo OE. (1989). Nematicidal potentials of pirimiphos-methyl and leaf extracts of *Ocimum gratissimum* in the control of *Meloidogyne incognita* on cowpea. M.Sc. Thesis. Department of Agric. Biology, University of Ibadan, Nigeria, p 205

Ashoub AH, Montasser SA, Mourad MH, Gamal M (2009) Impact of some fungi species against the root knot nematode, *Meloidogyne incognita*. Aust J Basic Appl Sci 3(4):3617–3624

Attitalla IH, Latiffah Z, Salleh B, Brishammar S (2011) Biology and partial sequencing of an endophytic Fusarium oxysporum and plant defense complex. Am J Biochem Mol Biol 1:121–144

Barker KR, Koenning SR (1998) Developing sustainable systems for nematode management. Annu Rev Phytopathol 36:165–205

Becker JO, Zavaleta-mejia E, Colbert SF, Schroth MN, Weinhold AR, Hancock JG, Van Gundy SD (1988) The effect of rhizobacteria on root-knot nematode and gall formation. Phytopathology 78:1466–1469

Bills GF, Gloer JB (2016) Biologically active secondary metabolites from the Fungi. Microbiol Spectr 4(6). https://doi.org/10.1128/microbiolspec.FUNK-0009-2016

Bogner CW, Kamdem RS, Sichtermann G, Matthaus C (2017) Bioactive secondary metabolites with multiple activities from a fungal endophyte. Microb Biotechnol 10:175–188

Brand D, Soccol CR, Sabu A, Roussos S (2009) Production of fungal biological control agents through solid state fermentation: a case study on Paecilomyces lilacinus against root-knot nematodes. Micol Appl Int 21(2):33–50

Cavoski I, Caboni P, Miano T (2011) Natural pesticides and future perspectives. In: Pesticides in the modern world – pesticides use and management. IntechOpen, Rijeka

Cavoski I, Al Chami Z, Bouzebboudja F, Sasanelli N (2012) Melia azedarach controls Meloidogyne incognita and triggers plant defense mechanisms on cucumber. Crop Prot 35:85–90

Chet I, Inbar J, Hadar Y (1997) Fungal antagonists and mycoparasitism. In: Wicklow DT, Söderström B (eds) The mycota. Volume IV: environmental and microbial relationships. Springer, Berlin, Heidelberg, pp 165–184

Cohen Y, Wang WQ, Ben-Daniel BH, Ben-Daniel Y (2006) Extracts of Inula viscosa control downy mildew of grapes caused by Plasmopara viticola. Phytopathology 96:417–424

Costa SD, Santos MSND, Ryan MF (2003) Effect of Artemisia vulgaris rhizome extracts on hatching, mortality and plant infectivity of Meloidogyne megadora. J Nematol 35:437–442

Degenkolb T, Nielsen KF, Dieckmann R, Branco-Rocha F, Chaverri P, Samuels GJ, Thrane U, von Döhren H, Vilcinskas A, Brückner H (2015) Peptaibol, secondary-metabolite, and hydrophobin pattern of commercial biocontrol agents formulated with species of the *Trichoderma harzianum* complex. Chem Biodivers 12:662–684

Dos Santos MA, Ferraz S, Muchovez JJ (1992) Evaluation of 20 species of fungi from Brazil for biocontrol of Meloidogyne incognita race-3. Nematropica 22:183–192

Dumas M, Boyonowski TNW (1992) Scanning electron microscopy of mycoparasitism of Armillaria rhizomorphs by species of Trichoderma. Eur J Forest Pathol 22:279–383

Fernández-García M, Martínez-Arias A, Hanson JC, Rodriguez JA (2004) Nanostructured oxides in chemistry: characterization and properties. Chem Rev 104(9):4063–4104

Furno F, Morley KS, Wong B, Sharp BL, Arnold PL, Howdle SM, Bayston R, Brown PD, Winship PD, Reid HJ (2004) Silver nanoparticles and polymeric medical devices: a new approach to prevention of infection. J Antimicrob Chemother 54:1019–1024

Galper S, Cohn E, Spiegel Y, Chet I (1990) Nematicidal effect of collagen-amended soil and the influence of proteasëand collagenase (l). Revue Néwzatol 13(1):67–71

Gan ZW, Yang JK, Tao N, Liang LM, Mi QL, Li J, Zhang KQ (2007a) Cloning of the gene Lecanicillium psalliotae chitinase Lpchi1 and identification of its potential role in the biocontrol of rootknot nematode Meloidogyne incognita. Appl Microbiol Biotechnol 76:1309–1317

Gan ZW, Yang JK, Tao N, Yu ZF, Zhang KQ (2007b) Cloning and expression analysis of a chitinase gene Crchi1 from the mycoparasitic fungus Clonostachys rosea (syn. Gliocladium roseum). J Microbiol 45:422–430

Gheysen G, Mitchum MG (2011) How nematodes manipulate plant development pathways for infection. Curr Opin Plant Biol 14(4):415–421

Gortari MC, Hours RA (2008) Fungal chitinases and their biological role in the antagonism onto nematode eggs. A review. Mycol Prog 7:221–238

Gozel U, Gozel C (2016) Entomopathogenic nematodes in pest management. In: Integrated pest management (IPM). IntechOpen, Rijeka. https://doi.org/10.5772/63894

Hardman R (2006) Toxicologic review of quantum dots: toxicity depends on physicochemical and environmental factors. Environ Health Perspect 114:165–172. https://doi.org/10.1289/ehp.8284

Hominick WM (2002) Biogeography. In: Gaugler R (ed) Entomopathogenic nematology. CABI Publishing, Wallingford, pp 115–143

Hominick WM, Briscoe BR (1990) Occurrence of entomopathogenic nematodes (Rhabditida: Steinernematidae and Heterorhabditidae) in British soil. Parasitology 100:295–302

Huang XW, Zhao NH, Zhang KQ (2004) Extracellular enzymes serving as virulence factors in nematophagous fungi involved in infection of the host. Res Microbiol 155:811–816

Huang XW, Tian BY, Niu QH, Yang JK, Zhang L, Zhang KQ (2005) An extracellular protease from *Brevibacillus laterosporus* G4 without parasporal crystals can serve as a pathogenic factor in infection of nematodes. Res Microbiol 156:719–727

Jahn SA (1989) *Moringa oleifera* for food and water purification—selection of clones and growing of annual short stems. Entwicklung Landlicher Raum 23:22–25

Jo Y, Byung K, Geunhwa J (2009) Antifungal activity of silver ions and nanoparticles on phytopathogenic fungi. Plant Dis 93:1037–1043

Kalimuthu K, Babu R, Venkataraman D, Bilal M, Gurunathan S (2008) Biosynthesis of silver nanocrystals by Bacillus licheniformis. Colloids Surf B: Biointerfaces 65:150–153. https://doi.org/10.1016/j.colsurfb.2008.02.018

Kalishwaralal K, Ramkumarpandian B, Deepak V, Gurunathan S (2008) Biosynthesis of silver nanocrystals by Bacillus licheniformis. Colloids Surf B: Biointerfaces 65(1):150–153

Keren-Zur M, Antonov J, Berconvitz A, Feldman K, Husid A, Keran G, Marcov N, Rebhun M (2000) Bacillus furmus formulations for the safe control of rootknot nematode. In: The BCPC Conf. Pests & Disease, Brighton, pp 47–52

Karwehl S, Stadler M (2017) Exploitation of fungal biodiversity for discovery of novel antibiotics. In: Stadler M, Dersch P (eds) How to overcome the antibiotic crisis-facts, challenges, technologies & future perspective, Current topics in microbiology and immunology, vol 398. Springer, Cham, Switzerland, pp 303–338

Khan MQ, Abbasi MW, Zaki MJ, Khan D (2011) Control of root-knot nematodes and amelioration of egg plant growth by the combined use of Bacillus Thuringiensis berliner and nematicides. Fuuast J Biol 1(2 December):83–86

Khattak B, Saifullah SM (2008) Effect of some indigenous isolates of *Trichoderma harzianum* on root knot nematode, *Meloidogyne javanica* (TREUB) Chitwood. Sarhad J Agric 24:285–288

Kim JS, Kuk E, Yu KN, Kim J-H, Park SJ, Lee HJ, Kim SH, Park YK, Park YH, Hwang C-Y, Kim Y-K, Lee Y-S, Jeong DH, Cho MH (2007) Antimicrobial effects of silver nanoparticles. Nanomedicine 3:95–101

Koppert Biological Systems. Mycotal (2014) http://www.koppert.fr/ravageurs/thrips/produits-contre-les-thrips/detail/mycotal-4

Kyndt T, Vieira P, Gheysen G, Almeida-Engler J (2013) Nematode feeding sites: unique organs in plant roots. Planta 238(5):807–818

Lim D, Roh J-Y, Eom H-J, Hyun JW, Choi J (2012) Oxidative stress-related PMK-1 P38 MAPK activation as a mechanism for toxicity of silver nanoparticles to reproduction in the nematode *Caenorhabditis elegans*. Environ Toxicol Chem 31:585–592

Linford MB, Yap F, Oliveira JM (1938) Reduction of soil population of root-knot nematode during decomposition of organic matter. Soil Sci 45:127–142

Lopez-Llorca LV (1990) Purification and properties of extracellular proteases produced by the nematophagous fungus *Verticillium suchlasporium*. Can J Microbiol 36:530–537

Lu D, Macchietto M, Chang D, Barros MM, Baldwin J, Mortazavi A (2017) Activated entomopathogenic nematode infective juveniles release lethal venom proteins. PLoS Pathog 13(4):e1006302

Luo LSR, Daniel JY, Richeng XA, Ole Becker J (2013) Effect of films on 1,3-dichloropropene and chloropicrin emission, soil concentration, and root-knot nematode control in a raised bed. J Agric Food Chem 61:10–2400

Maareg MF, Badr STA (2000) The effect of certain biocontrol organisms, oxamyl and their combination on Meloidogyne javanica infecting sugar beet. Egypt J Agronematol 4(1):95–104

Maareg MF, Badr STA, Allam AI (1999) Controlling of root-knot nematode, Meloidogyne incognita using organic soil amendments, nematicides and their mixtures in sugar beet. Egypt J Agronematol 3(1):75–94

Maareg MF, Gohar IMA, Tawfik SF (2008) Effect of certain organic soil amendments on sugarbeet (Beta vulgaris L.) infested with root-knot nematode, meloidogyne javanica under field conditions. Egypt J Biol Pest Control 18:235–241

Mahmoud MF (2009) Pathogenicity of three commercial products of entomopathogenic fungi, Beauveria bassiana, Metarhizum anisopilae and Lecanicillium lecanii against adults of olive fly, Bactrocera oleae (Gmelin) (Diptera:Tephritidae) in the laboratory. Plant Prot Sci 45(3):98–102

Meher HC, Walia S, Sethi CL (1988) Effect of steroidal and triterpenic saponins on the mobility of juveniles of *Meloidogyne incognita*. Indian J Nematol 18:244–247

Maistrello L, Henderson G, Laine RA (2003) Comparative effects of vetiver oil, nootkatone and diso-dium octaborate tetrahydrate on Coptotermes formosanus and its symbiotic fauna. Pest Manag Sci 59:58–68. https://doi.org/10.1002/ps.601

Maistrello L, Vaccari G, Sasanelli N (2010) Effect of chestnut tannins on the root-knot nematode Meloidogyne javanica. Helminthologia 47(1):48–57

Marchiol L (2018) Nanotechnology in agriculture: new opportunities and perspectives. In: New visions in plant science. IntechOpen, Rijeka

Mercer CF, Greenwood DR, Grant JL (1992) Effect of plant and microbial chitinases on the eggs and juveniles of *Meloidogyne hapla* Chitwood. Nematologica 38:227–236

Meyer SLF, Huettel RN, Sayre RM (1990) Isolation of fungi from Heterodera glycines and in vitro bioassays for their antagonism to eggs. J Nematol 22:532–537

Meyer LF, Huettel RN, Liu XZ, Humber RA, Jaba J, Nitao K (2004) Activity of fungal culture filtrates against soybean cyst nematode and root knot nematode egg hatch and juvenile motility. Nematology 6:23–32

Meyer JN, Christopher AL, Yang YX, Turner EA, Badireddyb AR, Marinakos SM, Chilkoti A, Wiesner MR, Auffanbc M (2010) Intracellular uptake and associated toxicity of silver nanoparticles in Caenorhabditis elegans. Aquat Toxicol 100(2):140–150

Mi QL, Yang JK, Ye FP, Gan ZW, Wu CW, Niu XM, Zou CG, Zhang KQ (2010) Cloning and overexpression of *Pochonia chlamydosporia* chitinase gene pcchi44, a potential virulence factor in infection against nematodes. Process Biochem 45:810–814

Moghadam M, Rohani EH, Rastegar MF (2009) Biological control of sugar beet cyst forming nematode with trichoderma under in vitro and green house condition. J Sci Technol Agric Nat Res 13(48):301–312

Muniasamy S, Pavaraj M, Rajan MK (2010) Efficacy of the fruit extract of *Citrullus colocynthis* (L.) on the root-knot nematode *Meloidogyne incognita* infecting *Vigna ungiculata* (L.). J Biopest 3(1):309–312

Naserinasab F, Sahebani N, Etebarian HR (2011) Biological control of *Meloidogyne javanica* by *Trichoderma harzianum* BI and salicylic acid on tomato. Afric J Food Sci 5:276–280

Nguyen KB, Buss EA (2011) *Steinernema phyllophagae* n. sp. (Rhabditida: Steinerne matidae), a new entomopathogenic nematode from Florida, USA. Nematology 13:425–442

Niu QH, Huang XW, Tian BY, Yang JK, Liu J, Zhang L, Zhang KQ (2006a) Bacillus sp. B16 kills nematodes with a serine protease identified as a pathogenic factor. Appl Microbiol Biotechnol 69:722–730

Niu QH, Huang XW, Zhang L, Li Y, Li J, Yang JK, Zhang KQ (2006b) A neutral protease from Bacillus nematocida, another potential virulence factor in the infection against nematodes. Arch Microbiol 185:439–448

Nordbring-Hertz B, Jansson HB, Tunlid A (2011) Nematophagous fungi eLS. John Wiley & Sons, Ltd, Chichester, pp 1–13

Novozymes (2014) Met 52, l'insecticide biologique pour le contrôle de l'otiorhynque de la vigne. Organic Materials Review Institute

Oka Y (2001) Nematicidal activity of essential oil components against the root-knot nematode *Meloidogyne javanica*. Nematology 3:159–164

Oka Y, Nacar S, Putievsky E, Ravid U, Yaniv Z, Spiegel Y (2000) Nematicidal activity of essential oils and their components against the root-knot nematodes. Phytopathology 90(7):710–715

Oka Y, Karssen G, Mor M (2003) Identification, host range and infection process of Meloidogyne marylandi from turf grass in Israel. Nematology 5(5):727–734. https://doi.org/10.1163/156854103322746904

Oka TA, Latha TKS, Mokbel AA, Shauthi A (2006) Infection of *Meloidogyne incognita Fusarium oxysporum* f. sp. *cubense* on tomato. Nematol Mediterr J 26:911

Orozco RA, Hill T, Stock SP (2013) Characterization and phylogenetic relationships of *Photorhabdus luminescens* subsp. *sonorensis* (gamma-Proteobacteria: Enterobacteriaceae), the bacterial symbiont of the entomopathogenic nematode *Heterorhabditis sonorensis* (Nematoda: Heterorhabditidae). Curr Microbiol 66:30–39

Parikh RY, Singh S, Prasad BL, Patole MS, Sastry M (2008) Extracellular synthesis of crystalline silver nanoparticles and molecular evidence of silver resistance from Morganella sp.: towards understanding biochemical synthesis mechanism. Chembiochem 9:1415–1422

Park HJ, Kim SH, Kim HJ, Choi SH (2006) A new composition of nanosized silica-silver for control of various plant diseases. Plant Pathol J 22:295–302

Pavaraj M, Karthikairaj K, Rajan MK (2010) Effect of leaf extract of *Ageratum conyzoides* on the biochemical profile of blackgram *Vigna mungo* infected by root-knot nematode, *Meloidogyne incognita*. J Biopest 3(1):313–316

Pavaraj J, Bakavathiappan M, Baskaran S (2012) Evaluation of some plant extracts for their nematicidal properties against root-knot nematode, Meloidogyne incognita. J Biopest 5:106–110

Pérez-de-Luque A (2017) Interaction of nanomaterials with plants: what do we need for real applications in agriculture? Front Environ Sci 5:12. https://doi.org/10.3389/fenvs.2017.00012

Plichta KL, Joyce SA, Clarke D, Waterfield N, Stock SP (2009) *Heterorhabditis gerrardi* n. sp (Nematoda: Heterorhabditidae): the hidden host of Photorhabdus asym- biotica (Enterobacteriaceae: Gamma-Proteobacteria). J Helminthol 83:309–320

Radwan MA, Abu-ELamayem MM, Farrag SAA, Ahmed NS (2011) Integrated management of Meloidogyne incognita infecting tomato using bio-agents mixed with either oxamyl or organic amendments. Nematol Mediterr 39:151–156

Raghavendra KV, Gowthami R, Shashank R, Harish Kumar S (2014) Panchagavya in organic crop production. Pop Kheti 2(2):233–236

Renčo M (2013) Organic amendments of soil as useful tools of plant parasitic nematodes control. Helminthologia 50(1):3–14

Renčo M, Sasanelli N, Papajovà I, Maistrello L (2012) The nematicidal effect of chestnut tannin solutions on the potato cyst nematode *Globodera rostochiensis* (Woll.) Barhens. Helminthologia 49:108–114. https://doi.org/10.2478/s11687-012-0022-1

Richardson PN (1990) Uses for parasitic nematodes in insect control strategies in protected crops. In: The exploitation of microorganisms in applied biology. Association of Applied Biologists, Wellesbourne, pp 205–210. Aspects of Applied Biology (24)

Rodriguez E (1988) Dithiopolyacetylenes as potential pesticides. In: Cutler HG (ed) Biologically active natural products: potential use in agriculture, 2nd edn. The Society, London, pp 432–437; ISBN: 9780841215566

Roh J-Y, Sim SJ, Yi J, Park K, Chung KH, Ryu D-Y, Choi J (2009) Ecotoxicity of silver nanoparticles on the soil nematode *Caenorhabditis elegans* using functional ecotoxicogenomics. Environ Sci Technol 43:3933–3940

Rouhi J, Mahmud S, Naderi N, Ooi CR, Mahmood MR (2003) Physical properties of fish gelatin-based bio-nanocomposite films incorporated with ZnO nanorods. Nanoscale Res Lett 8:364

Saharan BS, Nehra V (2011) Plant growth promoting Rhizobacteria: a critical review. Life Sci Med Res 21:1–30

Sahebani N, Hadavi N (2008) Biological control of the root-knot nematode Meloidogyne javanica by *Trichoderma harzianum*. Soil Biol Biochem 40:2016–2020

Saifullah, Thomas BJ (1996) Studies on the parasitism of *Globodera rostochiensis* by *Trichoderma harzianum* using low temperature scanning electron microscopy. Afro-Asian J Nematol 6:117–122

Salawu EO (1992) Effect of neem leaf extract and ethoprop singly and in combination on *Meloidogyne incognita* and growth of sugarcane. Pak J Nematol 10:51–56

Samuel AD, Murthy VN, Hengartner MO (2001) Calcium dynamics during fertilization in C. elegans. BMC Dev Biol 1:8. https://doi.org/10.1186/1471-213X-1-8

Sharma VP (1994) Potential of *Pleurotus sajor-caju* for biocontrol of *Aphelenchoides composticola* in *Agaricus bisporus* cultivation. Mushroom Res 3(1):15–20

Sharon E, Bar-Eyal M, Chet I, Herrera-Estrella A, Klei-feld O, Spiegel Y (2001) Biocontrol of the root-knot nematode *Meloidogyne javanica* by *Trichoderma harzianum*. Phytopathology 91:687–693

Siddiqui ZA, Mahmood I (1996) Biological control of plant parasitic nematodes by fungi: a review. Bioresour Technol 58:229–239

Soler-Serratosa A, Kokalis-Burelle N, Rodriguez-Kabana R, Weaver CF, King PS (1996) Allelochemicals for control of plant-parasitic nematodes. 1. In vivo nematicidal efficacy of thymol and thymol/benzaldehyde combinations. Nematropica 26:57–71

Stirling GR (1991) Biological control of plant-parasitic nematodes. CAB International, Wallingford, 282 pp

Stock SP, Pryor BM, Kaya HK (1999) Distribution of entomopathogenic nematodes (Steinernematidae and Heterorhabditidae) in natural habitats in California, USA. Biodivers Conserv 8:535–549

Thakur RK, Shirkot P (2017) Potential of biogold nanoparticles to control plant pathogenic nematodes. J Bioanal Biomed 9:4

Thoden TC, Hallmann J, Boppre M (2009) Effects of plants containing pyrrolizidine alkaloids on the northern root-knot nematode *Meloidogyne hapla*. Eur J Plant Pathol 123:27–36

Tian BY, Yang JK, Lian LH, Wang CY, Li N, Zhang KQ (2007) Role of an extracellular neutral protease in infection against nematodes by *Brevibacillus laterosporus* strain G4. Appl Microbiol Biotechnol 74:372–380

Tian B, Yang J, Zhang K (2007a) Bacteria used in the biological control of plant-parasitic nematodes: populations, mechanisms of action, and future prospects. FEMS Microbiol Ecol 61(2):197–213

Tian BY, Yang JK, Lian LH, Wang CY, Li N, Zhang KQ (2007b) Role of an extracellular neutral protease in infection against nematodes by Brevibacillus laterosporus strain G4. Appl Microbiol Biotechnol 74:372–380

Tikhonov VE, Lopez-Llorca LV, Salinas J, Jansson HB (2002) Purification and characterization of chitinases from the nematophagous fungi *Verticillium chlamydosporium* and *V. suchlasporium*. Fungal Genet Biol 35:67–78

Tiwari P, Jain P (2018) Evidence of isotropy at large-scale from polarizations of radio sources. Astronomy & Astrophysics manuscript no. agn_pol_align c ESO 2018 December 14

Taylor CE, Murant AF (1966) Nematicidal activity of aqueous extracts from raspberry canes and roots. Nematologica 12:488–494

Vetrivelkalai P, Sivakumar M, Jonathan EI (2010) Biocontrol potential of endophytic bacteria on *Meloidogyne incognita* and its effect on plant growth in bhendi. J Biopest 3(2):452–457

Wang KH, McGovern RJ, McSorley R (2004) Cowpea cover crop and solarization for managing root-knot and other plant-parasitic nematodes in herb and vegetable crops. Proc Soil Crop Sci Soc Fla 63:99–104

Whipps JM (2001) Microbial interactions and biocontrol in the rhizosphere. J Exp Bot 52(1):487–511

Woo-Jin J, Soon-Ju J, Kyu-Nam A, Yu-Lan J, Ro-Dong P, Kil-Yong K, Bo-Kyoon S, Tae-Hwan K (2002) Effect of chitinase-producing *Paenibacillus illinoisensis* KJA 424 on egg hatching of root knot nematode (*Meloidogyne incognita*). J Microbiol Biotechnol 12(6):865–871

Wright JB, Lam K, Hansen D, Burrell RE (1999) Efficacy of topical silver against fungal burn wound pathogens. Am J Infect Control 27:344–350

Xiang HQ, Feng ZX (2000) Effect of Pleurotus ostreatus on dynamics of Meloidogyne arenaria population and control effectiveness. Sci Agr Sinica 34:27–34

Yang JK, Tian BY, Liang LM, Zhang KQ (2007) Extracellular enzymes and the pathogenesis of nematophagous fungi. Appl Microbiol Biotechnol 75:21–31

Yang J, Liang L, Li J, Zhang K (2013) Nematicidal enzymes from microorganisms and their applications. Appl Microbiol Biotechnol 97:7081–7095

Yin HQ, Langford R, Burrell RE (1999) Comparative evaluation of the antimicrobial activity of ACTICOAT antimicrobial barrier dressing. J Burn Care Rehabil 20:195–200

Youssef MMA, El-Nagdi WMA, Eissa MFM (2014) Population density of root-knot nematode, Meloidogyne incognita infecting date palm under stress of aqueous extracts of some botanicals and a commercial bacterial byproduct. Middle East J Appl Sci 4:802–805

Youssef MMA, El-Nagdi WMA, Abd El-Fattah AI (2008) Efficacy of chicken compost, Bacillus thuringiensis and Pseudomonas fluorescens for biocontrolling Meloidogyne incognito infecting sugar beet. Int J Nematol 18(1):35–40

Chapter 16
Sustainable Management of Plant-Parasitic Nematodes: An Overview from Conventional Practices to Modern Techniques

Nishanthi Sivasubramaniam, Ganeshamoorthy Hariharan, and Mohamed Cassim Mohamed Zakeel

Abstract Plant-parasitic nematodes are microscopic roundworms that live in many habitats. They cause substantial problems to major crops throughout the world, including vegetables, fruits, and grain crops. These may become a major threat to the agricultural production system worldwide if management fails. This chapter reviews the economic importance and diagnostic methods of plant-parasitic nematodes, including a comprehensive account of existing strategies used for their management ranging from conventional to modern techniques. Some important genera of plant-parasitic nematodes such as *Meloidogyne* spp., *Heterodera* spp., and *Pratylenchus* spp. have been ranked uppermost in the list of the most economically and scientifically significant species of nematodes due to their complicated relationship with the host plants, wide host range, and the level of damage due to infection in crops. Further, obstacles encountered in parasitic nematode diagnosis by classical morphology-based methods have been resolved by the adoption of novel molecular techniques, which are rapid, precise, and cost-effective. As far as the existing cultural management techniques are concerned, crop rotation with non-host crops can suppress a wide range of nematode species effectively, followed by the use of organic soil amendments. Nematicide application is effective when speedy control of nematodes is required; however, the use is reappraised due to environmental concerns. Biological control of nematodes by fungi and bacteria is highly favored due to its environmentally friendly nature. In addition, bio-pesticides are becoming a promising option for the management of plant-parasitic nematodes. Biotechnology-, molecular biology-, and nanotechnology-based approaches have

N. Sivasubramaniam · G. Hariharan
Department of Agricultural Biology, Faculty of Agriculture, Eastern University Sri Lanka, Vantharumoolai, Sri Lanka

M. C. M. Zakeel (✉)
Department of Plant Sciences, Faculty of Agriculture, Rajarata University of Sri Lanka, Anuradhapura, Sri Lanka
e-mail: zakeel@agri.rjt.ac.lk

R. A. Ansari et al. (eds.), *Management of Phytonematodes: Recent Advances and Future Challenges*, https://doi.org/10.1007/978-981-15-4087-5_16

added a new dimension to nematode disease diagnosis and management. Identification of genes that reduce nematode's ability to reproduce has allowed the breeding of nematode-resistant plants. Marker-assisted selection, genetic engineering, and RNA interference to confer resistance in crop plants, nematode suppression using host plant proteinase inhibitors, and genome-editing technologies have helped tremendously in developing management strategies for plant-parasitic nematodes. In conclusion, a sustainable management of plant-parasitic nematodes is feasible when two or more compatible tactics are applied concurrently while appraising environmental protection.

Keywords Biological control · Cultural control · Integrated management strategies · Molecular techniques · Conventional practices

16.1 Introduction

Nematodes are microscopic roundworms that live in many habitats. They are ubiquitous, present in most moist to watery ecological niches such as cold oceans, hot springs, mountain peaks, soil ecosystem, plants, and animals (Nicol, 2002). Classification of nematodes is mainly based on their feeding types and sources of nutrition. The main group of nematodes are plant-parasitic nematodes, which can attack living plants and potentially are capable of penetrating roots and aboveground plant parts for their feeding and reproduction (Dong and Zhang 2006; Ansari and Khan 2012a, b; Ansari and Mahmood 2017b). Plant-parasitic forms comprise about 15% of all the forms of nematodes that exist in various habitats, exhibiting different feeding behaviors. So far, over 2500 species of plant-parasitic nematodes have been identified, characterized mainly by the presence of a stylet that is used for penetration of host plant tissues. Most of the nematodes attack roots and underground parts of plants, but some are able to feed on leaves and flowers (Nicol 2002). Nematodes parasitizing plants, animals, and humans can be found in all ecosystems, from the tropics to the Polar Regions, causing significant damage to the hosts (Mackenzie et al. 2017). Plant-parasitic nematodes cause colossal problems to major crops throughout the world, including vegetables, fruits, and grain crops (Shapiro-Ilan et al. 2015). Further, these nematodes signify serious risk to the world economy and are responsible for excessive losses in production systems worldwide (Abd-Elgawad and Askary 2015). Several vegetable crops, including bitter gourd, cabbage, carrot, cucumber, okra (Anwar et al. 2007; Singh and Kumar 2015), chili, lettuce, melon, mustard, pumpkin, sponge gourd, squash (Anwar et al. 2007), eggplant, potato (Anwar et al. 2007; Kathy 2000; Singh and Kumar 2015), and tomato (Anwar et al. 2007; Kathy 2000; Singh and Kumar 2015; Verdejo-Lucas et al. 1994) are affected by different species of plant-parasitic nematodes. The purpose of this chapter is to provide an insight into the economic importance of plant-parasitic nematodes, and the necessity and methods of nematode diagnosis with a special emphasis on the existing strategies used for the management of plant-parasitic nematodes from conventional measures to modern techniques.

16.2 Economic Importance of Plant-Parasitic Nematodes

Plant-parasitic nematodes are one of the critically limiting factors in the production of major crops worldwide. In general, they cause a projected $150 billion worth annual crop loss worldwide, which is on average 10–15% loss of crop yield (Abad et al. 2008; Barker 1998; Hugot et al. 2001; Maggenti 1981). Study of plant-parasitic nematodes becomes imperative, as the economic loss due to nematode problem in agriculture is significant. However, they are one of the hardest pest problems to identify, characterize, and manage, because most of them live in the soil (Stirling et al. 1999). Their negative impacts on the crop production are most of the time misjudged by farmers, agronomists, and related consultants, but it has been considered since recent past as one of the most important problems in agriculture (Whitehead 1998). Many nematodes have been identified in small grain cereals; however, only a few of them are economically important. They include cereal cyst nematodes (*Heterodera* spp.), root-knot nematodes (*Meloidogyne* spp.), root-lesion nematodes (*Pratylenchus* spp.), seed gall nematode (*Anguina tritici*), and stem nematode (*Ditylenchus dipsaci*) (Nicol 2002).

16.2.1 Cereal Cyst Nematodes

Heterodera spp., are a group of closely related species of cereal cyst nematodes. They are one of the most important groups of plant-parasitic nematodes in the world. The most frequently recorded species is *H. avenae*, which has been spotted in many wheat-growing countries, including the United States (Miller 1986), Canada, Israel, Australia, South Africa, Japan, most European countries (Kort 1972), India (Sharma and Swarup 1984; Sikora 1988), Algeria (Mokabli et al. 2001), and North African and West Asian countries, including Morocco, Tunisia, Pakistan, and Libya (Sikora 1988). Studies on this nematode have been extensively performed in Europe, Canada, Australia, and India (Swarup and Sosa-Moss 1990). *H. avenae* has been related with economic levels of damage exclusively in light soils. However, it can cause economic damage irrespective of soil type when the intensity of cereal cropping exceeds a certain limit (Kort 1972). Over 2.9 million hectares of wheat is sown annually in Australia, particularly in the states of Victoria and South Australia and 8% of the production is lost due to cyst nematodes (Eastwood et al. 1991). In Western Australia and southern New South Wales, smaller areas are infested, resulting in a 0.1% production loss on regional basis (Eastwood et al. 1991). Overall, 23–50% wheat and 20% barley in Australia (Meagher 1972), 15–20% wheat in Pakistan (Maqbool 1988), and 40–92% wheat and 17–77% barley in Saudi Arabia

(Ibrahim et al. 1999) were lost due to cyst nematodes. Cyst nematodes have been widely distributed in the main wheat-producing provinces of China and the magnitude of yield reduction has been significant, making the cyst nematode as a major biotic constraint to China's wheat production (Peng et al. 2009).

16.2.2 Root-Knot Nematodes

Root-knot nematodes (*Meloidogyne* spp.) pierce roots of certain plant species and lay their eggs inside the roots, giving the roots a "knotty" appearance and the whole plant a wilted or stunted look (Meadows et al. 2018). The most widespread and economically important species include *M. incognita*, *M. javanica*, *M. arenaria*, *M. hapla*, *M. chitwoodi*, and *M. graminicola* (Mitkowski and Abawi 2003). Damage caused by root-knot nematodes is very severe. As they feed on the roots, large galls or "knots" are formed throughout the root system of infected plants (Mitkowski and Abawi 2003). In addition to the yield loss, root-knot nematode infection severely affects the quality of the produce, particularly vegetables, making consumer acceptance very low (Mitkowski and Abawi 2003). Carrots, potato, tomato, onion, lettuce, and soybeans are important vegetable crops that are subject to the root-knot nematode problem (Ralmi et al. 2016). Among different species recorded in North Carolina in the United States, the southern root-knot nematode (*Meloidogyne incognita*) is the most common species (Meadows et al. 2018). Among many genera of nematodes of economic importance, *Meloidogyne* spp. are responsible for a larger proportion of annual agricultural production loss attributed to nematode damage (Ralmi et al. 2016). At least 5% of the world total crop production is destroyed by *Meloidogyne* spp. alone every year (Karajeh 2008). Root-knot nematodes have shown a wide host range (Olsen 2000), with over 2000 plant species being identified to date as hosts (Sasser 1980). Moreover, fertilizer and water use efficiency of severely root-knot damaged roots becomes poor, leading to additional losses for the growers (Trudgill and Phillips 1997).

16.2.3 Root Lesion Nematodes

Root-lesion nematodes (*Pratylenchus* spp.) are polycyclic, polyphagous migratory root endoparasites that are not limited to fixed places for their growth and reproduction. These parasites lay their eggs in soil and inside plant roots by invading and migrating into root tissues, and ultimately feeding on inside tissues of the roots. This invasion of nematodes causes characteristic dark brown or black lesions on the root surface, hence its common name (Nicol 2002). The species *P. thornei, P. neglectus, P. penetrans* are the most studied root-lesion nematodes, among which, *P. thornei* is considered the most economically important species in at least three countries (Eastwood et al. 1994; Nicol 1996; Orion et al. 1984; Taylor and McKay 1993;

Taylor et al. 1999; Thompson and Clewett 1986; Van Gundy et al. 1974). Wheat yield losses of 38–85%, 32–70% have been reported in Australia (Eastwood et al. 1994; Nicol 1996; Taylor and McKay 1993; Taylor et al. 1999; Thompson and Clewett 1986), Mexico (Van Gundy et al. 1974), and Israel (Orion et al. 1984), respectively.

16.2.4 Seed Gall Nematode

Seed gall nematodes, *Anguina tritici*, generally recognized as ear cockle, is regularly found on small grain cereals where farm-saved seed is sown without proper seed treatments (Nicol 2002). It was the first plant-parasitic nematode to be observed and characterized. John Turberville Needham reported this species in 1743 (Roe 1983). This was the first recorded microscopic observation of a pathogen causing diseases in plants (Lehman 1979). These nematodes form galls in wheat grains, causing enormous yield loss (Evans et al. 1993). In India, the annual crop loss caused by *A. tritici* ranges between 1% and 9%, which is equivalent to an economic loss of at least 70 million Indian rupees (Kaushal 1998). A survey has shown that the problem is very severe in Turkey with an infection rate of 1.5–55.2%, depending on the crop species and region (Elmali 2002).

16.2.5 Stem Nematode

Stem nematode (*Ditylenchus dipsaci*) is also an important plant-parasitic nematode particularly in horticultural crops (Subbotin et al. 2005). In the western United States, stem nematode caused a serious damage to alfalfa cultivation (Jordan 2017). It has a wide host range with over 500 plant species across 40 angiosperm families (Subbotin et al. 2005). However, some biological races of this nematode have a limited host range (Subbotin et al. 2005). As an endoparasite, *Ditylenchus dipsaci* mainly attack aerial parts of plants including stems, leaves, and flowers but also infect other parts like bulbs, tubers, and rhizomes (Subbotin et al. 2005). Significant crop loss due to stem nematode may occur when the nematode density exceeds 10 individuals per 500 g of soil (Seinhorst 1956).

16.3 Plant-Parasitic Nematode Identification

Plant-parasitic nematodes are mostly ectoparasitic, living in soil around roots; however, the rhizosphere also harbors many endoparasitic species in abundance (Siddiqi 1997). Some plant-parasitic nematodes do not cause economically significant damage to plants and therefore they are not considered pathogenic (Siddiqi

1997). When the population of plant nematodes inflates to reach economic threshold level, ectoparasitic forms become important (Siddiqi 1997). Fast, reliable, and apt diagnosis and specimen identification methods are vital for choosing appropriate management or control strategies for nematodes and also to prevent the spread of exotic nematodes in quarantine materials (Blok 2005; Blok and Powers 2009; Hunt and Handoo 2009; Castagnone-Sereno 2011). However, identification of nematodes using conventional methods is difficult due to their microscopic size and complex nature of key morphological characters under light microscope (Carneiro et al. 2004, 2008; Hunt and Handoo 2009; Castagnone-Sereno 2011; Oliveira et al. 2011). Some of the morphological and morphometric characters are indistinguishable and some are overlapping, leading to improper or erroneous identification nematode species (Hunt and Handoo 2009; Oliveira et al. 2011; Ye et al. 2015). This necessitates the availability of well-trained nematode taxonomist for proper morphological identification. However, use of modern biochemical and molecular tools has made the diagnosis and identification of plant parasitic nematode rapid and reliable (Blok 2005; Blok and Powers 2009; Castagnone-Sereno 2011; Hunt and Handoo 2009). Due to increased incidences of plant-parasitic nematodes and a lack of proper control strategies, a number of diagnostic laboratories catering services of detection and identification of nematode are on the rise recently (Lima et al. 2015).

16.3.1 Classical Morphology as a Tool for Plant-Parasitic Nematode Identification

Appropriate sampling and extraction procedures are crucial for detection and identification of any type of plant-parasitic nematodes. Ectoparasitic nematodes can be extracted from soil using a classical method of extraction described by Jenkins (1964), whereas endoparasitic nematodes can be extracted by macerating root or plant tissues in which the nematodes inhabit. Modified Baermann trays are used to separate ecto and endo plant-parasitic nematodes (Mekete et al. 2012). From the nineteenth century, morphological and morphometric features have been used to group nematodes within their respective genera (Roeber et al. 2013). Diagnosis and taxonomy have traditionally depended on morphological and anatomical characterization using light microscopy (Castillo and Vovlas 2007; Hunt and Handoo 2009). For morphological identification, features of adult stages and morphometric data such as body length, stylet length, etc. are necessary along with appropriate identification keys (Castillo and Vovlas 2007; Hunt and Handoo 2009). Axial stylet is a distinct morphological feature that can be used to differentiate plant-parasitic nematodes (the ones that possess axial stylet) from nonparasitic forms (Anon 2018; Cunha et al. 2018). For the identification of plant pathogenic nematodes, morphological and anatomical structures such as stylet morphology, tail type, dorsal esophageal gland orifice, esophageal lumen, median bulb, basal bulb and intestine, and reproductive structures of adult male (spicule) and female (vagina and its position) are highly useful (Cunha et al. 2018; Mattiucci and Nascetti 2008; Rahman and Mian 2010). Nematode identification based on classical morphology still remains as a key method due to the link between the function and morphology, and simplicity of the methods (Oliveira et al. 2011). Therefore, morphology-based identification methods are used in quantitative evaluations and population surveys of plant-parasitic nematodes (Oliveira et al. 2011). Nevertheless, some of the morphological features overlap interspecifically and some show intraspecific variation, making the identification tedious and necessitating many replicates of the sample for proper identification (Gasser et al. 2008; Hunt and Handoo 2009; Ye et al. 2015).

16.3.2 Biochemical and Molecular Techniques to Identify Plant-Parasitic Nematodes

Many different biochemical and molecular techniques, particularly protein-, DNA-, and RNA-based methods, are widely used for the identification of plant-parasitic nematodes. Nematode proteins, including isozymes, are extracted and separated for analysis and identification of nematodes. These techniques include separation of proteins by one-dimensional gel electrophoresis, isoelectric focusing, two-dimensional gel electrophoresis, and capillary gel electrophoresis (Abrantes et al. 2004). Some methods such as multi-locus enzyme electrophoresis (MEE or MLEE), also called isoenzyme phenotyping, have proven to be successful in the identification of *Meloidogyne* spp. (Esbenshade and Triantaphyllou 1985). The principle of this method is based on the relative mobility of enzymes extracted from mature female nematodes in gel electrophoresis (Blok and Powers 2009). Phenotyping with different isozymes such as superoxide dismutase and glutamate oxaloacetate transaminase has been widely used. Gel electrophoresis profiles of esterase and malate dehydrogenase have been used to distinguish *Meloidogyne* spp. (Carneiro et al. 2008, 2014, 2016, 2017). To quantify plant-parasitic nematodes, immunological methods using polyclonal and specific monoclonal antibodies have been used to differentiate *M. incognita* and *M. javanica* (Davies et al. 1996; Fujiwara et al. 1997; Kennedy et al. 1997; Masler 2002). Perera et al. (2005) conducted biochemical and spectrometry analyses, where they discovered a quick and easy method to diagnose plant-parasitic nematode *Ditylenchus dipsaci*, using protein profiles generated by matrix-assisted laser desorption/ionization time-of-flight mass spectrometry (MALDI-TOF/MS). Genomic or organelle DNA-based molecular techniques are highly robust to characterize inter- and intraspecific variability. With the availability of sequences in databases, polymerase chain reaction (PCR)-based molecular detection techniques have been widely developed and used for the diagnosis of plant-parasitic nematodes (Carneiro et al. 2017). Molecular markers such as Restriction Fragment Length Polymorphism (RFLP), Randomly Amplified Polymorphic DNA (RAPD), Amplified Fragment Length Polymorphism (AFLP), Microsatellites or Single Sequence Repeats (SSRs), and Sequence Characterized

Amplified Regions (SCARs) have been extensively used for molecular characterization of plant-parasitic nematodes (Correa et al. 2013, 2014; Curran et al. 1985; Fleming et al. 1993; Hahn et al.,1996; Han et al. 2004; Thiéry and Mugniéry 2000; Wang et al. 2001; Yu et al. 1998; Zijlstra et al. 2000).

Quantitative and multiplex PCR techniques are highly useful for detection and quantification of plant-parasitic nematodes (Braun-Kiewnick and Kiewnick 2018). Quantitative PCR (q-PCR) also known as 'Real-time PCR' is a fast and reliable technique used for detection and quantification of target sequences of plant-parasitic nematodes in real time (Braun-Kiewnick and Kiewnick 2018). Many qPCR protocols have been developed for the detection of many plant-parasitic nematodes in recent years. Berry et al. (2008) used q-PCR to detect and quantify *M. javanica*, *Pratylenchus zeae*, *Xiphinema elongatum* in sugarcane. Moreover, q-PCR was used for the identification of soybean cyst nematode (*Heterodera glycines*) (Ye 2012) and root-lesion nematode (*Pratylenchus penetrans*) (Sato et al. 2017). Multiplex PCR assay amplifies multiple gene targets using multiple primer sets in a single reaction, minimizing the amount of DNA template and PCR reagents necessary to complete an analysis. Several multiplex PCR protocols were developed to identify and detect many plant-parasitic nematodes including *M. incognita*, *M. enterolobii*, and *M. javanica* (Hu et al. 2011), *M. arenaria* (Kiewnick et al. 2013), and *Bursaphelenchus* spp. (Filipiak et al. 2017). Multiplex PCR has been successfully used for detection of two plant-parasitic nematodes *M. incognita* (Mi1: 5′-AAACG GCTGTCGCTGGTGTC-3′ and Mi2: 5′-CCGCTATAAGAGAAAATGACCC-3′) and *P. coffeae* (Pe1: 5′-ATGCCCACATTGCATTCAGC-3′ and Pc2: 5′-GAGAGA GAAACACCTCTCAC-3′) (Saeki et al. 2003). DNA barcoding is another molecular approach that can be used for identification to species. In this approach, PCR amplification, followed by sequencing of a short-conserved gene region and comparison of the sequence to a database of reference sequences is used to identify the organism (Hebert et al. 2003). Currently, many sequences such as 18S regions of rRNA gene, *ITS* regions, D2 and D3 expansion segments of 28S rRNA gene, and mitochondrial cytochrome c oxidase subunit 1 gene (*coxI*) are targeted to PCR amplify and sequence in order to identify the species and for phylogenetic studies (Kiewnick et al. 2014; Kumari et al. 2010; Powers et al. 2018). DNA microarrays have been used for the identification of *M. chitwoodi* (Françoisa et al. 2006) and *M. hapla* (van Doorn et al. 2007). Current developments in next-generation sequencing (NGS) permit assembly of DNA regions with high sequencing depth (Hahn et al. 2013). Besnard et al. (2014) followed a genome-skimming approach based on Illumina HiSeq platform (an NGS variant) to carry out an assembly of mitochondrial genome of *M. graminicola*. Despite much advancement in molecular techniques, integration of morphological, biochemical, and molecular techniques would warrant accurate diagnosis and identification of plant-parasitic nematodes (Carneiro et al. 2017).

16.4 Strategies Used for the Management and Control of Plant-Parasitic Nematodes

Considering the economic importance of plant-parasitic nematodes globally, control and management of plant-parasitic nematodes become an important crop management practice to reduce crop losses. The management strategies should focus on reducing parasitic nematode populations to a level below that of the economic threshold than complete destruction in the rhizosphere. Because of the microscopic nature and similar/overlapping taxonomical features of plant-parasitic nematodes, proper identification is cumbersome, making management difficult and expensive. This section discusses the cultural, chemical, biological and molecular strategies, integrated approaches, and innovative practices used for the management and control of plant-parasitic nematodes.

16.4.1 Cultural Methods

Cultural practices mainly target prevention or reduction of outbreaks of plant-parasitic nematodes; however, the results of these practices are often unseen and difficult to quantify. When methods specifically designed for nematode control can be integrated easily with existing practices, they are usually readily adopted by farmers (Luna and House 1990): cultural practices such as crop rotation, allowing fallow period, adding organic amendments to soil, planting trap crops, adjusting the time of planting, etc.

16.4.1.1 Crop Rotation

Crop rotation is an effective method to manage different types of plant-parasitic nematodes. Cultivation of non-host or resistant crops or cultivars may lower the density of nematodes, which can effectively mitigate the damage caused by them in the succeeding, susceptible crops. Fair understanding of the crops selected for the rotation is essential to achieve an effective nematode management (Kirkpatrick and Thomas 2007). Species such as *Meloidogyne*, which have a wide host range, often diminish the effectiveness of crop rotation (Sikora and Fernandez 2005). However, choice of maize as one of the rotational crop has been shown to reduce northern root-knot nematode *M. hapla* (Johnson 1982; Raymundo 1985). But, population densities of other plant-parasitic nematodes such as stubby-root, lesion, lance, and ring nematodes may increase when maize is planted continuously for many seasons without rotation or with short rotations with non-host or poor-host crops (Table 16.1) (Johnson 1982). The basic concept of crop rotation for nematode management is to reduce initial inoculums of damaging nematode species to levels that allow the following crop(s) to become established and complete early growth

Table 16.1 List of documented non-host plants used in crop rotations

Nematode species	Recommended non-hosts/ poor-hosts	References
Meloidogyne incognita	Maize, Soy bean	Kirkpatrick and Thomas (2007)
Rotylenchulus reniformis	Cotton, Soy bean, Rice, Maize	Kirkpatrick and Thomas (2007)
Heterodera glycines	Maize, Sorghum, Rice	Kirkpatrick and Thomas (2007)
Meloidogyne hapla	Sun grasses	Viaene and Abawi (1998)
Pratylenchus penetrans		Anon. (2002)
Heterodera glycines	Maize, cotton, cowpea, potato, small grains, grains, tobacco, most vegetables	Trivedi and Barker (1986)
Meloidogyne javanica	*Andropogon*, *Crotalaria* spp., cotton, groundnut, sorghum, velvet bean	Trivedi and Barker (1986)
Meloidogyne arenaria, *Meloidogyne incognita*, *Meloidogyne javanica*	*Mucuna pruriens* L., and *Crotalaria spectabilis*	Osei et al. (2010)

before being heavily attacked (Nusbaum and Ferris 1973). Crops such as wheat, barley, and other small grains can also be good choices for crop rotation to reduce *M. hapla* and certain cyst nematodes (*Heterodera* spp. and *Globodera* spp.) (McGawley 1986; Santo et al. 1980). If lesion nematodes (*Pratylenchus* spp.) are present at economic injury levels, rye is a promising rotational crop compared to wheat or oats to reduce *P. penetrans* populations significantly (Castillo and Vovlas 2007; Florini and Loria 1986). Although this practice enhances the population density of *P. crenatus*, it results in fewer damaging species of lesion nematodes (Florini and Loria 1986). Unusual rotation crops such as marigold (Huang 1994), *Crotalaria* (Brodie and Murphy 1975) and *Chrysanthemum* (Hackney and Dickerson 1975) have also been tried with varying degrees of success.

This approach of nematode management in crops is the oldest form. However, it is an important method for managing nematodes in annual crops (Chen and Tsay 2006; Thomason and Caswell 1987). Market value of the crop, local climate, availability of equipment, and cropping practices are the major factors that control the development of programs for crop rotations (Thomason and Caswell 1987). Ideally, in crop rotation, the preceding crop avoids damage to the subsequent crop by suppressing the target nematode population (Johnson 1985). However, under some crop rotation practices, increasing occurrences of fields with two or more species of plant-parasitic nematodes have been noticed (Fortnum et al. 2001). To be effective, the nematode population in the system should be evaluated every year (Kirkpatrick and Thomas 2007).

16.4.1.2 Fallow Period

Fallow is a simple starvation technique to reduce nematode population (Tyler 1933). In this practice, soil is prepared and kept free of any vegetation for a given period mainly through tillage to ensure that plant-parasitic nematodes do not find hosts (Tyler 1933). When compared to cooler regions that are wet due to heavy rainfall, this method is very effective in dry, arid climates (Barker 1997), particularly when nematodes are in the most critical stage of their life cycle (Hill 1988). However, nematode species like *Heterodera* spp. and *Globodera* spp. can survive in dry habitats through a process called anhydrobiosis and live for long periods in the absence of host plants (Grainger 1964). Further, it may affect physical properties of soil and increase soil erosion while reducing farm income (Barker 1997). A six-week fallow in late spring in Georgia between the harvesting of tomato and planting of a cover crop was as effective as continuous fallow in reducing population density of *Pratylenchus brachyurus* and *Trichodorus christiei* (Brodie and Murphy 1975).

16.4.1.3 Use of Organic Soil Amendments

Addition of organic amendments to soil improves soil structure and fertility, while managing plant-parasitic nematodes (Ansari and Mahmood 2017a, 2019a, b; Ansari et al. 2019). Poultry litter and composts add ample amount of ammonia to soil during the decomposition process due to their low carbon-to-nitrogen ratio (Rodriguez-Kabana 1986). These ammonium ions are converted into plant utilizable nitrates by microbes. Organic amendments improve plant growth and tolerance to nematodes, highlighting indirect contribution toward the management of nematodes. Organic wastes such as crop residues, livestock and poultry manures, compost, chitin, and oilseed cakes (Table 16.2) have shown promising effect on reducing the density of plant-parasitic nematodes (Akhtar and Mahmood 1996). Moreover, the soil amendments promote growth and activity of beneficial microbes in the rhizosphere, which can antagonize plant-parasitic nematodes (Bridge 1996). Decomposition of organic residues facilitates the accumulation of specific compounds in the soils that may be toxic to nematodes (Rodriguez-Kabana 1986; Rodriguez-Kabana and Morgan-Jones 1987; Sitaramaiah 1990). Steer and chicken manures reduced the numbers of cyst and citrus nematodes, while increasing yields of potato and citrus (Gonzalez and Canto-Sanenz 1993). Poultry manure-based soil amendments have shown effectiveness in controlling plant-parasitic nematodes in many vegetable crops (Akhtar and Mahmood 1997). Root-knot nematodes in carrot were minimized by addition of neem leaves and poultry litter to the soil and the efficacy was influenced by the amount of added (Agyarko et al. 2006). The nematicidal effect of organic amendments differs with the nematode species, type of amendment, and length of time after application (McSorley and Gallaher 1997). The large quantities of amendments that are required are the major disadvantage of this method (Bridge 1996).

16.4.1.4 Growing Antagonistic Plants and Trap Crops

There are numerous plants belonging to 57 families that have nematicidal properties (Sukul 1992). Among them, neem, sunn hemp (*Crotalaria juncea*), asparagus,

Table 16.2 Organic soil amendments used for the control of plant-parasitic nematodes

Nematode species	Recommended organic soil amendments	References
Meloidogyne graminicola	Decaffeinated tea waste and water hyacinth compost	Roy (1976)
Pratylenchus brachyurus	Cocoa pod husks and farmyard manure	Egunjobi and Larinde (1975)
Meloidogyne incognita	Oilseed cakes of castor, mustard, neem, and groundnut Cow dung, poultry manure	Ahmed (1988) Chindo and Khan (1990); D'Addabbo et al. (1997); Poswal and Akpa (1991)
Tylenchorhynchus brassicae	Oilseed cakes of castor, mustard, neem, and groundnut	Ahmed 1988
Hirschmanniella spp.	Neem cake and press mud	Johnathan and Pandiarajan (1991)
Pratylenchus spp	Yard-waste compost	McSorley and Gallaher (1995)
Meloidogyne javinca	Chicken litter	Marull et al. (1997)
Meloidogyne chitwoodi	Rapeseed amendments	Mojtahedi et al. (1993)
Globedera pallida	Steer and chicken manure	Gonzalez and Canto-Sanenz (1993)
Helicotylenchus multicinctus	Vermicompost, neem cake, poultry manure, distillery sludge	Sundararaju et al. (2002)
Rotylenchulus sp.	Municipal green compost, penicillin residues compost	Renco et al. (2009)
Xiphinema spp.	Decomposed and composed manure	Olabiyi et al. (2007)

mustard, *Tagetes* spp. and sesame are the most popular antagonistic plants among farmers. These plants often act as trap crops that reduce nematode population by allowing them to invade roots and interfere with their development (Bridge 1996). Thus, the nematodes suffer from incomplete development (Bridge 1996). Some legumes are antagonistic to nematodes, producing certain toxic compounds due to which nematodes that invade legumes do not migrate back to soil; rather, they are killed within the plant (Prot et al. 1992).

16.4.1.5 Use of Nematode-Free Seeds and Vegetative Planting Material

Plant-parasitic nematodes are mainly spread by vegetative planting materials, but hardly by true seeds (Bridge 1996). Therefore, use of seed as planting material is obviously favored to prevent the spread of nematodes (Bridge 1996; Khan et al. 2019). Plants propagated by stem and leaf cutting and via tissue culture, however, do not harbor root nematodes (Bridge 1996). Farmers who produce their own seedlings or rooted planting materials have low risk of nematode problem than farmers who depend on nurseries for their seedlings (Bridge 1996).

16.4.1.6 Managing Plant-Parasitic Nematodes by Adjusting the Time of Planting

Another promising approach to manage or minimize the problem of plant-parasitic nematodes is by altering the time of planting during seasons when the nematodes are less active (McSorley and Duncan 1995; Trivedi and Barker 1986; Wcischer 1994). This technique has been applied for wheat, which is a temperate crop and associated with many nematode diseases (Roberts et al. 1981).

16.4.2 Chemical Control of Plant-Parasitic Nematodes Using Fumigants and Non-fumigant Nematicides

Synthetic nematicides are fumigant or non-fumigant-based chemicals used to kill or reduce nematodes (Bridge and Starr 2007). Many different fumigant and non-fumigant nematicides are used to control plant-parasitic nematodes (Table 16.3). The first developed chemical to control plant-parasitic nematodes is carbon disulfide (CS_2), which was developed around the second half of the nineteenth century followed by halogenated hydrocarbons and other volatile compounds (Chitwood 2003). Fumigants have low molecular weight chemicals commonly used in gaseous or liquid form. As they volatilize, they can diffuse through the soil profile and kill nematodes in soil (Gan et al. 1998). Therefore, the rate of diffusion largely depends on soil physical properties (Gan et al. 1998). Traditionally, fumigants were applied to soil by shank injection, but their efficacy was low (Gan et al. 1998). Subsequently, drip-applied fumigants were developed, taking soil texture, fumigant distribution, and delivery methods into consideration (Ajwa and Trout 2004; Gan et al. 1998). Drip fumigation largely reduced the environmental impacts caused by shank injection, mainly due to uniform movement of the fumigants through soil substrata compared with shank application (Ajwa and Trout 2004; Gan et al. 1998; Westerdahl et al. 2003). Shank and drip-application of fumigants using 1,3-dichloropropene (1,3-D or Telone) with chloropicrin mixtures and iodomethane confirmed good efficacy in controlling *Tylenchulus semipenetrans* populations over a range of soil types, soil depths, and locations (Schneider et al. 2008). Drip application of chloropicrin and methyl bromide has shown very effective control of plant-parasitic nematodes (Schneider et al. 2008). A novel fumigation agent based on ammonium bicarbonate was a promising fumigant for the control of plant-parasitic nematodes under sealed conditions (Su et al. 2015). A mixture of lime and ammonium bicarbonate at a rate of 0.857 and 0.428 g kg^{-1}, respectively, showed stronger nematicidal activity against *Meloidogyne* sp. and *Rotylenchulus* sp. found in banana under pot and field experimental conditions (Su et al. 2015).

Fumigants were the chief method to control *M. incognita* during the last few decades; however, their uses have recently been limited. Soil fumigation with methyl iodide, propargyl bromide, 1, 3-dichloropropene, calcium cyanamide, and methyl

Table 16.3 Fumigants and non-fumigant nematicides used for the control of plant-parasitic nematodes

Chemical name	Trade name	Formulation
Fumigants		
DD Mixture	D-D, VIDDEN – D, Nemafene	Liquid
Methyl bromide*	Dowfume	Gas
1,3 dichloropropene	Telone/DD-95	Liquid
Ethylene dibromide*	Dowfume W-85	Liquid
Metam-sodium	Vapam	Gas
Dazomet	Basamid	Granular
Dibromochloropropane (DBCP)	Nemagon/Fumazone	Liquid
Methyl isothiocyanate	Di-Trapex	Liquid
Sodium tetrathiocarbonate/	Ezone	Liquid
Chloropicrin*	Larvacide	Liquid
Contact nematicide/non-fumigant nematicide		
(a) Organophosphates		
Thionazin	Nemafos	Granular or emulsifiable liquid
Ethoprophos	Mocap	Granular or emulsifiable liquid
Fenamiphos	Nemacur	Granular or emulsifiable liquid
Fensulfothion	Dasanit	Granular
Terbufos	Counter	Granular or emulsifiable liquid
Cadusafos/ Ebufos	Rugby	Granular/ Microencapsulated
Isazofos	Miral	Granular or emulsifiable liquid
Fosthiazate	Nemathorin/Cierto	Clay-based microgranule
(b) Carbamates		
Aldicarb	Temik	Granular
Aldoxycarb	Standak	Flowable
OxamyI	Vydate	Granular or emulsifiable liquid
Carbofuran	Furadan	Granular or Liquid
Cleothocarb	Lance	Granular
(c) Unclassified		
Tioxazafen	Nemastrike	Liquid

[This table was adopted from Gowen 1997, and updated with information sourced from Chitwood 2003 and Taylor 2003]

bromide, have been very effective for suppressing nematodes (Giannakou et al. 2002; Ibekwe 2004). Nevertheless, many products such as methyl bromide and calcium cyanamide are no longer available for farmers due to their toxic effects on nontarget organisms including fungi and bacteria (Ibekwe 2004) and impacts on the environment (Giannakou et al. 2002; Zasada et al. 2010). Studies have shown that free-living nematodes were significantly reduced by the fumigation of methyl bromide, metam sodium (Cao et al. 2004), chloropicrin (Okada et al. 2004), 1,3-Dichloropropene, and metam sodium (Collins et al. 2006). Currently, there are only a restricted number of pesticides registered to control plant-parasitic nematodes. Pre-plant soil fumigants were commonly used for nematode control until the end of the last century (King and Taberna Jr. 2013; Rich et al. 2004), but many of them are now phased out or facing significant regulatory pressure (Leader et al. 2010). Contact or non-fumigant nematicides used to control plant-parasitic nematodes fall into the chemical classes of organophosphates and carbamates (Wada and Toyota 2008). Organophosphorus nematicides such as fosthiazate and imicyafos showed trivial effects on soil-dwelling nontarget microbes (Wada and Toyota 2008). Application of non-fumigant nematicide imicyafos had a significant impact on the density of *Pratylenchus penetrans* in radish, but the effect on free-living soil nematodes was insignificant (Wada et al. 2011). Non-fumigant synthetic nematicides that are acetyl cholinesterase inhibiters will have mild impact on soil living organisms that do not have nervous systems (Osaki and Fukuchi 2010). A new class of seed treatment nematicide Tioxazafen was commercialized recently to offer consistent broad-spectrum control of plant-parasitic nematodes in corn, soybean, and cotton (Slomczynska et al. 2015). This new compound exhibits excellent activity against cyst, root-knot, and reniform nematodes in soy; lesion, root-knot, and needle nematodes in corn; and reniform and root-knot nematodes in cotton (Slomczynska et al. 2015). Due to the regulatory constraints and public resistance to synthetic nematicides, development of a viable toxin-free, ecofriendly approach for the control of plant-parasitic nematodes is indispensable.

16.4.3 Biological Methods of Plant-Parasitic Nematode Management

With growing environmental concerns and demands for organic crop production, there is a need to find alternative means of nematode control. In this regard, the use of biological control agents is of paramount importance and popular among crop producers. De Bach (1964) defined biological control as "the action of parasites, predators or pathogens in maintaining another organism's population density at a lower average than would occur in their absence." According to Stirling (1991), the biological control of nematodes is "the reduction of nematode populations which is accomplished through the action of living organisms other than the nematode-resistant host plant, which occur naturally or through the manipulation of the environment or the introduction of antagonists." There is plenty of research underway to identify organisms that can be applied to plants at any stage of cultivation to inhibit the growth of harmful nematodes. The expectation is that a specific organism should act rapidly to reduce nematode populations without causing any adverse effects to the host plant. Persistence and proliferation of the organism in the root zone is believed to be a beneficial attribute (Kerry 2000; Stirling 1991). Weller (1988) stated that rhizosphere microbes are ideal for use as biocontrol agents due to their natural frontline defense against root pathogens. Many organisms, such as bacteria, fungi, viruses, predatory nematodes, insects, mites, and some invertebrates have been identified to parasitize or prey on plant-parasitic nematodes (Stirling 1991). Four basic techniques are practised in biological control of plant-parasitic nematodes: (1) introduction, (2) augmentation, (3) inoculation, and (4) inundation (Kerry and Hominick 2002). Introduction is a classical method by which a beneficial organism is deliberately introduced into a place where it is needed and it becomes fully established (Kerry and Hominick 2002). This technique is usually practised against introduced plant-parasitic nematodes that will have no indigenous antagonist. Augmentation is the release of laboratory-grown microbial control agents to compensate for inadequate population level of microbial control agents of a particular plant-parasitic nematode (Kerry and Hominick 2002). Inoculation is the release of antagonistic organisms when indigenous antagonists are absent or introduced ones cannot survive permanently (Kerry and Hominick 2002). The antagonistic organisms are released at the beginning of the planting season and repeated each season (Kerry and Hominick 2002). Inundation is the mass cultivation and releases of antagonistic microbial agents against plant-parasitic nematodes to ensure rapid suppression of the nematodes when they exceed the threshold level (Kerry and Hominick 2002).

16.4.3.1 Fungi as Biological Control Agent of Plant-Parasitic Nematode

Among biological control agents against nematodes, fungi play a significant role in controlling plant-parasitic nematodes in agriculture (Jatala 1986; Stirling 1991). About 80% of the total microbial biomass in most of the soils is fungus (Clark and Paul 1970; Shield et al. 1973). Naturally, fungus residing in the rhizosphere has constant association with nematodes, thereby destroying parasitic-nematodes in most soils (Gams and Zare 2003; Jaffee 1992; Kerry 2000; Kerry and Hominick 2002; Siddiqui and Mahmood 1996). Antagonistic fungi represent a group that can utilize nematode biomass for their carbon, nitrogen, and other important elemental requirements (Siddiqui and Mahmood 1996). Two zoophoric fungi, *Catenaria auxiliaris* and *Nematophthora gynophila*, have a potential to regulate population dynamics of cyst nematodes, *Heterodera avenae* and *Heterodera schachtii* in some soils (Crump et al. 1983; Kerry and Crump 1980). Hallmann et al. (2009) classified fungi that antagonize nematodes into three major groups: nematophagous, saprophagous, and endophytic fungi, whereas Moosavi and Zare (2011) put them into nematophagous and endophytic groups. The term "nematophagous fungi" means fungi that can colonize on nematodes and fulfil nutritional requirements from them (Hallmann et al. 2009). Most of the nematophagous fungi are facultative saprophytes and some are obligate parasites (Hallmann et al. 2009). The role of nematophagous fungi in root zone is very crucial for the control of nematodes (Li et al. 2015). Plant exudates present in the root zone largely influence the rhizosphere ecosystem and provide nutrients for fungi in the rhizosphere for their growth and development (Lòpez-Llorca et al. 2008). These rhizosphere-associated fungi along with other myceliophagous fungal species effectively parasitize the nematodes in the hosts

(Lòpez-Llorca et al. 2008). Facultative nematophagous fungi produce external structures such as appressoria on specialized hyphae and specialized adhesive spores to infect various stages of nematodes, whereas obligate forms usually initiate their infection through spores (Barron 1977). Nematophagous fungi are capable of producing extracellular enzymes or proteins such as PII (Tunlid et al. 1995), Aoz1 (Zhao et al. 2004), Mlx (Wang et al. 2006a), Ds1 (Wang et al. 2006b), Hasp (Wang et al. 2009), P32 (Lòpez-Llorca and Claughe 1990; Lòpez-Llorca and Robertson 1992), CH143 (Tikhonov et al. 2002), which may have role in antagonistic activity and improving stress tolerance of fungi.

Endoparasitic fungi are obligate parasites, which complete their vegetative cycle inside the infected host. These fungi attack vermiform plant-parasitic nematodes through conidia or zoospores, which reach nematodes by ingestion or by means of a slender penetration tube with mechanical pressure (Dackman et al. 1992; Dijksterhuis et al. 1994). These spores sprout and grow inside the nematode and ultimately kill them (Dackman et al. 1992; Dijksterhuis et al. 1994). However, Jansson (1994) criticized the mode of entry by ingestion of spores and germination in intestines of nematodes; he stated that direct entry of conidia via cuticle is the only means of infection. Various fungal species such as *Pochonia chlamydosporia* and *Paecilomyces lilacinus* (Table 16.4) have been reported as effective biocontrol agents against nematodes (Siddiqui and Mahmood 1996).

16.4.3.2 Bacteria as Biocontrol Agent for Plant-Parasitic Nematode

Bacteria genera *Pasteuria, Pseudomonas, and Bacillus* are abundantly found in soil and have potential for use as biocontrol agent against plant-parasitic nematodes (Emmert and Handelsman 1999; Meyer 2003; Siddiqui and Mahmood 1999; Ansari et al. 2017a, b). These bacteria have been largely investigated for the last two decades for use as biocontrol agents against plant nematodes. Nematophagous bacteria are ubiquitous in distribution in nature, with different modes of action and a wide host range (Kerry 2000; Meyer 2003; Siddiqui and Mahmood 1999; Stirling 1991). A variety of nematophagous bacteria have been isolated from soil, host-plant parts, nematode cysts, and eggs of nematodes (Kerry 2000; Meyer 2003; Siddiqui and Mahmood 1999; Stirling 1991). The modes of action exhibited by nematophagous bacteria include the production of toxins, enzymes and antibiotics, disturbance caused in nematode-host plant recognition, stimulation of systemic resistance of host plants and competition for nutrients (Siddiqui and Mahmood 1999). *Pasteuria* spp. are obligate, mycelial, endospore-forming bacterial parasites of plant-parasitic nematodes (Bekal et al. 2001). Many species of this bacterial genus have exhibited huge potential for use as biocontrol agents of plant-parasitic nematodes (Chen and Dickson 1998). These bacteria can infect over 323 species of both plant-parasitic and free-living nematodes belonging to 116 genera (Chen and Dickson 1998). Many of these plant-parasitic nematodes parasitized by bacteria are economically important (Bird et al. 2003). Commercial formulations made of *Pasteuria nishizawa* have been in use for nematode control in soybean cultivation

Table 16.4 Fungal species used as biocontrol agents against plant parasitic nematodes

Fungal species	Antagonized plant-parasitic nematodes	References
Purpureocillium lilacinus along with *Verticilium chlamydosporium* and *Trichoderma harzianum*	*Meloidogyne javanica* in chick pea	Siddiqui and Mahmood (1996)
Pochonia chlamydosporia	*Meloidogyne incognita* in vegetable crops	Atkins et al. (2003)
Pochonia suchlosporia, P. rubescens, P. bulbillosa, and P. globispora	*Meloidogyne* spp., *Heterodera* spp. in green house and microplot trials	de Leij and Kerry (1991); Freire and Bridge (1985); Moosavi et al. (2010); Morgan-Jones et al. (1981)
Hirsutella minnesotensis	*Heterodera glycines* in soy bean	Jaffee and Muldoon (1989)
Hirsutella rhossiliensis	*Meloidogyne javanica, Heterodera avenae, Heterodera glycines, and Criconema xenoplax*	Sturhan and Schneider (1980)

(Tylka et al. 2015; Xiang et al. 2017). *Bacillus* spp. and *Pseudomonas* spp. are rhizosphere bacteria that are able to antagonize nematodes (Krebs et al. 1998). Some species of *Bacillus* could suppress nematodes and improve plant growth, whereas some species are plant pathogenic (Li et al. 2005). *Bacillus* spp. have shown antagonistic activity against *Meloidogyne*, *Heterodera*, and *Rotylenchulus* (Giannakou et al. 2002; Gokta and Swarup 1988; Insunza et al. 2002; Kloepper et al. 1992; Kokalis-Burelle et al. 2002; Madamba et al. 1999; Meyer 2003; Siddiqui and Mahmood 1999). Rhizosphere *Pseudomonas* strains have shown different pathogenic interactions with plant-parasitic nematodes (Andreogloua et al. 2003; Jayakumar et al. 2002; Kerry 2000; Siddiqui and Shaukat 2003; Siddiqui et al. 2005; Spiegel et al. 1991). Ammonia released due to the decomposition of nitrogenous organic material by bacteria reduced the amount of plant-parasitic nematodes in soil (Rodriguez-Kabana 1986). *Corynebacterium paurometabolu* is capable of producing hydrogen sulfide and chitinase enzyme, which potentially suppress the hatching of nematode eggs (Mena and Pimentel 2002). Some rhizobacteria are able to produce antibiotics and hydrogen cyanide, which inhibit organisms harmful to plants (Zuckerman and Jasson 1984). Six cry proteins (Cry5, Cry6, Cry12, Cry13, Cry14, and Cry21) in *Bacillus thuringiensis* (*Bt*) have been recognized as toxic to larval stages of certain free-living and parasitic nematodes (Chen et al. 2017; Kotze et al. 2005; Marroquin et al. 2000; Wei et al. 2003). An unknown Bt isolate has shown toxicity to root-lesion nematodes, but its pathogenic factor is still undiscovered (Bradfish et al. 1991). Some bacteria living in symbiotic association with plants have shown to suppress nematodes via production of certain compounds (Samaliev et al. 2000).

Table 16.5 Plant species used in various studies to extract botanical nematicides for the control of plant parasitic nematodes

Plant species	Plant parts used for extraction	Promising chemical compound with nematicidal property	Tested nematode/s	Host plants used	Biological activity or possible effects on tested nematode/s	References
Azadirachta indica, *Ipomoea fistulosa* and *Azolia pinnata*	Leaves	NI	*M. javanica*	Wheat	Reduce the number of knotted roots/plants	Hossain et al. (2002)
Azadiracta indica, *Curcuma longa* *Eucalyptus globulus* *Zingiber officinale*	Leaves	NI	*M. graminicola*	Rice	Reduce the number of galls and nematode population	Shukla and Chand (2018)
Aegle marmelos *Azadirachta indica* *Jatropha curcas* *Eucalyptus globus*	Leaves	NI	*M. graminicola*	Rice	Suppress the development of *M. graminicola* root galls and reduce the size of galls and the nematode population density	Dongre and Sobita (2013)
Myrtus communis	Leaves	NI	*M. javanica* *Tylenchulus semipenetrans*, *Ditylenchus dipsaci*	Cucumber Tomato	Reduce the gall index and number of nematode eggs and juveniles	Oka et al. (2012)
Nicotiana tabacum	Leaves	Eugenol, β-asarone	*M. incognita*	Tomato	Kill nematodes	Wiratno et al. (2009)
Syzygium aromaticum	Buds					
Piper betle	Leaves					
Acorus calamus	Rhizome					
Nepeta cataria, *Couroupita quianensis* and *Pentanema indicum*	Leaves	NI	*M. incognita.*	Tomato	Decrease egg hatching and larval formation, and increase nematode mortality	Pavaraj et al. (2012)

(continued)

Table 16.5 (continued)

Plant species	Plant parts used for extraction	Promising chemical compound with nematicidal property	Tested nematode/s	Host plants used	Biological activity or possible effects on tested nematode/s	References
Tagetes erecta	Leaves, flower, root and stem	NI	*M. incognita.*	Tomato	Decrease nematode population	Meena et al. (2010)
Canavalia ensiformis	Seeds	D-glucose, L-canavanine, xanthotoxin, *cis*-aconitic acid, *trans*-aconitic acid, malic acid, citric acid, palmiticacid, S-carboxymethylcysteine	*M. incognita.*	Tomato	Effective on infective stage (J2), nematicidal activity and reduce the number of egg masses	Rocha et al. (2017)
Hyoscyamus niger, Melia azedarach and *Xanthium strumarium*	Leaves	NI	*M. incognita.*	Tomato	Inhibit egg hatching and increase J2s mortality	Kepenekçi et al. (2016)
Tagetes erecta	Leaves	NI	*M. incognita*	Cucumber	Inhibit egg hatching, 100% juvenile mortality and reduce gall index	Taiwo et al. (2015)
Azadirachta indica	Seeds and leaves	NI	*M. javanica*	Tomato	Mortality of J2s, egg hatch inhibition, reduction in the infection rate and inhibition of gall formation	Moosavi (2012)
Chrysanthemum coronarium	Stem and leaves					
Nerium oleander	Leaves	NI		Tomato	Reduce number of juveniles	Salim (2016)
Ageratum conyzoides	Leaves and stem	*Alkaloid, tannins, phenol, saponins, glycosides, fl avonoids, carbohydrate, protein, mucilage/gum and phlobatannins*	*M. incognita*	–	Maximum egg inhibition and J2 juvenile mortality	Asif et al. (2017)
Coccinia grandis	Fruit					
Azadirachta indica	Leaves	NI	*M. incognita*	Tomato	Reduce population and the percentage of multiplication rate	Bawa et al. (2014)
Capsicum Annuum	Fruits					
Zingiber officinale	Rhizomes					
Curcuma longa	Rhizomes	NI	*M. incognita*	–	Induce mortality and inhibitory effect on egg hatching	Neeraj et al. (2017)

Ageratum conyzoide and Eichornia crassipes	Leaves	NI	*M. incognita*	–	Damage eggs, reduce the hatching of eggs and cause paralysis and death of second stage juveniles (J2)	Khan et al. (2017)
Allium sativumm and *Allium cepa*	Whole plant	NI	*M. incognita*	Tomato	Reduction in the number of egg-masses, galls and females/ root system	Bakr (2018)
Cymbopogon citratus	Leaves					
Tithonia diversifolia, Zeyheria montana and *Eclipta alba*	Leaves	Flavanols, tannins, and alkaloids	*Pratylenchus zeae* and *Pratylenchus jaehni*	–	Increase mortality rate with time and different concentrations	Slomp et al. (2009)
Tabernaemontana catharinensis	Stem bark					
Moringa oleifera	Leaves	NI	*M. javanica*	Eggplant	Egg hatching inhibition and increased juvenile mortality	Murslain et al. (2014)
Azadirachta indica	Leaves	NI	*M. incognita*	Soy bean	Reduce root galling and nematode reproduction	Atungwu et al. (2009)
Argemone maxicana and *Solanum xanthocarpum*	Leaves	NI	*M. incognita, Rotylenchulus reniformis* and *Tylenchorhynchus brassicae*	Eggplant and cauliflower	Reduce root-knot development and inhibition of nematode multiplication	Ajaz and Tiyagi (2003)

NI not identified in the particular investigation

16.4.3.3 Predatory Nematodes on Plant-Parasitic Nematodes

Predatory nematodes are valuable and widely accepted pest management agents worldwide. Predatory nematodes residing in soil have potentials for use as biological control agents of plant-parasitic nematodes (McSorley et al. 2006, 2008). Predatory and omnivorous nematodes are highly found in soils rich in organic matter (Ferris and Matute 2003; Wang et al. 2004) than agriculturally manipulated soils (Ferris and Matute 2003; Ferris et al. 2001; McSorley et al. 2008). Compared to vineyard, natural woodland showed presence of predatory nematodes in abundance (Sanchez-Moreno and Ferris 2007). Predatory and omnivorous nematodes are found in most soils, suppressing plant-parasitic nematodes (Bilgrami et al. 2005; Khan and Kim 2005; McSorley et al. 2006; Sanchez-Moreno and Ferris 2007), particularly the root-knot nematodes (Khan and Kim 2005; McSorley et al. 2006; Sanchez-Moreno and Ferris 2007).

16.4.3.4 Plant Extracts with Nematotoxic Compounds are Possible Bio-pesticides in Plant Parasitic Nematode Management

Gradual phase-out of nematicides due to environmental concerns and tight limits introduced for the registration of new biocidal molecules have led to an increasing demand for nonchemical alternatives for the control of plant-parasitic nematodes (Dutta et al. 2019). One such alternative are the plant extracts based nematode control agents generally known as botanical nematicides (Ntalli and Caboni 2012). Botanical nematicides are plant-derived bio-pesticides or botanical pesticides extracted from various plant parts with nematotoxic (nematicidal and/or nematostatic) properties and have been used as an excellent alternative to synthetic nematicides for crop protection (Sano 2005). In this context, nematode-antagonistic plants have received substantial consideration due to their significant capacity either to paralyze (nematostatic compounds) or to kill (nematicide compounds) nematodes within the plant tissues and in the soil in the close vicinity of roots, thus decreasing the population density of plant parasitic nematodes (Sano 2005). Botanical nematicides often comprise an array of secondary metabolite compounds and exhibit nematode-suppressive properties (Zhou et al. 2012). These compounds include attractants, repellents, hatching inhibitors or stimulants, and nematotoxicants, either constitutive or formed in response to the presence of nematodes (Chitwood 2002). Plant nematotoxic effects of diverse plant-based compounds such as phenolics, glucosides, flavonoids, tannins, saponins, alkaloids, thiophenics, glucosinolates, thienyls, diterpenes, thienyls isothiocyanates, acetylenes, etc. have been identified and extensively studied (Chitwood 2002; Ntalli and Caboni 2012).

The amount and quality of bioactive compounds extracted from plants greatly vary from plant species to species and depend on plant parts used for the extraction, ecology of the plant, type of solvents (methanol, ethanol, water, acetone, chloroform, etc.) with different degrees of polarity used for the extraction and extraction methods

(Asif et al. 2017; Rocha et al. 2017). Botanicals extracted from various plant species, entire plant or plant parts, have been tested against several *Meloidogyne* spp. and other plant-parasitic nematodes to investigate their biological activities and to identify active ingredients with nematotoxic properties (Table 16.5).

The botanical pesticides with nematotoxic properties possess many advantages over synthetic pesticides. Pest selectivity or specific activity of plant-derived bio-pesticides to plant parasitic nematodes is an excellent character compared to synthetic nematicides and fumigants in the protection of nontarget and nematode antagonistic microorganisms, including predatory nematodes (Oka et al. 2012). The effectiveness at lower dosage and low toxicity to humans, plants, domestic or raised livestock, and the environment make them better substitute for synthetic nematicides (Chitwood 2002). Using plant extracts to control nematodes is relatively economical, easy to apply, and capable of enriching soil physical and chemical properties, especially texture and fertility (Feizi et al. 2014; Mokrini et al. 2018). Some plant extracts showed allelopathic effects mostly on weeds, in addition to nematicidal potential. This is known as dual effect (El-Rokiek and El-Nagdi 2011). Leaf extract of *Eucalyptus citriodora* exhibited dual effect by controlling the weed purslane (*Portulaca oleracea* L.) and the nematode *M. incognita* in sunflower (El-Rokiek and El-Nagdi 2011). On the contrary, some plant-derived nematicidal compounds have shown allelopathy or phytotoxicity to the main crop with economic importance. Leaf powder extracts of *Myrtus communis* were slightly phytotoxic like *Inula viscosa*, reducing shoot fresh weight of tomato at higher doses (Oka et al. 2012). This effect can probably be prevented by delaying the planting after the incorporation of nematicidal compound to soil or finding appropriate rate of application of plant extracts (Oka et al. 2012).

16.4.4 Integrated Plant-Parasitic Nematode Management

Nematodes could be effectively controlled by use of integrated management strategies in which two or more management techniques are used together (Bridge 1996; Roberts 1993). Due to negative impacts on the environment, certain control measures applied to the soil have been gradually reduced (Stork and Eggleton 1992). Therefore, much focus has been paid for integrated approaches of nematode management, while restricting the use of synthetic chemical nematicides (Noe et al. 1991; Roberts 1993). Integrated nematode management is the use of two or more compatible nematode management methods concurrently, with the appraisal of suitability of the methods for the locality where it is applied. An effective technique in one locality may not be effective or maybe less effective in another locality (Duncan and Noling 1998; Roberts 1993). Integrated approaches are developed by combining methods representing cultural, chemical, and biological methods of nematode control or management (Robinson 2004; Stirling 1991). Moreover, integrated approaches have been recommended for obtaining optimum yield from high and low-value crops (Brown 1987; McKenry 1987).

16.5 Molecular-Based Strategies to Control Plant-Parasitic Nematodes

Biotechnology and molecular-based approaches have added a new dimension to nematode pest management. Identification of the genes that reduce nematodes' ability to reproduce has allowed the breeding of plants resistant to nematodes. Marker-assisted selection, genetic engineering, RNAi to impart resistance in crop plants, nematode suppression using host plant proteinase inhibitors, and genome-editing technologies to develop resistant plants to manage plant-parasitic nematodes are discussed.

16.5.1 Identification of Nematode-Resistant Genes, Marker-Assisted Selection (MAS), and Quantitative Trait Loci (QTL) Mapping for Breeding of Plants

Resistant trait against plant parasitic nematodes is not usually available in every economically important crop. Therefore, identification of resistant genes in plants against plant-parasitic nematode is an important step in breeding for resistant plants. Marker-assisted breeding methods are mainly used to identify sources of nematode resistance in order to use them in breeding to develop resistant cultivars against nematodes. This approach generally includes screening to identify and characterize resistant genes from resistant wild types and other cultivars, pre-breeding and development of linked markers, QTL mapping, positional cloning, and isolation and characterization of the genes conferring the resistance (Fosu-Nyarko and Jones 2015). This is a lengthy process; however, the use of molecular techniques for marker-assisted selection (MAS) of resistant cultivars as parents in breeding programs has significantly reduced the time needed for the whole process (Fosu-Nyarko and Jones 2015). Several naturally occurring resistant genes against plant-parasitic nematodes from various plant genetic resources have been identified with the characterization of resistant response to infection (Cai et al. 1997; Ganal and Tanksley 1996; Milligan et al. 1998; van der Voort et al. 1997). These genes include *Mi-1, 2,* and *9* from wild tomato (*Solanum peruvianum*) against *M. incognita* (Ammiraju et al. 2003; Ganal and Tanksley 1996; Milligan et al. 1998; Yaghoobi et al. 1995); *Hs1*pro-1 and *Hs2* from wild sugar beet (*B. procumbens*) against *H. schachtii* (Cai et al. 1997; Heller et al. 1996); *Rhg1* and *Rhg4* from soybean against *H. glycines* (Concibido et al. 2004; Webb et al. 1995); *Mae, Mag,* and *Rma* from peanut against *M. arenaria* (Chu et al. 2011; Garcia et al. 1996); and *Gpa2* and *Gro1-4* from potato against *G. pallida* and *G. rostochiensis* (Leister et al. 1996; van der Voort et al. 1997). Resistant genes against cereal cyst nematode *H. avenae* have been characterized in barley and wheat (Kretschmer et al. 1997; Lewis et al. 2009; Williams et al. 1996). Nematode resistance loci *Ha1* and *Ha2* (allelic to Ha3) on chromosome 2, *Ha4* (chromosome 5) in barley and *Cre1* locus on chromosome 2B,

Cre3 (*Ccn-D1*) from *Triticum tauschii* in wheat, and other *Cre* genes have been widely used in cereal-breeding programs using MAS and back-crossing into advanced breeding lines (Kretschmer et al. 1997; Lewis et al. 2009; Williams et al. 1996).

16.5.2 Development of Transgenic Plants with Resistant Genes Against Nematode

In transgenic plant production with resistance against nematodes, the resistant gene isolated from one species is introduced into another species to confer resistance against nematodes in the species that is otherwise susceptible (Fosu-Nyarko and Jones 2015). However, the effectiveness of these genes in heterologous systems depends on genotype or species and to induce hypersensitive response, effective signaling in the pathways triggered by several elements is required (Fosu-Nyarko and Jones 2015). Further, certain protein interactions needed to confer the resistance may not be available in the species, which receives the gene via gene transfer (Fosu-Nyarko and Jones 2015). Transfer of tomato *Hero A* gene into tomato cultivars confers desirable levels of resistance to potato cyst nematode in tomato, but not in potato (Sobczak et al. 2005). Tomato cultivars carrying the *Mi* gene show variation in resistance to *M. incognita*, attributed to their genotypic background (Jacquet et al. 2005). When the *Mi1* gene from tomato plant was introduced into a root-knot nematode susceptible cultivar of lettuce (*Lactuca sativa*) by *Agrobacterium*-mediated transformation method, the resultant transgenic lines showed resistance to *M. incognita* (Zhang et al. 2010). However, not every naturally existing nematode resistance is conferred by a single, dominant R-gene that can be manipulated with relative ease once identified (Tomczak et al. 2009). Some sources of nematode resistance are complex traits, inherited in a polygenic manner and such quantitative resistance is not readily amenable to transgenic manipulations (Tomczak et al. 2009).

16.5.3 RNA Interference to Suppress Plant-Parasitic Nematode

RNA interference (RNAi) has emerged as a valuable and robust gene-silencing tool for functional analysis of various genes by suppressing their expression in many organisms including plant nematodes (Rosso et al. 2009). A multitude of studies has described its application in plant-parasitic nematode management. In this posttranscriptional silencing process, small double-stranded RNA (dsRNA) molecules, also known as small interfering RNAs (siRNAs) with ribonuclease activity at homologous mRNA sequences cause degradation of mRNA, resulting in RNA silencing

(Agrawal et al. 2003). The second stage juvenile (J2) of a cyst nematode was able to take up dsRNA from a solution in which they were kept under *in vitro* condition (Urwin et al. 2002). Plants have also been genetically engineered to produce siRNA molecules that can potentially silence important genes in plant-parasitic nematodes (Fire et al. 1998; Huang et al. 2006). The molecules of siRNA that are taken into nematodes, when they feed on plant cytoplasm, cause endogenous RNAi activity in nematode, subsequently silencing the target gene involved in infection (Lilley et al. 2012). The gene *HgALD* encodes for fructose-1,6-diphosphate aldolase, which is important for the conversion of glucose into energy for actin-based motility of parasite during invasion into host (Youssef et al. 2013). A reduction of 58% in female produced by H. *glycines* was resulted by a transgenic soybean to express an RNAi construct targeting to silence *HgALD gene* (Youssef et al. 2013). A *PRP17* gene construct expressing a dsRNA in transgenic soybean showed 53% and 79% drop of infection and reproduction in *H. glycines*, respectively (Dutta et al. 2015). Another study revealed that transgenic potato (*Solanum tuberosum*) and *Arabidopsis thaliana* lines with an RNAi construct pART27(16D10i-2) to produce dsRNAs complementary to Mc16D10L effector gene in *M. chitwoodi* enhanced their resistance against the nematode (Dinh et al. 2014). As alternative to transgenic plants with nematode resistance, Tobacco Rattle Virus vector to deliver proteins that reduce expression of target genes in feeding cyst nematodes (Valentine et al. 2007). *In vitro* RNAi by dsRNA with synthetic neurotransmitants in plant-parasitic nematodes was successful. Nevertheless, delivery of dsRNAs for *in planta* RNAi is an innovative strategy with future prospect (Banerjee et al. 2017). Host delivered RNAi that targets splicing factors and integrase genes in *M. incognita* elevated the disease resistance of *A. thaliana* against nematodes (Kumar et al. 2017).

16.5.4 Proteinase Inhibitors to Evade Feeding by Plant-Parasitic Nematodes

Plant proteinase inhibitors are compounds produced by host plants. These compounds can degrade nematode proteases that break down food materials for nematodes to absorb the nutrient (Jaouannet et al. 2013). This mechanism of degradation of nematode proteases diminishes nutrient absorption by nematodes (Jaouannet et al. 2013). Resistance of durum wheat (*Triticum durum*) against *H. avenae* was improved by heterologous expression of a serine proteinase inhibitor, PIN2 from potato (Vishnudasan et al. 2005). Expression of the barley cystatin HvCPI-6 in maize inhibited cysteine protease activity of *T. urticae,* impairing its development and reproduction (Carrillo et al. 2011). Overexpression of phytocystatin gene *CeCPI* originated from taro (*Colocasia esculenta*) improved resistance in tomato to *M. incognita* (Chan et al. 2010). Many other proteinase inhibitors such as CpTI in sweet potato (*Ipomoea batatas*) (Cai et al. 2003), cystatin *Oc-I1D86* in rice (*Oryza sativa*) (Urwin et al. 1997, 1998), and cystatins in maize (*Zea mays*), taro (*Colocasia esculenta*) and sunflower (*Helianthus annuus*) (Chan et al. 2015; Fuller et al. 2008) have been studied extensively. Combinations of different proteinase inhibitors have shown resistance against a wide range of nematodes. A translational fusion protein made with two proteinase inhibitors, CpTI and Oc-I1186, in transgenic *Arabidopsis* exhibited additive influence against *H. schachtii* and *G. pallida* (Urwin et al. 1998). Transgenic expression of maize proteinase inhibitors, anti-feedant cystatin and anti-root invasion synthetic peptides transformed into plantain (*Musa* spp., cv. Gonja manjaya), separately and in combination showed enhanced yield and nematode resistance against *R. similis, H. multicinctus*, and *Meloidogyne* spp., confirming the potential of proteinase inhibitors for nematode resistance in crops (Tripathi et al. 2015).

16.5.5 CRISPR-Cas9 Genome Editing Technique with Potential for Nematode Control

CRISPR-Cas9 (Clustered Regularly Interspaced Short Palindromic Repeats-CRISPR associated protein9) is RNA-guided genome editing technology that exploits the CRISPR-Cas system to modify a genome in a target fashion (Khatodia et al. 2016). Guided by RNA, the Cas9 endonuclease cleaves the DNA at target sequence (Khatodia et al. 2016). Genome editing using CRISPR/Cas9 system has been well known in free-living nematode, *Caenorhabditis elegans* (Dickinson and Goldstein 2016; Paix et al. 2017). Currently, there are a number of genomes-editing protocols using CRISPR-Cas9 technology to edit the genome in *C. elegans* (Friedland et al. 2013). Several studies have recently outlined the mechanism of translation of CRISPR/Cas9 technology from *C. elegans* to *Strongyloides* spp., *Ascaris suum, Brugia malayi,* and *Haemonchus contortus* (Britton et al. 2016; Ward 2015; Zamanian and Andersen 2016). This advanced technology would help to characterize several vital genes involved in different physiological processes of nematodes; however, only a handful of studies have been performed on the application of CRISPR/Cas 9 system to study the resistance mechanism of plants against parasitic nematodes (Ali et al. 2019). By targeting the right gene to modify, it is possible to enhance plant defense or resistance against parasitic nematodes using CRISPR-Cas9 system of genome editing (Borrelli et al. 2018).

16.6 Nanotechnology Approaches for the Control of Plant-Parasitic Nematodes

Recent advances in materials science and chemistry have enabled the development, production, and utilization of nanoparticles for a wide range of applications in agriculture (Singh et al. 2015). "Particles having one or more dimensions of the order of 100 nm or less" are known as nanoparticles and they are used in nanoscience and nanotechnology (ISO 2008). Among different applications of nanoparticles, use as antimicrobial agent for plant disease management is promising. Despite limited studies on the use of nanoparticles to control plant nematodes, encouraging results are currently available and many studies have commenced in the recent past focusing in this area (Bhau et al. 2016).

Nanoparticles of gold (Thakur et al. 2018), silver (Cromwell et al. 2014; Taha and Abo-Shady 2016), silicon oxide, titanium oxide (Ardakani 2013), zinc oxide (Ma et al. 2009), silicon carbide (Banna et al. 2018), alumino silicate, etc. have been tested for their toxicity effects on *M. incognita, M. incognita,* and *C. elegans* (Pluskota et al. 2009). The effect of various nanoparticles on adult mortality, immobility, reduction egg masses, number of adult nematodes per root, and inhibition of egg hatchability have been studied in pot experiments (Abdellatif et al. 2016; Ardakani 2013). Functional genomics studies have shown that toxicity of silver nanoparticles increases expression of superoxide dismutases-3 (*sod-3*) and abnormal DAuer formation protein gene (*daf-12*) with significant drop in reproduction ability of *C. elegans* (Roh et al. 2010). However, the modes of action of other tested nanoparticles need to be investigated to find if those nanoparticles are involved in direct physical effects such as blocking sensory apertures and feeding structures or metabolic toxicity of ingested particles on plant-parasitic nematodes. The use of nanoparticles has shown some interesting effects on host plants as well. Titanium oxide at a concentration of 7×10^{-4} percent was the best in controlling *M. incognita* while improving tomato growth (Ardakani 2013). However, the treatments silver nanoparticles at 0.02, 0.01, and 0.005% and titanium oxide at 0.02% concentrations were toxic to *M. incognita* and host tomato plants (Ardakani 2013). Another approach of using nanoparticles for plant-parasitic nematode control is by enriching formulations of plant natural product to enhance their efficacy to control nematodes. Silver nanoparticle enriched formulations of *Urtica urens* extracts exhibited 11-fold more nematicidal activity compared to the corresponding raw extracts against root-knot nematode (Nassar 2016). Silver nanoparticles enriched leaf extracts of *Conyza dioscoridis*, *Melia azedarach*, and *Moringa oleifera* augmented the nematicidal activity by five-fold against J2 stage and two-fold against eggs of *M. incognita* (Abbassy et al. 2017). However, there is a potential to further enhance the biological activity of plant products against plant parasitic nematodes by the development and use of nano-sized formulations. Therefore, more research is needed to find possible ways to synthesize or modify nano-formulations of plant natural extracts.

16.7 Conclusions and Future Directions

There has been a wealth of information about plant-parasitic nematodes, their identification and management gathered from different field and laboratory experiments for the last three decades. Due to the microscopic nature of plant-parasitic nematodes, their identification is cumbersome, necessitating certain laboratory tools and techniques for their accurate identification. Diagnosis of plant nematodes is crucial in order to develop proper management strategy to suppress the buildup of nematode population. Many nematodes have been reported to cause significant diseases in plants; however, *Meloidogyne* spp., *Heterodera* spp., and *Pratylenchus* spp. are the most economically significant species due to their complicated relationships with host plants, wide host range, and the level of damage caused by infection. Development of plant-parasitic nematode management methods depends on the type and stage of crops affected, type of parasitic nematode, degree of symptoms and damages, geographical area and availability of agricultural inputs required. Existing cultural practices often prevent the upsurge of plant-parasitic nematodes, but the results of these practices are often invisible and difficult to quantify. Assessment of the effect and rate of different organic matters for the management of economically significant nematode species in different geographical regions is imperative. Different bacterial and fungal species such as *Pasteuria, Pseudomonas, Bacillus, Verticilium chlamydosporium, Trichoderma harzianum, Hirsutella minnesotensis, Hirutella rhossiliensis,* etc. have shown their potential for control of plant parasitic nematodes; however, their efficacy and other characters attributed to biocontrol agents need to be extensively assessed before their field applications. Information about predatory nematodes for controlling parasitic and free-living nematodes is meagre. Biotechnology, molecular-based and nanotechnology approaches have added a new dimension in nematode management. However, currently there is only limited information available about the application of CRISPR/Cas9 system to study the resistance mechanism of plants against plant-parasitic nematodes. A novel approach of using nanoparticles with plant extracts has shown promising results and could be effectively used for the control of various plant-parasitic nematodes. In summary, every method has merits and demerits. Therefore, integration of appropriate techniques is essential to achieve sustainability in plant-parasitic nematode management.

References

Abad P, Gouzy J, Aury J-M, Castagnone-Sereno P, Danchin EGJ, Deleury E, Perfus-Barbeoch L, Anthouard V, Artiguenave F, Blok VC, Caillaud M-C, Coutinho PM, Dasilva C, De Luca F, Deau F, Esquibet M, Flutre T, Goldstone JV, Hamamouch N, Hewezi T, Jaillon O, Jubin C, Leonetti P, Magliano M, Maier TR, Markov GV, McVeigh P, Pesole G, Poulain J, Robinson-Rechavi M, Sallet E, Ségurens B, Steinbach D, Tytgat T, Ugarte E, van Ghelder C, Veronico P, Baum TJ, Blaxter M, Bleve-Zacheo T, Davis EL, Ewbank JJ, Favery B, Grenier E, Henrissat B, Jones JT, Laudet V, Maule AG, Quesneville H, Rosso M-N, Schiex T, Smant G, Weissenbach J, Wincker P (2008) Genome sequence of the metazoan plant-parasitic nematode *Meloidogyne incognita*. Nat Biotechnol 26:909–915

Abbassy MA, Abdel-Rasoul MA, Nassar AMK, Soliman BSM (2017) Nematicidal activity of silver nanoparticles of botanical products against root knot nematode, *Meloidogyne incognita*. Arch Phytopathol Plant Prot 50(17-18):909–926. https://doi.org/10.1080/03235408.2017.1405608

Abd-Elgawad MM, Askary TH (2015) Impact of phytonematodes on agriculture economy. In: Askary TH, Martinelli PRP (eds) Biocontrol agents of phytonematodes. CAB International, Wallingford, UK, pp 3–49

Abdellatif KF, Abdelfattah RH, El-Ansary MSM (2016) Green nanoparticles engineering on root-knot nematode infecting eggplants and their effect on plant DNA modification. Iran J Biotechnol 14(4):250–259

Abrantes IMO, Santos MCVD, Conceição ILPM, Cunha MJM, Santos MSNA (2004) Biochemical and molecular characterization of plant-parasitic nematodes. Phytopathol Mediterr 43:232–258

Agrawal N, Dasaradhi PVN, Mohmmed A, Malhotra P, Bhatnagar RK, Mukherjee SK (2003) RNA interference: biology, mechanism and applications. Microbiol Mol Biol Rev 67(4):657–685. https://doi.org/10.1128/MMBR.67.4.657-685.2003

Agyarko K, Kwakye PK, Bonsu M, Osei BA, Fripong KA (2006) Effects of organic soil amendments on root-knot nematodes, soil nutrients and growth of carrot. J Agron 5(4):641–646

Ahmed S (1988) Farmer pest-control practices in India: some observations. Final workshop of the IRRI-ADB-EWC project on botanical pest control in rice-based cropping systems. IRRI, LosBãnos, Philippines

Ajaz S, Tiyagi SA (2003) Utilization of noxious weeds for the management of plant parasitic nematode infesting some vegetable crops. Arch Phytopathol Plant Prot 36:211–219

Ajwa HA, Trout T (2004) Drip application of alternative fumigants to methyl bromide for strawberry production. Hortic Sci 39:1707–1715

Akhtar M, Mahmood I (1996) Organic soil amendments in relation to nematode management with particular reference to India. Integr Pest Manag Rev 1(4):201–215

Akhtar M, Mahmood I (1997) Control of root-knot nematode *Meloidogyne incognita* in tomato plants by seed coating with suneem and neem oil. J Pestic Sci 22:37–38

Ali MA, Shahzadi M, Zahoor A, Dababat AA, Toktay H, Allah Bakhsh A, Nawaz MA, Li H (2019) Resistance to cereal cyst nematodes in wheat and barley: an emphasis on classical and modern approaches. Int J Mol Sci 20:432. https://doi.org/10.3390/ijms20020432

Ammiraju J, Veremis J, Huang X, Roberts P, Kaloshian I (2003) The heat-stable root knot nematode resistance gene Mi-9 from *Lycopersicon peruvianum* is localized on the short arm of chromosome 6. Theor Appl Genet 106:478–484

Andreogloua FI, Vagelasa IK, Woodb M, Samalievc HY, Gowena SR (2003) Influence of temperature on the motility of *Pseudomonas oryzihabitans* and control of *Globodera rostochiensis*. Soil Biol Biochem 35:1095–1101

Anon. (2018) Free-living and plant-parasitic nematodes (roundworms). https://www.apsnet.org/edcenter/disandpath/nematode/intro/Nematode/Pages/Background.aspx. Accessed on 30 Apr 2019

Anon. (2002) Agricultural statistics yearbook. Council of Agriculture, Executive Yuan, Taiwan, R. O. C.

Ansari RA, Mahmood I (2017b) Determination of disease incidence caused by *Meloidogyne* spp. and or *Fusarium udum* on pigeonpea in Aligarh district: a survey. Trends Biosci 10 (24):5239–5243

Ansari RA, Mahmood I (2019a) Plant health under biotic stress: volume 2: microbial interactions. Springer, Singapore. https://doi.org/10.1007/978-981-13-6040-4

Ansari RA, Mahmood I (2019b) Plant health under biotic stress: volume 1: organic strategies. Springer, Singapore. https://doi.org/10.1007/978-981-13-6043-5

Ansari RA, Khan TA (2012a) Parasitic association of root-knot nematode, *Meloidogyne incognita* on guava. e-J Sci Technol (e-JST) 5:65–67

Ansari RA, Khan TA (2012b) Diversity and community structure of phytonematodes associated with guava in and around Aligarh, Uttar Pradesh, India. Trends Biosci 5(3):202–204

Ansari RA, Mahmood I (2017a) Optimization of organic and bio-organic fertilizers on soil properties and growth of pigeon pea. Sci Hortic 226:1–9

Ansari RA, Mahmood I, Rizvi R, Sumbul A (2017b) Siderophores: augmentation of soil health and crop productivity. In: Kumar V, Kumar M, Sharma S, Prasad R (eds) Probiotics in agroecosystem. Springer, Singapore, pp 291–312

Ansari RA, Rizvi R, Sumbul A, Mahmood I (2017a) PGPR: current vogue in sustainable crop production. In: Kumar V, Kumar M, Sharma S, Prasad R (eds) Probiotics and plant health. Springer, Singapore, pp 455–472

Ansari RA, Sumbul A, Rizvi R, Mahmood I (2019) Organic soil amendments: potential tool for soil and plant health management. In: Ansari RA, Mahmood I (eds) Plant health under biotic stress. Springer, Singapore, pp 1–35

Anwar SA, Zia A, Hussain M, Kamran M (2007) Host suitability of selected plants to *Meloidogyne incognita* in the Punjab, Pakistan. Int Jf Nematol 17:144–150

Ardakani AS (2013) Toxicity of silver, titanium and silicon nanoparticles on the root-knot nematode, *Meloidogyne incognita*, and growth parameters of tomato. Nematology 15(6):671–677. https://doi.org/10.1163/15685411-00002710

Asif M, Tariq M, Khan A, Siddiqui MA (2017) Biocidal and antinemic properties of aqueous extracts of *Ageratum* and *Coccinia* against root-knot nematode, *Meloidogyne incognita in vitro*. J Agric Sci 12(2):108–122. https://doi.org/10.4038/jas.v12i2.8229

Atkins S, Hidalgo-Diaz L, Kalisz H, Mauchline T, Hirsch P, Kerry B (2003) Development of a new management strategy for the control of root-knot nematodes (*Meloidogyne* spp) in organic vegetable production. Pest Manag Sci 59:183–189. https://doi.org/10.1002/ps.603

Atungwu JJ, Ademola AA, Aiyelagbe IOO (2009) Evaluation of organic materials for inhibition of nematode reproduction in soybean. African Crop Sci J 17(4):167–173

Bakr RA (2018) Bionematicidal potential of some incorporating plants on *Meloidogyne javanica* control on tomato. Int J Curr Microbiol App Sci 7(5):1457–1464

Banerjee S, Banerjee A, Gill SS, Gupta OP, Dahuja A, Jain PK, Sirohi A (2017) RNA interference: a novel source of resistance to combat plant parasitic nematodes. Front Plant Sci 8:834. https://doi.org/10.3389/fpls.2017.00834

Banna L, Salem N, Ghrair AM, Habash SS (2018) Impact of silicon carbide nanoparticles on hatching and survival of soil nematodes *Caenorhabditis elegans* and *Meloidogyne incognita*. Appl Ecol Environ Res 16(3):2651–2662. https://doi.org/10.15666/aeer/1603_26512662

Barker KR (1997) Opportunities for integrated management of plant-parasitic nematodes in the Near East. In: Plant nematode problems and their control in the near east region. FAO Plant Production and Protection Paper, 144

Barker KR (1998) Introduction and synopsis of advancement in nematodes. In: Barker KR, Pederson GA, Windham GL (eds) Plant nematode interactions, 1st edn. American Society of Agronomy, Crop Science Society of America, and Soil Science Society of America, Madison, WI, pp 1–20

Barron GL (1977) The nematode-destroying fungi. Canadian Biological Publications Ltd., Guelph

Bawa JA, Mohammed I, Liadi S (2014) Nematicidal effect of some plants extracts on root-knot nematodes (*Meloidogyne incognita*) of tomato (*Lycopersicon esculentum*). World J Life Sci Med Res 3(3):81–87

Bekal S, Borneman J, Springer MS, Giblin-Davis RM, Becker JO (2001) Phenotypic and molecular analysis of a *Pasteuria* strain parasitic to the sting nematode. J Nematol 33:110–115

Berry SD, Fargette M, Spaull VW, Morand S, Cadet P (2008) Detection and quantification of root-knot nematode (*Meloidogyne javanica*), lesion nematode (*Pratylenchus zeae*) and dagger nematode (*Xiphinema elongatum*) parasites of sugarcane using real-time PCR. Mol Cell Probes 22:168–176

Besnard G, Jühling F, Chapuis E, Zedane L, Huillier E, Mateille T, Bellafiore S (2014) Fast assembly of the mitochondrial genome of a plant parasitic nematode (*Meloidogyne graminicola*) using next generation sequencing. C R Biol 337:295–301

Bhau BS, Phukon P, Ahmed B, Gogoi B, Borah B, Baruah J, Sharma DK, Wann SB (2016) A novel tool of nanotechnology: nanoparticle mediated control of nematode infection in plants. In:

Singh D, Singh H, Prabha R (eds) Microbial inoculants in sustainable agricultural productivity. Springer, New Delhi. https://doi.org/10.1007/978-81-322-2644-4_16

Bilgrami AL, Gaugler R, Brey C (2005) Prey preference and feeding behavior of the diplogasterid predator *Mononchoides gaugleri* (Nematoda: Diplogasterida). Nematology 7:333–342

Bird DM, Opperman CH, Davies KG (2003) Interaction between bacteria and plant-parasitic nematodes: now and then. Int J Parasitol 33:1269–1276

Blok VC, Powers TO (2009) Biochemical and molecular identification. In: Perry RN, Moens M, Star JL (eds) Root knot nematodes, 1st edn. CABI International, London, pp 98–112

Blok VC (2005) Achievements in and future prospects for molecular diagnosis of plant parasitic nematodes. Can J Plant Pathol 27:176–185

Borrelli VMG, Brambilla V, Rogowsky P, Marocco A, Lanubile A (2018) The enhancement of plant disease resistance using CRISPR/Cas9 technology. Front Plant Sci 9:1245; https://www.frontiersin.org/article/10.3389/fpls.2018.01245. Accessed on 04 Apr 2019

Bradfish GA, Hickle LA, Flores R, Schwab G (1991) Nematicidal *Bacillus thuringiensis* toxins: opportunities in animal health and plant protection. In: First international conference on *Bacillus thuringiensis*. St. Catherine's College, Oxford, UK, p 33

Braun-Kiewnick A, Kiewnick S (2018) Real-time PCR, a great tool for fast identification, sensitive detection and quantification of important plant parasitic nematodes. Eur J Plant Pathol 152 (2):271–283

Bridge J, Starr JL (2007) Plant nematodes of agricultural importance. Academic Press, San Diego

Bridge J (1996) Nematode management in sustainable and subsistence agriculture. Annu Rev Phytopathol 34:201–225

Britton C, Roberts B, Marks N (2016) Functional genomics tools for *Haemonchus contortus* and lessons from other Helminths. Adv Parasitol 93:599–623

Brodie BB, Murphy WS (1975) Population dynamics of plant nematodes as affected by combinations of fallow and cropping sequence. J Nematol 7:91–92

Brown RH (1987) Control strategies in low-value crops. In: Brown RH, Kerry BR (eds) *Principles and Practice of Nematode Control in Crops*. Academic Press, Sydney, Australia, pp 351–388

Cai D, Kleine M, Kifle S, Harloff HJ, Sandal NN, Marcker KA, Klein-Lankhorst RM, Salentijn EM, Lange W, Stiekema WJ, Wyss U, Grundler FM, Jung C (1997) Positional cloning of a gene for nematode resistance in sugar beet. Science 275:832–834

Cai D, Thurau T, Tian Y, Lange T, Yeh KW, Jung C (2003) Sporamin-mediated resistance to beet cyst nematodes (*Heterodera schachtii* Schm.) is dependent on trypsin inhibitory activity in sugar beet (*Beta vulgaris* L.) hairy roots. Plant Mol Biol 51:839–849. https://doi.org/10.1023/A:1023089017906

Cao ZP, Yu YL, Chen GK, Dawson R (2004) Impact of soil fumigation practices on soil nematodes and microbial biomass. Pedoshpere 14:387–393

Carneiro RMDG, Correa VR, Almeida RMA, Gomes ACMM, Deimi AM, Castagnone-Sereno P, Karssen G (2014) *Meloidogyne luci* n. sp. (Nematoda: Meloidogynidae), a root-knot nematode parasitising different crops in Brazil, Chile and Iran. Nematology 16(3):1–13

Carneiro RMDG, Dos Santos MFA, Almeida MRA, Mota FC, Gomes ACMM, Tigano MS (2008) Diversity of *Meloidogyne arenaria* using morphological, cytological and molecular approaches. Nematology 10:819–834

Carneiro RMDG, Lima FSO, Correia VR (2017) Methods and tools currently used for the identification of plant parasitic nematodes. doi: https://doi.org/10.5772/intechopen.69403. Accessed 28 Apr 2019

Carneiro RMDG, Monteiro JMS, Silva UC, Gomes G (2016) Gênero Meloidogyne: diagnose através de eletroforese de isoenzimas e marcadores SCAR. In: Oliveira CMG, Santos MA, Castro LHS (eds) Diagnose de fitonematoides, 3rd edn. Millennium Editora, Campinas, pp 47–72

Carneiro RMDG, Tigano MS, Randig O, Almeida MRA, Sarah JL (2004) Identification and genetic diversity of *Meloidogyne* spp. (Tylenchida: Meloidogynidae) on coffee from Brazil, Central America and Hawaii. Nematology 6:287–298

Carrillo L, Martinez M, Ramessar K, Cambra I, Castanera P (2011) Expression of a barley cystatin gene in maize enhances resistance against phytophagous mites by altering their cysteine-proteases. Plant Cell Rep 30:101–112

Castagnone-Sereno P (2011) Molecular tools for diagnosis. In: Jones J, Gheisen G, Fenoll C (eds) Genomics and molecular genetics of plant nematode interactions, 1st edn. Springer, New York, pp 443–464

Castillo P, Vovlas N (2007) *Pratylenchus* (Nematodea: Pratylenchidae): diagnosis, biology, pathogenicity and management, Nematology monographs and perspectives. Brill, Leiden, the Netherlands, p 529

Chan YL, He Y, Hsiao T-T, Wang CJ, Tian Z, Yeh KW (2015) Pyramiding taro cystatin and fungal chitinase genes driven by a synthetic promoter enhances resistance in tomato to root-knot nematode *Meloidogyne incognita*. Plant Sci 231:74–81. https://doi.org/10.1016/j.plantsci.2014.11.010

Chan Y, Yang A, Chen J, Yeh K, Chan M (2010) Heterologous expression of taro cystatin protects transgenic tomato against *Meloidogyne incognita* infection by means of interfering sex determination and suppressing gall formation. Plant Cell Rep 29:231–238. https://doi.org/10.1007/s00299-009-0815-y

Chen P, Tsay TT (2006) Effect of crop rotation on Meloidogyne spp. and Pratylenchus spp. populations in strawberry fields in Taiwan. J Nematol 38(3):339–344

Chen Q, Yang B, Liu X, Chen F, Ge F (2017) Long-term cultivation of Bt rice expressing the Cry1Ab/1Ac gene reduced phytoparasitic nematode abundance but did not affect other nematode parameters in paddy fields. Sci Total Environ 607:463–474

Chen ZX, Dickson DW (1998) Review of *Pasteuria penetrans*: biology, ecology, and biological control potential. J Nematol 30:313–340

Chindo PS, Khan FA (1990) Control of root-knot nematodes, *Meloidogyne* spp., on tomato, *Lycopersicon esculentum* Mill., with poultry manure. Trop Pest Manag 36:332–335

Chitwood DJ (2002) Phytochemical based strategies for nematode control. Annu Rev Phytopathol 40:221–249

Chitwood DJ (2003) Nematicides. In: Plimmer JR (ed) Encyclopedia of agrochemicals, vol 3. Wiley, New York, NY, pp 1104–1115

Chu Y, Wu CL, Holbrook CC, Tillman BL, Person G, Ozias-Akins P (2011) Marker assisted selection to pyramid nematode resistance and the high oleic trait in peanut. Plant Genome 4:110–117

Clark FE, Paul EA (1970) The microflora of grassland. Adv Agron 22:375–435

Collins HP, Alva A, Boydston RA, Cochran RL, Hamm PB, McGuire A, Riga E (2006) Soil microbial, fungal, and nematode responses to soil fumigation and cover crops under potato production. Biol Fertil Soils 42(3):247–257

Concibido VC, Diers BW, Arelli PR (2004) A decade of QTL mapping for cyst nematode resistance in soybean. Crop Sci 44:1121–1131

Correa VR, Mattos VS, Almeida MRA, Santos MFA, Tigano MS, Castagnone-Sereno P, Carneiro RMDG (2014) Genetic diversity of the root-knot nematode *Meloidogyne ethiopica* and development of a species-specific SCAR marker for its diagnosis. Plant Pathol 63(2):476–483

Correa VR, Santos MFA, Almeida MRA, Peixoto JR, Castagnone-Sereno P, Carneiro RMDG (2013) Species-specific DNA markers for identification of two root-knot nematodes of coffee: *Meloidogyne arabicida* and *M. izalcoensis*. Eur J Plant Pathol 137:305–313

Cromwell WA, Yang J, Starr JL, Jo YK (2014) Nematicidal effects of silver nanoparticles on root-knot nematode in bermudagrass. J Nematol 46(3):261–266

Crump DH, Sayre RM, Young LD (1983) Occurrence of nematophagous fungi in cyst nematode populations. Plant Dis 67:63–64

Cunha TG, Visôtto LE, Lopes EA, Oliveira CMG, God PIVG (2018) Diagnostic methods for identification of root-knot nematodes species from Brazil. Ciência Rural 48(2):1–11

Curran J, Baillie DL, Webster JM (1985) Use of genomic DNA restriction fragment length polymorphism to identify nematode species. Parasitology 90:137–144

D'Addabbo T, Fntanazza G, Lamberti F, Sasanelli N, Patumi M (1997) The suppressive effect of soil amendments with olive residues on *Meloidogyne incognita*. Nematol Mediterr 25:195–198
Dackman C, Jansson HB, Nordbring-Hertz B (1992) Nematophagous fungi and their activities in soil. In: Stotsky G, Bollag JM (eds) Soil biochemistry, vol 7. Marcel Dekker, New York
Davies KG, Curtis RH, Evans K (1996) Serologically based diagnostic and quantification tests for nematodes. Pestic Sci 47:81–87
De Bach P (1964) Biological control of insect pests and weeds, 1st edn. Chapman and Hall, London
de Leij FAAM, Kerry BR (1991) The nematophagous fungus *Verticillium chlamydosporium* as a potential biological control agent for *Meloidogyne arenaria*. Rev Nematol 14:157–164
Dickinson DJ, Goldstein B (2016) CRISPR-based methods for *Caenorhabditis elegans* genome engineering. Genetics 202:885–901
Dijksterhuis J, Sjollema KA, Veenhuis M, Harder W (1994) Competitive interactions between two nematophagous fungi during infection and digestion of the nematode *Panagrellus redivivus*. Mycol Res 98:1458–1462
Dinh P, Brown C, Elling A (2014) RNA interference of effector gene Mc16D10L confers resistance against *Meloidogyne chitwoodi* in *Arabidopsis* and potato. Phytopathology 104:1098–1106. https://doi.org/10.1094/PHYTO-03-14-0063-R
Dong LQ, Zhang KQ (2006) Microbial control of plant-parasitic nematodes: a five-party interaction. Plant and Soil 288(1–2):31–45. https://doi.org/10.1007/s11104-006-9009-3
Dongre M, Sobita S (2013) Efficacy of certain botanical extracts in the management of *Meloidogyne graminicola* of rice. Int J Agric Sci Res 3(3):91–98
Duncan LW, Noling JW (1998) Agricultural sustainability and nematode integrated pest management. In: Barker KR, Pederson GA, Windham GL (eds) Plant nematode interactions. American Society of Agronomy, Madison. pp 251–87
Dutta TK, Khan MR, Phani V (2019) Plant-parasitic nematode management via biofumigation using brassica and non-brassica plants: current status and future prospects. Curr Plant Biol 17:17–32. https://doi.org/10.1016/j.cpb.2019.02.001
Dutta T, Banakar P, Rao U (2015) The status of RNAi-based transgenics in plant nematology. Front Microbiol 5:760. https://doi.org/10.3389/fmicb.2014.00760
Eastwood RF, Lagudah ES, Appels R, Hannah M, Kollmorgen JF (1991) *Triticum tauschii*: a novel source of resistance to cereal cyst nematode (*Heterodera avenae*). Aust J Agr Res 42(1):69–77
Eastwood RF, Smith A, Wilson J (1994) *Pratylenchus thornei* is causing yield losses in Victorian Wheat Crops. Aust Nematol Newslett 5:2
Egunjobi OA, Larinde MA (1975) Nematodes and maize growth in Nigeria. II. Effects of some amendments on populations of *Pratylenchus brachyurus* and on growth and production of maize (*Zea mays*) in Nigeria. Nematol Mediterr 3:65–73
Elmali M (2002) The distribution and damage of wheat gall nematode [*A. tritici* (Steinbuch)] (Tylenchida: Tylenchidae) in western part of Anatolia. Turkish J Entomol 26(2):105–114
El-Rokiek KG, El-Nagdi WM (2011) Dual effects of leaf extracts of *Eucalyptus citriodora* on controlling purslane and root-knot nematode in sunflower. J Plant Prot Res 51(2):121–129
Emmert EAB, Handelsman J (1999) Biocontrol of plant disease: a (Gram1) positive perspective. FEMS Microbiol Lett 171:1–9
Esbenshade P, Triantaphyllou A (1985) Use of enzyme phenotypes for identification of *Meloidogyne* species. J Nematol 17:6
Evans K, Trudgil DL, Webster JM (1993) Extraction, identification and control of plant parasitic nematodes. In: Evans K, Trudgil DL, Webster JM (eds) Plant parasitic nematodes in temperate agriculture. CAB International Publishing, Wallingford, UK, pp 648–649
Feizi A, Mahdikhani-Moghadam E, Azizi M, Roohani H (2014) Inhibitory effect of *Allium cepa* var. *aggregatum*, *Salvia officinalis* and *Kelussiaodor atissima* essence on the root-knot nematode (*Meloidogyne javanica*) and extraction of active ingredients. J Plant Prot 28:220–225
Ferris H, Matute MM (2003) Structural and functional succession in the nematode fauna of a soil food web. Appl Soil Ecol 23:93–110

Ferris H, Bongers T, de Goede RGM (2001) A framework for soil food web diagnostics: extension of the nematode faunal analysis concept. Appl Soil Ecol 18:13–29

Filipiak A, Wieczorek P, Tomalak M (2017) Multiplex polymerase chain reaction for simultaneous detection and identification of *Bursaphelenchus xylophilus*, *B. mucronatus* and *B. fraudulentus*—three closely related species within the *Xylophilus* group. Nematology 19:1107–1116

Fire A, SiQun X, Montgomery M, Kostas S, Driver S, Mello C (1998) Potent and specific genetic interference by double-stranded RNA in *Caenorhabditis elegans*. Nature 391:806–811. https://doi.org/10.1038/35888

Fleming C, Shields R, Stratford R (1993). Identification of potato cyst nematodes using the polymerase chain reaction. In: Proceedings of crop protection in Northern Britain, pp 183–188

Florini DA, Loria R (1986) Changes in population densities of *Pratylenchus penetrans* and *Pratylenchus crenatus* on winter cover crops. J Nematol 18:607

Fortnum BA, Lewis SA, Johnson AW (2001) Crop rotation and nematicides for management of mixed populations of *Meloidogyne* spp. on tobacco. Suppl J Nematol 33:318–324

Fosu-Nyarko J, Jones MGK (2015) Application of biotechnology for nematode control in crop plants. Adv Botanical Res 73:340–376

Françoisa C, Kebdania N, Barkerb I, Tomlinson J, Boonham N, Castagnone-Sereno P (2006) Towards specific diagnosis of plant-parasitic nematodes using DNA oligonucleotide microarray technology: a case study with the quarantine species *Meloidogyne chitwoodi*. Mol Cell Probes 20(1):64–69

Freire FCO, Bridge J (1985) Parasitism of eggs, females and juveniles of *Meloidogyne incognita* by *Paecilomyces lilacinus* and *Verticillium chlamydosporium*. Fitopatol Braz 10:577–596

Friedland AE, Tzur YB, Esvelt KM, Colaiacovo MP, Church GM, Calarco JA (2013) Heritable genome editing in *C. elegans* via a CRISPR-Cas9 system. Nat Methods 10:741–743

Fujiwara R, Kurogi S, Nagatomo Y, Takaki H (1997) Identification of *Meloidogyne* spp. by enzyme linked immunosorbent assay using monoclonal antipbody. In: Proceedings of the association for plant protection of Hyushu, vol 43, pp 94–96.

Fuller VL, Lilley CJ, Urwin PE (2008) Nematode resistance. New Phytol 180:27–44. https://doi.org/10.1111/j.1469-8137.2008.02508.x

Gams W, Zare R (2003) A taxonomic review of the clavicipitaceous anamorphs parasitizing nematodes and other microinvertebrates. In: White JF Jr, Bacon CW, Hywel-Jones NL, Spatafora JW (eds) Clavicipitalean fungi: evolutionary biology, chemistry, and cultural impacts. Marcel Dekker, New York, pp 26–81

Gan J, Yates SR, Wang D, Ernst FF (1998) Effect of application methods on 1,3-dichloropropene volatilization from soil under controlled conditions. J Environ Qual 27:432–438

Ganal MW, Tanksley SD (1996) Recombination around the Tm2a and Mi resistance genes in different crosses of *Lycopersicon peruvianum*. Theor Appl Genet 92:101–108

Garcia GM, Stalker HT, Shroeder E, Kochert G (1996) Identification of RAPD, SCAR, and RFLP markers tightly linked to nematode resistance genes introgressed from *Arachis cardenasii* into Arachis hypogaea. Genome 39:836–845

Gasser RB, Bott NJ, Chilton NB, Hunt P, Beveridge I (2008) Toward practical, DNA based diagnostic methods for parasitic nematodes of livestock-bionomic and biotechnological implications. Biotechnol Adv 26:325–334

Giannakou IO, Sidiropoulos A, Prophetou-Athanasiadou D (2002) Chemical alternatives to methyl bromide for the control of root-knot nematodes in greenhouses. Pest Manag Sci 58:290–296

Gokta N, Swarup G (1988) On the potential of some bacterial biocides against root-knot cyst nematodes. Indian J Nematol 18:152–153

Gonzalez A, Canto-Sanenz M (1993) Comparison of live organic amendments for the control of *Globodera pallida* in microplots in Peru. Nematropica 23:133–139

Gowen SR (1997) Chemical control of nematodes: efficiency and side-effects. FAO Plant Production and Protection Paper, 144, pp 59–65. http://www.fao.org/3/v9978e/v9978e08.htm. Accessed 12 Apr 2019

Grainger J (1964) Factors affecting the control of eelworm diseases. Nematologica 10:5–20
Hackney RW, Dickerson OJ (1975) Marigold, castor bean, and chrysanthemum as controls of *Meloidogyne incognita* and *Pratylenchus alleni*. J Nematol 7:84–90
Hahn C, Bachmann L, Chevreux B (2013) Reconstructing mitochondrial genomes directly from genomic next-generation sequencing reads- a baiting and iterative mapping approach. Nucleic Acids Res 41(2013):e129
Hahn ML, Burrows PR, Wright DJ (1996) Genomic diversity between *Radopholus similis* populations from around the world detected by RAPD-PCR analysis. Nematologica 42:537–545
Hallmann J, Davies KG, Sikora RA (2009) Biological control using microbial pathogens, endophytes and antagonists. In: Perry RN, Moens M, Starr JL (eds) Root-knot nematodes. CABI, Wallingford
Han H, Cho MR, Jeon HJ, Lim CK, Jang HI (2004) PCR-RFLP identification of three major *Meloidogyne* species in Korea. J Asia Pac Entomol 7(2):171–175
Hebert PD, Cywinska A, Ball SL, deWaard JR (2003) Biological identifications through DNA barcodes. Proc Royal Soc London 270(1512):313–321. https://doi.org/10.1098/rspb.2002.2218
Heller R, Schondelmaier J, Steinrücken G, Jung C (1996) Genetic localization of four genes for nematode (*Heterodera schachtii* Schm.) resistance in sugar beet (*Beta vulgaris* L.). Theor Appl Genet 92:991–997
Hill SN (1988) Cultural practices for the management of plant-parasitic nematodes. Ornamentals Northwest Ach 12(4):7–9
Hossain M, Ahmadi MU, Ahmed N, Abulhossain M, Alim MA (2002) A study on control of root-knot nematode *(Meloidogyne javanica)* of wheat. Indian Agriculturist 46:121–128
Hu MX, Zhuo K, Liao JL (2011) Multiplex PCR for the simultaneous identification and detection of *Meloidogyne incognita*, *M. enterolobii* and *M. javanica* using DNA extracted directly from individual galls. Phytopathology 101(11):1270–1277
Huang G, Allen R, Davis E, Baum T, Hussey R (2006) Engineering broad root-knot resistance in transgenic plants by RNAi silencing of a conserved and essential root-knot nematodeparasitism gene. PNAS 103:14302–14306. https://doi.org/10.1073/pnas.0604698103
Huang SP (1994) Cropping effects of marigold, corn, and okra on population levels of *Meloidogyne javanica* and carrot yields. J Nematol 16:396–398
Hugot JP, Baujard P, Morand S (2001) Biodiversity in helminthes and nematodes as a field of study. Nematology 3:199–208
Hunt D, Handoo Z (2009) Taxonomy, identification and principal species. In: Perry RN, Moens M, Star JL (eds) Root knot nematodes (pp 55–88), 1st edn. CABI International, London
Ibekwe M (2004) Effects of fumigants on non-target organisms in soils. Adv Agron 83:1–35
Ibrahim AAM, Al-Hazmi AS, Al-Yahya FA, Alderfasi AA (1999) Damage potential and reproduction of *Heterodera avenae* on wheat and barley under Saudi field conditions. Nematology 1 (6):625–630
Insunza V, Alstrom S, Eriksson KB (2002) Root bacteria from nematicidal plants and their biocontrol potential against trichodorid nematodes in potato. Plant and Soil 241:271–278
ISO (2008) Nanotechnologies—terminology and definitions for nano objects - nanoparticle, nanofibre and nanoplate. http://www.iso.org/iso/iso_catalogue.htm. Accessed 12 Apr 2019
Jacquet M, Bongiovanni M, Martinez M, Verschave P, Wajnberg E, Catagnone-Sereno P (2005) Variation in resistance to the root-knot nematode Meloidogyne incognita in tomato genotypes bearing the Mi gene. Plant Pathol 54:93–99
Jaffee BA, Muldoon AE (1989) Suppression of cyst nematode by natural infestation of a nematophagous fungus. J Nematol 21:505–510
Jaffee BA (1992) Population biology and biological control of nematodes. Can J Microbiol 38:359–364
Jansson HB (1994) Adhesion of conidia of *Drechmeria coniospora* to *Caenorhabditis elegans* wild type and mutants. J Nematol 26:430–435

Jaouannet M, Magliano M, Arguel M, Gourgues M, Evangelisti E, Abad P, Rosso M (2013) The root-knot nematode calreticulin Mi-CRT is a key effector in plant defense suppression. Mol Plant Microbe Interact 26:97–105. https://doi.org/10.1094/MPMI-05-12-0130-R

Jatala P (1986) Biological control of plant parasitic nematodes. Annu Rev Phytopathol 24:453–489

Jayakumar J, Ramakrishnan S, Rajendran G (2002) Bio-control of reniform nematode, *Rotylenchulus reniformis* through fluorescent *Pseudomonas*. Pesttology 26:45–46

Jenkins WR (1964) A rapid centrifugal-flotation technique for separating nematodes from soil. Plant Dis Rep 48:692

Johnathan EI, Pandiarajan P (1991) Effect of soil amendments in controlling rice root nematodes. Int Rice Res Newslett 16:25

Johnson AW (1982) Managing nematode populations in crop production. In: Riggs Rd (ed) Nematology in the southern region of the United States, South. Coop. Ser. Bull. 276, pp 193–203

Johnson AW (1985) Specific crop rotation effects combined with cultural practices and nematodes. In: Sasser JN, Carter CC (eds) An advanced treatise on meloidogyne, Volume I. North Carolina State University Graphics, Raleigh, NC, pp 283–301

Jordan S (2017) Yield to the resistance: the impact of nematode resistant varieties on alfalfa yield. Nat Resour Model 31(2):e12150. https://doi.org/10.1111/nrm.12150

Karajeh M (2008) Interaction of root-knot nematode (*Meloidogyne Javanica*) and tomato as affected by hydrogen peroxide. J Plant Prot Res 48(2):2

Kathy M (2000) Root-parasitic nematode host range and damage levels on Oregon vegetable crops: a literature survey. Nematode Testing Service, Extension Plant Pathology Oregon

Kaushal KK (1998) Management of nematodes infecting wheat. Summer School Report, Division of Nematology, IARI, New Delhi, pp 1–8

Kennedy MJ, Schoelz JE, Donald PA, Niblack TL (1997) Unique immunogenic proteins in *Heterodera glycines* eggshells. J Nematol 29:276–281

Kepenekçi I, Erdoğuş D, Erdoğan P (2016) Effects of some plant extracts on root-knot nematodes *in vitro* and *in vivo* conditions. Türkish J Entomol 40(1):3–14. https://doi.org/10.16970/ted.84688

Kerry BR, Crump DH (1980) Two fungi parasitic on females of cyst nematodes (*Heterodera* spp.). Trans Br Mycol Soc 74:119–125

Kerry BR, Hominick WM (2002) Biological control. In: Lee DL (ed) The biology of nematodes. Taylor & Francis, London

Kerry BR (2000) Rhizosphere interactions and exploitation of microbial agents for the biological control of plant-parasitic nematodes. Annu Rev Phytopathol 38:423–441

Khan A, Tariq M, Asif M, Siddiqui MA (2017) Evaluation of botanicals toxicants against root knot nematode, Meloidogyne incognita in vitro. Asian J Biol 4(3):1–7. https://doi.org/10.9734/AJOB/2017/38376

Khan MR, Mukhopadhyay AN, Pandey RN, Thakur MP, Singh D, Siddiqui MA, Akram M, Mohiddin FA, Haque Z (2019) Biointensive approaches: application and effectiveness in the management of plant nematodes, insects and weeds. Indian Phytopathological Society, New Delhi, pp 1–698

Khan Z, Kim YH (2005) The predatory nematode, *Mononchoides fortidens* (Nematoda: *Diplogasterida*), suppresses the root-knot nematode, *Meloidogyne arenaria*, in potted field soil. Biol Control 35:78–82

Khatodia S, Bhatotia K, Passricha N, Khurana SMP, Tuteja N (2016) CRISPR/Cas genome-editing tool: application in improvement of crop. Front Plant Sci 7:506. https://doi.org/10.3389/fpls.2016.00506

Kiewnick S, Holterman M, van den Elsen S, van Megen H, Frey JE, Helder J (2014) Comparison of two short DNA barcoding loci (COI and COII) and two longer ribosomal DNA genes (SSU & LSU rRNA) for specimen identification among quarantine root-knot nematodes (*Meloidogyne* spp.) and their close relatives. Eur J Plant Pathol 140(1):97–110. https://doi.org/10.1007/s10658-014-0446-1

Kiewnick S, Wolf W, Willareth M, Frey J (2013) Identification of the tropical root-knot nematode species *Meloidogyne incognita*, M. javanica and M. arenaria using a multiplex PCR assay. Nematology 15:891–894. https://doi.org/10.1163/15685411-00002751

King BA, Taberna JP Jr (2013) Site-Specific management of *Meloidogyne chitwoodi* in Idaho potatoes using 1,3-Dichloropropene; approach, experiences, and economics. J Nematol 45:202–213

Kirkpatrick TL, Thomas AC (2007) Crop rotation for management of nematodes in cotton and soybean. Cooperative Extension Service, University of Arkansas, US Department of Agriculture, and county governments cooperating

Kloepper JW, Rodriguez-Kabana R, Mcinroy JA, Young RW (1992) Rhizosphere bacteria antagonistic to soybean cyst (*Heterodera glycines*) and root-knot (*Meloidogyne incognita*) nematodes: identification by fatty acid analysis and frequency of biological control activity. Plant and Soil 139:75–84

Kokalis-Burelle N, Vavrina CS, Rosskopf EN, Shelby RA (2002) Field evaluation of plant growth-promoting rhizobacteria amended transplant mixes and soil solarization for tomato and pepper production in Florida. Plant and Soil 238:257–266

Kort J (1972) Nematode diseases of cereals of temperate climates. In: Webster JM (ed) Economic nematology. Academic, New York, NY, pp 97–126

Kotze AC, O'Grady J, Gough JM, Pearson R, Bagnall NH, Kemp DH, Akhurst RJ (2005) Toxicity of *Bacillus thuringiensis* to parasitic and free-living life stages of nematodes parasites of livestock. Int J Parasitol 35:1013–1022

Krebs B, Hoeding B, Kuebart S, Workie MA, Junge H, Schmiedeknecht G, Grosch R, Bochow H, Hevesi M (1998) Use of *Bacillus subtilis* as biocontrol agent. I. Activities and characterization of Bacillus subtilis strains. J Plant Dis Prot 105:181–197

Kretschmer J, Chalmers K, Manning S, Karakousis A, Barr A, Islam A, Logue S, Choe Y, Barker S, Lance R (1997) RFLP mapping of the Ha 2 cereal cyst nematode resistance gene in barley. Theor Appl Genet 94(8):1060–1064

Kumar A, Kakrana A, Sirohi A, Subramaniam K, Srinivasan R, Abdin M, Jain PK (2017) Host-delivered RNAi-mediated root-knot nematode resistance in *Arabidopsis* by targeting splicing factor and integrase genes. J General Plant Pathol 83(2):91–97

Kumari S, Decraemer W, De Luca F, Tiefenbrunner W (2010) Cytochrome c oxidase subunit 1 analysis of *Xiphinema diversicaudatum*, *X. pachtaicum*, *X. simile* and *X. vuittenezi* (Nematoda, Dorylaimida). Eur J Plant Pathol 127:493–499

Leader AJ, Busacca J, Dawson J, Lyall T, Eelen H (2010) Is there a future for 1,3-dichloropropene and other chemical soil fumigants in European agriculture? Acta Hortic 883:67–73

Lehman PS (1979) Seed and leaf gall nematodes of the genus *Anguina* occurring in North America. Nematology Circular No. 55, September 1979

Leister D, Ballvora A, Salamini F, Gebhardt C (1996) A PCR-based approach for isolating pathogen resistance genes from potato with potential for wide application in plants. Nat Genet 14:421–429

Lewis JG, Matic M, McKay AC (2009) Success of cereal cyst nematode resistance in Australia: history and status of resistance screening systems. In: Riley IT, Nicol JM, Dababat AA (eds) Cereal cyst nematode: status, research and outlook, pp. 137–142, Proceedings of the first workshop of the international cereal cyst nematode initiative, Antalya, Turkey

Li B, Xie GL, Soad A, Coosemans J (2005) Suppression of *Meloidogyne javanica* by antagonistic and plant growth promoting rhizobacteria. J Zhejiang Univ Sci B 6:496–501

Li J, Zou C, Xu J, Ji X, Niu X, Yang J, Huang X, Zhang KQ (2015) Molecular mechanisms of nematode-nematophagous microbe interactions: basis for biological control of plant-parasitic nematodes. Annu Rev Phytopathol 53:67–95

Lilley C, Davis L, Urwin P (2012) RNA interference in plant parasitic nematodes: a summary of the current status. Parasitology 139:630–640. https://doi.org/10.1017/S0031182011002071

Lima FSO, Santos GR, Nogueira SR, Santos PRR, Correa VR (2015) Population dynamics of the root lesion nematode *Pratylenchus brachyurus* in soybean fields in Tocantins state and its effect to soybean yield. Nematropica 45(2):170–177

Lòpez-Llorca LV, Macia-Vicente JG, Jansson H-B (2008) Mode of action and interactions of nematophagous fungi. In: Ciancio A, Mukerji KG (eds) Integrated management and biocontrol of vegetable and grain crops nematodes. Springer, Dordrecht

Lòpez-Llorca LV, Claughe RD (1990) Appressoria of the nematophagous fungus *Verticillium suchlasporium*. Micron Microscopia Acta 21:125–130

Lòpez-Llorca LV, Robertson WM (1992) Immunocytochemical localization of a 32-kDa protease from the nematophagous fungus *Verticillium suchlasporium* in infected nematode eggs. Exp Mycol 16:261–267

Luna JM, House GJ (1990) Pest management in sustainable agricultural systems. In: Edwards CA, Lal R, Madden P, Miller RH, House G (eds) Miller Sustainable Agricultural Systems. Soil Water Conservation Society, Ankeny, IA, p 696

Ma H, Bertsch PM, Glenn TC, Kabengi NJ, Williams PL (2009) Toxicity of manufactured zinc oxide nanoparticles in the nematode Caenorhabditis elegans. Environ Toxicol Chem 28 (6):1324–1330

Mackenzie CD, Behan-Braman A, Hauptman J, Geary T (2017) Assessing the viability and degeneration of the medically important filarial nematodes. In: Nematology-concepts, diagnosis and control. IntechOpen

Madamba CP, Camaya EN, Zenarosa DB, Yater HM (1999) Screening soil bacteria for potential biocontrol agents against the root-knot nematode, *Meloidogyne* spp. Philippine Agric 82:113–122

Maggenti A (1981) General nematology, 1st edn. Springer-Verlag, New York, p 372. https://doi.org/10.1007/978-1-4612-5938-1

Maqbool MA (1988) Present status of research on plant parasitic nematodes in cereals and food and forage legumes in Pakistan. In: Saxena MC, Sikora RA, Srivastava JP (eds) Nematodes parasitic to cereals and legumes in temperate semi-arid regions. Aleppo, Syria, ICARDA, pp 173–180

Marroquin LD, Elyassnia D, Griffitts JS, Feitelson JS, Aroian RV (2000) *Bacillus thuringiensis* (Bt) toxin susceptibility and isolation of resistance mutants in the nematode *Caenorhabditis elegans*. Genetics 155:1693–1699

Marull J, Pinochet J, Rodrıguez-Kabana R (1997) Agricultural and municipal compost residues for control of root-knot nematodes in tomato and pepper. Compost Sci Util 5:6–15

Masler EP (2002) A polyclonal antiserum that recognizes the major female-specific proteins from the soybean cyst nematode Heterodera glycines. Int J Nematol 12:119–124

Mattiucci S, Nascetti G (2008) Chapter 2: advances and trends in the molecular systematics of anisakid nematodes, with implications for their evolutionary ecology and host—parasite co-evolutionary processes. In: Rollinson D, Hay SI (eds) Advances in parasitology. Academic Press, London, pp 47–148

McGawley EC (1986) Influence of crop rotation on nematode community structure. J Nematol 18:620

McKenry MV (1987) Control strategies in high-value crops. In: Brown RH, Kerry BR (eds) Principles and practice of nematode control in crops. Academic Press, Sydney, Australia, pp 330–349

McSorley R, Duncan LW (1995) Economic thresholds and nematode management. Adv Plant Pathol 11:147–170

McSorley R, Gallaher RN (1995) Effect of yard waste compost on plant-parasitic nematode densities in vegetable crops. Suppl J Nematol 27:545–549

McSorley R, Gallaher RN (1997) Effect of compost and maize cultivars on plant-parasitic nematodes. J Nematol 29:731–736

McSorley R, Wang RH, Kokalis-Burelle N, Church G (2006) Effects of soil type and steam on nematode biological control potential of the rhizosphere community. Nematropica 36:197–214

McSorley R, Wang RH, Kokalis-Burelle N, Church G (2008) Suppression of root-knot nematodes in natural and agricultural soils. Appl Soil Ecol 39:291–298
Meadows I, Averre C, Duncan H, Baker K (2018) Control of root-knot nematodes in the home vegetable garden. NC State Extension Publications
Meagher JW (1972) Cereal cyst nematode (Heterodera avenae Woll). Studies on ecology and content in Victoria. Technical Bulletin 24, Department of Agriculture, Victoria, Australia, pp 50
Meena KS, Sivakumar M, Jonathan EI, Devrajan K, Boopathi T (2010) Management of *Meloidogyne incognita* in tomato through dry powder and aqueous extracts of *Tagetes erecta* cv. Indian yellow. Pestology 34:28–30
Mekete T, Dababat A, Sekora N, Akyazi F, Abebe E (2012) Identification key for agriculturally important plant-parasitic nematodes prepared for the international nematode diagnosis and identification course 2012—a manual for nematology. CIMMYT, Mexico, DF
Mena J, Pimentel E (2002) Mechanism of action of *Corynebacterium pauronetabolum* strain C-924 on nematodes. Nematology 4:287
Meyer SLF (2003) United States Department of Agriculture—Agricultural Research Service research programs on microbes for management of plant-parasitic nematodes. Pest Manag Sci 59:665–670
Miller L (1986) Economic importance of cyst nematodes in North America. In: Lamberti F, Taylor CE (eds) Cyst nematodes. Plenum Press, New York, pp 373–385
Milligan SB, Bodeau J, Yaghoobi J, Kaloshian I, Zabel P, Williamson VM (1998) The root knot nematode resistance gene *Mi* from tomato is a member of the leucine zipper, nucleotide binding, leucine-rich repeat family of plant genes. Plant Cell 10:1307–1319. https://doi.org/10.1105/tpc.10.8.1307
Mitkowski NA, Abawi GS (2003) Root-knot nematodes. The Plant Health Instructor. https://www.apsnet.org/edcenter/disandpath/nematode/pdlessons/Pages/RootknotNematode.aspx. Accessed 11 Apr 2019
Mojtahedi H, Santo GS, Wilson JH, Hang AN (1993) Managing *Meloidogyne chitwoodi* on potato with rapeseed and green manure. Plant Dis 77:42–46
Mokabli A, Valette S, Gauthier J-P, Rivoal R (2001) Influence of temperature on the hatch of *Heterodera avenae* Woll. populations from Algeria. Nematology 3(2):171–178
Mokrini F, Janati S, Houari A, Essarioui A, Bouharroud R, Mimouni A (2018) Management of plant parasitic nematodes by means of organic amendment. Rev Mar Sci Agron Vét 6 (3):337–344
Moosavi MR, Zare R (2011) Fungi as biological control agents of plant-parasitic nematodes. Plant Defense: Biol Control:67–107. https://doi.org/10.1007/978-94-007-1933-0_4
Moosavi MR (2012) Nematicidal effect of some herbal powders and their aqueous extracts against *Meloidogyne javanica*. Nematropica 42:48–56
Moosavi MR, Zare R, Zamanizadeh HR, Fatemy S (2010) Pathogenicity of *Pochonia* species on eggs of Meloidogyne javanica. J Invertebr Pathol 104:125–133
Morgan-Jones G, Godoy G, Rodriguez-Kabana R (1981) *Verticillium chlamydosporium* fungal parasite of *Meloidogyne arenaria* females. Nematropica 11:115–119
Murslain M, Javed N, Khan SA, Khan HU, Abbas H, Munawar M (2014) Efficacy of moringa leaves and *Trichoderma harzianum* on the invasion and development of *Meloidogyne javanica*. Pak J Phytopathol 46(3):827–832
Nassar AM (2016) Effectiveness of silver nano-particles of extracts of *Urtica urens* (Urticaceae) against root-knot nematode *Meloidogyne incognita*. Asian J Nematol 5(1):14–19
Neeraj SG, Kumar A, Ram S, Kumar V (2017) Evaluation of nematicidal activity of ethanolic extracts of medicinal plants to *Meloidogyne incognita* (Kofoid and White) Chitwood under lab conditions. Int J Pure Appl Biosci 5(1):827–831. https://doi.org/10.18782/2320-7051.2525
Nicol JM (2002) Important nematode pests. Bread Wheat:345–366
Nicol JM (1996) The distribution, pathogenicity and population dynamics of *Pratylencus thornei* (Sher and Allen, 1954) on wheat in South Australia. Ph.D. thesis, The University of Adelaide, Adelaide, Australia

Noe JP, Sasser JN, Imbriani JL (1991) Maximizing the potential of cropping systems for nematode management. J Nematol 23:353–361

Ntalli NG, Caboni P (2012) Botanical nematicides: a review. J Agric Food Chem 60:9929–9940

Nusbaum CJ, Ferris H (1973) The role of cropping systems in nematode population management. Annu Rev Phytopathol 11:423–440

Oka Y, Ben-Daniel B, Cohen Y (2012) Nematicidal activity of the leaf powder and extracts of *Myrtus communis* against the root-knot nematode *Meloidogyne javanica*. Plant Pathol 61 (6):1012–1020. https://doi.org/10.1111/j.1365-3059.2011.02587.x

Okada H, Harada H, Kadota I (2004) Application of diversity indices and ecological indices to evaluate nematode community changes after soil fumigation. Jap J Nematol 34:89–98

Olabiyi TI, Akanbi WB, Adepoju IO (2007) Control of certain nematode pests with different organic manure on cowpea. Am-Eurasian J Agric Environ Sci 2:523–527

Oliveira CMG, Monteiro AR, Blok VC (2011) Morphological and molecular diagnostics for plant parasitic nematodes: working together to get the identification done. Trop Plant Pathol 36 (2):65–73

Olsen MW (2000) Root-knot nematode. University of Arizona, Arizona Cooperative Extension, AZ1187 (November), 1–3

Orion D, Amir J, Krikun J (1984) Field observations on *Pratylenchus thornei* and its effects on wheat under arid conditions. Rev Nematol 7:341–345

Osaki N, Fukuchi T (2010) Biological activities and characteristics of a novel nematicide, Imicyafos (NEMAKICK_). Plant Prot 64:333–337

Osei K, Gowen S, Pembroke B, Brandenburg R, Jordan D (2010) Potential of leguminous cover crops in management of a mixed population of root-knot nematodes (*Meloidogyne* spp.). J Nematol 42:173–178

Paix A, Folkmann A, Seydoux G (2017) Precision genome editing using CRISPR-Cas9 and linear repair templates in *C. elegans*. Methods 121–122:86–93

Pavaraj M, Bakavathiappan G, Baskaran S (2012) Evaluation of some plant extracts for their nematicidal properties against root-knot nematode, Meloidogyne incognita. J Biopest 5:106–110

Peng D, Nicol JM, Li H, Hou S, Li H, Chen S, Riley IT (2009) Current knowledge of cereal cyst nematode (*Heterodera avenae*) on wheat in China. Cereal cyst nematodes: status, research and outlook, pp 29–34

Perera MR, Vanstone VA, Jones MG (2005) A novel approach to identify plant parasitic nematodes using matrix-assisted laser desorption/ionization time-of-flight mass spectrometry. Rapid Commun Mass Spectrom 19(11):1454–1460

Pluskota A, Horzowski E, Bossinger O, von Mikecz A (2009) In Caenorhabditis elegans nanoparticle-biointeractions become transparent: silica-nanoparticles induce reproductive senescence. PLoS One 4:1–9

Poswal MAT, Akpa AD (1991) Current trends in the use of traditional and organic methods for the control of crop pests and diseases in Nigeria. Trop Pest Manag 37:329–333

Powers T, Harris T, Higgins R, Mullin P, Powers K (2018) Discovery and identification of *Meloidogyne* species using COI DNA barcoding. J Nematol 50(3):399–412

Prot J-C, Soriano IR, Matias DM, Savary S (1992) Use of green manure crops in control of *Hirschmanniella mucronata* and *H. oryzae* in irrigated rice. J Nematol 24:127–132

Rahman MM, Mian IH (2010) Isolation and identification of ten genera of plant parasitic nematodes. Bangladesh J Sci Industrial Res 45(3):267–270

Ralmi NHAA, Khandaker MM, Mat N (2016) Occurrence and control of root knot nematode in crops: a review. Aust J Crop Sci 11(12):1649

Raymundo SA (1985) Cropping systems research and root-knot nematode control. In: Sasser JN, Carter CC (eds) An advanced treatise on Meloidogyne. Vol I. Biology and control. N. C. State Univ. Graphics, Raleigh, pp 277–281

Renco M, Sasanelli N, Šalamun P (2009) The effect of two compost soil amendments based on municipal green and penicillin production wastes, on plant parasitic nematodes. Helminthologia 46(3):190–197. https://doi.org/10.2478/s11687-009-0035-6

Rich JR, Dunn RA, Noling JW (2004) Nematicides: past and present uses. In: Chen Z, Chen S, Dickson DW (eds) Nematology advances and perspectives, Vol. 2. Nematode management and utilization. CABI Publishing, Wallingford, UK, pp 1179–1200

Roberts PA (1993) The future of nematology: integration of new and improved management strategies. J Nematol 25:383–394

Roberts PA, VanGundy SD, McKinney HE (1981) Effects of soil temperature and planting date of wheat on *Meloidogyne incognita* reproduction. J Nematol 13:338–344

Robinson AF (2004) Nematode behavior and migrations through soil and host tissue. In: Chen ZX, Chen SY, Dickson DW (eds) Nematology advances and perspectives: Vol. I nematode morphology, physiology, and ecology. CAB International, Oxfordshire, UK, pp 330–405

Rocha TL, Soll CB, Boughton BA, Silva TS, Oldach K, Alexandre AP, Callahan FDL, Sheedy J, Silveira ER, Carneiro RMDG, Silva LP, Polez VLP, Pelegrini PB, Bacic A, Grossi-de-Sa MF, Roessner U (2017) Prospection and identification of nematotoxic compounds from *Canavalia ensiformis* seeds effective in the control of the root knot nematode *Meloidogyne incognita*. Biotechnol Res Innov 1(1):87–100

Rodriguez-Kabana R, Morgan-Jones G (1987) Biological control of nematodes: soil amendments and microbial antagonists. Plant and Soil 100:237–247

Rodriguez-Kabana R (1986) Organic and inorganic nitrogen amendments to soil as nematode suppressants. J Nematol 18:129–135

Roe SA (1983) John Turberville Needham and the generation of living organisms. Isis 74 (2):159–184. http://www.jstor.org/stable/233101. Accessed 10 Mar 2019

Roeber F, Jex AR, Gasser RB (2013) Next-generation molecular-diagnostic tools for gastrointestinal nematodes of livestock, with an emphasis on small ruminants: a turning point. Adv Parasitol 83:267–333

Roh JY, Park YK, Park K, Choi J (2010) Ecotoxicological investigation of CeO_2 and TiO_2 nanoparticles on the soil nematode *Caenorhabditis elegans* using gene expression, growth, fertility, and survival as endpoints. Environ Toxicol Pharmacol 29:167–172

Rosso MN, Jones JT, Abad P (2009) RNAi and functional genomics in plant parasitic nematodes. Annu Rev Phytopathol 47(1):207–232

Roy AK (1976) Effect of decaffeinated tea waste and water hyacinth compost on the control of *Meloidogyne graminicola* on rice. Indian J Nematol 6:73–77

Saeki Y, Kawano E, Yamashita C, Akao S, Nagatomo Y (2003) Detection of plant parasitic nematodes, *Meloidogyne incognita* and *Pratylenchus coffeae* by multiplex PCR using specific primers. Soil Sci Plant Nutr 49(2):291–295. https://doi.org/10.1080/00380768.2003.10410010

Salim H, Ali A, Abdalbaki A, Eshak H, Khamees K, Reski B (2016) Nematicidal activity of plant extracts against the root-knot nematode *Meloidogyne* sp. on tomato plants. J Biol Agric Healthcare 6(20):73–76

Samaliev HY, Andreoglou FI, Elawad SA, Hague NGM, Gowen SR (2000) The nematicidal effects of the bacteria *Pseudomonas oryzihabitans* and *Xenorhabdus nematophilus* on the root-knot nematode *Meloidogyne javanica*. Nematology 2:507–514

Sanchez-Moreno S, Ferris H (2007) Suppressive service of the soil food web: effects of environmental management. Agric Ecosyst Environ 119:75–87

Sano Z (2005) Cultural control of the nematode damage. In: Noubunkyo (ed) Large encyclopedia of environmental conservation agriculture, Tokyo, pp. 281–316. doi: https://doi.org/10.1177/147447400501200114

Santo GS, O'Bannon JH, Finley AM, Golden AM (1980) Occurrence and host range of a new root-knot nematode (*Meloidogyne chitwoodi*) in the Pacific Northwest. Plant Dis 64:951–952

Sasser JN (1980) Root-knot nematodes: a global menace to crop production. Plant Dis 64:36–45

Sato E, Min YY, Shirakashi T, Wada S, Toyota K (2017) Detection of the root-lesion nematode, *Pratylenchus penetrans* (Cobb), in a nematode community using real-time PCR. Jap J Nematol 37:87–92

Schneider SM, Ajwa HA, Trout TJ, Gao S (2008) Nematode control from shank- and drip-applied fumigant alternatives to methyl bromide. Hortic Sci 43(6):1826–1832

Seinhorst JW (1956) Population studies on stem nematodes (*Ditylenchus dipsaci*). Nematologica 1:159–164

Shapiro-Ilan DI, Hazir S, Lete L (2015) Viability and virulence of entomopathogenic nematodes exposed to ultraviolet radiation. J Nematol 47(3):184–189

Sharma SB, Swarup G (1984) Cyst forming nematodes of India, New Delhi. Indian Cosmo Publication 1:150

Shield JA, Paul EA, Lowe WE, Parkinson D (1973) Turnover of microbial tissue under field conditions. Soil Biol Biochem 5:753–764

Shukla A, Chand R (2018) Studies on the bio-efficacy of botanicals against rice root-knot nematode, *Meloidogyne graminicola* development. J Pharmacognosy Phytochem 7(6):1563–1565

Siddiqi MR (1997) Techniques and methodologies for nematode disease diagnosis and nematode identification. FAO Plant Production and Protection Paper 144:21–44

Siddiqui IA, Haas D, Heeb S (2005) Extracellular protease of *Pseudomonas fluorescens* CHA0, a biocontrol factor with activity against the root-knot nematode *Meloidogyne incognita*. Appl Environ Microbiol 7:5646–5649

Siddiqui IA, Shaukat SS (2003) Suppression of root-knot disease by *Pseudomonas fluorescens* CHA0 in tomato: importance of bacterial secondary metabolite 2,4diacetylphloroglucinol. Soil Biol Biochem 35:1615–1623

Siddiqui ZA, Mahmood I (1999) Role of bacteria in the management of plant parasitic nematodes: a review. Bioresour Technol 69:167–179

Siddiqui ZA, Mahmood I (1996) Biological control of plant parasitic nematodes by fungi: a review. Bioresour Technol 58:229–239

Sikora R, Fernandez E (2005) Nematode parasites of vegetables. In: Luc M, Sikora R, Bridge J (eds) Plant parasitic nematodes in subtropical and tropical. CABI, Wallingford, pp 319–392. https://doi.org/10.1079/9780851997278.0000

Sikora RA (1988) Plant parasitic nematodes of wheat and barley in temperature and temperate semi-arid regions - a comparative analysis. In: Saxena MC, Sikora RA, Srivastava JP (eds) Nematodes parasitic to cereals and legumes in temperate semi-arid regions. Aleppo, Syria, ICARDA, pp 46–48

Singh S, Singh BK, Yadav SM, Gupta AK (2015) Applications of nanotechnology in agriculture and their role in disease management. Res J Nanosci Nanotechnol 5(1):1–5

Singh R, Kumar U (2015) Assessment of nematode distribution and yield losses in vegetable crops of Western Uttar Pradesh in India. Int J Sci Res 4(5):2812–2816

Sitaramaiah K (1990) Mechanisms of reduction of plant parasitic nematodes in soils amended with organic materials. In: Saxena SK, Khan MW, Rashid A, Khan RM (eds) Progress in plant nematology. CBS, Delhi, India, pp 263–295

Slomczynska U, South MS, Bunkers GJ, Edgecomb D, Wyse-Pester D, Selness S, Ding Y, Christiansen J, Ediger K, Miller W, Charumilind P, Hartmann G, Williams J, Dimmic M, Shortt B, Haakenson W, Wideman A, Crawford M, Hresko M, James McCarter J (2015) Tioxazafen: a new broad-spectrum seed treatment nematicide. In: Discovery and synthesis of crop protection products; ACS symposium series. American Chemical Society, Washington, DC, pp 129–147

Slomp L, Pereira PS, França SC, Zingaretti S, Oliveira BOR (2009) *In vitro* nematocidal effects of medicinal plants from São Paulo state, Brazil. Pharm Biol 47(3):230–235

Sobczak M, Avrova A, Jupowicz J, Phillips MS, Ernst K, Kumar A (2005) Characterization of susceptibility and resistance responses to potato cyst nematode (*Globodera* spp.) infection of tomato lines in the absence and presence of the broad-spectrum nematode resistance *Hero* gene. Mol Plant-Microbe Interaction 18:158–168

Spiegel Y, Cohn E, Galper S, Sharon E, Chet I (1991) Evaluation of a newly isolated bacterium, *Pseudomonas chitinolytica* sp. nov., for controlling the root-knot nematode *Meloidogyne javanica*. Biocontrol Sci Technol 1:115–125
Stirling GR (1991) Biological Control of plant-parasitic nematodes: progress, problems, and prospects. CAB International, Wallingford, UK, p 282
Stirling GR, Nicol J, Reay F (1999) Advisory services for nematode pests: operational guidelines. Rural Industries Research & Development Corporation, ACT, Australia. Available http://nematologists.org.au/aan-rirdc.pdf. Accessed 12 Feb 2019
Stork NE, Eggleton P (1992) Invertebrates as determinants and indicators of soil quality. Am J Altern Agric 7:38–47
Sturhan D, Schneider R (1980) *Hirsutella heteroderae*, ein neuer nematodenparasitarer Pilz. Phytopathol Z 99:105–115
Su L, Ruan Y, Yang X, Wang K, Li R, Shen Q (2015) Suppression on plant-parasitic nematodes using a soil fumigation strategy based on ammonium bicarbonate and its effects on the nematode community. Sci Rep 5:17597. https://doi.org/10.1038/srep17597
Subbotin SA, Madani M, Krall E, Sturhan D, Moens M (2005) Molecular diagnostics, taxonomy, and phylogeny of the stem nematode *Ditylenchus dipsaci* species complex based on the sequences of the internal transcribed spacer-rDNA. Phytopathology 95(11):1308–1315
Sukul NC (1992) Plants antagonistic to plant parasitic nematodes. Indian J Life Sci 12:23–52
Sundararaju P, Mustaffa MM, Kumar V, Cannayane I, Tanuja Priya B (2002) Effect of organic farming on plant-parasitic nematodes infesting banana cv. Karpuravalli. Curr Nematol 13:39–43
Swarup G, Sosa-Moss C (1990) Nematode parasites of cereals. In: Luc M, Sikora RA, Bridge J (eds) Plant parasitic nematodes in subtropical and tropical agriculture. Wallingford, CAB International, pp 109–136
Taha EH, Abo-Shady NM (2016) Effect of silver nanoparticles on the mortality pathogenicity and reproductivity of entomopathogenic nematodes. Int J Zool Res 12(3–4):47–50
Taiwo BA, Fawole B, Cole AOC (2015) Nematicidal potential of extracts from some selected plants against the root-knot nematode *Meloidogyne incognita*. Agrotechnology 4(2):62. https://doi.org/10.4172/2168-9881.S1.015
Taylor AL (2003) Nematocides and nematicides—a history. *Nematropica* 33(2):225–232
Taylor SP, McKay A (1993) Sampling and extraction methods for *Pratylenchus thornei* and *P. neglectus*. In: Vanstone VA, Taylor SP, Nicol JM (eds) Proc. 9th biennial Australian plant pathology conference. Pratylenchus Workshop, Adelaide, Australia.
Taylor SP, Vanstone VA, Ware AH, McKay AC, Szot D, Russ MH (1999) Measuring yield loss in cereals caused by root lesion nematodes (*Pratylenchus neglectus* and *P. thornei*) with and without nematicide. Aust J Agr Res 50(4):617–622
Thakur RK, Dhirta B, Shirkot P (2018) Studies on effect of gold nanoparticles on *Meloidogyne incognita* and tomato plants growth and development. Ann Nanosci Nanotechnol 2(1):1–7. https://doi.org/10.1101/428144
Thiéry M, Mugniéry D (2000) Microsatellite loci in the phytoparasitic nematode *Globodera*. Genome 43:160–165
Thomason IJ, Caswell EP (1987) Principles of nematode control. In: Brown RH, Kerry BR (eds) Principles and practice of nematode control in crops. Academic Press, New York, pp 87–124
Thompson JP, Clewett TG (1986) Research on root-lesion nematode. In: Queensland Wheat Research Institute Biennial Report 1982–1984, Qld Dept. Primary Industries, Qld. Govt., Queensland, Australia, Toowoomba, Queensland, Australia, Wheat Research Institute, pp 32–35
Tikhonov VE, Lòpez-Llorca LV, Salinas J, Jansson HB (2002) Purification and characterization of chitinases from the nematophagous fungi *Verticillium chlamydosporium* and *V. suchlasporium*. Fungal Genet Biol 35:67–78
Tomczak AF, Danan S, van Dijk T, Beyene A, Bouwman L, Overmars H, Bakker E (2009) A high-resolution map of the Grp1 locus on chromosome V of potato harbouring broad-spectrum

resistance to the cyst nematode species *Globodera pallida* and *Globodera rostochiensis*. Theor Appl Genet 119(1):165–173
Tripathi L, Babirye A, Roderick H, Tripathi JN, Changa C, Urwin PE, Tushemereirwe WK, Coyne D, Atkinson HJ (2015) Field resistance of transgenic plantain to nematodes has potential for future African food security. Sci Rep 5:8127
Trivedi PC, Barker KR (1986) Management of nematodes by cultural practices. Nematropica 16:213–236
Trudgill DL, Phillips MS (1997) Nematode population dynamics, threshold levels and estimation of crop losses. http://www.fao.org/3/v9978e/v9978e07.htm. Accessed 14 Apr 2019
Tunlid A, Rosen S, Ek B, Rask L (1995) Purification and characterization of an extracellular serine protease from the nematode-trapping fungus *Arthrobotrys oligospora*. Microbiology 140:1687–1695
Tyler J (1933) The root-knot nematode. Berkeley, CA, USA, Circular No. 330, University of California College of Agriculture, Agricultural Experiment Station, p 33
Tylka GL, Marett CC, Robertson AE (2015) Field experiments show effects of clariva™ seed treatment in 2014. Integrated Crop Management News
Urwin PE, Lilley CJ, Atkinson HJ (2002) Ingestion of double-stranded RNA by preparasitic juvenile cyst nematodes leads to RNA interference. Mol Plant Microbe Interact 15:747–752
Urwin PE, Lilley CJ, Mcpherson MJ, Atkinson HJ (1997) Resistance to both cyst and root-knot nematodes conferred by transgenic *Arabidopsis* expressing a modified plant cystatin. Plant J 12:455–461. https://doi.org/10.1046/j.1365-313X.1997.12020455.x
Urwin PE, Mcpherson MJ, Atkinson HJ (1998) Enhanced transgenic plant resistance to nematodes by dual proteinase inhibitor constructs. Planta 204:472–479. https://doi.org/10.1007/s004250050281
Valentine TA, Randall E, Wypijewski K, Champman S, Jones J, Oparka KJ (2007) Delivery of macromolecules to plant parasitic nematodes using tobacco rattle virus vector. Plant Biotechnol J 5:827–834
van der Voort JR, Wolters P, Folkertsma R, Hutten R, Van Zandvoort P, Vinke H, Kanyuka K, Bendahmane A, Jacobsen E, Janssen R (1997) Mapping of the cyst nematode resistance locus Gpa2 in potato using a strategy based on comigrating AFLP markers. Theor Appl Genet 95 (5-6):874–880
van Doorn R, Szemes M, Bonants P, Kowalchuk GA, Salles JF, Ortenberg E, Schoen CD (2007) Quantitative multiplex detection of plant pathogens using a novel ligation probe-based system coupled with universal, high-throughput real-time PCR on OpenArrays™. BMC Genomics 8:276
Van Gundy SD, Jose Gustavo Perez B, Stolzy LH, Thomason IJ (1974) A pest management approach to the control of *Pratylenchus thornei* on wheat in mexico. J Nematol 6:107–116
Verdejo-Lucas S, Sorribas J, Puigdomenech P (1994) Yield losses caused by *Meloidogyne javanica* on lettuce and tomato in a plastic house. Invest Agrar Prod Prot Veg Suppl 2:395–400
Viaene NM, Abawi GS (1998) Management of *Meloidogyne hapla* on lettuce in organic soil with sudangrass as a cover crop. Plant Dis 82(8):945–952
Vishnudasan D, Tripathi MN, Rao U, Khurana P (2005) Assessment of nematode resistance in wheat transgenic plants expressing potato proteinase inhibitor (PIN2) gene. Transgenic Res 14:665–675
Wada S, Toyota K (2008) Effect of three organophosphorous nematicides on non-target nematodes and soil microbial community. Microbes Environ 23:331–336
Wada S, Toyota K, Takada A (2011) Effects of the nematicide imicyafos on soil nematode community structure and damage to radish caused by *Pratylenchus penetrans*. J Nematol 43 (1):1–6
Wang L, Wang Y, Yang B, Hu X, Yu S (2001) Studies on relationships of Bursaphelenchus xylophilus and B. mucronatus by RAPD. Acta Phytopathol Sinica 31:225–229

Wang B, Liu X, Wu W, Liu X, Li SH (2009) Purification, characterization, and gene cloning of an alkaline serine protease from a highly virulent strain of the nematode-endoparasitic fungus *Hirsutella rhossiliensis*. *Microbiol Res* 164:665–673

Wang KH, McSorley R, Marshall AH, Gallaher RN (2004) Nematode community changes associated with decomposition of *Crotalaria juncea* amendment in litterbags. Appl Soil Ecol 27:31–45

Wang M, Yang JK, Zhang KQ (2006a) Characterization of an extracellular protease and its cDNA from the nematode-trapping fungus *Monacrosporium microscaphoides*. Can J Microbiol 52:130–139

Wang RB, Yang JK, Lin C, Zhang Y, Zhang KQ (2006b) Purification and characterization of an extracellular serine protease from the nematode-trapping fungus *Dactylella shizishanna*. Lett Appl Microbiol 42:589–594

Ward JD (2015) Rapid and precise engineering of the *Caenorhabditis elegans* genome with lethal mutation co-conversion and inactivation of NHEJ repair. Genetics 199:363–377

Webb D, Baltazar B, Rao-Arelli A, Schupp J, Clayton K, Keim P, Beavis W (1995) Genetic mapping of soybean cyst nematode race-3 resistance loci in the soybean PI 437.654. Theor Appl Genet 91(4):574–581

Wei JZ, Hale K, Carta L, Platzer E, Wong C, Fang SC, Aroian RV (2003) *Bacillus thuringiensis* crystal proteins that target nematodes. PNAS 100:2760–2765

Weischer BA (1994) Where to go in phytonematode control? Phytoparasitica 22:95–99

Weller DM (1988) Biological control of soil-borne plant pathogens in the rhizosphere with bacteria. Annu Rev Phytopathol 26:379–407

Westerdahl B, Giraud D, Etter S, Riddle LJ, Radewald JD, Anderson CA, Darso J (2003) Management options for *Pratylenchus penetrans* in Easter lily. J Nematol 35:443–449

Whitehead AG (1998) *Plant nematode control*. CAB, Wallingford, UK

Williams KJ, Fisher JM, Langridge P (1996) Development of a PCR-based allelespecific assay from an RFLP probe linked to resistance to cereal cyst nematode in wheat. Genome 39:798–801

Wiratno D, Taniwiryono H, den Berg V, Riksen JAG, Rietjens IMCM, Djiwanti SR, Kammenga JE, Murk AJ (2009) Nematicidal activity of plant extracts against the root knot nematode, *Meloidogyne incognita*. *Open Nat Prod J* 2:77–85

Xiang N, Lawrence KS, Kloepper JW, Donald PA, McInroy JA, Zhang B (2017) Biological control of *Heterodera glycines* by spore- forming plant growth- promoting rhizobacteria (PGPR) on soybean. PLoS One 12(7):e0181201. https://doi.org/10.1371/journal.pone.0181201

Yaghoobi J, Kaloshian I, Wen Y, Williamson VM (1995) Mapping a new nematode resistance locus in *Lycopersicon peruvianum*. Theor Appl Genet 91:457–464

Ye W (2012) Development of prime time Real Time PCR for species identification of soybean cyst nematode (*Heterodera glycines* Ichinohe, 1952) in North Carolina. J Nematol 44(3):284–290

Ye W, Zeng Y, Kerns J (2015) Molecular characterization and diagnosis of root-knot nematodes (*Meloidogyne* spp.) from Turfgrasses in North Carolina, USA. PLoS One 10(11):e0143556. https://doi.org/10.1371/journal.pone.0143556

Youssef RM, Kim KH, Haroon SA, Matthews BF (2013) Posttranscriptional gene silencing of the gene encoding aldolase from soybean cyst nematode by transformed soybean roots. Exp Parasitol 134:266–274. https://doi.org/10.1016/j.exppara.2013.03.009

Yu S, Wang Y, Hu X, Bai W (1998) RAPD analysis of four most common *Meloidogyne* spp. Acta Phytopathol Sinica 28:359–365

Zamanian M, Andersen EC (2016) Prospects and challenges of CRISPR/Cas genome editing for the study and control of neglected vector-borne nematode diseases. FEBS J 283(17):3204–3221

Zasada IA, Halbrendt JM, Kokalis-Burelle N, LaMondia J, McKenry MV, Noling JW (2010) Managing nematodes without methyl bromide. Annual. Rev Phytopathol 48:311–328

Zhang LY, Zhang YY, Chen RG, Zhang JH, Wang TT, Li HX, Ye ZB (2010) Ectopic expression of the tomato *Mi-1* gene confers resistance to root knot nematodes in lettuce (*Lactuca sativa*). Plant Mol Biol Rep 28(2):204–211

Zhao M, Mo M, Zhang Z (2004) Characterization of a serine protease and its full-length cDNA from the nematode-trapping fungus *Arthrobotrys oligospora*. Mycologia 96:16–22
Zhou L, Wang J, Wang K, Xu J, Zhao J, Shan T, Luo C (2012) Secondary metabolites with antinematodal activity from higher plants. Stud Nat Prod Chem 37:67–114
Zijlstra C, Donkers-Venne DTHM, Fargette M (2000) Identification of *Meloidogyne incognita, M. javanica* and *M. arenaria* using sequence characterised amplified region (SCAR) based PCR assays. Nematology 2:847–883
Zuckerman BM, Jasson HB (1984) Nematode chemotaxis and possible mechanisms of host/prey recognition. Annu Rev Phytopathol 22:95–113

The manufacturer's authorised representative in the EU is Springer Nature Customer Service Centre GmbH, Europaplatz 3, 69115 Heidelberg, Germany. If you have any concerns regarding our products, please contact ProductSafety@springernature.com

Printed and bound by CPI Group (UK) Ltd, Croydon, CR0 4YY
30/09/2026
02225558-0016